Progress in the Chemistry of Organic Natural Products

The volumes of this classic series, now referred to simply as “Zechmeister” after its founder, Laszlo Zechmeister, have appeared under the Springer Imprint ever since the series’ inauguration in 1938. It is therefore not really surprising to find out that the list of contributing authors, who were awarded a Nobel Prize, is quite long: Kurt Alder, Derek H.R. Barton, George Wells Beadle, Dorothy Crowfoot-Hodgkin, Otto Diels, Hans von Euler-Chelpin, Paul Karrer, Luis Federico Leloir, Linus Pauling, Vladimir Prelog, with Walter Norman Haworth and Adolf F.J. Butenandt serving as members of the editorial board.

The volumes contain contributions on various topics related to the origin, distribution, chemistry, synthesis, biochemistry, function or use of various classes of naturally-occurring substances ranging from small molecules to biopolymers.

Each contribution is written by a recognized authority in the field and provides a comprehensive and up-to-date review of the topic in question. Addressed to biologists, technologists, and chemists alike, the series can be used by the expert as a source of information and literature citations and by the non-expert as a means of orientation in a rapidly developing discipline.

All contributions are listed in PubMed.

A. Douglas Kinghorn · Heinz Falk · Simon Gibbons · Yoshinori Asakawa · Ji-Kai Liu · Verena M. Dirsch
Editors

Phytochemistry of Bryophytes

Volume 126

By Yoshinori Asakawa

Springer

Editors
A. Douglas Kinghorn
College of Pharmacy
Ohio State University
Columbus, OH, USA

Heinz Falk
Institute of Organic Chemistry
Johannes Kepler University Linz
Linz, Austria

Simon Gibbons
Natural and Medical Sciences Research Center
University of Nizwa
Nizwa, Oman

Yoshinori Asakawa
Faculty of Pharmaceutical Sciences
Tokushima Bunri University
Tokushima, Japan

Ji-Kai Liu
School of Pharmaceutical Sciences
South Central University for Nationalities
Wuhan, China

Verena M. Dirsch
Department of Pharmaceutical Sciences
University of Vienna
Vienna, Austria

ISSN 2191-7043 ISSN 2192-4309 (electronic)
Progress in the Chemistry of Organic Natural Products
ISBN 978-3-031-77797-4 ISBN 978-3-031-77795-0 (eBook)
https://doi.org/10.1007/978-3-031-77795-0

This Springer imprint is published by the registered company Springer Nature Switzerland AG
The registered company address is: Gewerbestrasse 11, 6330 Cham, Switzerland

The author wishes to dedicate this book to his friend and colleague, the late Professor Dr. Jun'ichi Kobayashi, Professor Emeritus, Faculty of Pharmaceutical Sciences, Hokkaido University, Japan. He contributed greatly to the isolation, structure elucidation, and biological evaluation of a wide range of natural products, especially from marine organisms of Okinawan origin, and also investigated alkaloids from plants. Prof. Kobayashi published over 550 scientific reports and received several prestigious research honors, and among these were the Sumiki-Umezawa Memorial Award, the Japanese Society of Pharmacognosy Award, and the Pharmaceutical Society of Japan Award. In addition, he served as a Series Editor for "Progress in the Chemistry of Organic Natural Products" from volume 89 (2008) through to volume 114 (2021). Professor Kobayashi's legacy will continue to inspire future generations of scientists and he will be very greatly missed by all who knew him.

Yoshinori Asakawa

About This Book

This contribution discusses the phytochemistry of bryophytes, with a particular focus on their biologically active natural products and their potential applications in medicine, cosmetics, and as foods. Bryophytes, which include the mosses, liverworts, and hornworts, produce a plethora of terpenoids, phenolic compounds, and polyketides that exhibit diverse pharmacological activities, including anti-inflammatory, antiobesity, antioxidant, cytotoxic, and muscle-relaxant effects. The volume also explores the characteristic odors and flavors of bryophytes, as well as their possible use in the cosmetics industry, as food additives, and ultimately as medicinal drugs. Additionally, the biosynthesis pathways and synthesis of selected bioactive bryophyte compounds are discussed, highlighting the potential of these fascinating and ancient plants as a source of novel and valuable natural products.

Contents

Phytochemistry of Bryophytes: Biologically Active Compounds and Their Uses as Cosmetics, Foods, and in Drug Development

Yoshinori Asakawa

Contents

Y. Asakawa (✉)
Faculty of Pharmaceutical Sciences, Tokushima Bunri University, Yamashiro-Cho 180, Tokushima 770-8514, Japan
e-mail: asakawa@ph.bunri-u.ac.jp

A. D. Kinghorn, H. Falk, S. Gibbons, Y. Asakawa, J.-K. Liu, V. M. Dirsch (eds.),
Phytochemistry of Bryophytes, Progress in the Chemistry of Organic Natural Products 126, https://doi.org/10.1007/978-3-031-77795-0_1

1 Introduction

The bryophytes are found everywhere in the world, except in the sea. They grow on wet soil or rocks (Plate 1a), the trunks of trees (Plate 1b), and in lakes, rivers, and even in Antarctica.

The bryophytes are placed taxonomically between the algae and the pteridophytes. There are about 23,000 species globally and they are further divided into three phyla, the Bryophyta (mosses 14,000 species) (Plate 2), the Marchantiophyta (liverworts 6,000 species) (Plate 3a and b), and the Anthoceratophyta (hornworts 300 species) (Plate 4). They are considered to be the oldest terrestrial plants, although no strong scientific evidence for this has appeared in the literature. This hypothesis was mainly based on the resemblance of the present-day liverworts to the first land plant fossils, the spores of which date back to the Devonian period, i.e., almost 350–400 million years. Genetic studies seem even to point to their origin as early as the Ordovician, i.e., about 450 million years ago.

Plate 1 **a** Mosses on wet rocks (Tokushima, Japan); **b** Liverworts on the twigs of trees (Panama)

Plate 2 *Polytrichum commune* (moss)

Plate 3 **a** *Plagiochila sciophila* (liverwort); **b** *Wiesnerella denudata* (liverwort)

Among the bryophytes, almost all liverworts possess characteristic cellular oil bodies (Plate 5) that are peculiar, membrane-bound cell organelles, consisting of ethereal terpenoids and aromatic oils suspended in a carbohydrate- or protein-rich matrix, while the other two phyla do not. These oil bodies are very important biological markers for the taxonomy of the Marchantiophyta [1–34].

In the literature on Chinese medicinal spore-forming plants, 24 lichens, 74 sea algae, 22 mosses, five liverworts, 112 fungi, and 329 ferns have been listed with their Latin names, morphological characteristics, distribution locations, pharmacological activity, and biological effects, and their prescription uses in detail [35]. Several mosses have been used widely medicinally in China, to treat burns, bruises, external wounds, snake bites, pulmonary tuberculosis, neurasthenia, fractures, convulsions,

Plate 4 *Phaeoceros carolinianus* (hornwort)

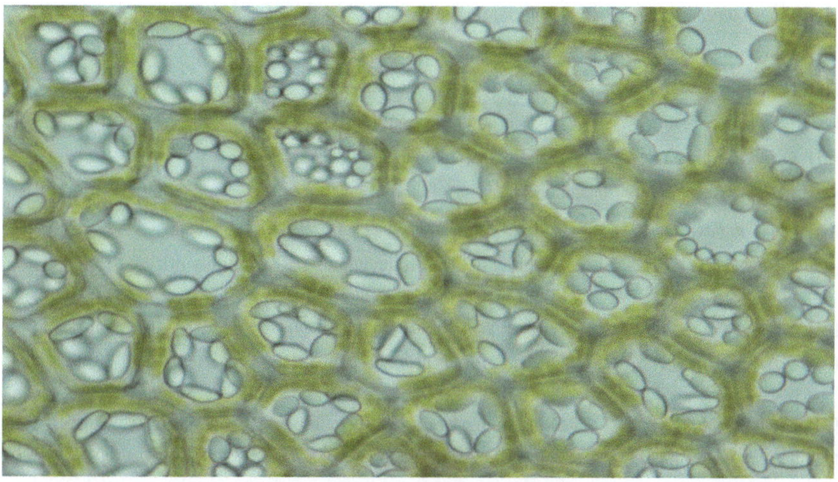

Plate 5 Oil bodies of *Bazzania japonica* (liverwort)

uropathy, pneumonia, and neurasthenia, and other conditions, as shown in Table 1. Glime in 2017 also summarized the medicinal uses of bryophytes, including as herbal medicines and medicinal teas, for liver, lung, skin, eye, and ear ailments, for gall stones, ringworm infections, heart and cardiovascular problems, nose bleeds, neurological conditions, inflammation, fever, and urinary and bowel disorders. Also, they have uses as disinfectants and in infections, for ailments of the nose and throat, for burns, hair treatments, as sedatives, antidotes, filters, for application to surgical and larger wounds, and for other purposes [36].

Table 1 Medicinal bryophytes and their biological activities, effects, and applications

Species name[a]	Biological activity, effects, applications	Refs.
[M] *Aerobryum lanosum*	Antibacterial, antidote, antirhinitis, antipyretic, for adenopharyngitis, bacteriosis, body aches, diarrhea, menstrual pain, wound healing, to treat bruises, burns, fever, fractures, intermittent fever; used as a medicinal plant (herbal medicine) and a filter for smoking	[36]
[L] *Bazzania pompeana*	Absorbs water up to four times its dry weight	[36]
[M] *Campylopus introfrexus*	Used as a medicinal plant (herbal medicine)	[36]
[H] *Ceratophyllum demersum*	Antipyretic, used as an astringent and for constipation	[36]
[L] *Conocephalum conicum*	Antidotal, antifungal (*Candida albicans*), antimicrobial, antipyretic, for cuts, burns, fractures, gallstones, poisonous snake bites, scalds, to treat boils, eczema, wounds	[35, 36]
[L] *Conocephalum japonicum*	Antidote, analgesic, for apocatastasis, bed sores, carbuncles, snake bites	[35, 36]
[M] *Cratoneuron filicinum*	To treat heart disease	[35, 36, 59]
[M] *Dawsonia longifolia*	Diuretic, for hair growth	[36]
[M] *Dawsonia superba*	Diuretic, for hair growth, colds	[36, 59]
[M] *Ditrichum pallidum*	For childhood convulsions	[35, 36]
[L] *Dumortiera hirsuta*	Used as sources of antibiotics	[59]
[M] *Entodon compressus*	Diuretic, for edema	[35]
[M] *Entodon concinnus*	Used as a filter for smoking	[36]
[M] *Entodon flavescens*	To treat earache, used as ear drops	[36, 59]
[M] *Eurhynchium striatum*	Potential use for producing unique and highly prized compounds for the pharmaceutical industry	[36]
[M] *Fissidens adianthoides*	For hair growth, to bandage wounds	[36]
[M] *Fissidens japonicum*	Diuretic, for burns, choloplania (jaundice, icterus), and growth of hair	[35]
[M] *Fissidens nobilis*	Diuretic, for hair growth, healing burns, jaundice	[36, 59]
[M] *Fissidens osmundoides*	Antibacterial, for swollen throats	[36]
[M] *Flatbergium sericeum*	To treat eye disease	[36]
[M] *Fontinalis antipyretica*	Hemostasis, to treat fever	[36]
[L] *Frullania ericoides*	For nourishing hair, to get rid of head lice	[36]
[L] *Frullania tamarisci*	Antiseptic	[35, 36]
[M] *Funaria hygrometrica*	Hemostasis, for athlete's foot dermatophytosis (dermatomycosis), bruises, pulmonary tuberculosis, skin infections, vomitus cruentus (hepatemesis)	[35, 36, 59]
[M] *Grimmia pilifera*	Antibacterial	[36]

(continued)

Table 1 (continued)

Species name[a]	Biological activity, effects, applications	Refs.
[M] *Haplocladium microphyllum* (= *H. capillatum*)	To treat bronchitis, fever, tonsillitis, tympanitis	[35, 36]
[L] *Haplomitrium mnioides*	Absorbs water up to 12 times its dry weight	[36]
[L] *Herbertus sendtneri*	Antiseptic, antidiarrheal, astringent, expectorant, used as a filter for smoking	[36]
[M] *Homalia trichomanoides*	Antifungal	[36]
[M] *Homalothecium sericeum*	To treat rhinorrhagia, whooping cough	[36]
[M] *Hydrogonicum amplexifolium*	Antimicrobial, antipyretic, for diarrhea	[35]
[M] *Hyophila attenuata*	To treat colds, coughs, neck pain	[35, 59]
[M] *Hyophila involuta*	For colds, coughs, sore throat	[36]
[M] *Hypnum cupressiforme*	Used as a filter for smoking and to stuff pillows	[36]
[M] *Leptodictyum riparium*	Used as antipyretic and for uropathy	[36, 59]
[M] *Leucobryum bowringii*	To treat pain	[36]
[M] *Loeskeobryum brevirostre*	To absorb water up to 9.8 times its dry weight	[36]
[M] *Loeskeobryum carvifolium*	To absorb water up to 9.8 times its dry weight	[36]
[L] *Marchantia convoluta*	To treat hepatitis, fever, gastric intolerance	[36]
[L] *Marchantia paleacea*	To treat fever, hepatitis, skin diseases, to reduce swelling	[36, 59]
[L] *Marchantia palmata*	Anti-inflammatory after burns	[36]
[L] *Marchantia polymorpha*	Antidotal, antifungal, antihepatic, antimicrobial, antipyretic, diuretic, for cuts, fractures, open wounds, scalds, swollen tissues, to treat abscesses, boils, burns, cut, eczema, jaundice of hepatitis, pulmonary tuberculosis, snake bites, wounds, an external cure to reduce inflammation, and to improve of cooling and cleansing of the liver	[35, 36]
[M] *Meteoriella soluta*	Antiinflammatory, hemostasis, for gastrointestinal hemorrhage, hemoptysis	[35]
[M] *Meteoriopsis squarosa*	Used as a filter for smoking	[36]
[M] *Mnium cuspidatum*	Hemostasis, for improper bleeding from sexual organs, nose bleeds	[35]
[M] *Mnium stellare*	For fractures, to reduce pain from bruises, burns, wounds	[36]
[M] *Octoblepharum albidum*	Anodyne, antipyretic, febrifuge, for body aches, to treat fever, external wounds	[35, 36, 59]
[M] *Oreas martiana*	Anodyne, hemostasis, for epilepsy, external wounds, menorrhagia, neurasthenia (nervosism, nervous exhaustion)	[35]
[L] *Pallavicinia* sp.	Used as antimicrobial agent	[59]

(continued)

Table 1 (continued)

Species name[a]	Biological activity, effects, applications	Refs.
[M] *Philonotis fontana*	Antidotal, antipyretic, for adenopharyngitis, fractures, to treat bruises, fever, skin diseases	[35, 36]
[M] *Physcomitrella patens*	To produce Factor H	[35, 36]
[L] *Plagiochasma appendiculatum*	Antifungal, to treat burns, boils, blisters, skin diseases (skin eruptions caused by sunlight)	[35, 59, 340]
[L] *Plagiochila beddomei*	For healing wounds	[36, 59]
[M] *Plagiomnium cuspidatum*	Hemostasis, to treat infectious swelling	[36]
[M] *Plagiomnium insigne*	Used as medicinal herb	[36]
[M] *Plagiomnium maximoviczii*	To absorb water up to 6–7 times its dry weight	[36]
[M] *Plagiopus oederi*	Sedative, for apoplexy, cardiopathy, epilepsy	[35]
[M] *Plagiopus oederi*	Used as sedative, to treat epilepsy	[35]
[M] *Pogonatum macrophyllum*	Inflammation, for fever, used as detergent, laxative, hemostatic agent	[36]
[M] *Pogonatum cirratum*	Hemostasis, laxative	[36]
[M] *Polytrichum commune*	Antidote, antipyretic, diuretic for hemostasis, laxative, for bleeding from gingivae, common cold, gynecological aid, hair growth cuts, hematemesis, inflammation, night sweats, pulmonary tuberculosis, to speed up the labor process during childbirth, uterine prolapse, to treat fever, to dissolve kidney and gall bladder stones, used as a herbal tea	[35, 36, 59]
[M] *Polytrichum juniperinum*	To treat prostate problems, painful urination in the elderly, obstruction or suppression, dropsy, skin ailments	[35, 36]
[M] *Polytrichum pallidisetum*	Cytotoxicity against human cancer cells	[36]
[M] *Polytrichum* sp.	Diuretic, hair growth	[35]
[L] *Reboulia hemisphaerica*	Hemostasis, for bruises, external wounds, skin blotches	[35, 36]
[M] *Rhodobryum cirratum*	For heart diseases	[36]
[M] *Rhodobryum giganteum*	Antihypertensive, antipyretic, diuretic, for cardiopathy, cuts, expansion of blood vessels, neurasthenia, psychosis, sedation	[35]
[M] *Rhodobryum giganteum*	Sedative, to cure angina, to treat cardiovascular diseases, fever, hypertension, nervous prostration, to reduce oxygen resistance by increasing the rate of flow in the aorta by over 30% to cuts, used as a herbal tea	[36]
[M] *Rhodobryum ontariense*	Antihypertension, sedative, for heart diseases	[36]
[M] *Rhodobryum cirratum*	For heart diseases	[36]
[M] *Rhodobryum giganteum*	Antihypertensive, antipyretic, diuretic, for cardiopathy, cuts, expansion of blood vessels, neurasthenia, psychosis, sedation	[35]

(continued)

Table 1 (continued)

Species name[a]	Biological activity, effects, applications	Refs.
[M] *Rhodobryum giganteum*	Sedative, to cure angina, to treat cardiovascular diseases, fever, hypertension, nervous prostration, to reduce oxygen resistance by increasing the rate of flow in the aorta by over 30% to cuts, used as herbal tea	[36]
[M] *Rhodobryum ontariense*	Antihypertension, sedative, for heart diseases	[36]
[M] *Rhodobryum roseum*	Sedative, for cardiopathy	[35]
[M] *Rhodobryum roseum*	For cardiovascular diseases, nervous prostration	[36]
[L] *Riccia austinii*	To cure ringworm	[36, 59]
[L] *Scapania gracilis*	Used as a filter for smoking	[36]
[M] *Sematophyllum adnatum*	Used as a herbal tea	[36]
[M] *Sphagnum centrale*	Induction of allergy and lymphocutaneous sporotrichosis, to bandage wounds, make salves, treat boils, used as medicinal plant in Malaysia (suppository)	[36]
[M] *Sphagnum cymbifolium*	Substitute for cotton	[35]
[M] *Sphagnum girgensohnii*	Substitute for cotton	[35]
[M] *Sphagnum magellanicum*	Substitute for cotton	[35]
[M] *Sphagnum palustre*	For edema and ulcers	[35]
[M] *Sphagnum sericeum*	Used to dress wounds, with antimicrobial properties, for skin ailments (insect bites, scabies, acne), hemorrhoids, to treat eye disease	[35]
[M] *Sphagnum* sp.	To relieve itching from mosquito bites, to treat skin problems (acne, chilblains, dandruff, eczema, insect bites, painful inflammation of the small blood vessels in the skin, ringworm), used as a contraceptive to block the entry of sperm, and as a heart tonic tea	[36]
[M] *Sphagnum squarrosum*	Substitute for cotton	[35]
[M] *Sphagnum teres*	Antipyretic, for conjunctivitis, cataracts and other eye disorders	[35, 36, 59]
[L] *Targionia hypophylla*	To treat skin diseases	[36]
[M] *Taxiphyllum taxirameum*	Antiphlogistic, antiinflammatory, for hemostasis, external wounds	[35, 36]
[M] *Tetraplodon bryoides* (= *T. mnioides*)	Sedative, for heart failure, Hansen's disease	[35]
[M] *Thuidium recognitum*	Antibacterial, antiinflammation	[36]
[M] *Timmiella barbuloides*	Used as medicinal plant in Ancient Egypt	[36]
[M] *Trachycystis macrophylla*	Absorbs water up to 3.2 times its dry weight	[36]
[M] *Weissia controversa*	Antidote, to treat colds, intermittent fever, rhinitis	[35, 36]
[M] *Weissia viridula*	Antidotal, antipyretic, for rhinitis	[35]
[L] *Wiesnerella denudata*	To treat gallstones	[36]

[a] [L] liverwort, [M] moss, [H] hornwort

Many species of liverworts possess characteristic fragrant odors and/or an intense pungent, sweet, or bitter taste. Generally, bryophytes are not damaged by bacteria and fungi, insect larvae and adults, snails, slugs, and other small mammals. Furthermore, some liverworts cause intense allergenic contact dermatitis and exhibit allelopathy.

Although liverworts possess various biologically active compounds, their isolation was neglected for almost a century. The present author and his associates have been interested in the application of bryophytes for human or domestic animal uses and in the isolation and the structural elucidation of their biologically active compounds. Since 1972, Asakawa and his coworkers have collected more than 1,000 species of bryophytes from around the world and have analyzed them with respect to their chemistry, pharmacology, and for possible application as sources of cosmetic ingredients and in the human diet, and as medicinal or agricultural agents. The biological activities of liverworts are due mainly to the terpenoids and aromatic compounds that are present in the oil bodies in each species [6–34, 37–51].

Herein, the biological and chemical diversity of bryophytes and the structural diversity of the terpenoids, aromatic compounds, and polyketides found in bryophytes, along with their biological and pharmacological activities, including their characteristic odors and tastes, as well as the possibility of using bryophytes as cosmetics, foods, and medicines, are surveyed. Selected biosynthesis pathways and the semi- and total synthesis of selected biologically active compounds of liverwort constituents are also discussed.

Asakawa and his coworkers have published widely on the details of, compound isolation and structure elucidation as well as of their biological activities in addition to the chemosystematics of bryophytes [1–34, 37–51], and many review articles of phytochemicals found in bryophytes and their potential application for pharmaceutical uses have been published [52–72].

Recently, Lou and his group joined this area to work on phytochemicals of Chinese liverworts, of which many are similar to Japanese taxa, and they have published various new terpenoids and aromatic compounds having pharmacological activity, as shown later on in Table 9.

2 Biodiversity of Bryophytes

The Marchantiophyta (liverworts) include two subclasses, the Jungermanniidae and Marchantiidae, and six orders, 49 families, 130 genera, and 6,000 species. There are 54 endemic genera in the southern hemisphere, such as from New Zealand and South America [73]. In Asia, including Japan, a relatively large number of endemic genera (21) has been recorded. However, Africa, Madagascar, North America, and Europe are very poor regions of endemic genera [74]. The richness of the endemic genera of bryophytes in the southern hemisphere suggests that the bryophytes might have originated in the Antarctic region from 350–400 million years ago, and developed and spread to the northern hemisphere during a long evolutionary process. In Southeast Asia and South America, there are rainforests where so many liverwort species have been found, although some species, like those in the Lejeuneaceae, are intermingled with other species, and it is tedious and time-consuming work to differentiate each of these.

3 Chemical Diversity of Bryophytes

Extraction of the oil bodies with *n*-hexane or ether, using ultrasonic apparatus is facile to perform for stem-leafy liverworts, in order to yield a large amount of crude extract. In the case of thalloid liverworts, the specimens are ground mechanically and then extracted with non-polar solvents. Several hundred new compounds have been isolated from the liverworts (Marchantiophyta) and more than 40 new carbon skeletal terpenoids and aromatic compounds, such as the bis-bibenzyls marchantin A (**1**) and riccardin A (**2**), which are very rare natural products, have been found [1–3].

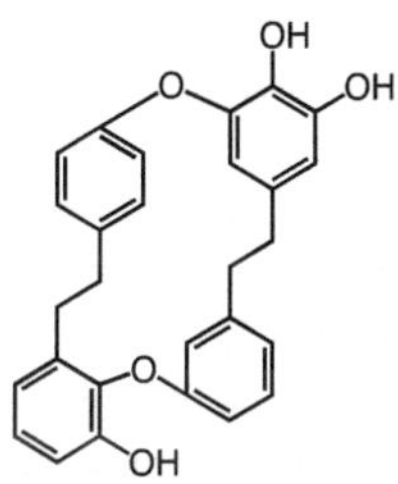

1 (marchantin A)

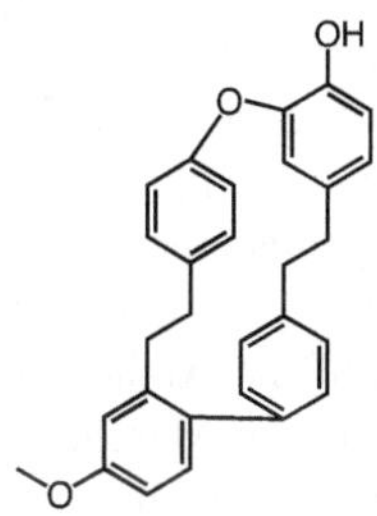

2 (riccardin A)

Most of the liverworts studied elaborate characteristic scents, having pungent, and bitter-tasting constituents, of which many show antimicrobial, antifungal, antiviral, allergenic contact dermatitis, cytotoxic, insecticidal, anti-HIV, superoxide anion radical release, plant growth regulatory, neurotrophic, NO production inhibitory, muscle relaxant, antiobesity, piscicidal, and nematocidal effects, as well as numerous other activities [1–3, 75, 76]. The biological activities ascribed to the liverworts are due mainly to lipophilic sesqui- and diterpenoids, phenolic compounds, and polyketides, which are the principal constituents of the oil bodies. The most characteristic chemical phenomenon of the liverworts is that most of the sesqui- and diterpenoids are enantiomers of those found in higher plants, although there are a few exceptions, such as the guaiane-type sesquiterpenoids. It is highly noteworthy that different liverwort species of the same genus, such as *Frullania tamarisci* and *F. dilatata* (Frullaniaceae), or *Lepidozia* species (Lepidoziaceae), may produce sesquiterpene lactone enantiomers, e.g., (+)-frullanolide (**3**) and (–)-frullanolide (**4**) [1–3].

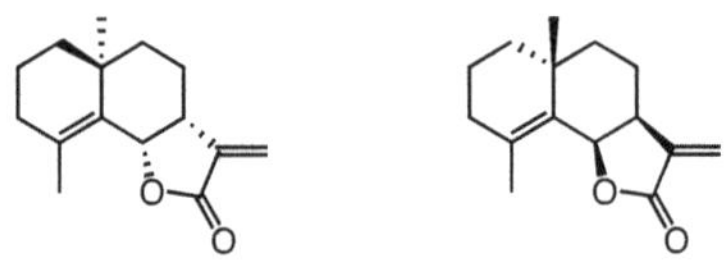

3 ((+)-frullanolide) **4** ((–)-frullanolide)

In general, the presence of nitrogen- or sulfur-containing compounds in the bryophytes is very rare [2, 3, 75]. However, recently, several nitrogen- and sulfur-containing compounds, (*E*)-coriandrin (**5**), (*Z*)-coriandrin (**6**), (*E*)-*O*-methyltridentatol (**7**), and (*Z*)-*O*-methyltridentatol (**8**), were isolated from the Mediterranean liverwort, *Corsinia coriandrina* (Corsiniaceae, Marchantiales) [77] and two prenylated indole derivatives (**9** and **10**) from *Riccardia* species (Aneuraceae, Metzgeriales) [2]. Skatole (**11**) was isolated from or detected in two liverwort species, namely, the Malaysian *Asterella* or *Mannia* (Aytoniaceae) [78] and the Tahitian *Cyathodium foetidissimum* (Cyathodiaceae) [79], and isotachin A (**12**), isotachin B (**14**), and isotachin C (**13**) from the Isotachidaceae [2].

Highly evolved liverworts belonging to the Marchantiaceae produce phytosterols, such as campesterol, stigmasterol, and sitosterols [1–3]. Almost all liverworts elaborate α-tocopherol (**15**) and squalene. The characteristic components of the Bryophyta are highly unsaturated fatty acids, and alkanones, such as 5,8,11,14,17-eicosapentaenoic acid, 7,10,13,16,19-docosapentaenoic acid, and 10,13,16-nonadecatrien-7-yn-2-one. The neolignans are one of the most important chemical markers of the Anthocerotophyta [2]. The presence of hydrophobic terpenoids is very rare in the Marchantiophyta. A few bitter-tasting kaurene diterpene glycosides have been found in *Jungermannia* species and a number of flavonoid glycosides were detected in liverworts and mosses [1–3].

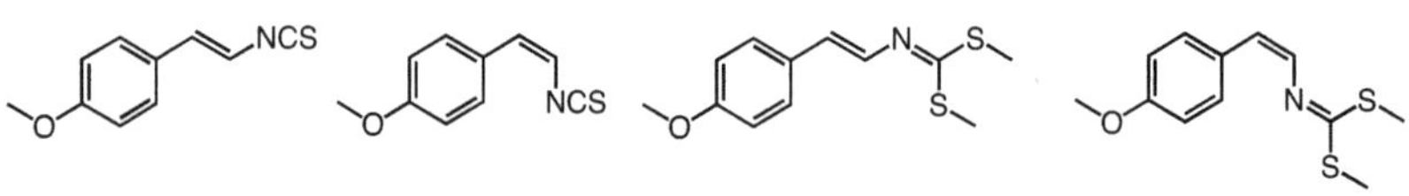

5 ((*E*)-coriandrin) **6** ((*Z*)-coriandrin) **7** ((*E*)-*O*-methyltridentalol) **8** ((*Z*)-*O*-methyltridentalol)

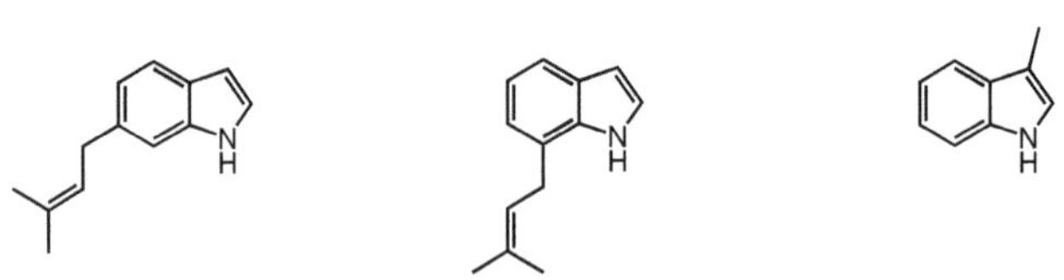

9 (6-(3-methyl-2-butenyl)indole) **10** (7-(3-methyl-2-butenyl)indole) **11** (skatole)

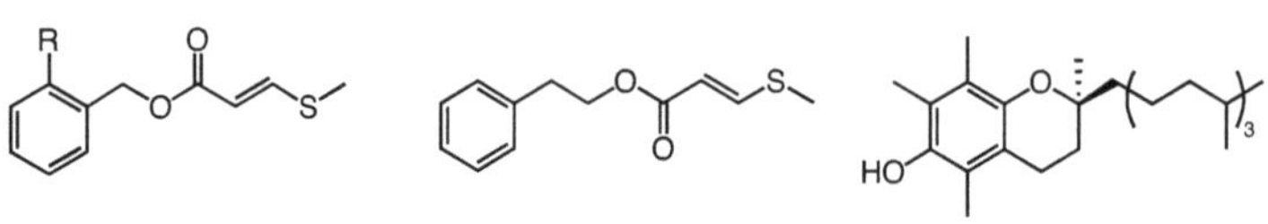

12 (isotachin A) R = H
13 (isotachin C) R = OMe
14 (isotachin B)
15 (α-tocopherol)

4 Biological Activity of Bryophytes

The bryophytes have been explored more as possible medicinal agents rather than items in the human diet, since they elaborate a number of biologically active secondary metabolites. The biological characteristics of the secondary metabolites obtained from bryophytes discussed in this contribution have been investigated in terms of their following effects: (1) characteristic odor, (2) pungency, bitterness, and sweetness, (3) allergenic contact dermatitis, (4) cytotoxicity (liverworts), (5) cytotoxicity (mosses), and (6) antimicrobial, (7) antifungal, (8) antiviral, (9) insect antifeedant, mortality and nematocidal, (10) antitrypanosomal, antileishmanial and antitrichomonal, (11) brine shrimp lethality, (12) piscicidal (13) 5-lipoxidase, calmodulin, hyaluronidase, cyclooxygenase, DNA polymerase β, α-amylase and α-glucosidase inhibitory and quinone reductase-inducing, (14) melanin production inhibition and sedative, (15) antidiabetic nephropathic, (16) superoxide release inhibitory, (17) antioxidant, (18) anti-inflammatory and NO production inhibitory, (19) acetylcholinesterase and tyrosinase inhibitory, (20) neurotrophic, (21) muscle relaxant and calcium inhibitory, (22) cardiotonic and vasopressin antagonistic, (23) liver X receptor α agonism and liver X receptor β antagonism, (24) cathepsins B and L inhibition, (25) antiplatelet, antithrombin and thromboxane synthase inhibition, (26) farnesoid X-receptor activating, (27) vasorelaxant, (28) psychological, (29) diuretic, (30) skin irritancy and tumor promoting, (31) sex pheromonal of brown algae, and (32) plant growth-regulating activities.

4.1 Characteristic Odors

Almost all liverworts emit a characteristic odor when crushed. In Table 2, representative liverworts that emit both pleasant and unpleasant odors are listed. Lipophilic terpenoids and aromatic constituents in the oil bodies are responsible for the resultant intense sweet-woody, turpentine, sweet-mossy, fungal-, carrot-, mushroom-, or seaweed-like odors [38, 80, 81]. Almost all liverworts that emit a mushroom-like odor produce oct-1-en-3-ol (**16**) and its acetate (**17**).

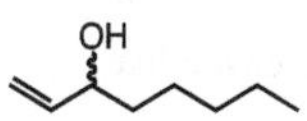

16 (oct-1-en-3-ol)

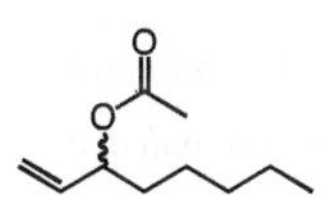

17 (oct-1-en-3-yl acetate)

One of the most characteristic liverworts possessing a pleasant odor is the Japanese *Leptolejeunia elliptica* (Plate 6), which is a minute species that grows on the leaves of *Camellia chinensis* and *Illicium anisatum*. It emits a complex citrus, minty, floral, and anise-like odor in the presence of water and the remaining scent exhibits a moldy odor.

Table 2 Characteristic odors of liverworts and their evaluation [38, 79]

Species name[a]	Property of odor	Fragrant (+) and bad odor (▲)[b]
[L] *Asterella* sp. (Japanese)	*Houttuynia cordata* like	▲
[L] *Asterella* sp. (Malaysian)	Fecal smell	▲▲
[L] *Bazzania japonica*	Sweet, balsamic, tree moss	+++
[L] *Bazzania pompeana*	Oak moss-like	+++
[L] *Chandonanthus hirtellus*	Ocean, stink bug smell	+++
[L] *Cheilolejeunea imbricata*	Milky smell	++
[L] *Chiloscyphus pallidus*	Stink bug smell	▲▲▲
[L] *Chiloscyphus profundus*	Camphoraceous	+++
[L] *Conocephalum conicum*[c]	Camphoraceous, sweet lactone	+++
[L] *Conocephalum conicum*[c]	Mossy	+++
[L] *Conocephalum conicum*[c]	Strong mossy	++
[L] *Conocephalum conicum*[c]	Strong mossy	++
[L] *Conocephalum conicum*[d]	Strong mushroom like	++++
[L] *Conocephalum japonicum*[c]	Carbonyl	++
[L] *Conocephalum japonicum*[c]	Carbonyl	++
[L] *Corsinia coriandrina*	Sulfur	▲▲
[L] *Cyathodium foetidissimum*	Fecal smell	▲▲▲▲
[L] *Fossombronia angulosa*	Ocean smell	+
[L] *Frullania davurica*	Mossy	+++
[L] *Frullania diversitexta*	Ocean smell	++
[L] *Frullania tamarisci* subsp. *tamarisci*	Carnation like	+++
[L] *Heteroscyphus coalitus*	Stink bug	▲▲
[L] *Jungermannia obovata*	Carrot like	++
[L] *Leptolejeunea elliptica*	Mold like, cresol like	+++
[L] *Leptolejeunea epiphylla*	Mold like, cresol like	+++
[L] *Lophocolea heterophylla*	Earthy	+++
[L] *Makinoa crispata*	Rooty, earthy, woody, amber	+++
[L] *Marchantia paleacea* subsp. *diptera*[c]	Perillaldehyde like	++++
[L] *Marchantia paleacea* subsp. *paleacea*[c]	Perillaldehyde like	++
[L] *Mannia fragrans*	Mossy	++
[L] *Metzgeria lindbergii*	Ocean smell, mossy	++

(continued)

Table 2 (continued)

Species name[a]	Property of odor	Fragrant (+) and bad odor (▲)[b]
[L] *Pallavicinia subciliata*	Earthy, woody, amber	++
[L] *Pellia endiviifolia*	Ocean smell	++
[L] *Plagiochila sciophila*	Sweet mossy, hay	+++
[L] *Plagiochila ovalifolia*	Mossy	+++
[L] *Porella densifolia*	Mossy	+++
[L] *Porella gracillima*	Mossy	+++
[L] *Porella japonica*	Earthy, mossy, root like, woody	++
[L] *Porella vernicosa*	Pungent	+++
[L] *Ptychanthus striatus*	Woody, earthy, mossy	+
[L] *Radula perrottetii*	Castoreum animal like	+++
[L] *Riccardia multifida* subsp. *decrescens*	Ocean smell	++
[M] *Takakia lepidozioides*	Coumarin like	+++
[L] *Targionia hypophylla*	Resinous smell	+++
[L] *Wiesnerella denudata*	Earthy, mossy, green, citrus	++

[a] [L] liverwort, [M] moss
[b] + and ▲: strength of odor
[c] Species collected in different places
[d] Cultured species

Plate 6 *Leptolejeunea elliptica* (liverwort)

In order to identify the components responsible for its characteristic odor, fresh *L. elliptica* was analyzed by headspace-solid-phase microextraction gas chromatography-mass spectrometry (HS-SPME-GC/MS). The volatile fraction contained 1-ethyl-4-hydroxybenzene (**18**, 13.4%), ethyl-4-methoxybenzene (**19**, 51.7%), and 1-ethyl-4-acetoxybenzene (**20**, 4.4%), with a mixture of these compounds exhibiting the typical odor of this liverwort. As the minor volatiles present, nine aromatic compounds also occurred, 1,2-dimethoxy-4-ethyl benzene (**21**), 4-vinylphenyl acetate (**22**), phenol (**23**), 4-ethylguaiacol (**24**), 4-vinylguaiacol (**25**), 5-vinylguaiacol (**26**), 4-vinylphenol (**27**), *m*- (**28**) and *p*-cresol (**29**), and two alkanes, *n*-pentadecane and *n*-heptadecane. In addition, methyl (*E*)-cinnamate (**30**), 4-methoxystyrene (**31**) and 13 sesquiterpenoids, α-barbatene (**32**), β-barbatene (**33**), *cis*-calamenene (**34**), β-caryophyllene (**35**), α-cedrene (**36**), β-elemene (**37**), elemol (**38**), germacrene A (**39**) (6*E*)-nerolidol (**40**), α-selinene (**41**), β-selinene (**42**), γ-selinene (**43**), and thujopsene (**44**), were detected by gas chromatographic olfactometry, together with acetic acid (**45**), linalool (**46**), isovaleric acid (**47**) and 2-acetyl-1-pyrroline (**48**), as the volatile compounds present. All of the identified compounds originated from *L. elliptica* since the detected products were not found in the host plants by the same analytical procedures as used for the fresh liverwort [82, 83].

18 (1-ethyl-4-hydroxybenzene)

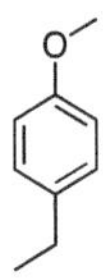

19 (1-ethyl-4-methoxybenzene)

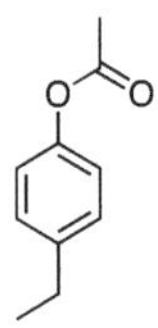

20 (1-ethyl-4-acetoxybenzene)

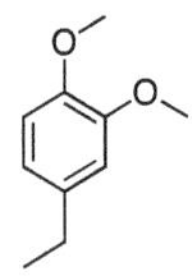

21 (1,2-dimethoxy-4-ethyl benzene)

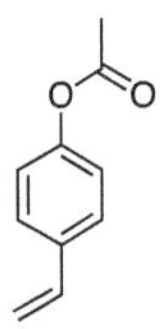

22 (4-vinylphenyl acetate)

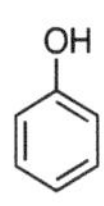

23 (phenol)

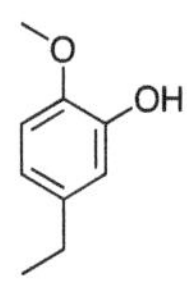

24 (4-ethylguaiacol)

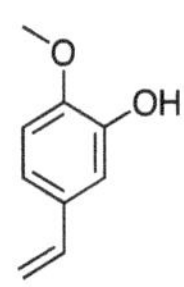

25 (4-vinylguaiacol)

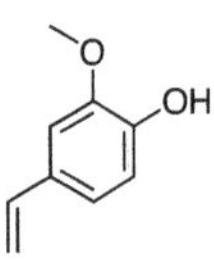

26 (5-vinylguaiacol)

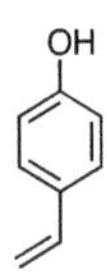

27 (4-vinylphenol)

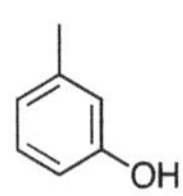

28 (*m*-cresol)

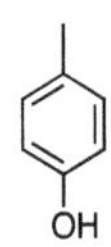

29 (*p*-cresol)

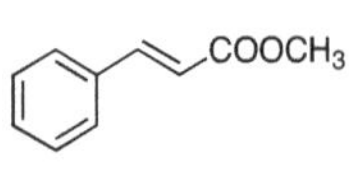

30 (methyl (*E*)-cinnamate)

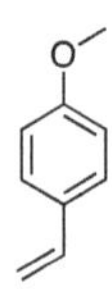

31 (4-methoxystyrene)

32 (α-barbatene)

33 (β-barbatene)

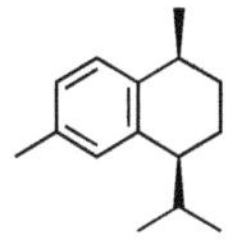
34 (*cis*-calamenene)

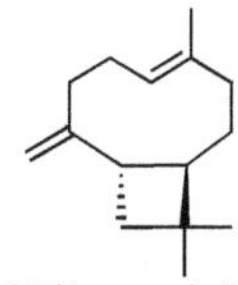
35 (β-caryophyllene)

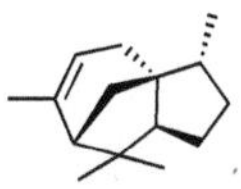
36 (α-cedrene)

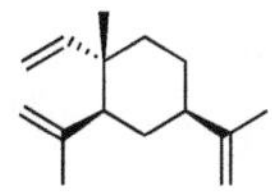
37 (β-elemene)

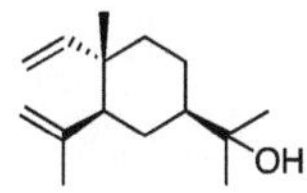

38 (elemol)

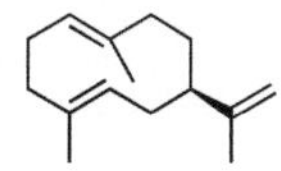
39 (germacrene A)

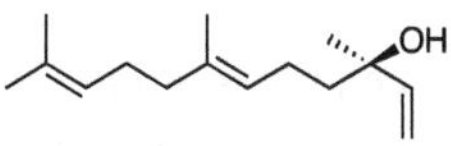

40 ((6*E*)-nerolidol)

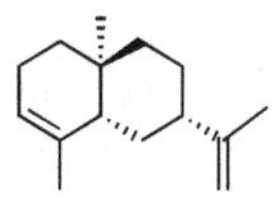
41 (α-selinene)

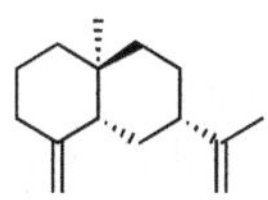
42 (β-selinene)

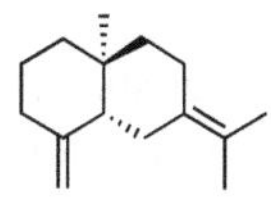
43 (γ-selinene)

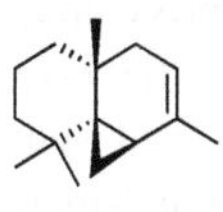
44 (thujopsene)

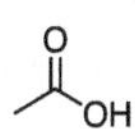

45 (acetic acid)

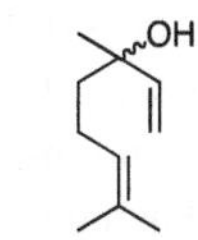

46 (linalool)

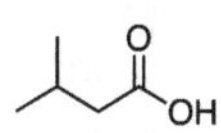

47 (isovaleric acid)

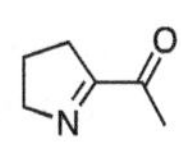

48 (2-acetyl-1-pyrroline)

In New Caledonia, there are two *Leptolejeunea* species, *L. epiphylla* and *L. leratii* (= *Colura leratii*), which both grow on the leaves of the banana and certain ferns. The two liverworts emit a strong mold-like smell and a myrrh-like odor, respectively. The former species contains 1-ethyl-4-hydroxybenzene (**18**) as the major component, with 1-ethyl-4-methoxybenzene (**19**) being absent in these New Caldonian collections. The major component of the latter species was found to be 4-methoxystyrene (**31**), having a myrrh-like odor [83].

Cheilolejeunea imbricata emits an intense milky smell. A mixture of (*R*)-dodec-2-en-1,5-olide (**49**) and (*R*)-tetradec-2-en-1,5-olide (**50**) is responsible for the characteristic smell of the liverwort [82].

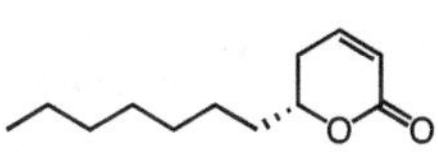

49 ((*R*)-dodec-2-en-1,5-olide)

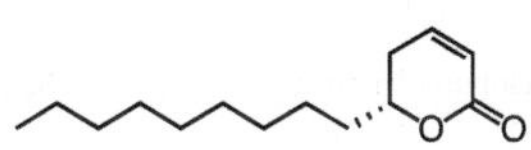

50 ((*R*)-tetradec-2-en-1,5-olide)

Plagiochila sciophila elaborates bicyclohumulenone (**51**), which possesses an aroma reminiscent of a variety of scents based on a strong woody note, resembling the odor of patchouli, vetiver, cedar wood, iris, moss, and carnation [2]. There are about 300–375 species of *Frullania* belonging to the Frullaniaceae, which are epiphytic liverworts. Tamariscol (**52**), isolated from *Frullania tamarisci* subsp. *tamarisci, F. tamarisci* subsp. *obscura, F. nepalensis,* and *F. asagrayana*, similarly possesses a

Plate 7 **a** *Conocephalum conicum* (Type 2; liverwort); **b** *C. conicum* (cultured for 6 months); **c** *C. conicum* (cultured for 9 months)

characteristic aroma reminiscent of the woody and powdery green notes of mosses, hay, costus, violet leaves, and seaweeds. Both compounds **51** and **52** are important in commerce, where they are used as perfumes as such, and as fragrance components in powdery floral-, oriental bouquet-, fantastic chypre-, fancy violet-, and white rose-components in various cosmetics. The synthetic compound epoxytamariscol (**53**) also emits the same odor as tamariscol. It is noteworthy that the *Frullania* species producing **52** only grow in high mountains [84, 85]. The total synthesis of (±)-tamariscol (**52**) was accomplished using commercially available 4-methoxy-acetophenone in 13 steps [86]. A synthetic minitamariscol, 1-hydroxy-1-(2-methyl-1-propenyl)-cyclohexane (**54**), has a sweet mossy aroma, similar to that of **52** itself [85].

There are three chemotypes of the liverwort, *Conocephalum conicum* (Plate 7). Types 1, 2, and 3 emit (–)-sabinene (**55**), (+)-bornyl acetate (**56**), and methyl (*E*)-cinnamate (**30**) as the major components, respectively, which are responsible for the characteristic odors of each type [87]. An overview of the main liverwort compounds is provided in Table 3.

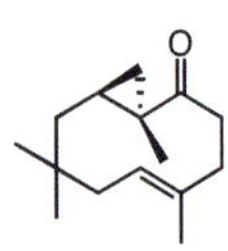

51 (bicyclohumulenone)

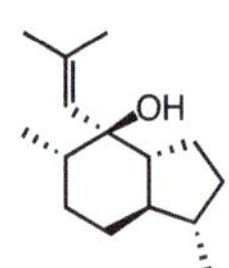

52 (tamariscol)

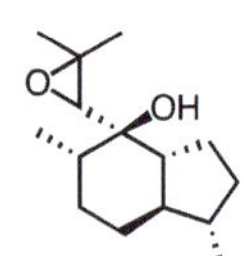

53 (epoxytamariscol)

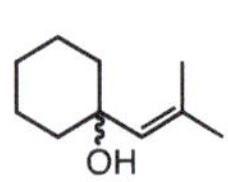

54 (1-hydroxy-1-(2-methyl-1-propenyl)-cyclohexane)

55 ((–)-sabinene)

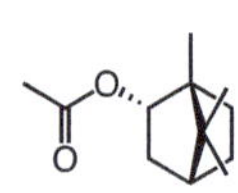

56 ((+)-bornyl acetate)

When *C. conicum* (Type 2) (Plate 7a) was stored in a glass vessel or plastic box for 1 to 9 months, the liverwort dramatically changes morphologically to a fine form

Table 3 Odiferous compounds from liverworts

Species name	Compound	Odor property	Refs.
Asterella species	Skatole (**11**)	Fecal smell	[78]
Chandonanthus hirtellus	Dictyotene (**92**), ectocarpene (**95**)	Ocean smell	[101]
Cheilolejeunea imbricata	(*R*)-Dodec-2-en-1,5-olide (**49**), (*R*)-Tetradec-2-en-1,5-olide (**50**)	Milky	[82]
Chiloscyphus pallidus	(*E*)-Dec-2-enal (**97**), (*Z*)-dec-2-enal (**98**), (*E*)-pent-2-enal (**59**), (*Z*)-pent-2-enal (**100**)	Stink bug	[102]
Colura leratii (= *Leptolejeunea leratii*)	4-Methoxystyrene (**31**)	Myrrh-like	[83]
Conocephalum conicum	1-Octen-3-ol (**16**) 1-Octen-3-yl acetate (**17**) Methyl (*E*)-cinnamate (**30**)	Strong mushroom like Medicinal	[87, 88]
Corsinia coriandrina	(*E*)-Coriandrin (**5**), (*Z*)-coriandrin (**6**), (*E*)-*O*-methyltridentatol (**7**), (*Z*)-*O*-methyltridentatol (**8**)	Faint sulfur	[77]
Cyathodium foetidissimum	Skatole (**11**), 4-methoxystyrene (**31**), 3,4-dimethoxystyrene (**96**), bicyclogermacrene (**101**), isolepidozene (**102**), lunularin (**103**), 2-aminoacetophenone (**123**)	Fecal smell Myrrh like	[103, 104]
Drepanolejeunea madagascariensis	α-Phellandrene, β-phellandrene (**69**), β-pinene (**72**), terpinen-4-ol, β-elemene (**37**), bicyclogermacrene (**101**), β-bourbonene (**1284**), δ-cadinene, α-copaene, α-cubebene, germacrene D (Herbaceous effects) Limonen-10-ol (**60**), *p*-menth-1-en-9-yl acetate, dill ether, limonene (**104**), *p*-methen-1-en-9-ol (major in three samples of *D. madagascariensis*)	Sweet, warm, woody-spicy, and herbaceous fragrance, slightly reminiscent of dill	[2]
Fossombronia angulosa	Dictyotene (**92**), multifidene (**93**), dictyopterene (**94**)	Ocean smell	[100]
Frullania tamarisci subsp. *tamarisci*, *F. tamarisci* subsp. *obscura*	Tamariscol (**52**) Epoxytamariscol (**53**) (synthetic)	Carnation like	[84–86]
Gackstroemia decipiens	(–)-13-Hydroxybergamota-2,11-diene (**78**) β-Santalene derivatives	Fruity citrus smell	[96]
Jungermannia obovata	4-Hydroxy-4-methylcyclohex-2-en-1-one (**62**)	Carrot	[90, 91]

(continued)

Table 3 (continued)

Species name	Compound	Odor property	Refs.
Leptolejeunea elliptica	1-Ethyl-4-hydroxybenzene (**18**), 1-ethyl-4-methoxybenene (**19**), 1-ethyl-4-acetoxybenzene (**20**), 1,2-dimethoxy-4-ethylbenzene (**21**), 4-ethylguaiacol (**24**)	Moldy, medicinal	[82, 83]
	4-Vinylguaiacol (**25**)	Grocery	
	5-Vinylguiaoacol (**26**)	Acid	
	4-Vinylphenol (**27**)	Green, floral	
	p-Cresol (**29**)	Sweaty	
	m-Cresol (**28**)	Waxy, clay	
	Methyl (*E*)-cinnamate (**30**)	Acetic	
	Acetic acid (**45**)	Phenolic, Medicinal	
	Linalool (**46**)	Medicinal	
	Isovaleric acid (**47**)	Medicinal	
	trans-4,5-Epoxy-(2*E*)-decenal, 2-acetyl-1-pyrroline (**48**)	Moldy, medicinal	
Lophocolea bidentata	(–)-2-Methylisoborneol (**63**), geosmin (**64**)	Earthy	[92]
Lophocolea heterophylla	(–)-2-Methylisoborneol, geosmin (**64**)	Earthy	[92]
Mannia fragrans	Grimaldone (**65**)	Strong mossy	[94]
Marchantia paleacea subsp. *diptera* *M. paleacea* subsp. *paleacea*	(–)-(*S*)-Perillaldehyde (**57**)	*Perilla frutescens*-like smell	[89]
Metzgeria species *Pallavicinia* sp. *Pellia endiviifolia* *Riccardia* sp.	Dimethyl sulfide (**91**)	Algae	[2]
Pellia epiphylla	Isoafricanol (**83**)	Sharp, aromatic	[97]
Plagiochila sciophila	Bicyclohumulenone (**51**)	Strong mossy	[2]
Porella vernicosa complex	Polygodial (**90**)	Strong pungent	[99]
Symphyogyna brongniartii	Geosmin (**64**)	Earthy	[93]
Takakia lepidozioides	Coumarin (**84**)	Strong vanillin	[98]
Targionia hypophylla	*cis*-Pinocarveyl acetate (**66**), *trans*-pinocarveyl acetate (**67**)	Resinous	[95]

(continued)

Table 3 (continued)

Species name	Compound	Odor property	Refs.
Trichocolea tomentella	Sulfur	Volcanic odor (hydrogen sulfide)	[2]
Wiesnerella denudata	1-Octen-3-ol (**16**), 1-octen-3-yl acetate (**17**), nerol (**75**), neryl acetate (**76**), linalyl acetate (**77**)	Sweet citrus	[2]

with thin gametophytes as shown in Plate 7b, c. This gives the appearance of a totally different plant compared to the original one. The stressed *C. conicum* also shows a drastically changed chemical composition. A chemical marker, (+)-bornyl acetate (**56**), of this type gradually decreased in its concentration levels and then methyl (*E*)-cinnamate (**30**) (35.3, 50.0, and 50.2%) was newly produced after 3, 6, and 9 months, respectively [88]. The chemical profile of the volatile components after 6 to 9 months cultivation of *C. conicum* was almost the same as that of the most expensive Japanese mushroom, *Tricholoma matsutake*, which has been used in Japanese cuisine, for example in consommé soup. Thus, the significant flavor of *T. matsutake* can be obtained in limitless quantities for the whole year. The Japanese *C. japonica* was classified into the same genus of *C. conicum*, although the volatile components did not change when it was cultured by the same method as that for treated *C. conicum* [88].

The large thalloid liverwort *Marchantia paleacea* subsp. *diptera* (Plate 8a) was cultured on Nipp soil or half strength of Gamborg's B5 medium including 2% sucrose in a growth cabinet at 24°C for 16 h (light) and 22°C for 8 h (dark). (–)-(*S*)-Perillaldehyde (**57**) was identified, which is the most important aroma constituent of *Perilla frutescens* (Lamiaceae) and used in Japanese cuisine and herbal medicines, and not present in the original liverwort. Compound **57** was present at 50% yield, along with 1-perillyl alcohol (**58**), *cis*-shisool (**59**), *trans*-shisool (**60**), and *p*-menth-1-en-10-al (**61**). Considering the biosynthesis pathway, only (–)-(*S*)-limonene-7-hydroxygenase may play an important role in the formation of perillaldehyde (**57**) [89]. When the present author and his colleagues cultured the same species in a similar manner to that mentioned earlier without any other medium or sucrose, (–)-(*S*)-perillaldehyde (**57**) was generated, although this took a few months (Plate 8b). *M. paleacea* subsp. *paleacea* also produces the same aldehyde when it was cultured in a sealed plastic box. Thus, (–)-(*S*)-perillaldehyde (**57**) can be produced readily for a year in this simple manner [89].

The intense, carrot-like odor *Jungermannia obovata* is due to the presence of 4-hydroxy-4-methyl-cyclohex-2-en-1-one (**62**) [90, 91]. The strong, and distinct mossy odor of *Lophocolea heterophylla* and *L. bidentata* is due to a mixture of (–)-2-methylisoborneol (**63**) and geosmin (**64**) [92]. The latter compound was also found in the cultured liverwort, *Symphyogyna brongniartii* [93].

Plate 8 **a** *Marchantia paleacea* subsp. *diptera* (liverwort); **b** *M. paleacea* subsp. *diptera* (cultured for 3 months)

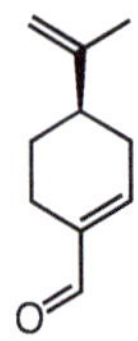

57 ((–)-(*S*)-perillaldehyde)

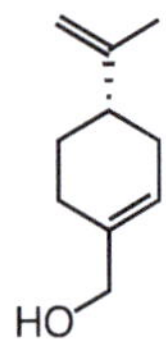

58 (1-perillyl alcohol)

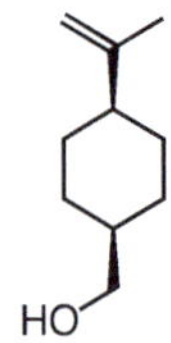

59 (*cis*-shisool)

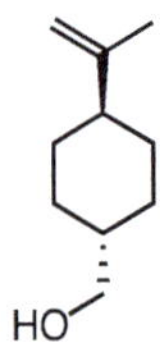

60 (*trans*-shishool)

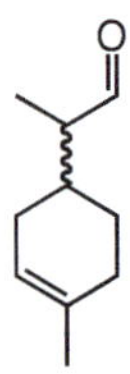

61 (*p*-menth-1-en-10-al)

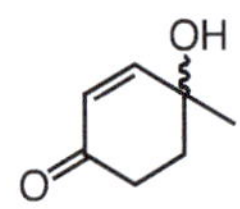

62 (4-hydroxy-4-methyl-cyclohex-2-en-1-one)

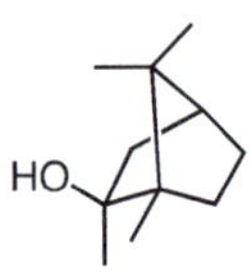

63 ((–)-2-methyl-*iso*-borneol)

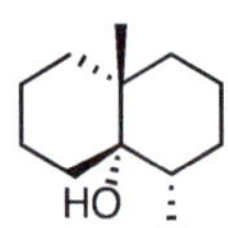

64 (geosmin)

The strong, sweet mossy note of *Mannia fragrans* (Plate 9) is attributable to grimaldone (**65**) [94]. In turn, the sweet turpentine-like odor of the French *Targionia hypophylla* is due to a mixture of *cis*- and *trans*-pinocarveyl acetates (**66**, **67**) [95]. The strong sweet-mushroom-like scent of the ether extract of *Wiesnerella denudata* comes from (+)-bornyl acetate (**56**) and a mixture of the monoterpene hydrocarbons, α-terpinene (**68**), β-phellandrene (**69b**), terpinolene (**70**), α-pinene (**71**), β-pinene

Plate 9 *Mannia fragrans* (liverwort)

(**72**), and camphene (**73**). The odor of the steam distillate of *W. denudata* is weaker than that of its ether extract. The steam distillate contains γ-terpinene (**74**, 31%), nerol (**75**, 14%), neryl acetate (**76**, 27%), and linalyl acetate (**77**), but the content of 1-octen-3-ol (**16**, 7%) and its acetate (**17**, 2%) is lower than that of the steam distillate of *Conocephalum conicum* [2].

65 (grimaldone)

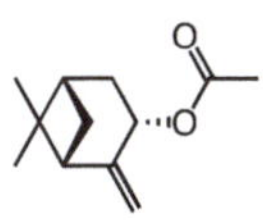

66 (*cis*-pinocarveyl acetate)

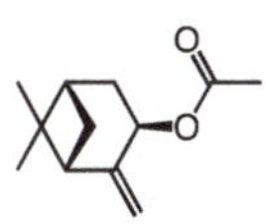

67 (*trans*-pinocarveyl acetate)

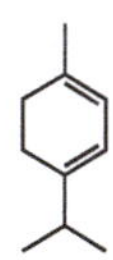

68 (α-terpinene)

69 (β-phellandrene)

70 (terpinolene)

71 (α-pinene)

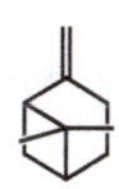

72 (β-pinene)

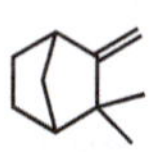

73 (camphene)

74 (γ-terpinene)

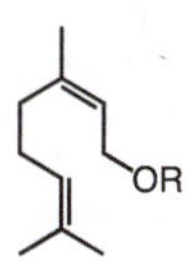

75 (nerol, R = H)
76 (neryl acetate, R = Ac)

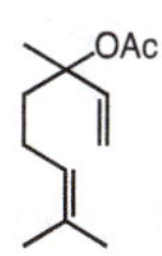

77 (linalyl acetate)

Gackstroemia decipiens emits a characteristic scent that is due to the presence of a mixture of (–)-13-hydroxybergamota-2,11-diene (**78**) and four santalane derivatives **79–82**. These compounds were characterized by their olfactory effects [96]. Isoafricanol (**83**), isolated from *Pellia epiphylla*, is responsible for the typical odor of its sporophyte [97].

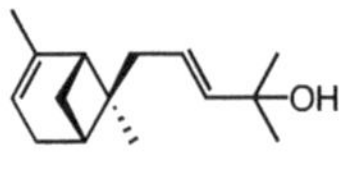

78 ((–)-13-hydroxy-bergamota-2,11-diene)

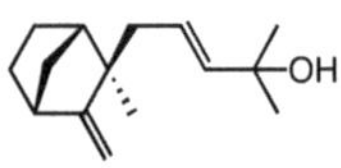

79 (β-santalene derivative)

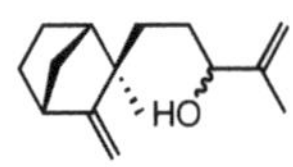

80 (β-santalene derivative)

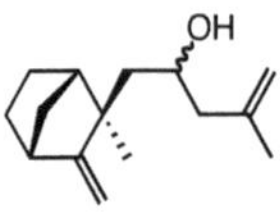

80 (β-santalene derivative)

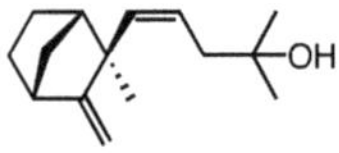

79 (β-santalene derivative)

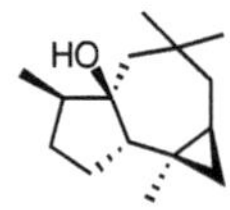

83 (isoafricanol)

Takakia lepidozioides found in Japan was thought originally to be a liverwort, but, from a recent morphological study, it has been classified in the Bryophyta. This fresh or dried moss emits a powerful coumarin-like odor when crushed. In fact, the fresh leaves contain coumarin (**84**) and 1,4-dihydroquinone (**85**) as the major components, along with α-tocopherol (**15**), dihydrocoumarin (**86**), 1,4-benzoquinone (**87**), dihydrobenzofuran (**88**), and α-asarone (**89**) as the minor components [98].

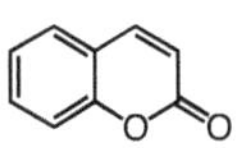

84 (coumarin)

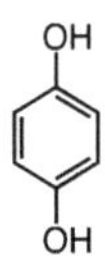

85 (1,4-dihydroquinone)

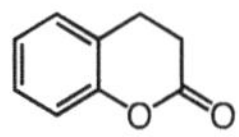

86 (dihydrocoumarin)

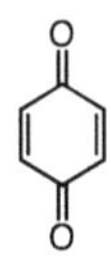

87 (1,4-benzoquinone)

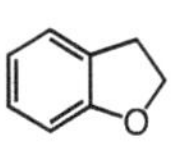

88 (dihydrobenzofuran)

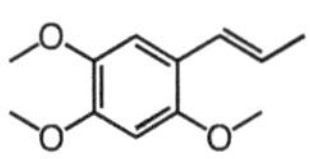

89 (α-asarone)

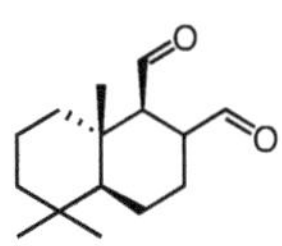

90 (polygodial)

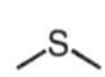

91 (dimethyl sulfide)

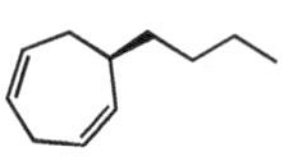

92 (dictyotene)

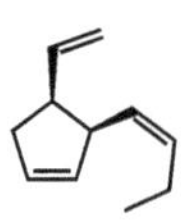

93 (multifidene)

94 (dictyopterene)

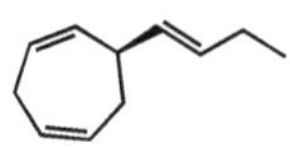

95 (ectocarpene)

When the fresh or dried liverwort, *Porella vernicosa* complex, is crushed, immediately an intensely pungent odor is emitted. This odor is due to a mixture of the pungent polygodial (**90**) and other unidentified volatile compounds. When *Porella* *vernicosa* was cultured on certain media, **90** is also created readily [99]. This pungent compound shows significant biological activities, as described later.

Liverworts belonging to the Metzgeriales emit an ocean or marine algal-like smell. When they are dried, immediately one detects this characteristic smell. Their GC/MS analysis indicated that *Blasia, Metzgeria, Pallavicinia*, and *Riccardia* contain dimethyl sulfide (**91**), which is responsible for the typical odor of some sea algae used in Japanese cuisine [2].

The Greek *Fossombronia angulosa* (Plate 10) is chemically very characteristic, since it elaborates dictyotene (**92**), multifidene (**93**), and dictyopterene (**94**), which were all previously isolated from a brown algal source [100]. The French Polynesian liverwort, *Chandonanthus hirtellus*, also was found to produce dictyotene (**92**) and ectocarpene (**95**) [101]. These hydrocarbons are sex pheromones of marine brown algae, such as *Dictyota* species. Dictyotene (**92**) is responsible for the ocean-like smell.

Trichocolea tomentella and similar species from New Zealand emit a sulfur smell [2]. In fact, pure sulfur crystals were isolated from this liverwort. The Mediterranean thalloid liverwort *Corsinia coriandrina* also emits a faint horseradish-like odor. It is noteworthy that this liverwort biosynthesizes nitrogen- and sulfur-containing aromatic compounds, namely, (*E*)- (**5**) and (*Z*)-coriandrins (**6**), and (*E*)- (**7**) and (*Z*)-*O*-methyltridentatols (**8**), which emit only a weak sulfur-like odor [77]. The characteristic odiferous compounds of the remaining liverworts listed in Table 2 have not yet been isolated.

Plate 10 *Fossombronia angulosa* (liverwort)

While many liverworts emit a pleasant odor, others have an incredibly intense bad smell. An unidentified Malaysian liverwort (*Asterella* or *Mannia*) was found to contain only two volatiles, skatole (**11**, 20%), which is responsible for its unpleasant fecal-like smell, and 3,4-dimethoxystyrene (**96**, 80%) [78]. The New Zealand *Chiloscyphys pallidus* emits a very strong stink bug-like smell. In fact, this stem-leafy liverwort produces exactly the same (*E*)- (**97**) and (Z)-dec-2-enal (**98**) and (*E*)- (**99**) and (Z)-pent-2-enal (**100**) constituents as those from some stink bugs. It is the first example of a liverwort found to biosynthesize insect defense compounds [102].

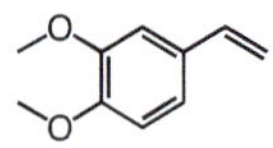

96 (3,4-dimethoxystyrene)

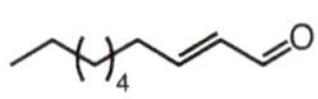

97 ((*E*)-dec-2-enal)

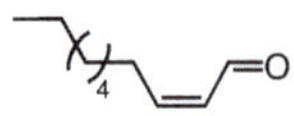

98 ((*Z*)-dec-2-enal)

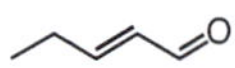

99 ((*E*)-pent-2-enal)

100 ((*Z*)-pent-2-enal)

The liverwort, *Cyathodium foetidissimum* belonging to the Marchantiaceae, is one of the most powerfully unpleasant odor emitting species, which grows in Panama and in several southern hemisphere islands, as for example, Tahiti. When the Tahitian *C. foetidissimum* (Plate 11) was stored within three layers of zipped bags, its characteristic fecal-like odor permeated even through the double layered bag after the outer bag was removed [103].

Plate 11 *Cyathodium foetidissimum* (liverwort)

In 2018, Sakurai et al., analyzed the chemical constituents of the volatiles of this taxon using solid-phase microextraction to identify 4-methoxystyrene (**31**, 24%), 3,4-dimethoxystyrene (**96**, 29%), skatole (**11**, 16%), bicyclogermacrene (**101**) and its diastereoisomers, isolepidozene (**102**), and lunularin (**103**) as the predominant components, among which **11** was recognized as being responsible for the bad odor of this species.

Cyanthodium foetidissimum elaborates not only the above major volatiles, but also the monoterpenoids linalool (**46**), limonene (**104**), 1,8-cineole (**105**), menthol (**106**), and α-terpineol (**107**), the sesquiterpenoids β-elemene (**37**), β-caryophyllene (**35**), α-selinene (**41**), δ-elemene (**108**), longifolene (**109**), calarene (**110**), and β-cadinene (**111**), and the polyketides *n*-hexanal (**112**), isobutyl butyrate (**113**), methyl hexanoate (**114**), butyl butyrate (**115**), isoamyl isovalerate (**116**), 6-methyl-5-hepten-2-one (**117**), nonanal (**118**), tetradecane, 2-ethyl-hexanol (**119**), decanal (**120**), the hydrocarbons pentadecane, dodecane, and tridecane, and the aromatic compounds, *m*- (**28**) and *p*-cresol (**29**), benzyl alcohol (**121**), phenol (**23**), 2-phenylethanol (**122**), 2-aminoacetophenone (**123**), 4-acetylacetophenone (**124**), and veratraldehyde (**125**) as minor components [103].

The characteristic odor of *C. foetidissimum* is reminescent of an aged "old person" smell, or old "chest of drawers", or "nostalgic" odor. Skatole (**11**) has been used as an alternative aroma having a civet-like scent after its animal of origin was prohibited in certain countries. 4-Methoxystyrene (**31**) shows the characteristic aroma of the Ayrshire Splendens and Anbridge Rose varieties of rose with a myrrh-type smell, and of *Rosa hurtula* (white rose) that possesses a 1,3-dimethoxy-5-methylbenzene (**126**)-like odor. The GC/O analysis of their volatiles showed a nostalgic and old-armoire odor. 4-Methoxystyrene (**31**) also has a ten times more potent skin-whitening effect than kojic acid and possesses the characteristic smell of the field rose (*Rosa arvensis*). The pleasant and nostalgic old person's odor of *C. foetidissimum* is due to the presence of appropriate proportions of 4-methoxystyrene (**31**) to skatole (**11**) and 3,4-dimethoxystyrene (**96**). A few alkyl aldehydes were detected in *C. foetidissimum*, however, (2*E*)-nonenal (**127**), which possesses an aged person smell, was not detected in the volatile fraction. Gas chromatography-olfactometry analysis of this showed the presence of 2-aminoacetophenone (**123**), which contributed considerably to the characteristic scent of this rare liverwort as well as skatole (**11**) [103]. A Costa Rican sample of a volatile fraction of *C. foetidissimum* was analyzed by GC-MS to detect 50% of skatole (**11**) [104].

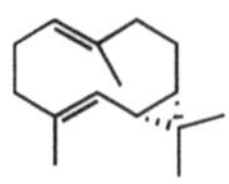
101 (bicyclogermacrene)

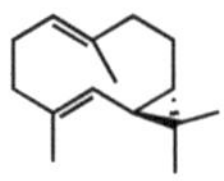
102 (isolepidozene)

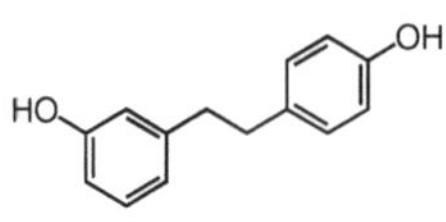

103 (lunularin)

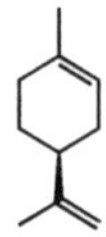
104 (limonene)

105 (1,8-cineole)

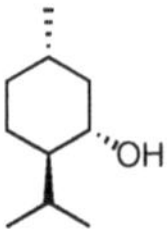

106 (menthol)

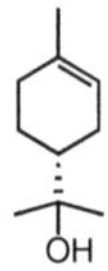

107 (α-terpineol)

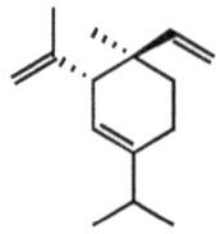
108 (δ-elemene)

109 (longifolene)

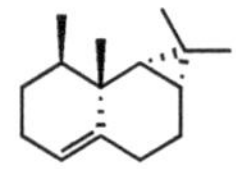
110 (calarene)

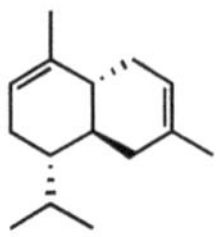
111 (β-cadinene)

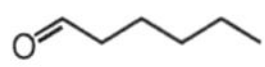

112 (*n*-hexanal)

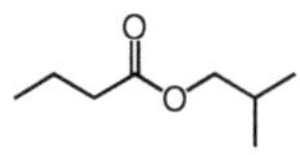

113 (isobutyl butyrate)

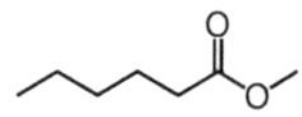

114 (methyl hexanoate)

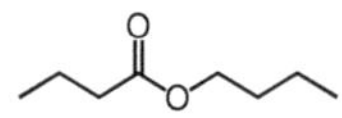

115 (butyl butyrate)

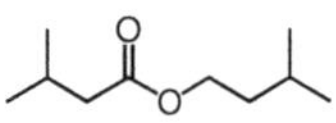

116 (isoamyl isovalerate)

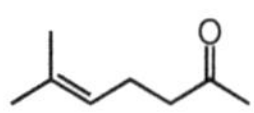

117 (6-methyl-5-hepten-2-one)

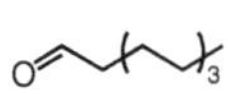

118 (nonanal)

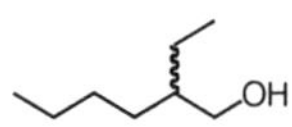

119 (2-ethyl-hexanol)

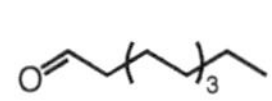

120 (decanal)

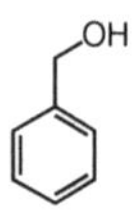

121 (benzyl alcohol)

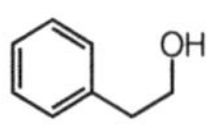

122 (2-phenylethanol)

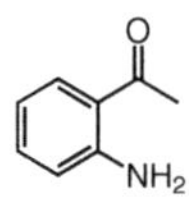

123 (2-amino-acetophenone)

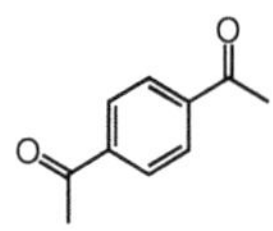

124 (4-acetylaceto-phenone)

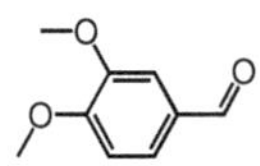

125 (veratraldehyde)

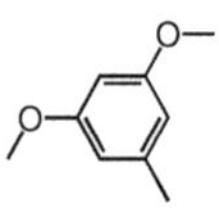

126 (1,3-dimethoxy-5-methylbenzene)

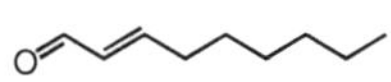

127 ((2*E*)-nonenal)

Drepanolejeunea madagascariensis emits a pleasant, sweet, warm, woody-spicy, and herbaceous fragrance. Using headspace solid-phase microextraction α- (**69a**) and β-phellandrene (**69b**) and several sesquiterpene hydrocarbons, such as β-elemene (**37**), bicyclogermacrene (**101**), and β-bourbonene (**1284**) could be characterized [2].

In contrast to the liverworts, almost no mosses emit any characteristic odor, since they do not possess any cellular oil bodies. When one crushes mosses, their smell may be detected and usually has a green note, similar to *n*-hexanol or *n*-hexanal. When a large moss sample is collected, a strong sweet fungal odor may be apparent on the back of those specimens that are attached to the soil. However, this odor does not originate from the mosses per se, but rather from geosmin (**64**) [2], which is biosynthesized by some actinomycete species in the soil.

4.2 Pungency, Bitterness, and Sweetness

Some genera of liverworts, such as the *Hymenophyton, Lobatiriccardia, Pallavicinia, Pellia, Porella, Trichocoleopsis,* and *Wiesnerella* species elaborate potent pungent constituents [2, 79, 105], which exhibit interesting biological activities described in subsequent sections. An overview of the particular species concerned and their tastes is provided in Table 4. Most North American liverworts produce unpleasant-tasting compounds, of which some taste like immature green peas or pepper [106]. Species in the genera *Anastrepta, Lophozia,* and *Scapania,* and many other stem-leafy liverworts produce intensely bitter principles [2, 3].

Members of the *Porella vernicosa* complex (*P. arboris-vitae, P. fauriei, P. gracillima, P. obtusata* subsp. *macroloba, P. roellii,* and *P. vernicosa*) contain intensely pungent compounds [1–3]. *Jamesoniella autumnalis* affords an intensely bitter principle for which the taste resembles that of the leaves of lilac (*Syringa vulgaris*), or the roots of *Gentiana scabra* var. *orientalis.* The strong hot taste of the *Porella vernicosa* complex is due to the presence of the drimane sesquiterpene dialdehyde, (–)-polygodial (**90**) [1, 107, 108], which is a major component of the Japanese medicinal plant, *Polygonum hydropiper,* the Malaysian *P. minus,* and the Argentinean *P. punctatum* var. *punctatum* (Polygonaceae) [108].

The Malaysian *P. minus* and Japanese *P. hydropiper* have been used as vegetables. Polygodial (**90**) is found not only in higher plants and liverworts, but also in some ferns, the Argentinean *Thelypteris hispidula* and the New Zealand *Blechnum fluviatile,* as the major components [2].

The sacculatane diterpene dialdehyde, sacculatal (**128**), 1β-hydroxysacculatal (**129**), together with several sacculatane-type diterpenoids, 18- (**130**) and 19-hydroxysacculatal (**131**), and two eudesmanolides, *ent*-diplophyllolide (**132**), *ent*-7α-hydroxydiplophyllolide (**133**), and the germacranolide, tulipinolide (**134**), which possess potent pungent effects, were isolated from *Pellia endiviifolia* (Plate 12), *P. epiphylla, Trichocoleopsis sacculata, Chiloscyphus polyanthos, Diplophyllum albicans,* and *Wiesnerella denudata,* respectively [1–3, 109].

Table 4 Pungent, bitter, and sweet-tasting bryophytes, and their strength of taste [38, 79]

Species name[a]	Pungency (◆), bitterness (✱), sweetness (❖)	Strength
[L] *Anastrepta orcadensis*	✱	+++
[L] *Barbilophozia lycopodioides*	✱	+++
[L] *Gymnocolea inflata*	✱	++++
[L] *Chiloscyphus polyanthos*	◆	+
[L] *Diplophyllum albicans*	◆	+
[L] *Hymenophyton flabellatum*	◆	+
[L] *Jamesoniella autumnalis*	✱	++++
[L] *Jungermannia inflata*	✱	+++
[L] *Lobatiriccardia yakushimensis*	◆	++
[L] *Pallavicinia levieri*	◆	++
[L] *Pellia endiviifolia*	◆	++
[L] *Plagiochila asplenioides*	◆	++
[L] *Plagiochila fruticosa*	◆	++
[L] *Plagiochila hattoriana*	◆	++
[L] *Plagiochila ovalifolia*	◆	+++
[L] *Plagiochila pulcherrima*	◆	+
[L] *Plagiochila semidecurrens*	◆	++
[L] *Plagiochila yokogurensis*	◆	++
[L] *Porella acutifolia* subsp. *tosana*	◆	+
[L] *Porella arboris-vitae*	◆	++++
[L] *Porella fauriei*	◆	++
[L] *Porella gracillima*	◆	+++
[L] *Porella obtusata* var. *macroloba*	◆	++
[L] *Porella roellii*	◆	+++
[L] *Porella vernicosa*	◆	+++
[L] *Scapania undulata*	✱	+
[L] *Trichocoleopsis sacculata*	◆	++
[L] *Wiesnerella denudata*	◆	+
[M] *Fissidens* species	❖	+++
[M] *Rhodobryum gigantea*	❖	+++

[a] [L] liverwort, [M] moss

Plate 12 *Pellia endiviifolia* (liverwort)

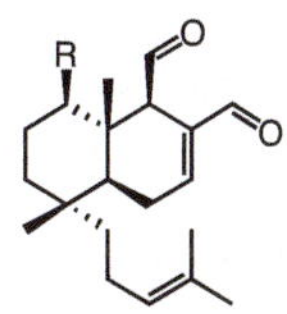

128 (sacculatal, R = H)
129 (1β-hydroxysacculatal, R = OH)

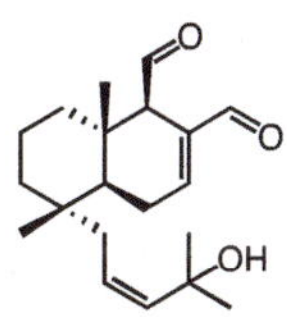

130 (18-hydroxysacculatal)

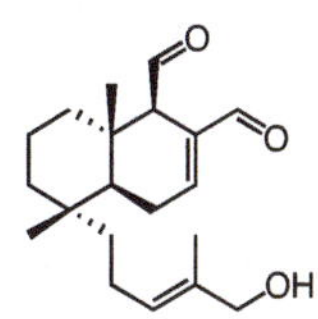

131 (19-hydroxysacculatal)

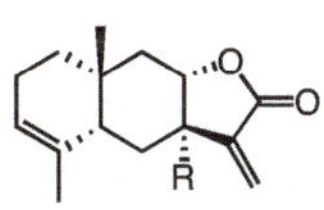

132 (*ent*-diplophyllolide, R = H)
133 (*ent*-7β-hydroxydiplophyllolide, R = OH)

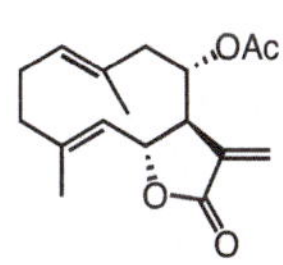

134 (tulpinolide)

The hot tastes of two *Fossombronia* species, *F. alaskana* and *F. wondraczeki*, and of *Pallavicinia levieri* and *Lobatiriccardia yakushimensis* (= *Riccardia lobata* var. *yakushimensis*), which belong to the Metzgeriales, are also due to sacculatal (**128**) [110]. Polygodial (**90**) and **128** have been obtained from cell suspension cultures of *Porella vernicosa* and *Pellia endiviifolia*, respectively [2, 3, 111, 112].

When one chews a whole plant of the stem-leafy liverworts, *Plagiochila asplenioides, P. fruticosa, P. hattoriana, P. ovalifolia, P. pulcherrima* (Plate 13), or *P. yokogurensis*, which contain the 2,3-*seco*-aromadendrane sesquiterpene hemiacetals plagiochiline A (**135**) and plagiochiline I (**136**), a potent pungent taste slowly develops. It is suggested that both compounds might be converted into a pungent unsaturated dialdehyde by human saliva. In fact, as shown in Scheme 1, treatment

Plate 13 *Plagiochila pulcherrima* (liverwort)

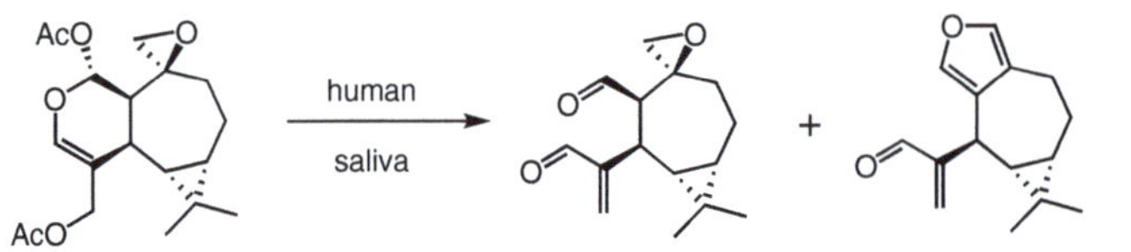

135 (plagiochiline A) **137** (plagiochilal B) **138** (furanoplagiochilal)

Scheme 1 Formation of two pungent 2,3-*seco*-aromadendranes, plagiochilal B (**137**) and furanoplagiochilal (**138**) from plagiochiline A (**135**) by human saliva

of **135** with amylase in phosphate buffer or with human saliva produced two strong pungent components, plagiochilal B (**137**), for which the partial structure is similar to that of the pungent drimane-type sesquiterpene dialdehyde, (–)-polygodial (**90**), and furanoplagiochilal (**138**) [113].

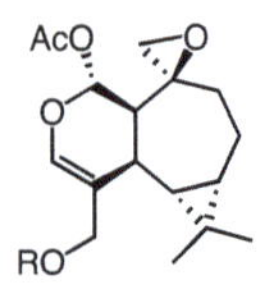

135 (plagiochiline A, R = Ac)
136 (plagiochiline I, R = H)

The pungent taste of *Porella acutifolia* subsp. *tosana* is due to the presence of the hydroperoxysesquiterpene lactones, 1α- (**139**), and 1β-hydroperoxy-4α,5β-epoxygermacra-10(14),11(13)-dien-12,18α-olide (**140**) [114].

The New Zealand liverwort, *Hymenophyton flabellatum,* produces a different pungently tasting compound from the other aforementioned liverworts, namely, 1-(2,4,6-trimethoxyphenyl)-but-(2*E*)-en-1-one (**141**) [115].

Most of the species belonging to the Lophoziaceae produce bitter compounds. *Jamesoniella autumnalis* is the epitome of bitter-tasting liverworts, since when one chews it, the intense bitter taste remains on the tongue for at least five hours. This taste is due to a complex mixture of clerodane diterpenoids, such as the furanoclerodane lactones jamesoniellides A, D, E, F, H, and I (**142–147**) [2, 3, 75].

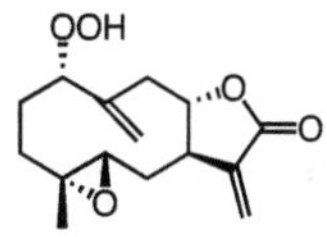

139 (1α-hydroperoxy-4α,5β-epoxygermacra-10(14),11(13)-dien-12,18α-olide)

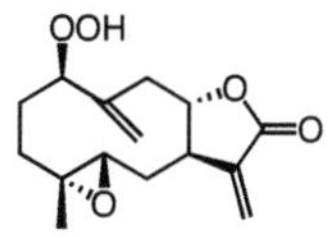

140 (1β-hydroperoxy-4α,5β-epoxygermacra-10(14),11(13)-dien-12,18α-olide)

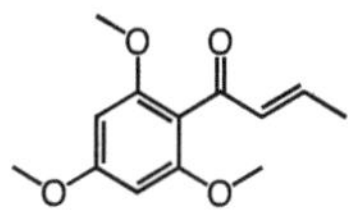

141 (1-(2,4,6-trimethoxyphenyl)-but-(2*E*)-en-1-one)

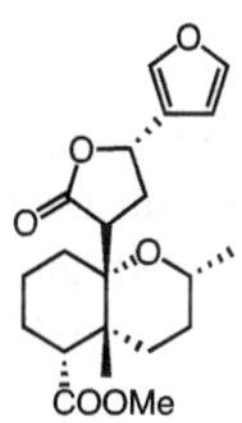

142 (jamesoniellide A)

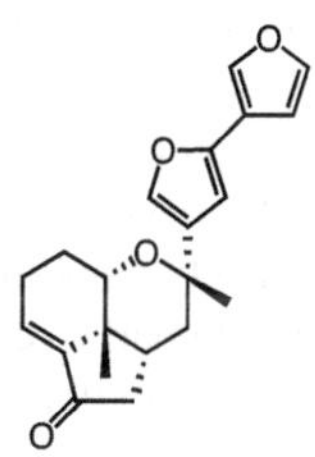

143 (jamesoniellide D)

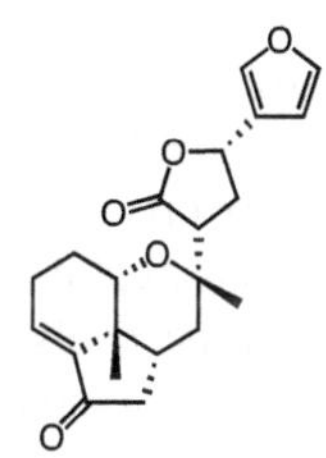

144 (jamesoniellide E)

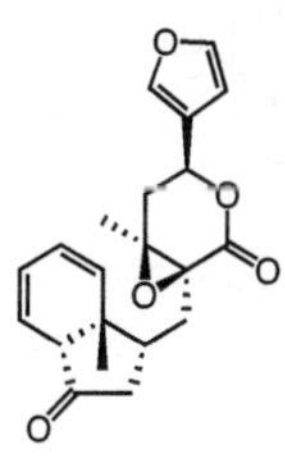

145 (jamesoniellide F)

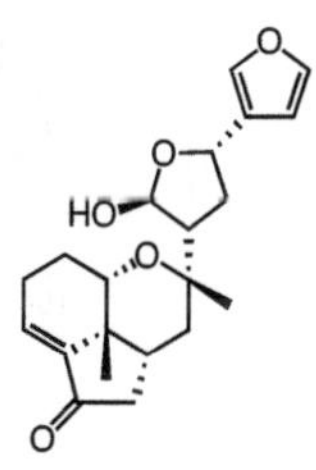

146 (jamesoniellide H)

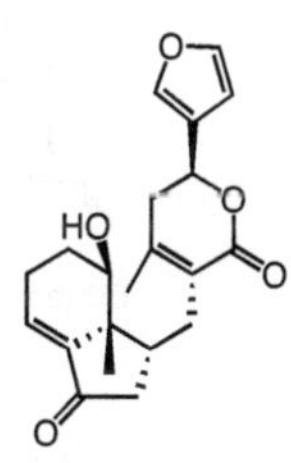

147 (jamesoniellide I)

Gymnocolea inflata is also persistently bitter, and induces vomiting when a few leaves are chewed for several seconds. This surprisingly intense bitterness is due to its clerodane diterpene lactone constituent, gymnocolin A (**148**) [116].

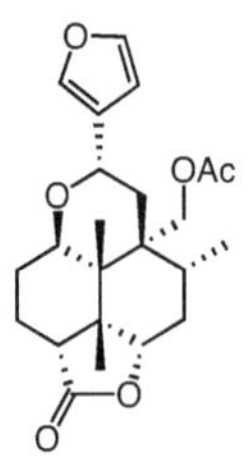

148 (gymnocolin A)

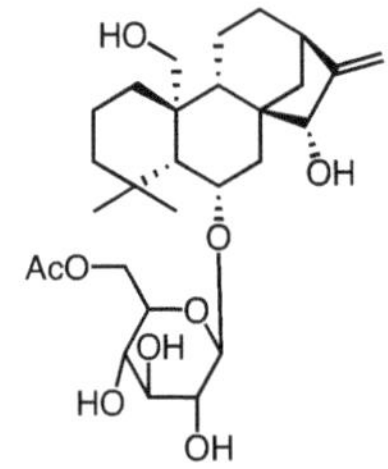

149 (infuscaside A)

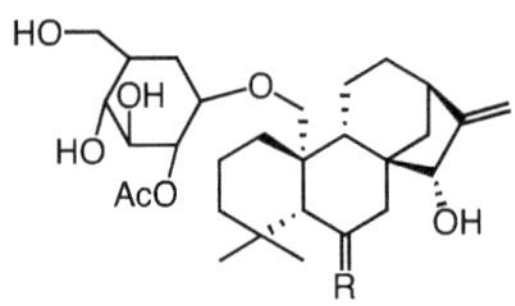

150 (infuscaside B, R = O)
151 (infuscaside C, R = H_2)

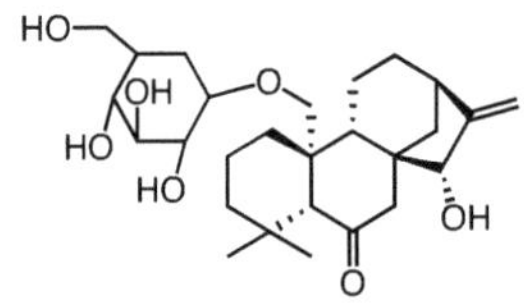

152 (infuscaside D)

Jungermannia infusca has an intensely bitter taste due to the presence of infuscasides A–D (**149–152**) and E [117]. These metabolites were the first diterpene glycosides isolated from liverworts.

Anastrepta orcadensis, *Barbilophozia lycopodioides,* and *Scapania undulata* are also bitter liverworts, from which were isolated the highly oxygenated bitter-tasting diterpenoids anastreptin A (**153**), barbilycopodin (**154**) [91, 118], and scapanin A (**155**) [119], respectively. 10-Deacetoxybarbilycopodin (**156**) from *Barbilophozia floerkei* also shows a moderately bitter taste [91, 118].

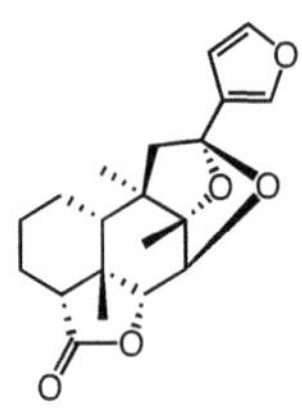

153 (anastreptin A)

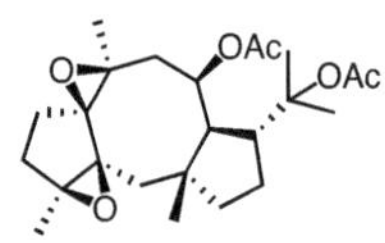

154 (barbilycopodin)

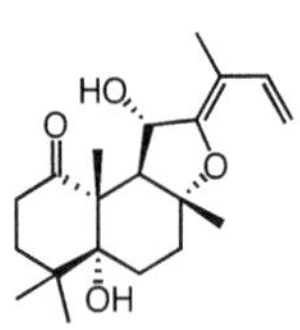

155 (scapanin A)

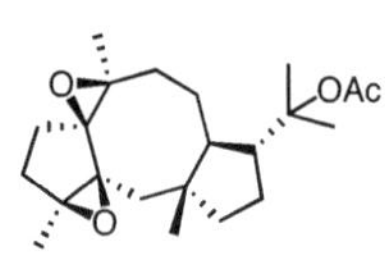

156 (10-deacetoxy-barbilycopodin)

Plagiochiline B (**157**) from *Plagiochila hattoriana* and perrottetianal A (**158**) from *Porella perrottetiana* are structurally similar to the pungent plagiochiline A (**135**) and sacculatal (**128**), respectively. However, the sesquiterpene hemiacetal **157** and the diterpene dialdehyde **158** show only a moderate bitter taste [1, 38]. An overview of the taste profiles of the compounds isolated from bryophytes is provided in Table 5.

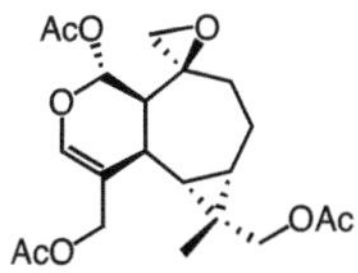

157 (plagiochiline B)

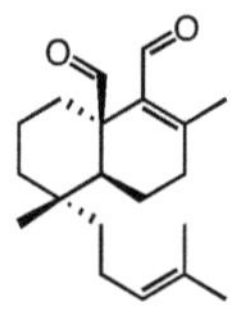

158 (perottetianal A)

There are several sweet-tasting mosses, such as *Fissidens* and *Rhodobryum* species. Almost all *Fissidens* species produce an unpleasant but persistent sweet taste. When one chews the fresh mosses, *Fissidens japonica* or *Rhodobryum giganteum*, this persistent sweet taste, which is not eliminated from the tongue, occurs. However, the individual sweet principles have not yet been isolated and identified in these particular mosses except for the presence of sucrose as a minor component [75].

4.3 Allergenic Contact Dermatitis

Several mosses and liverworts induce potent or weak allergenic contact dermatitis. Such species are shown in Table 6. In particular, two epiphytic liverworts, *Frullania dilatata* and *F. tamarisci* subsp. *tamarisci,* contain powerful allergens in their oil bodies. The allergy-inducing compounds of the first mentioned species are the sesquiterpene lactones, (+)-frullanolide (**3**), (+)-β-cyclocostunolide (**159**), (+)-oxyfrullanolide (**160**), (+)-eremofrullanolide (**161**), with (–)-frullanolide (**4**), the enantiomer of **3**, being obtained from the latter species [120], as shown in Table 7 and illustrated in Plate 14.

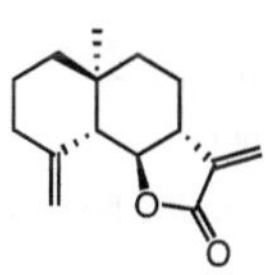

159 ((+)-β-cyclocostunolide)

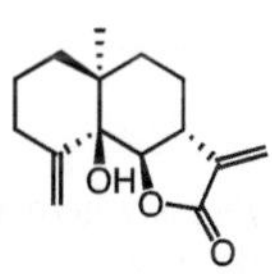

160 ((+)oxyfrullanolide)

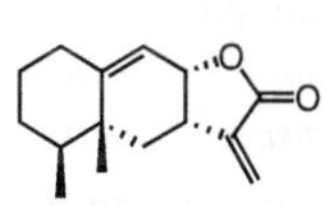

161 (eremofrullanolide)

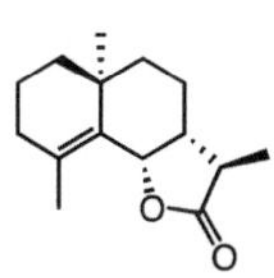

162 (*ent*-dihydrofrullanolide)

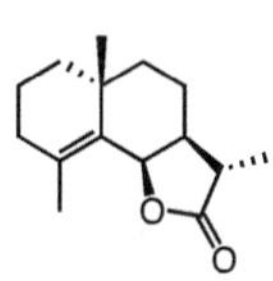

163 (dihydrofrullanolide)

Table 5 Pungent, bitter, and sweet-tasting substances from bryophytes

Species name (liverwort, moss)	Compound or solvent extract	Taste	Refs.
Chiloscyphus polyanthos	*ent*-Diplophyllolide (**132**), *ent*-7α-hydroxydiplophyllolide (**133**)	Weakly hot	[1–3]
Diplophyllum albicans	*ent*-Diplophyllolide (**132**)	Weakly hot	[1–3]
Fossombronia alaskana *F. wondraczeki*	Sacculatal (**128**)	Persistent hot	[3]
Hymenophyton flabellatum	1-(2,4,6-Trimethoxyphenyl)-1-but-(2*E*)-en-1-one (**141**)	Weakly hot	[115]
Lobatiriccardia yakushimensis (= *Riccardia lobata* var. *yakushimensis*)	Sacculatal (**128**)	Persistent hot	[110]
Pallavicinia levieri	Sacculatal (**128**)	Persistent hot	[2]
Pellia endiviifolia	Sacculatal (**128**), 1β-hydroxysacculatal (**129**), 18-hydroxysacculatal (**130**), 19-hydroxysacculatal (**131**)	Persistent hot	[1–3, 109]
Pellia epiphylla	Sacculatal (**128**)	Persistent hot	[3]
Plagiochila adiantoides *P. fruticosa* *P. hattoriana* *P. ovalifolia* *P. pulcherima* *P. yokogurensis* and many other pungent *Plagiochila* spp.	Plagiochiline A (**135**), plagiochilal B (**137**)[a], furanoplagiochilal (**138**)[a]	Persistent hot	[1–3, 113]
Plagiochila yokogurensis	Plagiochiline I (**136**)	Persistent hot	[2]
Porella acutifolia subsp. *tosana*	1α-Hydroperoxy-4α,5β-epoxy-germacra-10(14), 11(13)-dien-12,8α-olide (**139**), 1β-hydroperoxy-4α,5β-epoxy-germacra-10(14), 11(13)-dien-12,8α-olide (**140**)	Weakly hot	[114]

(continued)

Table 5 (continued)

Species name (liverwort, moss)	Compound or solvent extract	Taste	Refs.
P. arboris-vitae *P. canariensis* *P. fauriei* *P. gracillima* *P. obtusata* subsp. *macroloba* *P. roellii* *P. vernicosa*	(–)-Polygodial (**90**)	Persistent hot	[1, 107, 108]
Wiesnerella denudata	Tulipinolide (**134**)	Weakly hot	[2]
Anastrepta orcadensis	Anastreptin A (**153**)	Potent bitterness	[91]
Barbilophozia attenuata *B. barbata* *B. floerkei* *B. hatcheri* *B. lycopodioides* *Chandonanthus setiformis*	Barbilycopodin (**154**)	Potent bitterness	[2, 3, 52, 90]
Barbilophozia floerkei	10-Deacetoxybarbilycopodin (**156**)	Moderately bitter	[91, 118]
Gymnocolea inflata	Gymnocolin A (**148**)	Potent and persistent bitterness	[116]
Jamesoniella autumnalis	Jamesoniellides A, D, E, F, H, I (**142–147**)	Potent and persistent bitterness	[2, 3, 75]
Jungermannia infusca	Infuscaside A (**149**), infuscaside B (**150**), infuscaside C (**151**), infuscaside D (**152**), infuscaside E	Potent persistent bitterness	[119]
Plagiochila hattoriana	Plagiochiline B (**157**)	Moderate bitterness	[38]
Porella perrottetiana	Perrottetianal A (**158**)	Moderate bitterness	[1, 38]
Scapania undulata	Scapanin A (**155**)	Potent bitterness	[119]
Fissidens sp. (moss)	EtOH	Strong sweetness	[75]
Rhodobryum giganteum (moss)	EtOH	Moderate sweetness	[75]

[a] Pungent derivatives from plagiochiline A (**72**) produced in human saliva

Table 6 Allergy-inducing bryophytes [38]

Species name (liverwort)	Species name (moss)
Frullania asagrayana	*Eurhynchium oreganum*
Frullania bolanderi	*Isothecium stoloniferum*
Frullania dilatata	*Leucolepis menziesii*
Frullania eboracensis	*Mnium cuspidatum*
Frullania franciscana	*Mnium undulatum*
Frullania inflata	*Polytrichum juniperinum*
Frullania kunzei	*Rhytidiadelphus loreus*
Frullania nisquallensis	*Sphagnum palustre*
Frullania riparia	
Frullania tamarisci subsp. *tamarisci*	
Frullania tamarisci subsp. *obscura*	
Marchantia polymorpha	
Metzgeria furcata	
Radula complanata	

The compounds *ent*-dihydrofrullanolide (**162**) and its enantiomer dihydrofrullanolide (**163**), with an α-methyl-γ-butyrolactone unit as isolated from the above mentioned liverworts, do not cause an allergenic response.

A patch test indicated that both (+)-frullanolide (**3**), (–)-frullanolide (**4**), and (+)-oxyfrullanolide (**160**) are the most potent liverwort allergens so far known. Young males (aged 1–17 years) were found to be more sensitive than women (aged 18–23 years) to these sesquiterpene lactones, as shown in Table 8 [120].

Frullania asagrayana, F. bolanderi, F. brasiliensis, F. brocheri, F. eboracensis, F. franciscana, F. inflata, F. kunzei, F. nisquallensis, F. riparia, and other *Frullania* species that contain the sesquiterpenes **164–169** also with an α-methylene-γ-butyrolactone unit, also cause potent allergenic contact dermatitis [49] (Table 7). Also the pungent polygodial found in the *Porella vernicosa* complex has been found to cause an allergic effect on the skin of the guinea pig [38].

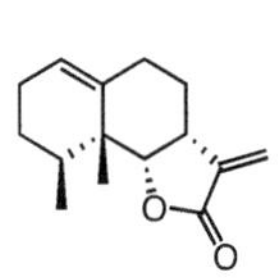

164 (5-*epi*-dilantanolide)

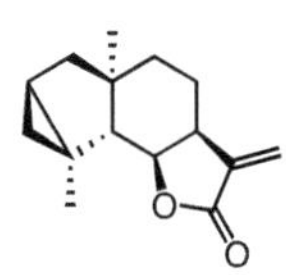

165 (nepalensolide C)

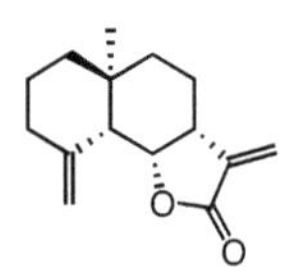

166 (β-frullanolide)

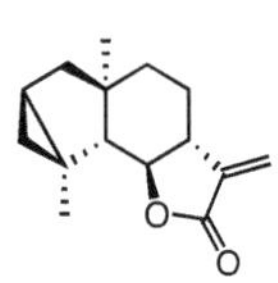

167 (brothenolide)

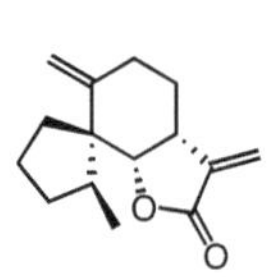

168 (spirodilatanolide)

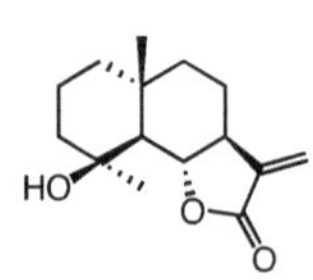

169 (4-*epi*-arbusculin A)

Table 7 Allergy-inducing compounds from liverworts and their potency

Species name	Compound	Potency	Refs.
Frullania dilatata	(+)-Frullanolide (**3**)	++++	[120]
	(+)-β-Cyclocostunolide (**159**)	++	
	(+)-Oxyfrullanolide (**160**)	+++	
	Eremofrullanolide (**161**)	++	
	ent-Dihydrofrullanolide (**162**)	Inactive	
Frullania apiculata *Frullania asagrayana* *F. nisquallensis* *F. tamarisci* subsp. *tamarisci*	(–)-Frullanolide (**4**)	++++	[1, 2, 120]
	ent-Dihydrofrullanolide (**162**)	Inactive	
Frullania brasiliensis *F. brocheri* *F. dilatata* *F. nepalensis*	5-*epi*-Dilatanolide (**164**), nepalensolide C (**165**), β-frullanolide (**166**), brothenolide (**167**), spirodilatanolide (**168**), 4-*epi*-arbusculin A (**169**)	Allergenic (potential)	[49]
Porella vernicosa complex	Polygodial (**90**)	+	[38]
Schistochila appendiculata	3-Undecyl-, 3-tridecyl-, 3-pentadecyl-, and 3-heptadecyl phenols (**170–173**)	+	[121]
	6-Undecyl-, 6-tridecyl-, and 6-pentadecyl salicylic acids (**174–176**)	+	
	Potassium 6-undecyl-, 6-tridecyl-, and 6-pentadecyl salicylates (**177–179**)	+	
	6-Undecyl catechol (**180**)	++	

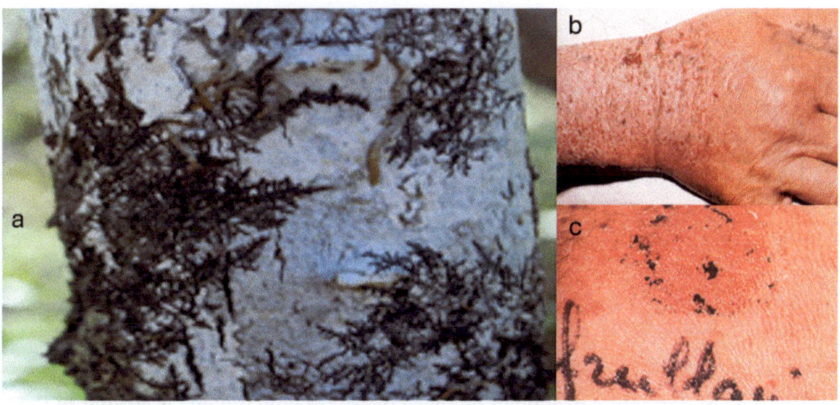

Plate 14 **a** *Frullania tamarisci* subsp. *tamarisci*, **b** allergenic contact dermatitis on the human hand, and **c** on the back skin caused by frullanolide (**3**)

Table 8 Patch testing of selected sesquiterpene lactones isolated from the European liverwort *Frullania dilatata* [120]

Compound tested (EtOH soln. 1‰)	Patient number[a]								
	1–5	6–14	15	16	17	18, 19	20, 21	22	23
(+)-Frullanolide (**3**)	+	+	+	+	+	0	0	0	0
(–)-Frullanolide (**4**)[b]	+	+	+	+	+	0	0	0	0
(+)-β-Cyclocostunolide (**159**)	+	+	0	0	0	+	0	0	+
(+)-Oxyfrullanolide (**160**)	+	+	+	+	+	+	+	0	0
(+)-Eremofrullanolide (**161**)	–	x	+	x	0	+	+	+	x

[a] +… active, 0… inactive, x… not tested
[b] isolated from *F. tamarisci* subsp. *tamarisci*

The New Zealand liverwort *Schistochila appendiculata*, having a length of about one meter, is a stem-leafy liverwort, which causes allergic effects. The allergens of this taxon are a mixture of long-chain alkylphenols, such as 3-undecyl- (**170**), 3-tridecyl- (**171**), 3-pentadecyl- (**172**), and 3-heptadecyl-phenol (**173**), the long-chain alkyl salicylic acids, 6-undecyl- (**174**), 6-tridecyl- (**175**), 6-pentadecyl-salicylic acid (**176**), and their potassium salts, **177–179**, as well as 6-undecyl-catechol (**180**) [121] (Table 7).

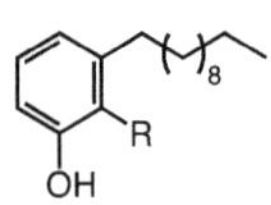

170 (3-undecyl phenol, R = H)
174 (6-undecyl salicylic acid, R = COOH)
177 (potassium 6-undecyl salicylate, R = COOK)

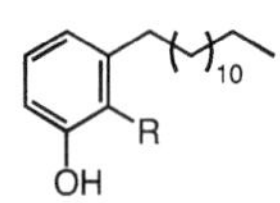

171 (3-tridecyl phenol, R = H)
175 (6-tridecyl salicylic acid, R = COOH)
178 (potassium 6-tridecyl salicylate, R = COOK)

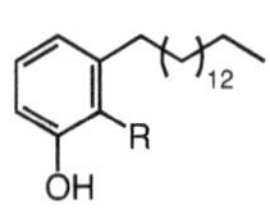

172 (3-pentadecyl phenol, R = H)
176 (6-pentadecyl salicylic acid, R = COOH)
179 (potassium 6-pentadecyl salicylate, R = COOK)

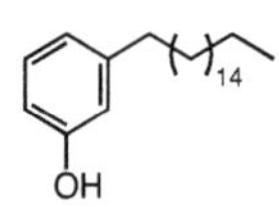

173 (3-heptadecyl phenol)

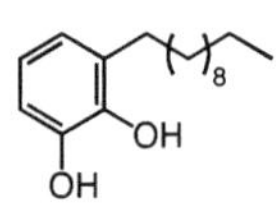

180 (6-undecyl catechol)

The type of dermatitis produced by these liverwort constituents is similar to those caused by the long-chain alkylphenols of the fruits of *Ginkgo biloba* and by plants in the Anacardiaceae, such as *Toxicodendron vernicifluum* and *Rhus succedanea* [122].

4.4 Cytotoxic Activity (Liverworts)

4.4.1 Bis-Bibenzyls

Since Asakawa for the first time reported in 1982 that the Japanese liverworts, *Marchantia polymorpha* (Plate 15) and *Riccardia multifida* produce the very rare bis-bibenzyl derivatives, marchantin A (**1**) and riccardin A (**2**), more than 70 macrocyclic and acyclic bis-bibenzyls have been isolated from many liverworts and their structures established [2, 3, 31–51, 75, 76, 123]. Furthermore, not only the above-mentioned macrocyclic and acyclic bis-bibenzyls, but also various sesqui- and diterpenoids, and bibenzyls have been isolated from liverworts and their cytotoxicity against human tumor cells evaluated. Although none of these natural products has yet led to any as clinically useful drugs, many of them have been chemically modified to produce more potently active cytotoxic compounds than their original natural forms (Table 9).

While the presence of macrocyclic bis-bibenzyls has been known only in liverworts, an acyclic bis-bibenzyl, perrottetin H (**181**) was isolated from the Japanese fern, *Hymenophyllum barbatum* [124]. Also, riccardin C (**182**) has been isolated from the Japanese liverworts, *Reboulia hemisphaerica*, and various *Blasia, Dumortiera, Plagiochasma, Jungermannia, Mastigophora*, and *Plagiochila* species, and perrottetin E (**183**) obtained from the liverwort *Radula perrottetii* [3].

Russian phytochemists identified riccardin C (**182**) and perrottetin E (**183**) and their analogues from the higher plant, *Primula macrocalyx* (Primulaceae) [125, 126]. In a further investigation in 2021 of the distribution of bis-bibenzyls in higher plants, Bukvicki et al. isolated riccardin C (**182**) from the rhizomes of *Primula veris* subsp. *macrocalyx*. Riccardin C (**182**), and its structurally related bis-bibenzyls, marchantin A (**1**) and the acyclic bis-bibenzyls, perrottetins E (**183**) and F (**184**), were evaluated against three human cancer cell lines, namely, A549 non-small cell lung carcinoma, SW480 colon carcinoma, and CH1/PA-1 ovarian teratocarcinoma, respectively. Riccardin C (**182**) and perrottetin E (**183**) showed weak growth inhibitory activities at concentrations between 23–36 μ*M*, with a low dependence on the intrinsic resistance of the cell lines [127].

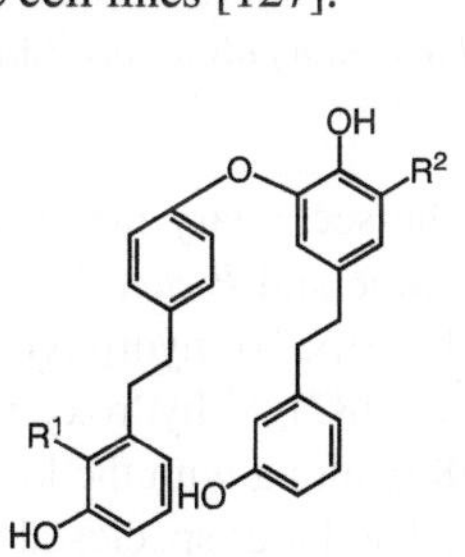

181 (perrottetin H, R^1 = OH, R^2 = H)
183 (perrottetin E, R^1 = R^2 = H)
184 (perrottetin F, R^1 = H, R^2 = OH)

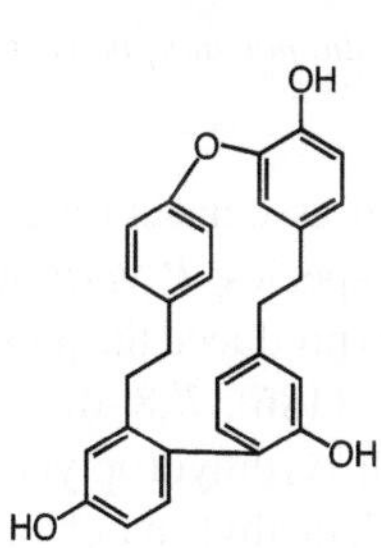

182 (riccardin C)

Plate 15 **a** *Marchantia polymorpha* (liverwort); **b** *Marchantia polymorpha* (dried)

Novakovic and associates (2023) studied the secondary metabolites of the two Serbian *Primula* species, *P. veris* subsp. *columnae* and *P. acaulis*, and showed that the former species produced the previously undescribed 6′-hydroxyriccardin C (**185**), 8′-oxoriccardin C (**186**), 7′,8′-dioxoriccardin C (**187**), 8′-hydroxy-isomarchantin C (**188**), and 8′-hydroxydihydroptychantol A (**189**), along with the known riccardin C (**182**) and 11-*O*-demethyl-marchantin I (**192**). The latter species also contained the known isoperrottetin A (**190**) and isoplagiochin E (**191**) as found in liverworts, as well as compounds **182**, **186**, **188**, and **189**, as mentioned above, together with the two acetophenone derivatives **193** and **194**. Among the isolated compounds, riccardin C (**182**) was the predominant bis-bibenzyl in both *Primula* species. The cytotoxicity

Table 9 Cytotoxic compounds and solvent extracts from bryophytes and their activity against various cancer cell lines and plant cells

Species name (Liverwort)	Compound or solvent extract	Activity value Cell lines[a]	Refs.
Asterella angusta	Marchantin M (**217**)	*IC*$_{50}$ (μ*M*) Du145: 10 for inhibition of growth Du145: 20 for antiproliferation Du145: 10 for 40% induction of apoptosis	[146]
Asterella angusta	Marchantin M (**217**)	Induction of autophagic cell death in prostate cancer cells (PCa)	[147]
Asterella angusta	Marchantin C (**264**)	*IC*$_{50}$ (μ*M*) A-172: 8–12 HeLa: 8–24	[149]
	Dihydroptychantol A (**228**)	U205: 21.2–29.6 U87: 23.7–24.7 U205: induced autophagy and apoptosis in human osteosarcoma U20S cells	
Asterella angusta	Riccardin C (**182**)	*IC*$_{50}$ (μ*M*) PC3: 3.22	[145]
	Marchantin M (**217**)	PC3: 5.45	
Bazzania albifolia	Albifolione (**489**)	*IC*$_{50}$ (μ*M*) MCF-7: PC3: 37.73 59.1	[193]
	Methyl 2-oxo-aromadendra-1(10),3-dien-12-oate (**400**)	64.8 57.9	
	δ-Cuparenol (**401**)	62.4 53.3	
	Chiloscyphenol A (**402**)	5.6 29.6	
	Chiloscyphenol B (**403**)	14.8 38.5	
	Fusicoauritone (**490**)	24.7 43.1	

(continued)

Table 9 (continued)

Species name (Liverwort)	Compound or solvent extract	Activity value Cell lines[a]	Refs.
Bazzania albifolia	Chiloscyphenol A (**402**)	IC_{50} (μ*M*) HBE HeLa 12.78 1.3	[193]
Bazzania novae-zelandiae	Naviculyl caffeate (**398**)	GI_{50} (μg/cm^3) P388: 0.8–1.1	[192]
	Bazzanenyl caffeate (**397**)	P388: > 25	
	Naviculol (**399**)	Inactive	
Blasia pusilla	Pusilatin B (**250**)	IC_{50} (μ*M*) KB: 7.1–15.3	[157]
	Pusilatin C (**251**)	KB + VLB: 7.1–15.3 KB – VLB: 7.1–15.3	
Chandonanthus birmensis	Chandonanone C (**526**)	IC_{50} (μ*M*) PC12: 40.6 NCI-H292: 48.7 NCI-H1299: 40.3 A172: 37.3	[216]
Chandonanthus hirtellus	Chandolide (**366**)	IC_{50} (μg/cm^3) HL-60: 5.3 KB: 11.2	[184] [214]
	13,18,20-*epi*-iso-Chandonanthone (**501**)	18.1	
	Fusicoauritone 6*α*-methyl ether (**504**),	11.2	
	Anadensin (**499**)	HL-60: 17.0	
Chandonanthus hirtellus	Anadensin (**499**)	IC_{50} (μ*M*) PC12: 31.1 NCI-H1299: 34.7	[216]
	(8*E*)-4*α*-Acetoxy-12α,13*α*-epoxycembra-1(15),8-diene (**530**)	PC12: 47.1 NCI-H1299: 41.6	

(continued)

Table 9 (continued)

Species name (Liverwort)	Compound or solvent extract	Activity value Cell lines[a]	Refs.
	Chandonanthone (**519**) Isochandonanthone (**520**)	NCI-H1299: 48.1 PC12: 36.4	
	Chandonanone A (**524**)	DU145: 43.7 PC3: 24.5 A549: 31.1 PC12: 24.5 NCI-H292: 21.4 NCI-H1299: 17.2 A172: 23.1	
	Chandonanone B (**525**)	PC3: 44.9 PC12: 43.8 NCI-H292: 38.9 NCI-H1299: 19.5 A172: 31.9	
	Chandonanone D (**527**)	A549: 21.5 PC12: 30.2 NCI-H1299: 20.9	
	Chandonanone E (**528**)	PC12: 41.1 NCI-H1299: 28.3	
	Chandonanone F (**529**)	PC12: 37.1 NCI-H1299: 23.5	
Chiloscyphus polyanthos *Diplophyllum albicans*	Diplophyllin (**411**)	ED_{50} (μg/cm^3) KB: 2.1	[1–3]
Chiloscyphus rivularis	13-Hydroxy-chiloscyphone (**371**)	IC_{50} (μg/cm^3) A-549: 2.0	[186]
Chiloscyphus polyanthos var. *rivularis*	(3*S*,5*S*,7*R*,10*S*)-3-Hydroperoxy-7-hydroxyeudesm-4(15)-ene (**372**)	IC_{50} (μM) A549: 27.7	[187]

(continued)

Table 9 (continued)

Species name (Liverwort)	Compound or solvent extract	Activity value Cell lines[a]	Refs.
Chiloscyphus subporosa *Clasmatocolea vermicularis*	Diplophyllolide (**132**)	IC_{50} (μg/cm^3) P388: 0.4 Monkey kidney cell (BSC): cytotoxic zone 75% of well	[189]
Conocephalum conicum	Lunularic acid (**196**)	IC_{50} (μg/cm^3) HepG2: 7.5	[176]
	2α,5*β*-Dihydroxy-bornane-2-cinnamate (**765**)	4.5	
Conocephalum conicum	(+)-Bornyl acetate (**56**)	*Marchantia polymorpha* apoptosis inducing	[252]
Conocephalum conicum	Lepidoza-1(10),4-dien-14-ol (**484**) *rel*-(1(10))*Z*,4*S*,5*E*,7*R*)-Germacra-1(10),6-diene-11,14-diol (**485**), *rel*-(1(10)*Z*,4*S*,5*E*,7*R*)-Humula-1(10),5-diene-7,14-diol (**486**) Conocephalenol (**487**)	Rat splenocyte: 1000–500 mM (cell death)	[212]
	rel-(1(10))*Z*,4*S*,5*E*,7*R*)-Germacra-1(10),6-diene-11,14-diol (**485**), *rel*-(1(10)*Z*,4*S*,5*E*,7*R*)-Humula-1(10),5-diene-7,14-diol (**486**)	Concanavalin A-stimulated rat splenocyte: immunosuppressive effect	
Denotarisia linguifolia	*n*-Hexane	IC_{50} (μg/cm^3) MCF-7 42.79	[262]
	$CHCl_3$	51.92	
	EtOAc	79.47	
	EtOH	> 100	
Dumortiera hirsuta	Riccardin D (**253**)	IC_{50} (μ*M*) KB: MCF-7: PC3: 7.1 6.6 10.1	[158]
	12-*N,N*-Dimethylamino-methoxymethylriccardin D (**254**) (synthetic)	10.7 13.3 9.7	

(continued)

Table 9 (continued)

Species name (Liverwort)	Compound or solvent extract	Activity value Cell lines[a]	Refs.
	10-Bromoriccardin D (**255**) (synthetic)	5.9 5.4 5.6	
Dumortiera hirsuta	Riccardin D (**253**)	IC_{50} (μ*M*) A549: MCF-7: K562: 28.14 18.31 17.54	[164]
	13,1′,12′-tri-*N,N*-dimethylamino-ethoxyriccardin D (**257**) (synthetic)	0.51 0.23 0.190	
	11,6′-Bis(pyrrolidin-1-yl) methylriccardin D (**258**) (synthetic)	0.43 0.50 0.99	
Dumortiera hirsuta	Plagiochin E (**219**)	IC_{50} (μ*M*) KB: MCF-7: PC3: 40.1 34.9 28.0	[140]
	10-*N,N*-Dimethylamino-plagiochin E (**260**) (synthetic)	15.1 33.8 25.1	
	12-Bromoplagiochin E (**261**) (synthetic)	8.2 6.3 9.3	
	12,12′-Dibromo-plagiochin E (**262**) (synthetic)	9.7 8.4 9.2	
Dumortiera hirsuta	Riccardin D (**253**)	IC_{50} (μ*M*) HL60: K562: K562A02: 25.2 21.7 27.2 10–60 10–60 10–60 90.6% 77.2 (antiproliferation) IC_{50} (μ*M*) 1.25–10 27.8 28.0 20.3 apoptosis (due to DNA topoisomerase-II inhibition, %)	[158]
Dumortiera hirsuta	Riccardin D (**253**)	(1) Decreases of vascular endothelial growth factor (VEGF) in H460 human lung cancer cells (2) Delay in growth of H469 xenografts	[160]

(continued)

Table 9 (continued)

Species name (Liverwort)	Compound or solvent extract	Activity value Cell lines[a]	Refs.
Dumortiera hirsuta	Riccardin D (**253**)	(1) Prevention of intestinal adenoma (polyps) in mice (2) Angiogenesis decrease in intestinal polyps	[162]
Dumortiera hirsuta	Riccardin D (**253**)	PCa cells: (1) Induces cellular senescence (2) Diminishes colonogenic potential and induce cyclic arrest (3) Provokes DNA damage PC3 xenograft: triggers cellular senescence	[163]
Dumortiera hirsuta	Lunularin (**103**)	IC_{50} (μg/cm^3) HepG2: 7.4	[175]
Dumortiera hirsuta	Isoriccardin C (**942**)	IC_{50} (μM) KB HL60 38 38	[168]
	Marchantin A 1′-methyl ether (**263**)	33 37	
	Isomarchantin C (**264**)	35 36	
Frullania dilatata	(+)-Oxyfrullanolide (**160**)	ED_{50} (μg/cm^3) KB: 0.80	[38]
	Eremofrullanolide (**161**)	1.70	
	(+)-Epoxyfrullanolide (**1228**) (synthetic)	2.65	
Frullania inouei	Brittonin A (**282**)	IC_{50} (μM) KB: 3.55 KB/VCR: 5.3 K562: 1.5 K562/A02: 4.7	[175]

(continued)

Table 9 (continued)

Species name (Liverwort)	Compound or solvent extract	Activity value Cell lines[a]	Refs.
	Brittonin B (**283**)	KB: 2.02 KB/VCR: 2.0 K562: 5.4 K562/A02: 6.0	
	3,3′,4,4′-Tetramethoxy-bibenzyl (**284**)	KB: 9.4 KB/VCR: 7.3 K562: 1.1 K562/A02: 1.4	
	Chrysotobibenzyl (**285**)	KB: 29.4 KB/VCR: 23.8 K562: 15.4 K562/A02: 11.4	
Frullania muscicola	Muscicolin (**698**)	IC_{50} (μg/cm^3) KB: 20–50 PG: 20–50 HT-29: 20–50 BEL-7402: 20–50	[243]
Frullania species (Indonesian unidentified *Frullania* sp.)	Et_2O	IC_{50} (μ*M*) HL-60: KB: 6.7 3.3	[82, 183]
	(+)-3α-(4-Methoxy-benzyl)-5,7-dimethoxy-phthalide (**353**)	0.9 1.0	
	(–)-3α-(3′-Methoxy-4′,5′-methylenedioxybenzyl)-5,7-dimethoxy-phthalide (**354**)	6.3 5.5	
	3-Methoxy-3′,4′-methylene-dioxybibenzyl (**355**)	96.6 122.1	
	2,3,5-Trimethoxy-9,10-dihydrophenanthrene (**357**)	5.5 124.3	
Tahitian unidentified *Frullania* sp.	Et_2O	IC_{50} (μ*M*) HL-60: KB: 2.4 32.6	[82, 183]

(continued)

Table 9 (continued)

Species name (Liverwort)	Compound or solvent extract	Activity value Cell lines[a]	Refs.
	Tulipinolide (**134**)	4.59 32.7	
Frullania nisquallensis	Costunolide (**360**)	LD_{50} (μg/cm^3) A549: 12 RS321N (mutant yeast strain): 50 RS322YK (mutant yeast strain): 150 RS167K (mutant yeast strain): 330	[191]
Frullania tamarisci subsp. *obscura*	4-*epi*-Arbusculin A	ED_{50} (μg/cm^3) KB: 0.50	[38]
Gottschelia schizopleura	*cis*-Cleroda-3,13-dien-16,18-dihydroxy-15-oic acid-15,16-olide (**699**)	IC_{50} (μ*M*) MDA-MB-435: 0.1 LOVO: 0.1	[244]
Gottschelia schizopleura	Schizopleurolide A (**704**) Schizopleurolide B (**705**)	HL-60: B16-F10: (No active values described)	[245]
Hepatostolonophora paucistipula	(–)-*ent*-Arbusculin B (**395**)	IC_{50} (μg/cm^3) P-388: 1.1	[190]
	(–)-*ent*-Costunolide (**396**)	0.7	
Heteroscyphus tener		IC_{50} (μ*M*)	[217]
	Anadensin (**499**)	PC3: > 40 DU145: > 40 K562: > 40 A549: > 40 NCI-H252: 39.9 NCI-H1299: 27.0 NCI-H446: > 40 RWPE-1: > 40 HBE: > 40	

(continued)

Table 9 (continued)

Species name (Liverwort)	Compound or solvent extract	Activity value Cell lines[a]	Refs.
	Heteroscyphin C (**996**)	PC3: 20.2 DU145: 18.9 K562:34.0 A549: 21.5 NCI-H252: 10.7 NCI-H1299: 21.9 NCI-H446: 16.7 RWPE-1: 26.4: HBE: 53.2	
	Isomanool (**536**)	PC3: 26.6 DU145: 28.5 K562:15.8 A549: > 40 NCI-H252: 17.1 NCI-H1299: 35.1 NCI-H446: 39.5 RWPE-1: 81.0 HBE: 53.9	
	ent-Labdanes, heteroscyphin A–D (**994–997**), fusicoauritone (**490**), 4*α*-hydroxyfusicocca-3(7)-en-6-one (**498**)	Antiproliferative effects (cytotoxicity) against human cancer cell lines: 10.7–40.0 μ*M* and normal epithelial cell lines	
Jungermannia (New Zealand sp.)	*ent*-11*α*-Hydroxykaur-16-en-15-one (**550**)	IC_{50} (μg/cm^3) P-388: 0.48	[220]
	ent-1*β*-Hydroxy-9(11),16-kaura-dien-15-one (**551**)	IC_{50} (μ*M*) HL-60: 7.0	
	ent-9(11),16-Kauradiene-12,15-dione (**552**)	0.6	
	ent-6*β*-Hydroxykaur-16-en-15-one (**555**)	0.4	

(continued)

Table 9 (continued)

Species name (Liverwort)	Compound or solvent extract	Activity value Cell lines[a]	Refs.
Jungermannia fauriana	Jungermannenone A (**553**)	IC_{50} (μM) HL-60: 0.3	[220, 226, 227]
	ent-11α-Hydroxy-kauran-15-one (**554**)	> 100	
	Jungermannenone B (**556**)	1.2	
	Jungermannenone C (**557**)	1.3	
	Jungermannenone D (**558**)	0.8	
Jungermannia fauriana *J. hyaline*	*ent*-11α-Hydroxykaur-16-en-15-one (**560**)	IC_{50} (μM) PC3: 1.9 DU145: 4.5 LNCaP: 2.5 RWPE1: 5.2 MDA-MB-231: 5.2 NCI-H1299: 4.8 SKOV3: 2.6 LOVO:1.4 T24: 2.1 K562: 1.5 HL-60: 0.9 SH-SY5Y: 0.7 A-172: 3.2 Saos-2: 1.4	[228]

(continued)

Table 9 (continued)

Species name (Liverwort)	Compound or solvent extract	Activity value Cell lines[a]	Refs.
	13,18-Dihydroxy-*ent*-9(11),16-kauradien-15-one (**567**)	PC3: 5.5 DU145: 5.3 LNCaP: 11.6 RWPE1:11.7 MDA-MB-231: 12.3 NCI-H1299: 10.7 SKOV3: 8.6 LOVO: 9.0 T24: 10.4 K562: 10.2 HL-60: 3.7 SH-SY5Y: 4.4 A172: 4.4 Saos-2: 7.8	
	13-Hydroxy-*ent*-9(11),16-kauradien-18-al-15-one (**568**)	PC3: 11.4 DU145: 11.6 LNCaP: 9.5 RWPE1: 12.7 MDA-MB-231: 21.6 NCI-H1299: 12.2 SKOV3: 7.4 LOVO: 9.3 T24: 5.9 K562: 6.3 HL-60: 4.4 SH-SY5Y: 4.2 A-172: 4.4 Saos-2: 7.3	

(continued)

Table 9 (continued)

Species name (Liverwort)	Compound or solvent extract	Activity value Cell lines[a]	Refs.
	13,19-Dihydroxy-*ent*-9(11),16-kauradien-15-one (**569**)	PC3: 6.6 DU145: 8.3 LNCaP: 12.8 RWPE1: 13.2 MDA-MB-231: 15.5 NCI-H1299: 19.0 SKOV3: 11.3 LOVO: 18.4 T24: 13.2 K562: 10.4 HL-60: 6.7 SH-SY5Y: 4.9 A-172: 13.0 Saos-2: 15.8	
	18-Hydroxy-*ent*-9(11),16-kauradien-15-one (**570**)	PC3: 21.1 DU145: 31.0 LNCaP: 16.3 RWPE1: 19.3 MDA-MB-231: 32.0 NCI-H1299: 14.5 SKOV3: 22.3 LOVO: 11.2 T24: 17.7 K562: 13.8 HL-60: 4.7 SH-SY5Y: 9.8 A-172: 20.3 Saos-2: 14.0	

(continued)

Table 9 (continued)

Species name (Liverwort)	Compound or solvent extract	Activity value Cell lines[a]	Refs.
	13-Hydroxy-9(11),16-kauradien-15-one (**571**)	PC3: 3.2 DU145: 5.5 LNCaP: 3.2 RWPE1: 6.2 MDA-MB-231: 4.2 NCI-H1299: 2.4 SKOV3: 1.8 LOVO: 1.8 T24: 1.7 K562: 1.9 HL-60: 1.0 SH-SY5Y: 1.2 A-172: 3.8 Saos-2: 1.3	
	13-Hydroxyjungermannenone B (**572**)	PC3: 2.2 DU145: 3.2 LNCaP: 2.8 RWPE1: 6.3 MDA-MB-231: 3.4 NCI-H1299: 2.6 SKOV3: 2.9 LOVO: 2.3 T24: 2.3 K562: 2.2 HL-60: 1.9 SH-SY5Y: 1.6 A-172: 2.7 Saos-2: 2.4	

(continued)

Table 9 (continued)

Species name (Liverwort)	Compound or solvent extract	Activity value Cell lines[a]	Refs.
	13,18-Dihydroxy-jungermannenone B (**573**)	PC3: 2.2 DU145: 3.2 LNCaP: 2.8 RWPE1: 6.3 MDA-MB-231: 3.4 NCI-H1299: 2.6 SKOV3: 2.9 LOVO: 2.3 T24: 2.3 K562: 2.2 HL-60: 1.9 SH-SY5Y: 1.6 A-172: 2.7 Saos-2: 2.4	
	18-Hydroxy-jungermannenone B (**574**)	PC3: – DU145: – LNCaP: – RWPE1: – MDA-MB-231: – NCI-H1299: – SKOV3: – LOVO: 35.2 T24: 49.1 K562: 41.7 HL-60: 23,6 SH-SY5Y: 33.8 A-172: – Saos-2: 35.5	

(continued)

Table 9 (continued)

Species name (Liverwort)	Compound or solvent extract	Activity value Cell lines[a]	Refs.
	ent-13-Hydroxy-kauren-15-one-19-oic acid (= steviol 15-one) (**606**)	PC3: 30.1 DU145: 48.5 LNCaP: NT RWPE1: 27.0 MDA-MB-231: 37.2 NCI-H1299: 32.8 SKOV3: NT LOVO: NT T24: 2.3 K562: 9.9 HL-60: 7.9 SH-SY5Y: NT A-172: NT Saos-2: NT	
	13-Hydroxy-kauren-15-one-19-oic acid methyl ester (= steviol 15-one methyl ester (**607**)	PC3: 6.1 DU145: 10.8 LNCaP: NT RWPE1: 7.0 MDA-MB-231: 5.2 NCI-H1299: 9.2 SKOV3: NT LOVO: NT T24: NT K562: 4.1 HL-60: 3.0 SH-SY5Y: NT A-172: NT Saos-2: NT	

(continued)

Table 9 (continued)

Species name (Liverwort)	Compound or solvent extract	Activity value Cell lines[a]	Refs.
	ent-13-Hydroxy-kauren-15-one-19-oic acid *N*-methyl-piperazine ethyl ester (= steviol-15-one *N*-methylpiperazine ethyl ester) (synthetic) (**608**)	PC3: 8.2 DU145: 14.9 LNCaP: NT RWPE1: 29.3 MDA-MB-231: 27.5 NCI-H1299: 10.9 SKOV3:NT LOVO: NT T24: NT K562: 2.2 HL-60: 1.9 SH-SY5Y: NT A-172: NT Saos-2: NT	
Jungermannia fauriana	13-Hydroxy-jungermannenone B (**572**)	IC_{50} (μM) PC3: 1.3 DU145: 5.01 LNCaP: 2.8 K562: 2.2 A549: 8.6 NCI-H1299: 2.6 NCI-H446: 1.2 MCF-7: 18.3 HepG2: 5.3 RWPE1: 5.1	[229]

(continued)

Table 9 (continued)

Species name (Liverwort)	Compound or solvent extract	Activity value Cell lines[a]	Refs.
	13,18-Dihydroxyjungermannone B (**573**)	PC3: 4.9 DU145: 5.5 LNCaP: 3.2 K562: 1.9 A549: 5.3 NCI-H1299: 2.4 NCI-H446: 7.9 MCF-7: 14.2 HepG2: 6.0 RWPE1: 18.2	
	13-Hydroxyjungermannenone B (**572**)	IC_{50} (μ*M*) PC3 cell apoptosis induction: 1.5	
	13,18-Dihydroxyjun B (**573**)	5.0	
Jungermannia tetragona	*ent*-11*α*-Hydroxy-19,20-diacetoxy-kaur-16-en-15-one (**609**)	IC_{50}(μ*M*) HBE HepG2 A549 A2780 2.4 5.5 > 20 18.8 CVCL2780 > 20	[230]
	ent-11*α*,20-Dihydroxy-19-acetoxy-kaur-16-en-15-one (**610**)	15.0 6.0 > 20 19.5 CVCL2780 > 20	
	ent-20-Hydroxy-kaur-16-en-15-one (**611**)	> 20 4.2 13.0 > 20 CVCL2780 > 20	
	ent-11*α*-Hydroxy-19-acetoxy-(16*S*)-kauran-15-one (**614**)	20.0 > 20 > 20 > 20 CVCL2780 > 20	

(continued)

Table 9 (continued)

Species name (Liverwort)	Compound or solvent extract	Activity value Cell lines[a]	Refs.
	ent-11α,20-Dihydroxy-19-acetoxy-(16*S*)-kauran-15-one (**615**)	> 20 > 20 9.8 > 20 CVCL2780 3.8	
	ent-11α-Hydroxy-19,20-diacetoxy-(16*R*)-kauran-15-one (**616**)	> 20 > 20 > 20 > 20 CVCL2780 4.3	
	ent-11α,19-Dihydroxy-20-acetoxy-(16*S*)-kauran-15-one (**622**)	> 20 > 20 > 20 > 20 CVCL2780 1.5	
	ent-11α-Hydroxy-19-acetoxy-kaur-16-en-15-one (**612**)	8.1 15.5 19.0 2.4 CVCL278 8.0	
	ent-11α-Hydroxy-20-acetoxy-kaur-16-en-15-one (**613**)	9.6 8.0 4.0 2.6 CVCL278 8.8	
	ent-Kaur-11α-hydroxy-16-en-15-one (**550**)	5.1 4.2 3.8 0.9 CVCL278 4.9	
Jungermannia truncata		IC_{50} (μ*M*)	[221, 222]
	ent-11α-Hydroxykaur-16-en-15-one (**560**)	HL-60: 0.5–0.8 GI_{50} (μg/cm^3) P388: 0.5 GI_{50} (μ*M*) P388: 1.6	
Lepidolaena clavigera	Clavigerins A–D (**475–478**)	30 μg/disk BSC cells	[209]
Lepidolaena clavigera		IC_{50} (μg/cm^3)	[250]

(continued)

Table 9 (continued)

Species name (Liverwort)	Compound or solvent extract	Activity value Cell lines[a]	Refs.
	Atisane diterpenoid-1 (**739**)	P388: less active than compound **372**	
	Atisane diterpenoid-2 (**740**)	P388: 16	
Lepidolaena taylorii *L. palpebrifolia*	Et_2O	IC_{50} (µg/cm^3) P-388: 1.3	[232, 233]
		ID_{50} (µg/cm^3) GI_{50} (µg/cm^3) GI_{50} (µM)	
	Rabdoumbrosanin (**634**)	P-388: 0.06 P-388: 0.1 P-388: 0.3 IC_{50} (µM) five leukemia cells: 0.4 seven colon cancer cells: 0.6	
L. taylorii	16,17-Dihydrorabdoumbrosanin (**635**)	ID_{50} (µg/cm^3) GI_{50} (µg/cm^3) GI_{50} (µM) P-388: 0.8 P-388: 1.9 P-388: 5.9	
L. taylorii *L. palpebrifolia*	8,14-Epoxyrabdoumbrosanin (**636**)	ID_{50} (µg/cm^3) GI_{50} (µg/cm^3) GI_{50} (µM) P-388: 0.3 P-388: 1.2	
L. taylorii	11*β*-Hydroxyrabdoumbrosanin	GI_{50} (µM) P-388: 0.8	
Lepidozia borneensis	MeOH	Antiproliferative activity	[253]

(continued)

Table 9 (continued)

Species name (Liverwort)	Compound or solvent extract	Activity value Cell lines[a]	Refs.
Lepidozia fauriana	Lepidozenolide (**367**)	ED_{50} (μg/cm^3) P-388: 2.1	[185]
Lepidozia reptans	Lepidozin A (**748**)	IC_{50} (μM) PC-3: 6.5 A549: 7.0 H3255: 9.4 H446: 7.5	[252]
	Lepidozin F (**752**)	PC-3: 8.6 A549: 8.8 H3255: > 10 H446: 9.6	
	Lepidozin G (**753**)	PC-3: 4.2 A549: 4.4 H3255: 5.7 H446: 5.0	
	Riccardin D (**253**)	NSCLC H460: induction of apoptosis (human non-small cell lung cancer) A549 and H460: induction of apoptosis (due to DNA topoisomerase-II inhibition)	[159]
	Riccardin D (**253**)	Inhibits proliferation of human umbilical vascular endothelial cells (HUVEC) Decreases motility and migration of HUVEC cells Decreases vascular endothelial growth factor (VEGF) against human lung cancer H469 cells	[160]

(continued)

Table 9 (continued)

Species name (Liverwort)	Compound or solvent extract	Activity value Cell lines[a]	Refs.
	Riccardin D (**253**)	Prevents intestinal adenoma (polyp) formation in $APC^{Min/+}$ mice (80 mg/kg was given to mice) Decreases of β-catenin and cyclin D1 expression Prevents proliferation of intestinal polyps and trigger apoptosis via caspase-dependent pathway Decreases angiogenesis in intestinal polyps	[151]
Lunularia cruciata	Riccardin G (**244**)	IC_{50} (μM) A549: MRC5: 2.5 7.5	[152]
	1,2,7-Trihydroxy-3-(3′-hydroxy-1-benzyl-oxy)-phenanthrene (**242**)	5.0 5.0	
	2-Hydroxydehydro-cavicularin-3-methyl ether (**239**)	5.0 3.0	
Marchantia convoluta	Petroleum ether	IC_{50} (μg/cm^3) H1299: HepG2: > 500 > 500	[207, 208]
	EtOAc	100 50	
	n-BuOH	> 200 > 200	
Marchantia emarginata subsp. *tosana*	Marchantin A (**1**)	IC_{50} (μg/cm^3) MCF-7: 4.0	[134]

(continued)

Table 9 (continued)

Species name (Liverwort)	Compound or solvent extract	Activity value Cell lines[a]	Refs.
Marchantia polymorpha wild specimen	*n*-Hexane	IC_{50} (μg/cm^3) Breast cancer Head cancer Neck cancer Colon (DL-1) Colon (SW-620)	[148]
	H_2O	Cell proliferation for all cancer cell lines	
Cultured specimen	*n*-Hexane	MDA-MB-231 SW-620 FaDu	
	EtOAc	MDA-MB-231: 100	
	H_2O	Cell proliferation for all cancer cell lines	
Marchantia polymorpha	*n*-Hexane	Inhibition (%) for MCF-7 8.6	[261]
	$CHCl_3$	65.6	
	EtOAc	48.9	
	EtOH	4.6	
	Camptothecin (control)	65.1	
Marchantia polymorpha	Marchantin A (**1**)	ED_{50} (μg/cm^3) KB: 8.4	[38]
	Marchantin B (**195**)	10.0	
	Marchantin C (**197**)	10.0	

(continued)

Table 9 (continued)

Species name (Liverwort)	Compound or solvent extract	Activity value Cell lines[a]	Refs.
Marchantia polymorpha *M. paleacea* subsp. *diptera* *M. emarginata* subsp. *tosana* *Radula perrottetii* *M. paleacea* subsp. *diptera* *Blasia pusilla*	Marchantin A (**1**)	IC_{50} (μ*M*) KB KB+VLB KB–VLB 3.7 1.3 2.7	[26, 38, 49]
	Marchantin B (**195**)	3.2 9.3 ›4.0	
	Marchantin D (**198**)	10.8 > 20 > 20	
	Marchantin E (**199**)	7.0 1.6 2.9	
	Perrottetin F (**184**)	> 20 12.6 9.5	
	Paleatin B (**202**)	> 20 > 20 > 20	
	Pusilatin B (**250**)	13.1 15.3 11.9	
	Pusilatin C (**251**)	13.8 7.1 11.7	
Marchantia polymorpha	Marchantin A (**1**)	IC_{50} (μ*M*) A256: 5.5 T47D: 15.3 MCF7: 11.5	[135]
Marchantia polymorpha	Marchantin A (**1**)	IC_{50} (μg/cm^3) A375: 7.5–12.0	[136]
Marchantia polymorpha	VCR	IC_{50} (μ*M*) KB KB/VCR 1483.5	[137]
	VCR + marchantin C (**197**) (16 μ*M*)	478.8	
	VCR + marchantin C dimethyl ether (**204**) (synthetic) (16 μ*M*)	197.0	
	VCR + 7,8-dehydromarchantin C (**205**) (synthetic) (16 μ*M*)	259.3	
Marchantia polymorpha	Marchantin C (**197**)	IC_{50} (μ*M*) U87: 12 (inhibition of proliferation) T98G: 12 (inhibition of proliferation	[138]

(continued)

Table 9 (continued)

Species name (Liverwort)	Compound or solvent extract	Activity value Cell lines[a]	Refs.
		IC_{50} (μ*M*)	
	Marchantin C (**197**)	U87: 10 T98G: 6.8 (decreases glioma cell migration and invasiveness) inhibition of angiogenesis induced by T98G cells (glioma cells) U87: 71.2% T98G: 69.3%: (decreases migration ability glioma cells) Inhibition of angiogenesis for chicken embryo chorioallantoic membrane: 55.9% at 10 μ*M* Inhibition of phosphorylation of the ERK.MAPK signaling pathway Inhibition of tumor growth with induction of lung cancer cells (A549, H1299)	
Marchantia polymorpha	Marchantin C (**197**)	IC_{50} (μ*M*) KB: MCF-7: PC3: 15.3 12.8 15.8	[240]
	10-Bromomarchantin C (**208**)[a]	11.0 6.3 15.6	
	11-Bromomarchantin C (**209**)[a]	11.5 15.3 15.7	
	12-Bromomarchantin C (**210**)[a]	11.3 12.1 8.5	
	Marchantin C5 dimer (**211**)[a]	13.8 16.4 27.2	

(continued)

Table 9 (continued)

Species name (Liverwort)	Compound or solvent extract	Activity value Cell lines[a]	Refs.
	10-*N*,*N*-Dimethylamino-marchantin C (**207**)[a] [a](synthetic)	16.8 17.5 12.6	
Marchantia polymorpha	Marchantin C (**197**)	IC_{50} (μ*M*) A172 and HeLa: 8 (microtubule inhibition)	[142, 143]
	Marchantin C (**197**)	A172: Cell growth inhibitory and colony formation ability Induction of apoptosis	
Marchantia polymorpha	Marchantin C (**197**)	Inhibition of lung cancer growth via the induction of cancer cell senescence	[144]
Marchantia polymorpha	Plagiochin E (**219**)	IC_{50} (μ*M*) PC3: 6.0 PCa: Induction of apoptosis Induction of apoptosis of prostate cancer cells (PC3) through endoplasmic reticulum stress	[145]
Marchantia polymorpha	Dihydroptychantol A (**228**)	MDR reversal activity (multidrug resistance)	[149]
Marsupella emarginata	Marsupellone (**414**)	IC_{50} (μ*M*) P-388: 1.0	[198]
	Acetoxymarsupellone (**415**)	P-388: 1.0	
Mastigophora diclados	EtOH	IC_{50} (μg/cm^3) MCF-7: 29.2	[196]
Mastigophora diclados	Et_2O	IC_{50} (μg/cm^3) HL-60: KB: 2.4 14.6	[195]

(continued)

Table 9 (continued)

Species name (Liverwort)	Compound or solvent extract	Activity value Cell lines[a]	Refs.
	MeOH	13.1 32.5	
	(–)-Diplophyllolide (**132**)	1.7 3.3	
	(–)-α-Herbertenol (**405**)	12.8 12.5	
	(–)-Herbertene-1,2-diol (**406**)	1.4 11.8	
	(–)-Mastigophorene C (**407**)	2.4 14.8	
	(–)-Mastigophorene D (**408**)	2.5 14.2	
	1-Hydroxy-2-methoxy-herbertene (**409**)	2.6 11.0	
	Herbertene-1,2-diaetate (**410**)	15.0 14.5	
Marchantia polymorpha	Plagiochin E (**219**)	IC_{50} (μg/cm^3) K562/A02: 2–6	[167]
Notoscyphus lutescens		IC_{50} (μ*M*)	[246]
	Notolutesin A (**707**)	PC3: 6.2 A549: > 10 DU145 > 10 H2688; > 10	
Notoscyphus collenchymatosus	Notolutesin K (**720**)	IC_{50} (μ*M*) MCF-7 HCC-1428 HT-29 14.97 > 40 > 40	[247]
	Notolutesin L (**721**)	26.9 30.8 27.3	
	Notolutesin M (**722**)	30.0 12.4 16.4	
	Notolutesin N (**723**)	> 40 > 40 > 40	
	Notolutesin O (**724**)	> 40 > 40 > 40	
	Notolutesin P (**725**)	5.2 4.8 3.5	

(continued)

Table 9 (continued)

Species name (Liverwort)	Compound or solvent extract	Activity value Cell lines[a]	Refs.
	Notolutesin A (**707**)	22.5 30.7 35.7	
	Notolutesin C (**709**)	2.2 38.2 19.8	
	Notolutesin F (**712**)	22.5 34.9 20.0	
	Notolutesin H (**714**)	14.9 20.8 33.0	
	Notolutesin I (**715**)	21.5 32.4 > 40	
	(–)-Pimara-9(11),15-dien-19-ol (**719**)	14.1 12.8 16.8	
Paraschistochila pinnatifolia	*ent*-1β-Hydroxykauran-12-one (**632**)	IC_{50} (μg/cm^3) P-388: 15	[231]
Pallavicinia ambigua	Pallambin A (**661**)	Reversal hold values (at 10 μ*M*) 4.3	[239]
	Pallambin B (**662**)	1.9	
	Pallambin C (**663**)	2.9	
	Pallambin D (**664**)	1.9 (Ability to reverse the adriamycin-induced resistance of K562/A02 cells)	
Pellia endiviifolia	Sacculatal (**128**)	IC_{50} (μ*M*) Human melanoma cells: 2–4 Lu-1: 5.7 KB: 3.2 KB-V: 2.7 LNCaP: 7.6 ZR-75–1: 7.6	[75]
Pellia endiviifolia	Perrottetin E (**183**)	IC_{50} (μ*M*) HL-60 U-937 K-562 14.2 50.5 37.2	[154, 171]

(continued)

Table 9 (continued)

Species name (Liverwort)	Compound or solvent extract	Activity value Cell lines[a]	Refs.
	8-Hydroxyperrottetin E	> 100 38.5 > 100	
	10,10′-Dihydroxyperrottetin E (**248**)	> 100 > 100 > 100	
	Perrottetin F (**184**)	68.5 46.8 59.2	
	8-Hydroxyperrottetin F (**270**)	> 100 62.2 91.5	
	Perrottetin F-6′-sulfate (**272**)	> 100 > 100 > 100	
	1,2,6-Trihydroxy-2-(3′-hydroxy-1-benzyloxy)-phenanthrene (**242**)	39.2 13.2 35.5	
	Lunularic acid (**196**)	> 100 > 100 > 100	
	Perrottetin E (**183**)	NT2/D1 A-172 U251 < 10 13.5 15.2	
	8-Hydroxyperrottetin E	15.5 26.2 47.2	
	10,10′-Dihydroxyperrottetin E (**248**)	6.8 53.8 54.8	
	Perrottetin F (**184**)	< 10 8.5 60.2	
	8-Hydroxyperrottetin F (**270**)	< 5.5 10.5 80.2	
	Perrottetin F-6′-sulfate (**272**)	> 100 > 100 > 100	
	1,2,6-Trihydroxy-2-(3′-hydroxy-1-benzyloxy)-phenanthrene (**242**)	< 10 13.5 15.2	
	Lunularic acid (**196**)	56.2 56.4 > 100	
Pellia endiviifolia	Perrottetin E (**183**)	μ*M* Human lymphocytes genotoxicity Genotoxicity 25–100	[154]
	10′-Hydroxyperrottetine E	Apoptotic potential 25–100	
	10,10′-Dihydroxyperrottetin E (**248**)	Redox modulating 25–100	
	Perrottetin E (**183**)	Induction of proapoptosis: 1	
	MeOH	1	

(continued)

Table 9 (continued)

Species name (Liverwort)	Compound or solvent extract	Activity value Cell lines[a]	Refs.
Plagiochasma intermedium	Pakyonol (**218**)	IC_{50} (μ*M*) PC3: 7.98	[156]
Plagiochasma intermedium	Riccardin F (**243**) Pakyonol (**218**)	LD_{50} (μg/cm^3) K562 K562/A02 In the presence of 3 μg/cm^3 riccardin F or pakyonol, the IC_{50} of ADR against K562/A02 cells decreased by 2.51- and 4.78-fold, respectively	[156]
Plagiochila disticha	Plagiochiline A (**135**)	GI_{50} (μ*M*) A549: 9.0 DU145[a]: 1.4 DU145[b]: 4.7 H460: 4.8 HCT116:4.7 HeLa 16.8 HT-29: 1.9 K562: 4.8 LNCaP:12.8 M-14: 3.9 MCF-7: 6.8 3T3: 1.7 U937: 2.6 VERO: 4.7	[200]

(continued)

Table 9 (continued)

Species name (Liverwort)	Compound or solvent extract	Activity value Cell lines[a]	Refs.
	Plagiochiline I (**136**)	DU145: 1.4 H460: 15.5 HT-29: 6.1 K562: 8.7 M-14: 6.4 MCF-7: 9.0 3T3: 3.8 VERO: 6.8 H460: 69.2	
	Plagiochiline R (**419**)	DU145: 6.1 HT-29: 25.3 K562: 30.0 M-14: 9.0 MCF-7: 55.0 3T3: 4.8 VERO: 85.2 [a]Human prostate carcinoma [b]Human hormone-refractory prostate cancer cell	
Plagiochila fasciculata	2-Hydroxy-4,6-di-methoxyacetophenone (**767**)	IC_{50} (μg/cm^3) P-388: > 50	[258]
	2-Hydroxy-3,4,6-tri-methoxyacetophenone (**766**)	P-388: > 50	
Plagiochila fruticosa	Plagiochiline A (**135**)	ED_{50} (μg/cm^3) KB: 3.0	[38]
Plagiochila fruticosa	Isoplagiochin A (**265**)	IC_{50} (μM) KB: 50 (tubulin polymerization inhibition)	[169]
	Isoplagiochin B (**266**)	KB: 25 (tubulin polymerization inhibition)	

(continued)

Table 9 (continued)

Species name (Liverwort)	Compound or solvent extract	Activity value Cell lines[a]	Refs.
Plagiochila nitens	Plagiochilarins A–G (**652–658**), anadensin (**499**), cinnamolide (**659**), loliolide (**600**)	IC_{50} (μ*M*) Hepa-1c1c7: > 50.0	[236]
Plagiochila ovalifolia	Plagiochiline A (**135**)	IC_{50} (μg/cm^3) P-388: 3.0	[199]
	Plagiochiline A 15-yl octanoate (**460**)	0.05	
	14-Hydroxyplagiochiline A-15-yl-(2*E*,4*E*)-dodecadienoate (**465**)	0.05	
Plagiochila porelloides	Et_2O	IC_{50} (μg/cm^3) 1774: 1.3 W138: 3.0	[247]
	Essential oil	28.8 28.9	
Plagiochila pulcherrima		IC_{50} (μg/cm^3)	[201]
	Ethoxyplagiochiline A (**453**)	HeLa: 24.6 A172: 15.2 H460: 7.3	
	Methoxyplagiochiline A1 (**454**)	HeLa: 15.2 A172: 17.9 H460: 26.0	
	Methoxyplagiochiline A2 (**455**)	HeLa: 7.2 A172: 6.4 H460: 6.7	
	Plagiochiline C (**420**)	HeLa: 18.4 A172: 4.3 H460: 13.1	
Plagiochila sciophila	Fusicosciophin B (**227**) Fusicosciophin D (**229**)	ED_{50} (μ*M*) HL-60: 31.2 59.1	[248]

(continued)

Table 9 (continued)

Species name (Liverwort)	Compound or solvent extract	Activity value Cell lines[a]	Refs.
Plagiochila stephensoniana	3-Methoxy-4′-hydroxy-bibenzyl (**286**)	IC_{50} (μg/cm^3) P-388: > 25	[177]
	3-Methoxy-4-hydroxy-(*E*)-stilbene (**287**) (synthetic)	9.9	
	3-Methoxy-4-hydroxy-(*Z*)-stilbene (**288**) (synthetic)	16.3	
Porella perrottetiana	Et_2O	IC_{50} (μ*M*) HL-60: 3.2 KB: 15.1	[183]
	MeOH	14.8 2.7	
	Acutifolone A (**362**)	> 177 46.6	
	4*α*,5*β*-Epoxy-8-*epi*-inunolide (**363**)	8.5 52.4	
	7*α*-Hydroxypinguisenol-12-methyl ester (**365**)	IC_{50} (μg/cm^3) HL-60: P-388: KB: 2.7 1.1 177	
Porella platyphylla	*n*-Hexane	% inhibition (μg/cm^3) HeLa: 31.9 (10), 79.2 (30) A2689: 48.2 (10), 83.3 (30) TC7D: 48.9 (10), 64.4 (30)	[263]
	$CHCl_3$	HeLa: 35.7 (10), 47.4 (30) A2689: 41.9 (30) TC7D: 29.3 47.9 (30)	
	MeOH	HeLa: 41.7 (30) TC7D: 27.7 (30)	

(continued)

Table 9 (continued)

Species name (Liverwort)	Compound or solvent extract	Activity value Cell lines[a]	Refs.
Porella perrottetiana	Perrottetianal E (**649**) Oplopanone C (**651**) Perrottetianal A (**158**) Perrottetianal B (**650**) 13^2-Hydroxy-(13^2-*R*)-phaeophytin A (**892**) 13^2-Hydroxy-(13^2-*S*)-phaeophytin A (**891**)	IC_{50} (μ*M*) KB: 0.6–102 Lu-1: 0.6–102 Hep-G2: 0.6–102 Huh7: 0.6–102 HT29: 0.6–102	[235]
Porella viridissima	Perrottetianal A (**158**)	IC_{50} (μg/cm^3) A2780: 1.6	[234]
Ptilidium pulcherrimum	Ursolic acid (**762**)	IC_{50} (μ*M*) PC3: 11.9	[256]
	2*α*,3*β*-Dihydroxyurs-12-en-28-oic acid (**764**)	39.7	
	Acetyl ursolic acid (**763**)	10.9	
	Dihydroptychantol A (**228**) with a thiazole moiety (**230–233**) DHA2 (**229**) (synthetic)	IC_{50} (μ*M*) KB/VCR: 0.1–9.2 K562/A02: MRD reversal activity by increasing adriamycin Cytotoxicity against ovarian cancer cells, SKOV3: 3.44 Decreases polymerization rate of tubulin	[150]
Radula amoena	2-Carbomethoxy-3,5-dihydrostilbene (**317**)	IC_{50}(μg/cm^3) HepG-2: 22.5 SMMC-7721: 27.6 A549: 18	[179]

(continued)

Table 9 (continued)

Species name (Liverwort)	Compound or solvent extract	Activity value Cell lines[a]	Refs.
	3,5-Dimethoxybibenzyl (**318**)	HepG-2: 19.1 SMMC-7721: 18.5 A549: 16.6	
Radula apiculata		*IC*$_{50}$ (μ*M*)	[178]
	Radulapin A (**289**)	PC-3: 5.4 A549: 5.8 MCF-7: 4.6 NCI-H12199: 7.2	
	Radulapin B (**290**)	PC-3: 8.2 MCF-7: 9.8	
	Radulapin C (**291**)	PC-3: 4.1 A549: 4.5 MCF-7: 4.9 NCI-H12199: 6.6	
	Radulapin D (**292**)	PC-3: 3.2 A549: 3.6 MCF-7: 4.5 NCI-H12199: 6.2	
	Radulapin E (**293**)	PC-3: 3.5 A549: 3.1 MCF-7: 6.8 NCI-H12199: 6.7	
	Radulapin F (**294**)	PC-3: 1.4 A549: 2.6 MCF-7: 5.4 NCI-H12199: 5.8	

(continued)

Table 9 (continued)

Species name (Liverwort)	Compound or solvent extract	Activity value Cell lines[a]	Refs.
	Radulapin G (**295**)	PC-3: 1.5 A549: 2.8 MCF-7: 3.6 NCI-H1219: 6.9	
	Radulapin H (**296**)	PC-3: 8.2 A549: 6.5 MCF-7: 5.7 NCI-H12199: 8.2	
Radula constricta	2-Carbomethoxy-3,5-dihydroxy-4-(3-methyl-2-butenyl)-bibenzyl (**334**)	IC_{50} (μ*M*) NCI-H1299: 5.1 A549: 6.0 HepG-2: 6.3 SMMC7721: 8.3 U251: 9.0 SW620: 6.3 HT29: 5.0 KB: 6.7	[180]
	Bibenzyl/*o*-cannabicyclol hybrid (**341**)	NCI-H1299: 7.5 A549: 9.8	
Radula perrottetii	Perrottetin E (**183**)	ED_{50} (μg/cm^3) KB: 12.5	[38]

(continued)

Table 9 (continued)

Species name (Liverwort)	Compound or solvent extract	Activity value Cell lines[a]	Refs.
Radula sumatrana	Bibenzyl/*o*-cannabicyclol hybrid (**341**)	IC_{50} (μM) MCF-7: 14.0 PC-3: 19.6 SMMC-7721: 14.5	[181]
	Rasumatranin B (**343**)	MCF-7: 38.2 PC-3: 40 SMMC-7721: 14.5	
	5-Oxoradulanin A (**349**)	MCF-7: 3.9 PC-3: 6.6 SMMC-7721: 3.6 MCF-7: 24.6	
	Radulanin I (**351**)	PC-3: > 40 SMMC-7721: 34.2	
Reboulia hemisphaerica *Targionia lorbeeriana*	α-Zeorin (**761**)	IC_{50} (μg/cm^3) P-388: 1.1	[3, 254]
Riccardia crassa		μg/disk	[194]
	Riccardiphenol C (**404**)	BSC-1: 60	
Riccardia multifidi	Riccardin A (**2**)	ED_{50} (μg/cm^3) KB: 10.0	[38]
	Riccardin B (**216**)	KB: 12.0	
Riccia billardierei	Phytol/caryophyllene 1:1	mg/cm^3 HT-29: 1 ($p < 0.0001$) HCT-116: 1 ($p < 0.0001$)	[211]

(continued)

Table 9 (continued)

Species name (Liverwort)	Compound or solvent extract	Activity value Cell lines[a]	Refs.
Scapania irrigua	Scapairrin G (**671**)	IC_{50} (μ*M*) A-549: > 10 MDA-MB-231: 4.1 A2780: 5.3 HeLa: 6.6 HT-29: 5.8 HUVEC: > 10	[240]
	Scapairrin H (**672**)	A549: > 10 MDA-MB-231: 5.4 A2780: 4.4 HeLa: 5.5 HT-29: 7.0 HUVEC: > 10	
	Scapairrin I (**673**)	A549: > 10 MDA-MB-231: 5.0 A2780: 7.8 HeLa: 8.7 HT-29: 6.8 HUVEC: > 10	
	Scapairrin J (**674**)	A549: > 10 MDA-MB-231: 4.2 A2780: 8.6 HeLa: 8.3 HT-29: 5.2 HUVEC: > 10	
Scapania parva	Scaparvin A (**691**)	IC_{50} (μ*M*) KB: HeLa: MCF-7: PC-3: > 10	[242]
	Scaparvin B (**692**)	KB: HeLa: MCF-7: PC-3: > 10	
	Scaparvin C (**693**)	KB: HeLa: MCF-7: PC-3: > 10	

(continued)

Table 9 (continued)

Species name (Liverwort)	Compound or solvent extract	Activity value Cell lines[a]	Refs.
	Scaparvin D (**694**)	KB: HeLa: MCF-7: PC-3: > 10	
	Scaparvin E (**695**)	KB: HeLa: MCF-7: PC-3: > 10	
	Parvitexin B (**696**)	KB: HeLa: MCF-7: PC-3: > 10	
	Parvitexin C (**697**)	KB: HeLa: MCF-7: PC-3: > 10	
	Adriamycin (control)	KB: 0.6 HeLa: 0.8 MCF-7: 1.2 PC-3: 1.5	
Scapania undulata	Scapaundulin A (**685**)	IC_{50} (μ*M*) A546: > 50 K562: > 50 A2780: > 50 HeLa: > 50 HT29: > 50	[241]
	Scapaundulin C (**687**)	A546: > 50 K562: > 50 A2780: > 49.4 HeLa: > 50 HT29: > 50	
	5*α*,8*α*,9*α*-Trihydroxy-(13*E*)-labden-12-one (**688**)	A546: 37.6 K562: 36.8 A2780: 19.5 HeLa: > 50 HT29: > 50	
	5*α*,8*α*-Dihydroxy-(13*E*)-labden-12-one (**689**)	A546: 39.2 K562: 37.1 A2780: 16.8 HeLa: > 50 HT29: > 50	

(continued)

Table 9 (continued)

Species name (Liverwort)	Compound or solvent extract	Activity value Cell lines[a]	Refs.
	(13*S*),15-Hydroxylabd-8(17)-en-19-oic acid (**690**)	A546: > 50 K562: > 50 A2780: > 50 HeLa: > 50 HT29: > 50	
Scapania verrucosa	Et_2O (a mixture of volatile terpenoids and fatty acids)	IC_{50} (μg/cm^3) A549: > 100 LOVO: > 100 HL-60: 42.9 QGY: 90.8	[206]
Schistochila acuminata	*cis*-3,14-Clerodadien-13-ol (**737**)	IC_{50} (μg/cm^3) B16-F10: 40	[249]
	12*β*-Hydroxy-dolabella-(3*E*,7*E*)-diene (**738**)	0: 62	
Schistochila glaucescens	Marchantin C (**197**)	IC_{50} (μg/cm^3) P-388:18	[141]
	Neomarchantin A (**212**)	7.6	
	Neomarchantin B (**213**)	8.5	
	Glaucescens-bis-bibenzyls GBB A (**214**) + GBB B (**215**)	10.3	
	Glaucescenolide (**359**)	2.3	
	Riccardin D-26 (=10-*N*,*N*-dimethyl methyl riccardin D) (**254**) (synthetic)	IC_{50} μ*M* KB: KB/VCR: 3.06 3.61 (anti-proliferation) 5.2% 48.8% cancer growth inhibition in mice (20 mg/kg injection of **254**) 52.4% 48.6% Induction of apoptosis (8 μ*M* of **254**: flow cytometry analysis)	[165]

(continued)

Table 9 (continued)

Species name (Liverwort)	Compound or solvent extract	Activity value Cell lines[a]	Refs.
Trichocolea hatcheri	Methyl 4-[(2*E*)-3,7-dimethyl-2,6-octadienyl]oxy-3-hydroxybenzoate (**770**)	IC_{50} μ*M* KB: > 100 SK-MEL-3: > 100 NTT3T3: > 100	[260]
Trichocolea mollissima	1*α*-Hydroxy-*ent*-sandaracopimara-8(14),15-diene (**633**)	μg/cm^3 P-388: > 25	[231]
Trichocolea tomentella	Tomentellin (**768**) Demethyltomentellin (**769**)	μg/cm^3 African green monkey kidney epithelial cells (BSC-1): 15	[259]
Trocholejeunea sandvicensis	Dehydropinguisenol (**361**)	ED_{50} (μg/cm^3) KB: 12.55	[38]
Wiesnerella denudata	Tulipinolide (**134**)	ED_{50} (μg/cm^3) KB: 0.45	[38]
	Zaluzanin D (**368**)	1.50	
	8*α*-Acetoxyzaluzanin C (**369**)	1.60	
	8*α*-Acetoxyzaluzanin D (**370**)	2.50	
Species name (Moss)	Compound, essential oils or solvent extract	Cell lines Activity value	Refs.
Aulacomnium palustre	EtOH	IC_{50} (μg/cm^3) C6: 51 A431: 63 A549: 100 B16-F10: ND MCF-7: 100 CaCo-2: 84	[265]
Barbula javanica	MeOH (16-Methyl-heptadecanoic acid methyl ester, 10-methyl-heptadecanoic acid methyl ester, 2-hydroxy-1-(hydroxy-methyl)ethyl ester, hardwickiic acid-rich)	IC_{50} (μg/cm^3) HT-29: 100 HCT-6: 1000	[277]

(continued)

Table 9 (continued)

Species name (Moss)	Compound, essential oils or solvent extract	Cell lines Activity value	Refs.
Bryum capillare	EtOH	Viability (%) (500/1000 μg/cm^3) SKBR-3: 69/40 HeLa: 67/24 MCF-12A: 100/82	[278]
Climacium dendroides	EtOH	IC_{50} (μg/cm^3) C6: 43 A431: 64 A549: 57 B16-F10: 98 MCF-7: not determined CaCo-2: 66	[265]
Dicranum polysetum	EtOH	IC_{50} (μg/cm^3) C6: 27 A431: 57 A549: 57 B16-F10: 12 MCF-7: 67 CaCo-2: 45	[265]
Hedwigia ciliata	EtOH EtOH/H_2O (1:1) EtOAc	% (inhibition) MDA-MB-231: 50	[279]
Hylocomium splendens	EtOH	IC_{50} (μg/cm^3) C6: 3 A431:27 A549: 12 MCF-7: 67 CaCo-2: 45	[265]

(continued)

Table 9 (continued)

Species name (Moss)	Compound, essential oils or solvent extract	Cell lines Activity value	Refs.
Hypnum cupressiforme	EtOH/H_2O/:1:1 EtOAc H_2O (Gallic (**796**), protocatechuic (**797**), *p*-hydroxybenzoic (**799**), caffeic (**800**), *p*-coumaric acids (**801**), kaempferol (**807**)-containing)	(μg/cm^3) HCT-116: 10 MDA-MB-231: 10	[272]
Hypnum plumaeforme	Momilactone B (**776**)	IC_{50} (μ*M*) HT-29: 1 SW620: 1	[266, 276]
Isothecium subdiversiforme	15-Methoxy-ansamitocine (**771**)	IC_{50} (μg/cm^3) P388: 0.002	[38]
Paraleucobryum longifolium	Leucobryn A (**790**)	IC_{50} (μ*M*) SiHa: 45.6 HeLa: 60.4 A2780: 86.4 MDA-MB-231:50.8	[271]
	Leucobryns B–E (**788–794**)	(**401** and **402** are less active than **400**)	
Phyllogonium viride	Essential oil (β-bazzanene (**854**), β-caryophyllene (**474**), β-chamigrene (**855**) and germacrene B-rich)	IC_{50} (μg/cm^3) MCF-7: 50–250 (inactive) HCT-116: 50–250 (inactive) HaCaT:50–250 (inactive)	[280]
Plagiomnium acutum	Essential oil (*ent*-dolabella-3,7-dien-18-ol-rich)	IC_{50} (μg/cm^3) A549: 23.6 HepG2: 25.8	[273, 274]

(continued)

Table 9 (continued)

Species name (Moss)	Compound, essential oils or solvent extract	Cell lines Activity value	Refs.
Pleurozium schreberi	EtOH	IC_{50} (μg/cm^3) C6: 5 A431: 28 A549: 55 B16-F10: 39 MCF-7: > 100 CaCo-2: 61	[265]
Polytrichum commune	EtOH	IC_{50} (μg/cm^3) C6: 53 A431: 23 A549: not determined B16-F10: 24 MCF-7: 47 CaCo-2: 24	[265]
Polytrichum commune	Ohioensin A (**777**)	ED_{50} (μg/cm^3) 9KB: > 10 9PS: 1.0 A549: > 10 MCF-7: 9.0 HT-29: > 10	[268]
	Ohioensin B (**778**)	9KB: 9.7 9PS: 10.0 A549: > 10.0 MCF-7: 3.4 HT-29: 4.3	
	Ohioensin C (**779**)	9KB: > 10 9PS: 1.0 A549: 8.7 MCF-7: 6.7 HT-29: > 10	

(continued)

Table 9 (continued)

Species name (Moss)	Compound, essential oils or solvent extract	Cell lines Activity value	Refs.
	Ohioensin D (**780**)	9KB: > 10 9PS: 1.0 A549: > 10 MCF-7: > 10 HT-29: > 10	
	Ohioensin E (**781**)	9KB: > 10 9PS: 1.0 A549: 6.2 MCF- 7: > 10 HT-29: > 10	
Polytrichum commune	Communin A (**787**) Communin B (**788**) Ohioensin H (**789**)	IC_{50} (μg/cm^3) A549: 0.10 HepG2: 1.7 LOVO: 0.1 MDA-MB-435: 0.1 5 T-(CEM): 0.002	[270]
Polytrichum juniperinum	EtOH	IC_{50} (μg/cm^3) C6: 68 A431: 84 A549: 24 B16-F10: not determined MCF-7: > 100 CaCo-2: 42	[265]
Polytrichum pallidiscetum	Pallidisetin A (**782**)	IC_{50} (μg/cm^3) RPMI-7951: 1.0 U-251MG: 2.0	[269]
	Pallidisetin B (**783**)	RPMI-7951: 1.0 U-251MG: 2.0	

(continued)

Table 9 (continued)

Species name (Moss)	Compound, essential oils or solvent extract	Cell lines Activity value	Refs.
	1-*O*-Methylohioensin B (**784**)	HT-29: 1.0 RPMI-7951: 1.0 U-251MG: 2.0	
	1-*O*-Methyldihydro-ohioensin B (**785**)	U-251MG: 0.8	
	1,14-Di-*O*-methyl-dihydroohioensin B (**786**)	A549: 1.0 RPMI-7951: 1.0	
Ptilium crista-castrensis	EtOH	IC_{50} (μg/cm^3) C6: 23 A431: 77 A549: 72 B16-F10: 79 MCF-7: > 100 CaCo-2: 42	[265]
Rhytidiadelphus triquetrus	EtOH	IC_{50} (μg/cm^3) C6: 86 A431: 40 A549: 98 B16-F10: 76 MCF-7: not determined CaCo-2: 49	[265]
Sphagnum fallax	EtOH	IC_{50} (μg/cm^3) C6: 27 A431: 57 A549: > 100 B16-F10: 69 MCF-7: not determined CaCo-2: 65	[265]

(continued)

Table 9 (continued)

Species name (Moss)	Compound, essential oils or solvent extract	Cell lines Activity value	Refs.
Sphagnum magellanicum	EtOH	IC_{50} (μg/cm^3) C6: 0.9 A431: 13 A549: 44 B16-F10: 89 MCF-7: not determined CaCo-2: 43	[265]
Sphagnum magellanicum	EtOH/H_2O (1/1)	% (control cell viability) Normal human dermal fibroblasts: (NHDF) 70%	[281]
Sphagnum rubellum	EtOH	IC_{50} (μg/cm^3) C6: 53 A431: 72 A549: > 100 B16-F10: not determined MCF-7: 70 CaCo-2: 42	[265]
Sphagnum tenellum	EtOH	IC_{50} (μg/cm^3) C6: 26 A431: 62 A549: > 100 B16-F10: ND MCF-7: > 100 CaCo-2: 41	[265]
Ceratodon purpureus *Dryptodon pulvinatus* *Hypnum cupressiforme* *Rhytidiadelphus squarrosus* *Tortula muralis*	EtOH	mg/cm^3 L929: 0.5	[275]
T. muralis	EtOH	HFF-1 (anti-proliferation)	

(continued)

Table 9 (continued)

Species name (Moss)	Compound, essential oils or solvent extract	Cell lines Activity value	Refs.
Rhytidiadelphus triquetrus	*n*-Hexane	IC_{50}/IC_{75}(μg/cm^3) HeLa: 17.9/48.2	[276]
	$CHCl_3$	57.1/81.6	
	EtOAc	not determined /31.5	
	MeOH	112.3/120.2	
	H_2O	80.8/97.9	
	H_2O/EtOAc	65.0/124.1	
	H_2O/*n*-BuOH	Values could not be determined	
Tortella tortuosa	*n*-Hexane $CHCl_3$ EtOAc MeOH H_2O H_2O/EtOAc H_2O/*n*-BuOH	33.3/57.3 94.6/104.3 55.6/72/8 138.6/122.2 70.8/85.2 59.5/76.7 139.2/139.9	
Sphagnum magellanicum	EtOH/H_2O (1:1)	(%) (μg/cm^3) Normal human dermal fibroblasts: 70 (500)	[281]
Species name (Higher plant)	Compound or solvent extract	Cell lines Activity value	Refs.
Primula veris subsp. *macrocalyx*	Riccardin C (**182**)	IC_{50} (μ*M*) A549 SW480 GH1/PA1 34 1.3 1.0	[127]
	Perrottetin E (**183**)[a]	34 33 27	
	Perrottetin F (**184**)[a]	5.3 4.6 1.9	
	Marchantin A (**1**)[a] [a]Originated from liverworts	31 36 23	

(continued)

Table 9 (continued)

Species name (Higher plant)	Compound or solvent extract	Cell lines Activity value	Refs.
Primula veris subsp. *columnae*	Riccardin C (**182**)	IC_{50} (μM) A549 MRC5 23 15	[128]
	6′-Hydroxyriccardin C (**185**)	23 21	
	8′-Oxoriccardin C (**186**)	18 11	
	7′,8′-Dioxoriccardin C (**187**)	45 27	
	8′-Hydroxy-isomarchantin C (**188**)	23 13	
	8′-Hydroy-dihydroptychantol A (**189**)	23 11	
	11-*O*-Demethyl-marchantin I (**192**)	12 8	

(continued)

Table 9 (continued)

Species name (Higher plant)	Compound or solvent extract	Cell lines Activity value	Refs.
P. acaulis	Riccardin C (**182**)	18 11	
	8′-Oxoriccardin C (**186**)	12	
	8′-Hydroxy-isomarchantin C (**188**)	23 21	
	8′-Hydroxydihydro-ptychantol A (**189**)	– –	
	Isoperrottetin A (**190**)	45 42	
	Isoplagiochin E (**191**)	45 40	

[a] The following abbreviations are used for cancer and normal (non-tumorogenic) cell lines: 3T3 (murine embryonic fibroblast cells), 9KB (human epithelial carcinoma cells), 9PS (murine P388 leukemia cells), A-172 (human glioblastoma cell; human astrocyte cells), A256 (human breast cancer cells), A2780 (human ovarian cancer cells), A375 (human malignant melanoma cells), A431 (human epidermoid carcinoma cells), A549 (human non-small-cell lung carcinoma cells), B16-F10 (mouse musculus skin melanoma cells), BSC1 (African green monkey kidney epithelial cells), BEL-7402 (hepatoma cells), C6 (rat glioma cells), CaCo-2 (human colorectal carcinoma cells), CCRF-CEM (human T lymphoblast leukemia cells), DLD-1(colon cancer cells), DU145 (human hormone independent prostate carcinoma cells), FaDu (human head and neck cancer cells), H1688 (human lung carcinoma cells), H3255 (human prostate carcinoma cells), H446 (human prostate carcinoma cells), H460 (lung carcinoma epithelial cells), HaCaT(human immortalized keratinocytes), HBE(normal human bronchial epithelium cells), HCT116 (human colon cancer cells), HBE (human normal bronchial epithelial cells), HFF-1 (human foreskin fibroblast cells), HL-60 (acute promyelocytic leukemia cells), HT (Human B cell lymphoma cells), HT-29 (human colon adenocarcinoma cells), HUVEC (normal human vascular cells), HaCaT (immortalized human keratinocytes), HeLa-60 (human acute promyelocytic leukemia cells), Hepa-1C1C7 (cultured murine hepatoma cells), HepG2 (human epithelial tumor cells), Hepatoblastoma (derived cell from human liver cancer cells), HuCCA-1 (human cholangiocarcinoma cells), K562 (human chronic myelogenous leukemia cells), K562/A02 (multidrug-resistant leukemia cells), KB (human epithelial carcinoma cells), KB/VCR (vincristine-resistant KB cells), L929 (mouse fibroblasts cells), LNCaP (human prostate carcinoma cells), LOVO (human colon cancer cells), LU-1 (human adenocarcinoma cells), M-14 (human melanoma cells), MCF-7 (human breast adenocarcinoma cells), MDA-MB231 (human breast carcinoma cells), MOLT-3 (hyper tetraploid human cells), MRC5 (human lung fibroblast cell), MDA-MB-435 (human breast carcinoma cells), NCI60 (The US National Cancer Institute (NCI) 60 human tumor cell line panel), NCI-H1299 (human non-small-cell lung cancer cells), NCI-H292 (human non-small-cell lung cancer cells), NCI-H446 (human small-cell lung cancer cells), NCI-H522 (human non-small cell lung cancer cells), NIH3T3 (murine embryonic fibroblast cells), NSCLC H460 (human non-small cell lung cancer cells), NT2/D1 (human embryonal teratocarcinoma cells), P388 (murine leukemia cells), PCa (human prostate cancer cells) PC-12 (rat pheochromo-cytoma cells), PC-3 (human hormone independent prostate carcinoma cells), PG (PG3: human fibroblast-like cells), QGY (colon cancer cells), RPMI-7951(human melanoma cells), RWPE-1 (human prostate epithelial cells), Saos-2 (human osteosarcoma cells), SH-SYSY (human astrocytoma cells), SiHa (human cervical carcinoma cells), SKBR-3 (human breast cancer cells), SK-ME-3 (human melanoma cells), SKOV3 (human ovarian cancer cells), SW620 (human colon carcinoma cells), SMMC-7721 (human hepatocellular carcinoma cells), T24 (human urinary bladder carcinoma cells), T47D (human breast cancer cells), T98G (human glioblastoma multiforme tumor cells), U205 (human osteosarcoma cells), U-251 (human glioblastoma cells), U-251MG (human glioblastoma cells), U-937 (acute monocytic leukemia cells), U87 (human glioblastoma cells), VERO (normal African green monkey kidney epithelial cells), ZR-75–1 (human breast cancer cells)

of all of the isolated products was evaluated against the A549 human lung cancer cell line, and non-tumorigenic human MRC5 human lung fibroblast cells. The bis-bibenzyls **182**, **185**, **186**, **187**, **188**, **189**, and **191**, as well as **190** showed cytotoxicity against the A549 cell line and MRC5 cells, with IC_{50} values of 12–45 and 8–42 μM, respectively [128].

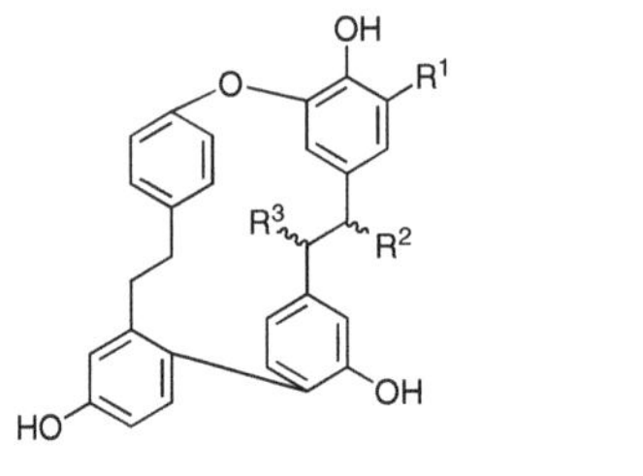

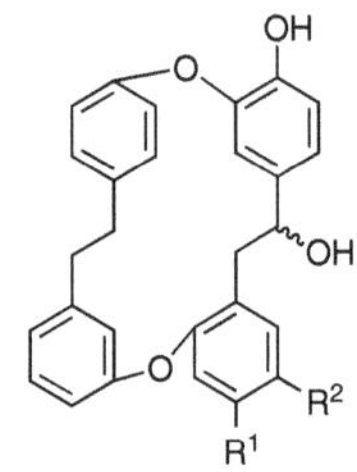

185 (6'-hydroxyriccardin C, R^1 = OH, R^2 = R^3 = H) **188** (8'-hydroxyisomarchantin C, R^1 = OH, R^2 = H)
186 (8'-oxoriccardin C, R^1 = R^2 = H, R^3 = O) **189** (8'-hydroxydihydroptychantol A, R^1 = H, R^2 = OH)
187 (7',8'-dioxoriccardin C, R^1 = H, R^2 = R^3 = O)

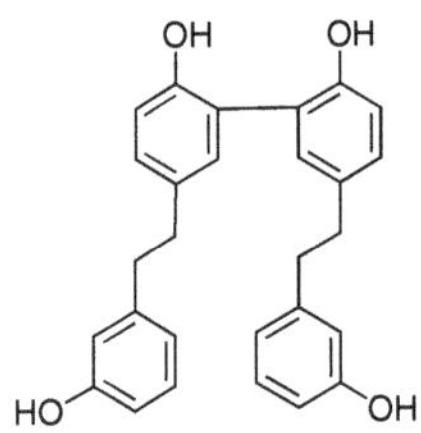

190 (isoperrottetin A)

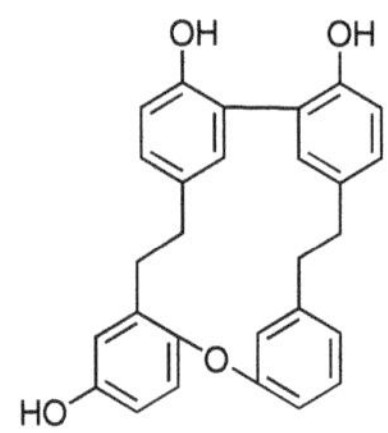

191 (isoplagiochin E)

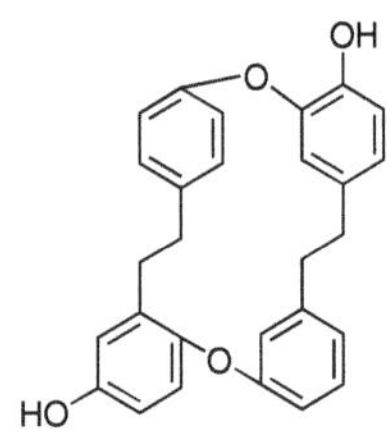

192 (11-*O*-demethyl-marchantin I)

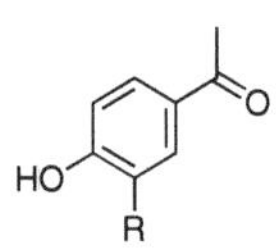

193 (4-hydroxyphenylmethylketone, R = H)
194 (4-hydroxy-3-methoxy-phenylmethyl ketone, R = OMe)

Cyclic bis-bibenzyls, such as those in the marchantin (e.g. marchantin (**1**) or marchantin B (**195**)) and riccardin series (e.g. (**2**)) might be biosynthesized from bibenzyls that correspond chemically to dihydrostilbenes [129]. This hypothesis was substantiated by feeding experiments using radioactive and ^{13}C-labeled precursors, including L-[U-^{14}C]-phenylalanine, [U-^{14}C]-dihydro-*p*-coumaric acid, [2-^{13}C]-acetate, and L-[$^{13}COOH$]-phenylalanine [130, 131]. Thus, marchantin C (**197**) is biosynthesized by the coupling of two lunularic acid (**196**) moieties, catalyzed by the cytochrome P-450, marchantin C hydroxylase to afford marchantin A (**1**), as shown Scheme 2.

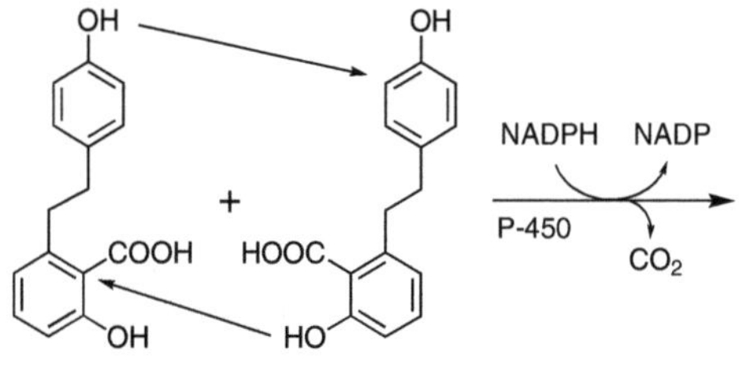

Scheme 2 Biosynthesis route to marchantin C (**197**) in the liverwort, *Marchantia polymorpha*

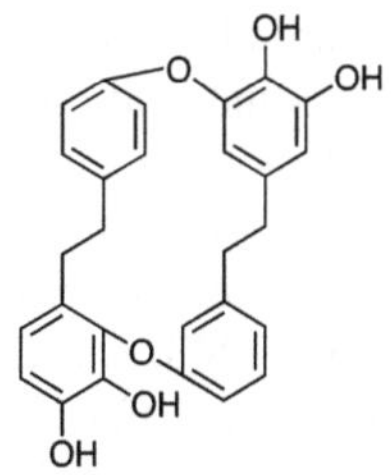

195 (marchantin B)

Macrocyclic bis-bibenzyls produced in liverworts possess various biological activities, in addition to cytotoxicity, including antimicrobial, antifungal, antiviral, antitrypanosomal, enzyme and NO production inhibitory, antioxidant, and muscle relaxant effects, among many others, as shown in Tables 9, 10, 11, 12, 13, 14, and also in Tables 17, 20, 21, 25, 27, 29, 30, 31, and 33, 34, 35, 36, 37.

Jantwal et al., [132] described in a review article titled "Pharmacological potential of genus *Marchantia*", the species names and phytochemical uses, inclusive of the marchantin series of compounds. *Marchantia* species are thalloid liverworts, and abundant sources of macrocyclic bis-bibenzyls, such as marchantin A (**1**) [1–3, 75, 76]. Thus, a methanol extract (105 g) of a Japanese specimen of *M. polymorpha* was chromatographed over silica gel and Sephadex LH-20 to yield a large amount of the cyclic bis-bibenzyl, marchantin A (**1**, 30 g), with its analogues, marchantin B (**195**), C (**197**), D (**198**), E (**199**), G (**200**), and J (**201**) also obtained. The yield of **1** is dependent upon the particular *Marchantia* species investigated. Large quantities of pure **1** (80–120 g) were isolated from 6.7 kg of the dried *M. paleacea* subsp. *diptera*. [76]. This liverwort elaborates other compounds in the marchantin series, including marchantins B (**195**), D (**198**), and E (**199**), and the acyclic bis-bibenzyls, perrottetin F (**184**) and paleatin B (**202**). Marchantins A (**1**), B (**195**), D (**198**), perrottetin F (**184**), and paleatin B (**202**) showed cytotoxicity against KB cells (IC_{50} range 3.7–20 μ*M*), in addition to KB + VLB (1.3–20 μ*M*) and KB − VLB (2.7–20 μ*M*) cells [26, 38, 49].

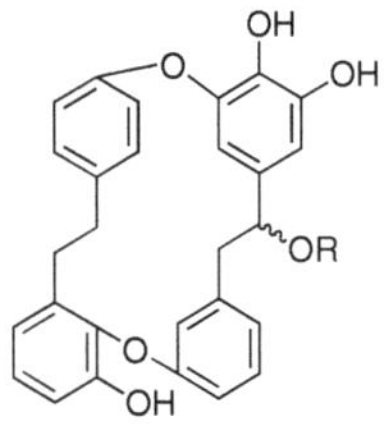
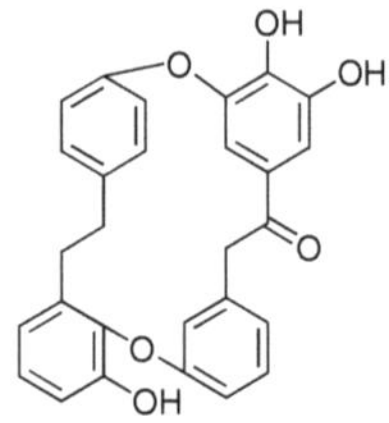
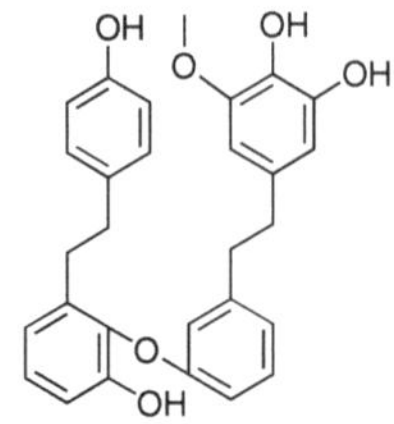

198 (marchantin D, R = H)
199 (marchantin E, R = Me)
201 (marchantin J, R = Et)

200 (marchantin G)

202 (paleatin B)

To obtain novel bis-bibenzyls on a laboratory scale, Sabovljević et al., have produced an axenically farmed culture of *M. polymorpha* and compared the phytochemicals between their cultured mass and field-collected *M. polymorpha*, both qualitatively and quantitatively. The methanol extracts of both samples were analyzed by LC/MS. The only similarity between the two extracts was the presence of marchantin A (**1**). In the cultured extract, the presence of marchantins E (**199**) and G (**200**), and/or C (**197**), and dehydromarchantin A (**203**) was determined. Thus, the axenically farmed liverwort biosynthesized numerous different secondary metabolites not found in naturally grown populations [133].

Marchantin A (**1**), isolated from *M. emarginata* subsp. *tosana*, induced cell growth inhibition in MCF-7 human breast cancer cells, with an IC_{50} value of 4.0 μg/cm^3. Fluorescence microscopy and a Western blot analysis indicated that **1** induces apoptosis of MCF-7 cells through a caspase-dependent pathway. The phenolic hydroxy groups at C-1′ and C-6′ were deemed responsible for the cytotoxic and antioxidant activities of compound **1** [134].

Jensen et al., reported in 2012 [135] that marchantin A (**1**) induced reductions in the cell viability of the three breast cancer cell lines, Walker A256, MCF-7, and T47D, with IC_{50} values of 5.5, 11.5, and 15.3 μ*M*, respectively. These effects increased dramatically for all cell lines in a synergistic manner when Aurora-A kinase, which inhibits MLN8237, was added simultaneously [135]. Marchantin A (**1**) was found to possess inhibitory activity (IC_{50} 7.5–12.0 μg/cm^3) against the A375 human malignant melanoma cell line [136].

Marchantin C (**197**), from *M. polymorpha* and other *Marchantia* species, its dimethyl ether **204**, and 7′,8′-dehydromarchantin C (**205**), and its dimethyl ether **206**, were synthesized and their possible modulatory effects on P-glycoprotein in vincristine-resistant KB/VCR cells were investigated [137]. The results indicated that **197** was the most potent inhibitor of cell proliferation in both KB and vincristine-resistant KB/VCR cells among these four compounds, while the three synthetic derivatives of **197** did not show any antiproliferative activity. Apoptosis in KB/VCR cells was induced by treatment with a 16 μ*M* solution of the dimethyl ether (**204**) of marchantin C (**197**) and vincristine for 48 h [137].

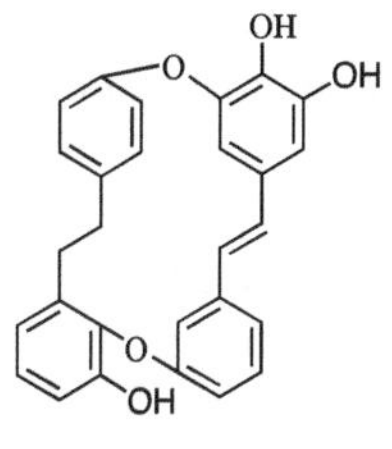

203 (7',8'-dehydromarchantin A)

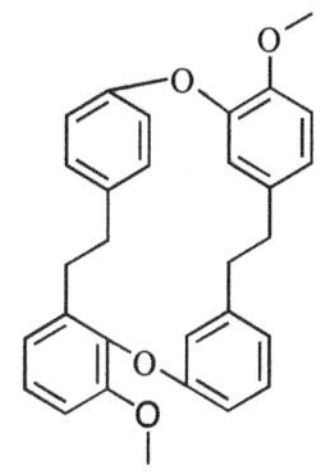

204 (marchantin C dimethyl ether)

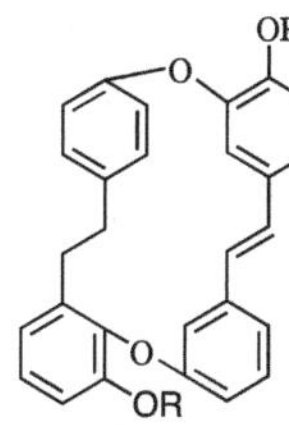

205 (7',8'-dehydromarchantin C, R = H)
206 (7',8'-dehydromarchantin C diemethyl ether, R = Me)

Shen et al., studied in 2010 the effect of marchantin C (**197**) on the inhibition of migration in T98G and U87 glioblastoma cells. At a concentration of 12 μ*M*, the viability of both these cell lines was reduced. Marchantin C (**197**) decreased the migration and invasion abilities in U87 and T98G glioma cells at final concentrations of 6.8 and 10 μ*M*, respectively. Marchantin C (**197**) also decreased the invasion ability in glioma cells at 6–10 μ*M*, and inhibited phosphorylation of the ERK/MAPK signaling pathway and 55.9% of angiogenesis at 10 μ*M* in a chorioallantoic membrane assay. The effect of marchantin C (**197**) on angiogenesis induced by T98G or THP cells was determined by a tube formation assay using human umbilical vascular endothelial cells cultured with a conditioned medium of a marchantin C-treated cell. Marchantin C reduced T98G cell-induced tube formation by 57% while THP1 cell-induced tube formation was not affected. Marchantin C (**197**) also increased the sFlt-1 level in the supernatant of T98G cells by about two-fold compared to a control group [138]. These results suggest that **197** can inhibit angiogenesis induced by glioma cells or monocytes. Up-regulation of sFlt-1 played a significant role in this process [139].

A nitrogen-containing marchantin C derivative (12-*N,N*-di-methyl-amino-marchantin C (**207**)), in addition to aminated, and brominated derivatives, as well as a dimer, have been synthesized, as illustrated in Scheme 3. The products **208–210** prepared from marchantin C (**197**), as shown in Scheme 3, possessed cytotoxicity against KB, MCF-7 and PC3 cells (IC_{50} range 6.3–27.2 μ*M*). 10-Bromo- (**208**), 11-bromo- (**209**) and 12-bromomarchantin C (**210**) showed slightly higher cytotoxic activities against all of the tested human cancer cells, when compared with marchantin C (**124**). The dimer (**211**) and 12-*N,N*-dimethylaminomarchantin C (**207**) were less active than the brominated products [140].

Marchantin C (**197**), neomarchantin A (**212**) and B (**213**), and a mixture of sesquiterpene/bis-bibenzyl dimers, GBB A (**214**) and GBB B (**215**), from *Schistochila glaucescens*, showed growth inhibitory activities against the P-388 murine lymphocytic leukemia cell line, with IC_{50} values of 18.0, 7.6, 8.5, and 10.3 μg/cm^3, respectively [141].

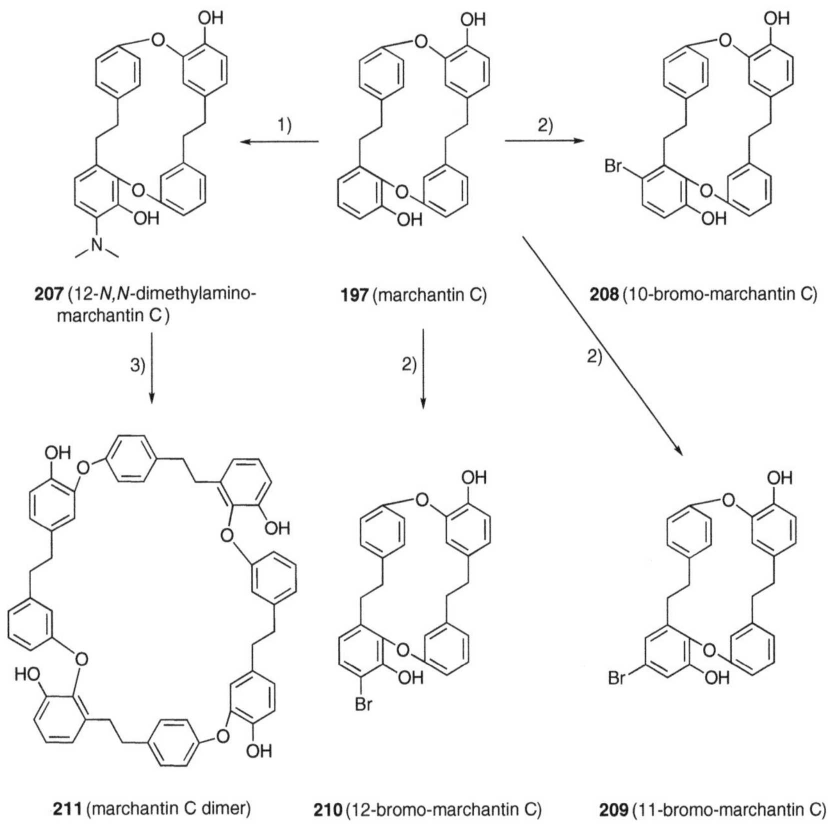

Scheme 3 Preparation of aminomarchantin C (**207**), bromomarchantins C (**208–210**), and dimeric marchantin C (**211**) from marchantin C (**197**). (1) Me_2NH, HCHO, CH_3OH, reflux, (2) NBS, CH_3CN, 0°C, (3) NaOMe, DCM

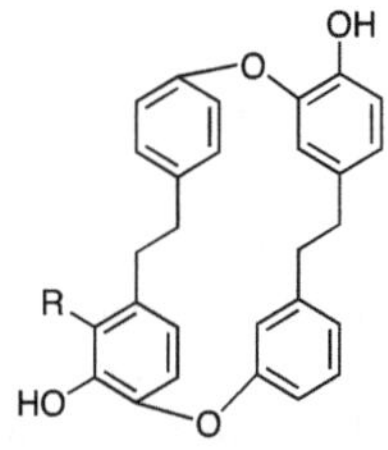

212 (neomarchantin A, R = H)
213 (neomarchantin B, R = OH)

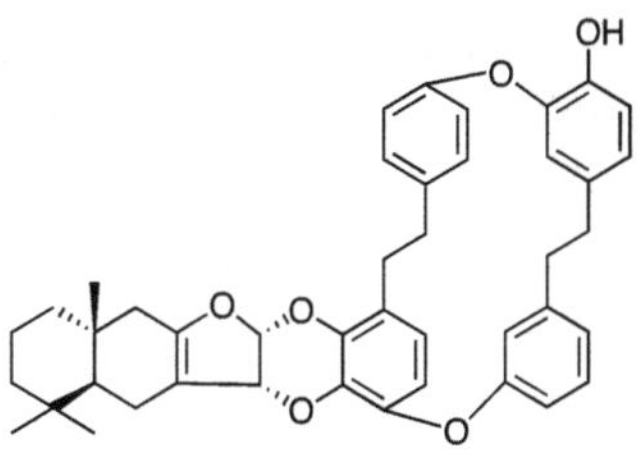

214 (GBB A)

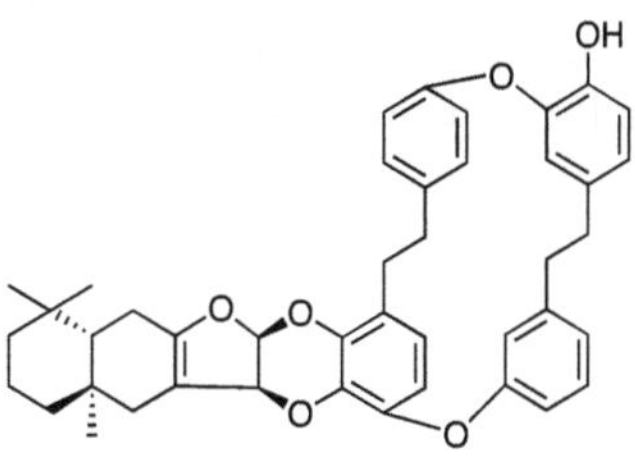

215 (GBB B)

Marchantin C (**197**) inhibited the growth of HeLa human cervical tumor xenografts in a nude mouse model at doses of 10 and 20 mg/kg, when administered by subcutaneous injection. It also decreased the quantity of microtubules in a time- and dose-dependent manner at the G2/M phase in HeLa cells at 8–16 μ*M* [142, 143]. This same compound decreased the polymerization rate of gross tubulin, in a similar manner to the potent microtubule depolymerizer, vincristine, at 8–24 μ*M*. Compound **197** is a novel microtubule inhibitor that induces mitotic arrest of tumor cells and suppresses tumor cell growth. The structure of marchantin C (**197**) is distinct from classical microtubule inhibitors like colchicine, podophyllotoxin, vinblastine, and vincristine. Therefore, it may be regarded as a possible antitumor agent lead compound for further in vivo studies in additional antitumor models, as a result of its ability to inhibit microtubule polymerization [142, 143].

As marchantin C (**197**) functioned as a microtubule inhibitor to induce apoptosis of A172 human glioblastoma cells and HeLa cervical carcinoma cells with low cytotoxicity [144], Zhang et al., in 2019, investigated the induction of cancer cellular senescence at a low concentration of **197** below its IC_{50} value, and found a similar amelioration of the promoting effects of the senescence-associated secretory phenotype on tumor cell proliferation to other genotoxic chemotherapies. The results showed that **197** can selectively inhibit tumor growth via the induction of cancer cell senescence with little cytotoxicity, further indicating the potential of **197** as a promising anticancer agent lead compound [144].

Riccardins A (**2**), B (**216**), and C (**182**) were the first riccardin-type bis-bibenzyls isolated from the Japanese liverwort, *Riccardia multifida* subsp. *decrescens.* The former two compounds were found to inhibit KB cells with ED_{50} values of 10 and 12 μg/cm^3, respectively [38].

In 2010, Xu et al., studied the cytotoxic effects of four cyclic bis-bibenzyls, riccardin C (**182**), marchantin M (**217**), pakyonol (**218**), and plagiochin E (**219**), obtained from *Asterella angusta*, *Plagiochasma intermedium* and *Marchantia polymorpha.* Cell growth was evaluated by the MTT method, and the apoptosis-related proteins BCL-2 and Bax, as well as poly(ADP-ribose) polymerase, were examined by Western blotting using PC3 cells, and also were evaluated with flow cytometry and by morphological examination. These tested bis-benzyls inhibited proliferation and elicited cell death in a dose- and time-dependent manner with IC_{50} values of 3.2, 8.1, 5.5, and 6.0 μM for riccardin C (**182**), marchantin M (**217**), pakyonol (**218**) and plagiochin E (**219**), respectively. Exposure to these bis-benzyls caused a decrease in the antiapoptotic protein Vcl-2 and an increase in proapoptotic Rax expression. Poly(ADP-ribose)-polymerase cleavage and caspase-3 activity were also observed [145].

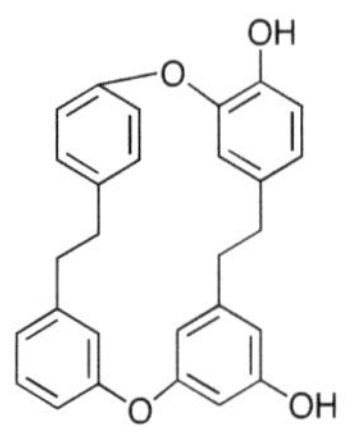

216 (riccardin B)

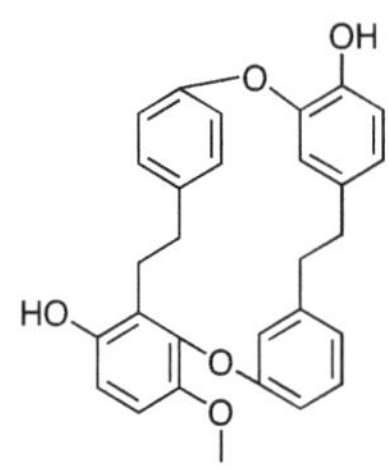

217 (marchantin M)

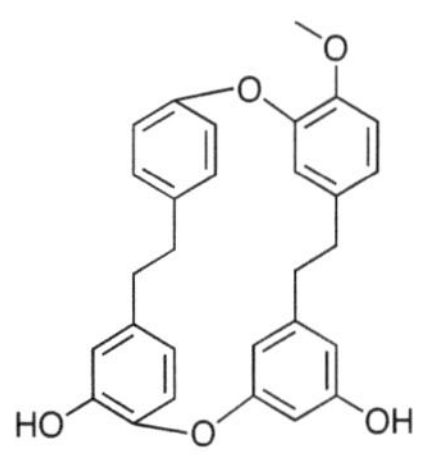

218 (pakyonol)

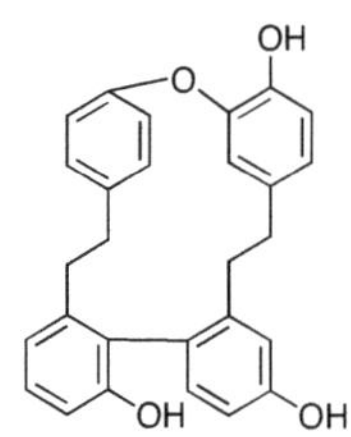

219 (plagiochin E)

Zhang et al., reported in 2015 that marchantin M (**217**) inhibited the growth of PC-3 human prostate cancer cells and also up-regulated the expression of CHOP and GRP78. The same authors indicated that **217** limited the proliferation and promoted apoptosis of DU145 human prostate cancer cells in a time- and dose-dependent manner [146].

Jian et al., found in 2013 that marchantin M (**217**) induced autophagy-dependent cell death, which was accompanied with the induction of endoplasmic reticulum stress and the inhibition of proteasome activity in PC3 and DU145 prostate cancer cells [147].

n-Hexane and ethyl acetate extracts of wild and cultured Indian *Marchantia polymorpha* were analyzed by GC/MS and 55 volatile components were identified. The major components of the *n*-hexane extract from the wild specimen were aromadendrene oxide (**220**), 1,2-dihydrocuparene (**221**), cedrene epoxide (**222**), and geranyl isovalerate (**223**), while palmitic, α-linolenic, and oleic acids, campesterol, and stigmasterol were confirmed in the cultured plant as the major components. The ethyl acetate extract from the wild sample contained α- (**224**) and β-santalol (**225**), and shikimic (**226**) and palmitic acids as significant components, while that from the cultured sample contained succinic (**227**), shikimic (**226**), and palmitic acids as predominant components. These chemical profiles were different from those of European and Japanese *M. polymorpha* specimens, except for the presence of fatty acids and steroids [3]. The cytotoxic effects of each crude extract were studied against four cancer cell lines. Only the *n*-hexane extract of the wild specimen showed some cytotoxic activity against the cell lines used, while the *n*-hexane and ethyl acetate extracts of the cultured specimen were also somewhat cytotoxic. The water extracts from both wild and cultured *M. polymorpha* samples showed no discernible cancer cell antiproliferative activity. No purified constituents were evaluated against the cancer cell line panel in this investigation [148].

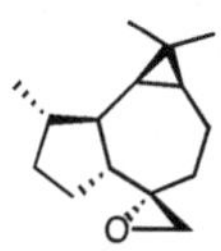

220 (aromadendrene oxide)

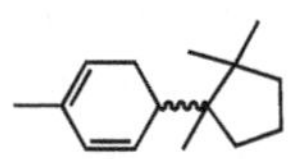

221 (1,2-dihydrocuparene)

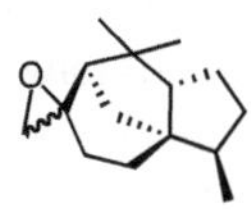

222 (cedrene epoxide)

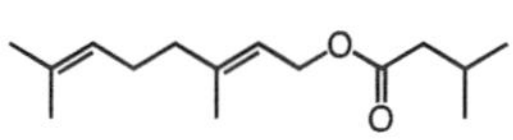

223 (geranyl isovalerate)

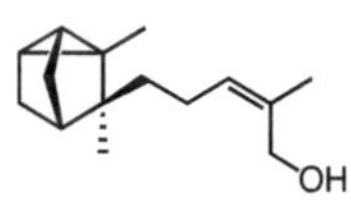

224 (α-santalol)

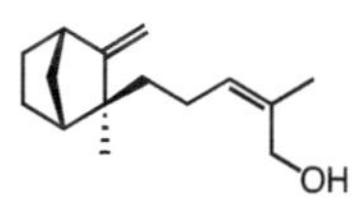

225 (β-santalol)

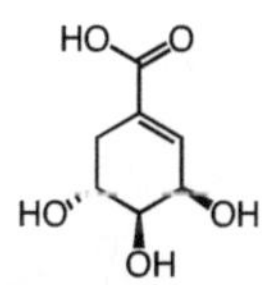

226 (shikimic acid)

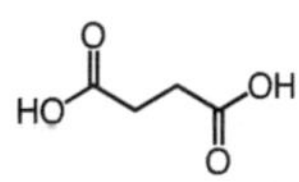

227 (succinic acid)

Dihydroptychantol A (DHA) (**228**) and DHA-2 (**229**) were isolated from *Asterella angusta* and tested for their cytotoxic effects against two different cancer cell lines. Dihydroptychantol A (**228**) showed growth inhibitory effects against both the U2O2 human osteosarcoma and U87 human gliobastoma cells. The IC_{50} values of this compound when administered for 24 and 48 h to U2OS and U87 cells were 29.6, 24.7, and 21.2, 23.7 μ*M*, respectively. In U2O2 cells, dihydroptychantol A (**228**) induced autophagy, followed by apoptotic cell death accompanied with G2/M-phase cell cycle arrest [149].

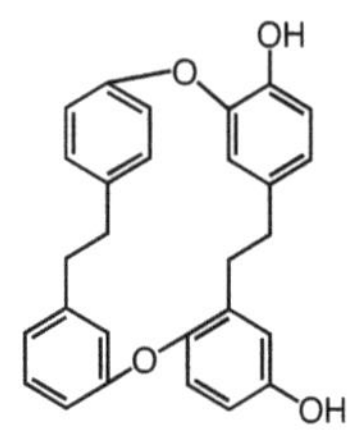

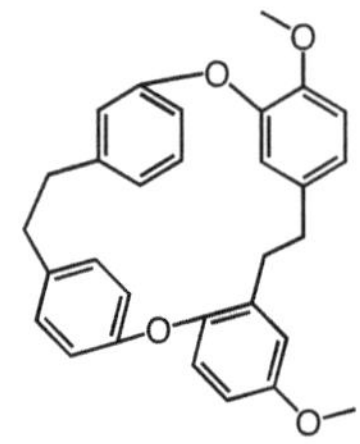

228 (dihydroptychantol A, DHA) **229** (DHA-2)

In order to obtain more potently cytotoxic bis-bibenzyl derivatives, dihydroptychantol A (**228**) was modified to produce four thiazole ring system compounds, **230–233**, as shown in Scheme 4. These were tested for reversal of multidrug resistance against the adriamycin-resistant K562/A02 and vincristine-resistant KB/VCR cell lines. Among the compounds tested, **231** showed the most potent multidrug-resistance-reversal activity. When co-administered with each derivative, there was an increase in the cytotoxicity of vincristine toward KB/VCR cells, and its IC_{50} values ranged from 0.14 to 0.18 μ*M*, with the multidrug resistance-reversal-fold of these compounds ranging from 10.5 to 13.8 μ*M*, which was 3.2–4.2-fold more potent than the parent compound, dihydroptychantol A (**228**) [150].

Pang et al., in 2014 also evaluated dihydroptychantol A (**228**) and its synthetic derivative, DHA2 (**229**) against ovarian cancer cells. The exposure of the SK-Ov-3 ovarian cancer cell line to DHA2 resulted in the downregulation of the anti-apoptotic

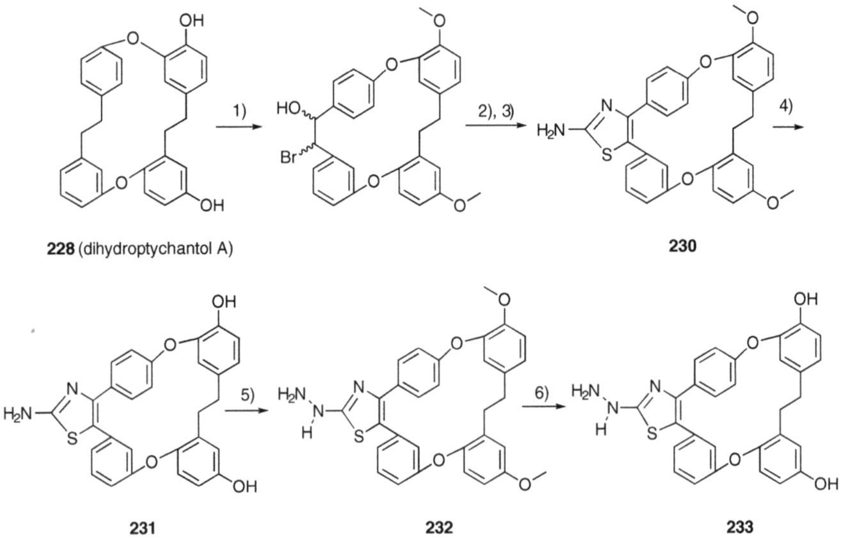

Scheme 4 Preparation of thiazole ring-containing bis-bibenzyls **230–233** from dihydroptychantol A (**228**). (1) NBS, THF/H_2O, 45°C, (2) DMSO, TFAA, CH_2Cl_2, – 78°C, (3) thiourea, DMF, 60°C, (4) BBr_3, CH_2Cl_2, (5) thiosemicarbazide, DMF, 75°C, (6) thioacetamide, K_2CO_3, DMF, 80°C

X-linked inhibition of apoptosis protein and BCL-2 and also caspase-independent cell death. Moreover, DHA2 suppressed tumor growth in vivo leading to autophagy in a SK-OV-3 murine xenograft model, when administered intraperitoneally at 15 and 30 mg/kg [151].

The thalloid liverwort, *Lunularia cruciata* (Plate 16), when grown in Japan, was investigated chemically by Asakawa et al., in 1995. The isolated compounds included the bis-bibenzyls, perrottetin F (**184**) and 7′,8′-dehydroperrottetin F (**234**) along with the bis-bibenzyl dimer, cruciatin (**235**). In turn, fractionation of a methanol extract of an Italian specimen of this species led to the isolation of seven new bis-bibenzyls, 6′-hydroxyriccardin F (**236**), riccardin G-10,13-quinone (**237**), riccardin G-10,13′-quinone (**238**), 2-hydroxydehydro-cavicularin-3-methyl ether (**239**), 7′-methoxyperrottetin F (**240**), 2-(3′-hydroxy-1-bibenzyloxy)-3,7-dihydroxydihydrophenanthrene (**241**), and 1,2,7-trihydroxy-3(3′-hydroxy-1-bibenzyloxy)-phenanthrene (**242**), along with lunularin (**103**), and five known bis-bibenzyls, perrottetins E (**183**) and F (**184**), and riccardins C (**182**), F (**243**), and G (**244**). The stereostructures of the new compounds were elucidated on the basis of 2D-NMR spectroscopic analysis [152]. It was suggested that 7′-methoxyperrottetin F (**240**) might be an extration artifact of perrottetin F (**184**) obtained during its isolation procedure, as seen for the bis-bibenzyl marchantin G (**200**) [3].

Plate 16 *Lunularia cruciata* (liverwort)

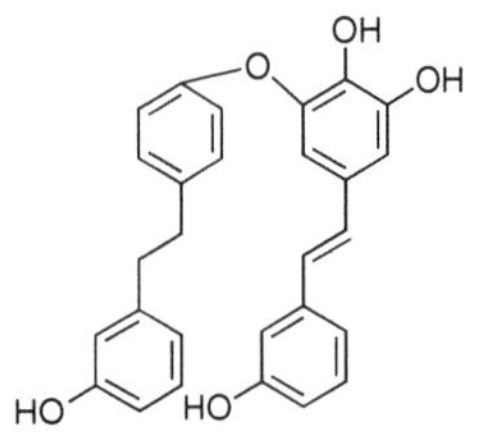

234 (7',8'-dehydroperrottetin F)

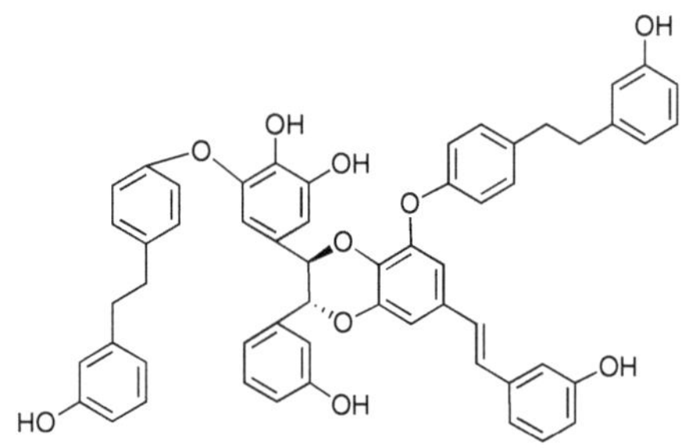

235 (cruciatin)

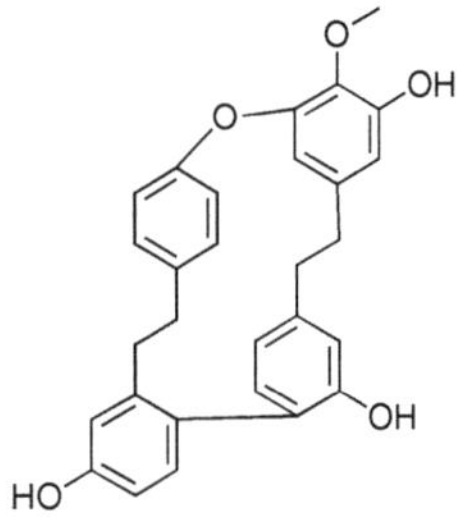

236 (6'-hydroxyriccardin F)

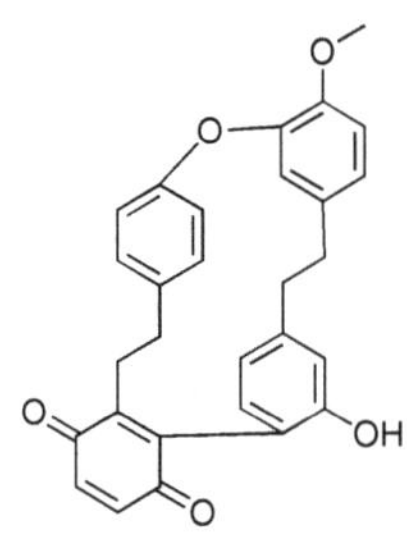

237 (riccardin G 10,13-quinone)

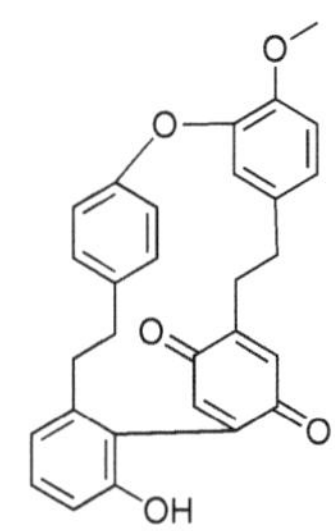

238 (riccardin G 10',13'-quinone)

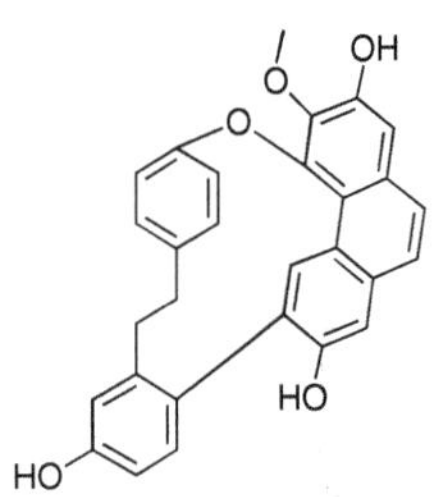

239 (2-hydroxy-dehydro-cavicularin-3-methyl ether)

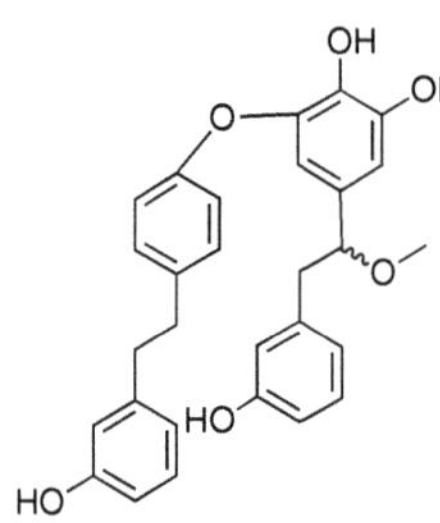

240 (7'-methoxy-perrottetin F)

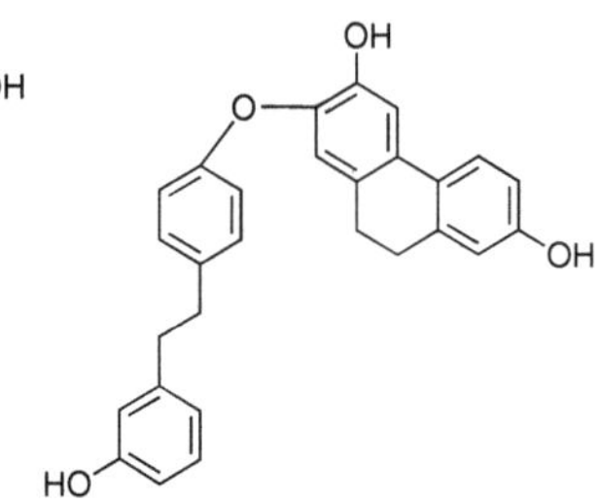

241 (2-(3'-hydroxy-1-bibenzyloxy)-3,7-dihydroxydihydrophenanthrene)

A similar dihydrophenanthrene-containing bis-bibenzyl has been found in the methanol extract of the Ecuadorian liverwort *Frullania convoluta*, along with lunularin (**103**) and lunularic acid (**196**) [26]. Compound **239** is structurally similar to the phenanthrene bis-bibenzyl angustin A (**245**) isolated from the Chinese liverwort, *Asterella angusta* [153] and cavicularin (**246**) from the Japanese liverwort *Cavicularia densa* [3]. Similar *p*-quinone bis-bibenzyls have been found in the liverworts *Marchantia paleacea* and *Reboulia hemisphaerica* [3]. It should be noted that most of the isolated bis-bibenzyls are optically active. Compound **239** showed the highest optical rotation, $[\alpha]_D - 33°dm^{-1}\ cm^3g^{-1}$, although there is no chiral center present in this molecule, thus suggesting that it possesses a highly strained conformation. This phenomenon has already been documented for the bis-bibenzyl cavicularin (**246**) [3]. Plausible biogenetic pathways for the new bis-bibenzyls **239**, **241**, and **242** were proposed [152]. Compounds **236**, **242**, and **244** showed cytotoxic activity against the A549 human lung cancer cells lines and MRC5 normal lung fibroblast cells, in the IC_{50} range 2.5–7.5 μ*M* [152].

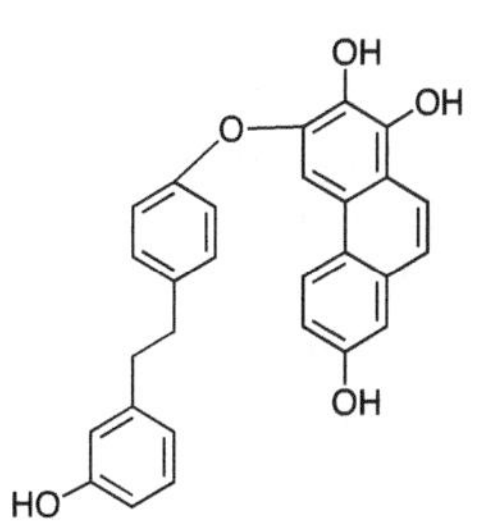

242 (1,2,7-trihydroxy-3(3'-hydroxy-bibenzyloxyphenanthrene)

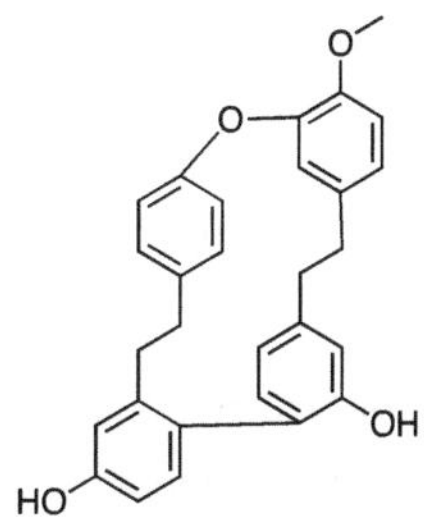

243 (riccardin F)

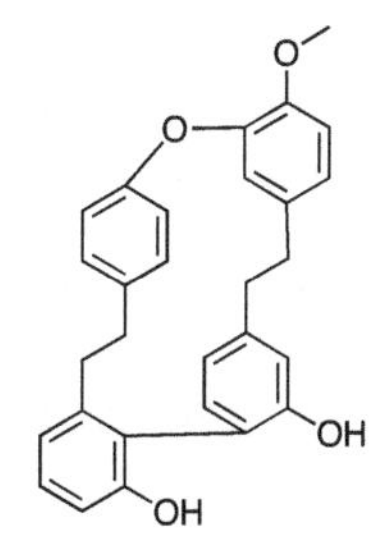

244 (riccardin G)

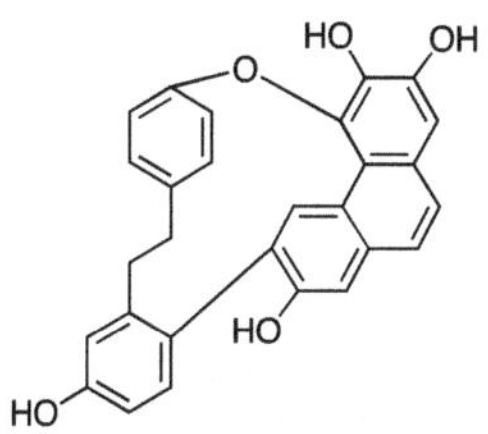

245 (angustin A = 6'-hydroxy-7',8'-dehydrocavivularin)

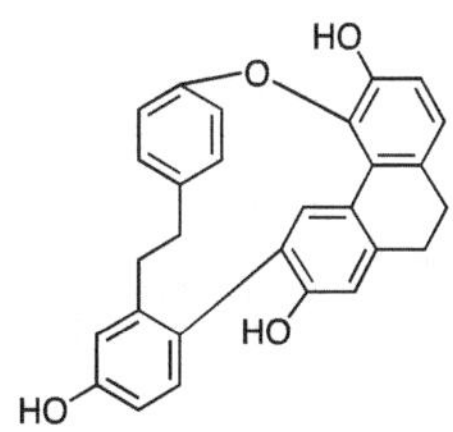

246 (cavicularin)

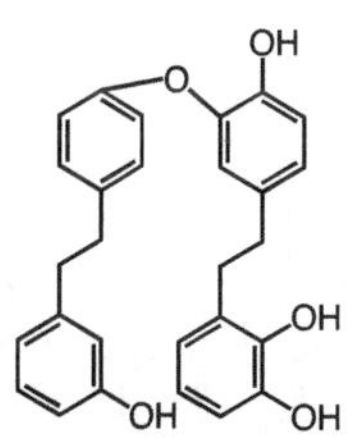

247 (10'-hydroxy-perrottetin E)

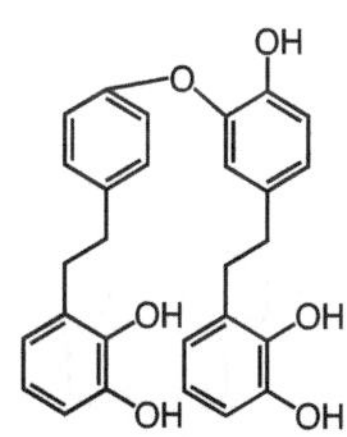

248 (10,10'-dihydroxy-perrottetin E)

Pellia endiviifolia elaborates not only a large amount of the pungent diterpene dialdehyde, sacculatal (**128**), but also acyclic bis-bibenzyls [3]. Fractionation of a methylene chloride-methanol extract of the Serbian *Pellia endiviifolia* led to the isolation of three bis-bibenzyls, perrottetin E (**183**), 10′-hydroxyperrottetin E (**247**), and 10,10′-dihydroxyperrottetin E (**248**). These three compounds and the perrottetin F phenanthrene derivative **242**, which was isolated from *Lunularia cruciata*, were evaluated for their cytotoxic effects against six cancer cell lines. Perrottetin E (**183**) showed cytotoxic activities against HL-60, U-937, K562, NT2/D1, A-172, and U-251 cells with an IC_{50} value range of 6.8–54.8 μ*M*. The latter two compounds isolated in general were less active, but **247** displayed more potent activity than compound **248** against NT2/D1 cells [154].

The genus *Plagiochasma* belongs to the family Aytoniaceae, and is an abundant source of bis-bibenzyl derivatives [3, 75, 123]. In 1999, Bardon et al., isolated pakyonol (**218**) and riccardin F (**243**) from *P. intermedium* [155], and in 2011

Ji et al., evaluated their cytotoxic activities by employing K562 leukemia cells and adriamycin-resistant K562/A02 cells. A P-glycoprotein 3-(4,5-dimethylthiazol-2-yl)-2,5-diphenyl-tetrazolium bromide assay indicated that pakyonol (**218**) and riccardin F (**243**) at concentrations ranging up to 6 μg/cm^3 exhibited no inhibitory effects on the growth of these two cell lines. However, in the presence of a non-cytotoxic concentration of 3 μg/cm^3 of either **218** or **243**, the IC_{50} of adriamycin against K562/A02 cells was modified by 2.5- and 4.8-fold, respectively. Flow cytometry showed that **218** and **243** enhanced the accumulation of adriamycin in K562/A02 cells [156].

Blasia pusilla elaborates not only bis-bibenzyls, but also the bis-bibenzyl dimers, pusilatins A–D (**249–252**), of which pusilatins B (**250**) and C (**251**) both exhibited moderate cytotoxic potencies against KB, KB+VLB, and KB–VLB cells (IC_{50} range 7.1–15.3 μM) [157].

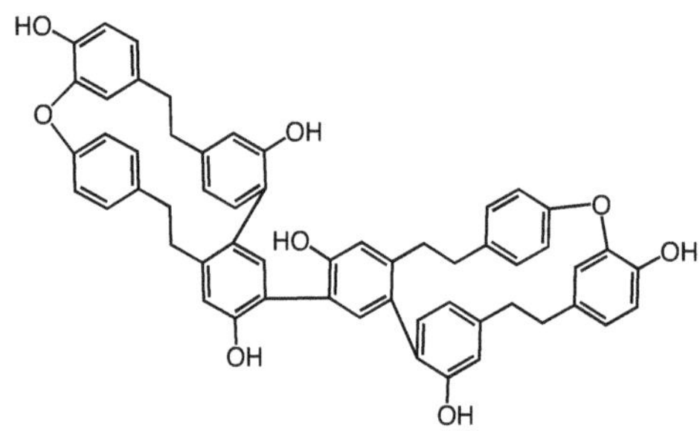

249 (pusilatin A)

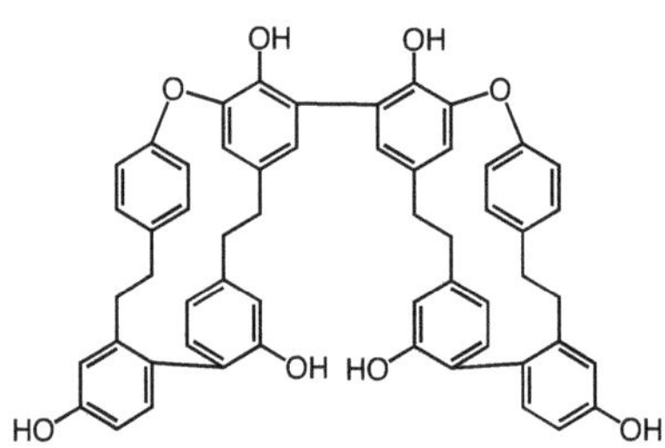

250 (pusilatin B)

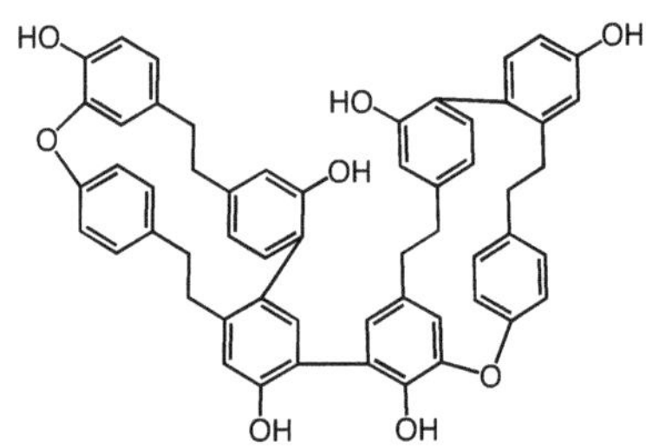

251 (pusilatin C)

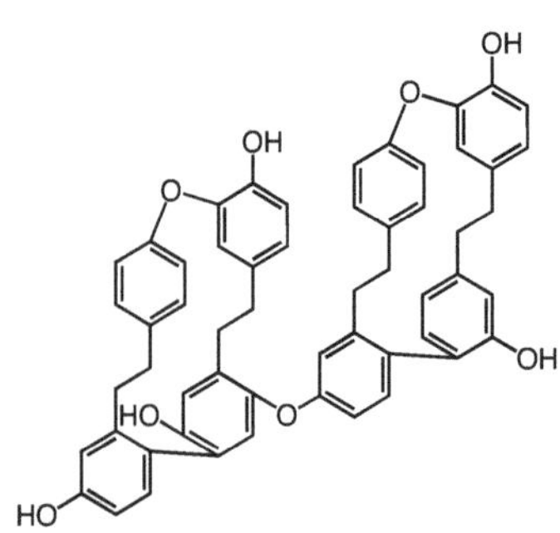

252 (pusilatin D)

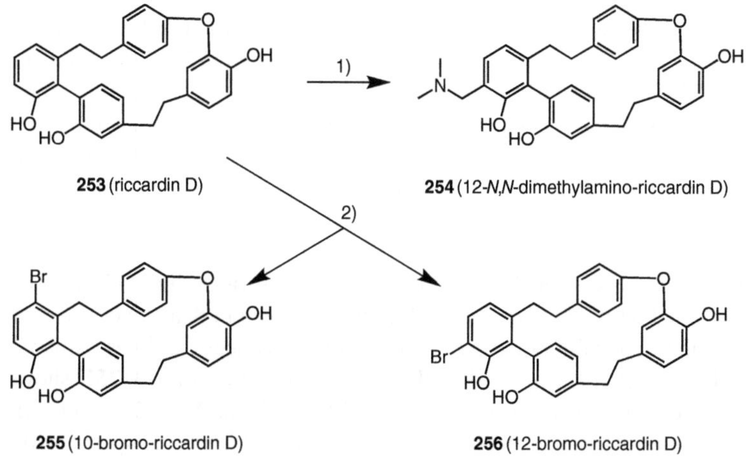

Scheme 5 Preparation of 12-*N,N*-dimethylaminomethylriccardin D (**254**), 10- (**255**), and 12-bromoriccardin D (**256**) from riccardin D (**253**). (1) Me_2NH, HCHO, MeOH, reflux, (2) NBS, CH_3CN, 0°C

Dumortiera hirsuta produces not only dumortane sesquiterpenoids but also the bis-bibenzyls, plagiochin E (**219**) and riccardin D (**253**), which inhibited the growth of KB, MCF-7, and PC3 cells, with IC_{50} values in the ranges 28.0–40.0 and 6.6–10.1 and μM, respectively [158].

The synthesis of 12-*N,N*-dimethylaminomethylriccardin D (= riccardin D-26) (**254**), and 10- (**255**), and 12-bromoriccardin D (**256**) derived from riccardin D (**253**), is shown in Scheme 5. The newly prepared products also showed somewhat improved cytotoxic activities against KB, MCF-7, and PC3 cells [140]. Xue et al., indicated in 2012 that riccardin D (**253**) also displayed cytotoxic effects against HL-60, K562, and K562/A02 cells (IC_{50} range 1.3–21.7 μM) [158].

Riccardin D (**253**) also demonstrated induction of apoptosis against the H460 and A549 human non-small cell lung cancer cell lines, due to DNA topoisomerase-II inhibition [159]. This bis-bibenzyl (**253**) inhibited the proliferation of human umbilical vascular endothelial cells, and decreased their motility and and migration. In addition, riccardin D (**253**) reduced the expression vascular endothelial growth factor (VEGF) and other cellular factors in human umbilical vascular endothelial cells. Compound **253** inhibited the growth of H460 human lung cancer tumors in a murine xenograft model, when administered via a tail injection at a dose of 20 mg/kg [160]. Riccardin D (**253**) also inhibited intestinal adenoma (polyp) formation in $APC^{Min/+}$ mice, decreased β-catenin and cyclin D1 expression, prevented the proliferation of intestinal polyps, and triggered apoptosis via the caspase-dependent pathway and decreased angiogenesis in intestinal polyps [161].

Riccardin D (**253**) significantly inhibited the growth of HT-29 human colon cancer cells. The cDNA expression of cyclooxygenase-1 and the protein expression and activity of NF-κB and tumor necrosis factor-α were downregulated. Riccardin D

(**253**) induced apoptosis in HT-29 cells, which may be associated with the blocking of the NF-κB signaling pathway [162]. Hu et al., reported in 2014 that this same bis-bibenzyl (**253**) exhibited cytotoxic activity by induction of apoptosis and inhibition of angiogenesis and topoisomerase-2. This group confirmed that apoptosis was not the sole mechanism by which riccardin D (**253**) inhibited tumor cell growth, because a low concentration of **253** caused cellular senescence in PCa prostate cancer cells [163].

In 2017, Sun et al., designed and synthesized a novel series of nitrogen-containing macrocyclic bis-bibenzyl derivatives from riccardin D (**253**), including 4′-((piperidin-1-yl)methyl)-riccardin D (**259**), as shown in Scheme 6, and evaluated them for their antiproliferative effects against three cancer cell lines. Among these new products, 13,1′,12′-tri-*N,N*-dimethylamino-ethoxyriccardin D (**257**) showed the most potent cytotoxic activity against the A549, K562, and MCF-7 cancer cell lines, with IC_{50} values of 0.5, 0.2, and 0.2 μ*M*, respectively, which were superior to those of the parent compound, riccardin D (**253**). 1,6′-Bis-(pyrrolidin-1-yl)methyl-riccardin D (**258**) also showed cytotoxicity against these cancer cell lines with IC_{50} values in the range 0.5–1.0 μ*M* [164].

In 2013, Yue et al., reported that riccardin D-26 (**254**) derived from riccardin D (**253**) significantly inhibited cancer growth in both KB and KB/VCR xenografts without significant systemic toxicity, when administered inttravenously at 20 mg/kg. It also inhibited cancer cell growth by inducing apoptosis in the activation of the mitochondrial mediated intrinsic apoptosis pathway, and was more potent than compound **253** in this regard [165].

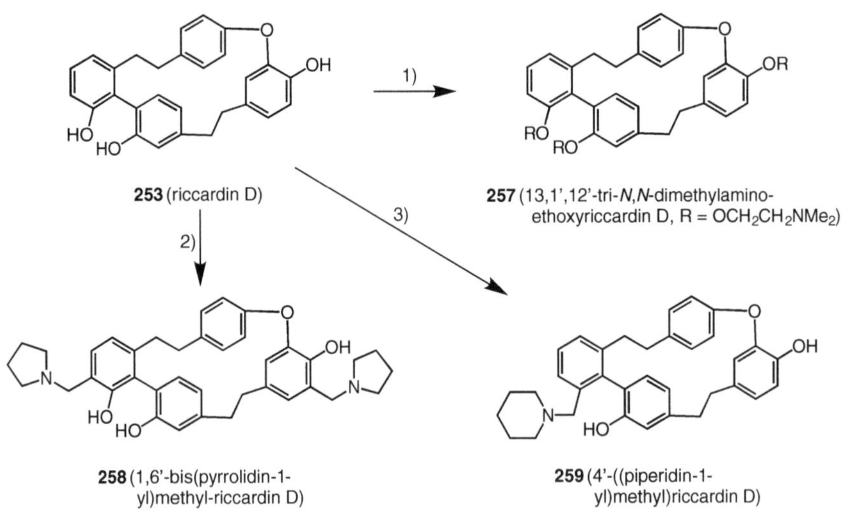

Scheme 6 Preparation of 13,1′,12′-*N,N*-dimethylamino-ethoxy-riccardin D (**257**), 1,6′-bis(pyrolidin-1-yl)methyl-riccardin D (**258**), and 4′-((piperidin-1-yl)methyl)-riccardin D (**259**) from riccardin D (**253**). (1) piperidine, HCHO, EtOH, reflux, (2) $ClCH_2NMe_2$, K_2CO_3, acetone, 60°C, (3) pyrrolidine, HCHO, EtOH, reflux

Plagiochin E (**219**), obtained from *Monoclea forsteri* [2] and *Marchantia polymorpha* [166, 167], showed cytotoxic effects against K562/A02 cells, having IC_{50} values from 2–6 μg/cm^3. When compound **219** was combined with adriamycin, it enhanced the sensitivity of K562/A02 cells by increasing the intracellular accumulation of adriamycin. The inhibitory effect of **219** on P-glycoprotein activity was the major cause of increased retention of adriamycin inside K562/A02 cells, indicating that **219** may effectively reverse the multidrug resistance in the above-mentioned cell line, via inhibiting expression and drug transport function of P-glycoprotein [167].

12-*N,N*-Dimethylaminoplagiochin E (**260**), 12-bromoplagiochin E (**261**), and 12,12′-dibromoplagiochin E (**262**), derived from plagiochin E (**219**) as indicated in Scheme 7, were found to possess cytotoxic effects against KB, MCF-7, and PC-3 cells (IC_{50} 9.7–15.1, 6.3–33.8 and 9.2–25.1 μ*M*, respectively). Compound **260** showed almost the same potency as that of compound **219**, while the two brominated derivatives were 2–6 times more active than the parent compound against these cancer cell lines [158].

Dumortiera hirsuta contains riccardin C (**182**), marchantin A 1′-methylether (**263**) and isomarchantin C (**264**), which displayed cytotoxic activities against KB and HL-60 cells, with IC_{50} values between 33 and 38 μ*M* [168].

Isoplagiochins A (**265**) and B (**266**) isolated from *Plagiochila fruticosa* inhibited the polymerization of tubulin, having IC_{50} values of 50 and 25 μ*M*. The dihydro

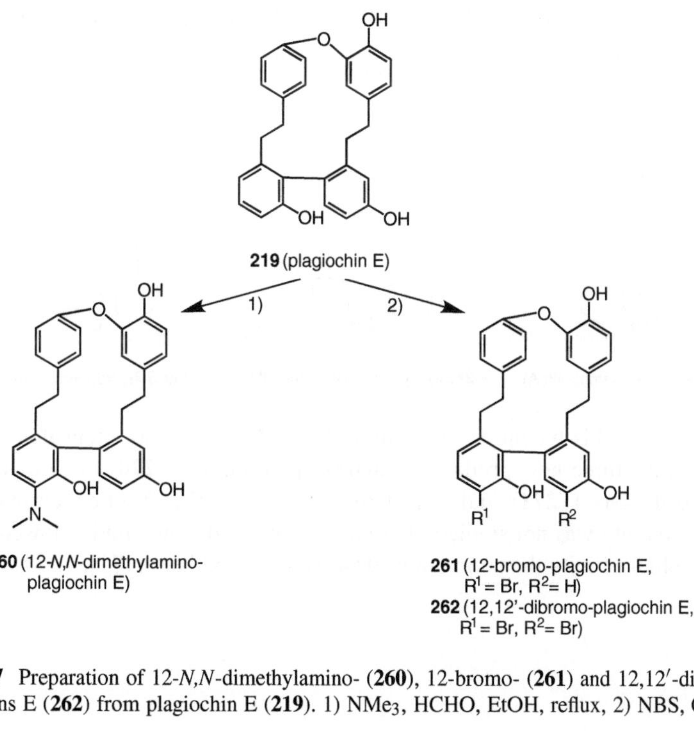

Scheme 7 Preparation of 12-*N,N*-dimethylamino- (**260**), 12-bromo- (**261**) and 12,12′-di-bromo-plagiochins E (**262**) from plagiochin E (**219**). 1) NMe_3, HCHO, EtOH, reflux, 2) NBS, CH_3CN, 0°C

derivatives of both **265** and **266** were inactive ($IC_{50} > 100\ \mu M$), and when compared with the parent compounds, indicated that a rotationally restricted biaryl ring system is favorable for tubulin binding. A Monte Carlo search showed that the presence of two aromatic rings connected by a two-carbon bridge with a double bond may serve to maintain the backbone conformation [169]. The same phenomenon has been encountered for the bis-bibenzyl, marchantin C (**197**), as mentioned above [142, 143].

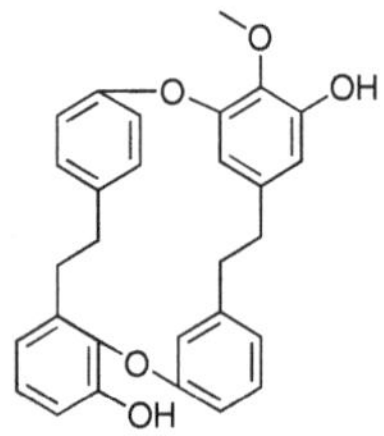

263 (marchantin A 1′-methyl ether)

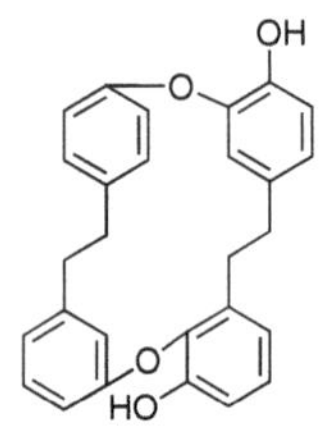

264 (isomarchantin C)

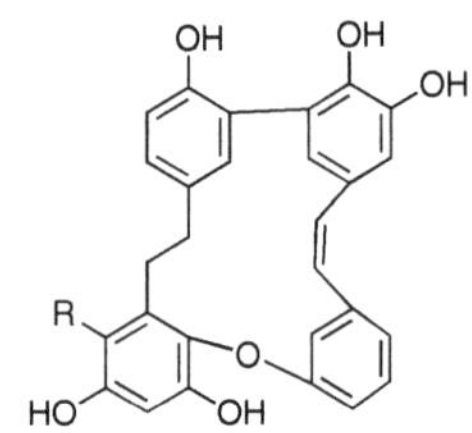

265 (isoplagiochin A, R = OH)
266 (isoplagiochin B, R = OH)

In order to obtain different biologically active compounds from a bis-bibenzyl, marchantin A (**1**) was biotransformed by the fungi, *Aspergillus niger* and *Neurospora crassa.* Marchantin A (**1**) was converted by the former fungus to give 10-hydroxymarchantin A (**267**) in 33% yield, while the latter fungus converted **1** to two hydroperoxy bis-bibenzyls, 3′-hydroperoxymarchantin A (**268**) and 5′-hydroperoxymarchantin A (**269**) in 10 and 23% yields, respectively. Both fungi showed regiospecificity in the addition of an oxygen atom to the B and C rings in the marchantin molecules investigated [170].

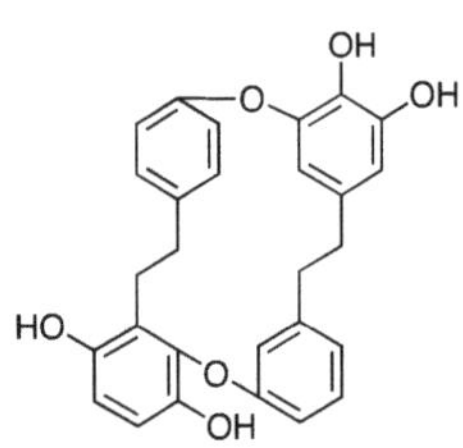

267 (10-hydroxymarchantin A)

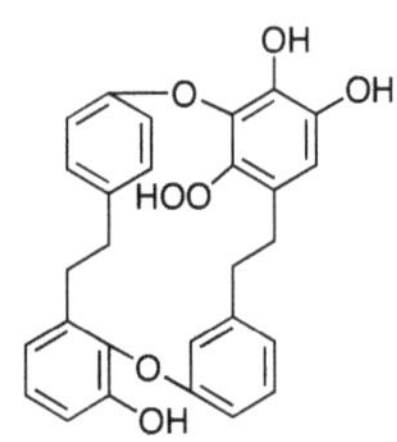

268 (3′-hydroperoxymarchantin A)

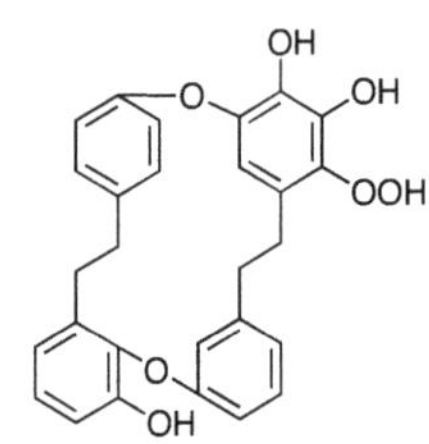

269 (5′-hydroperoxymarchantin A)

An acyclic bis-bibenzyl, perrottetin F (**184**), was also incubated with *A. niger* to give three new products, 8-hydroxyperrottetin F (**270**), 3,4-dihydroxy-5-benzyloxybibenzyl (**271**), and perrottetin F 6′-sulfate (**272**). While their cancer cell line cytotoxicity was not studied, these newly obtained compounds showed weaker antimicrobial activity than compound **184**, as discussed later [171].

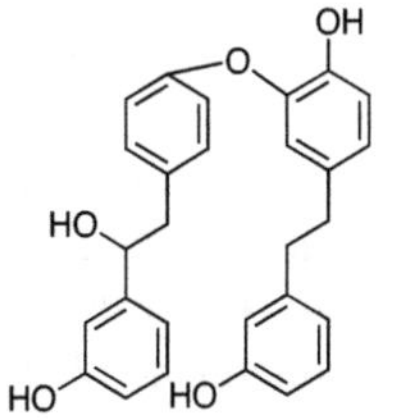

270 (8-hydroxyperrotetin F)

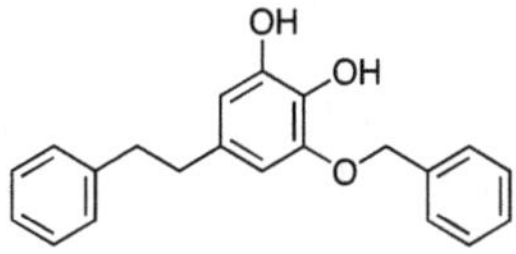

271 (3,4-dihydroxy-5-benzyloxybibenzyl)

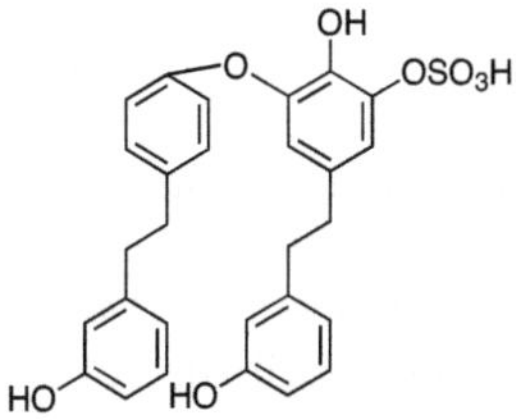

272 (perrottetin F 6'-sulfate)

Novakovic et al., obtained different esters from marchantin A (**1**) by chemical and enzymatic reactions. Chemical synthesis gave three peresterified marchantin A derivatives, the 13,1′,6′-tripropanoyl-, tributanoyl, and trihexanoyl esters (**273**–**275**), while enzymatic treatment using lipase, regioselectively gave the monoester derivatives regioselectively, the 13-propanoyl-, butanoyl, and hexanoyl marchantin A derivatives **276**–**278**.

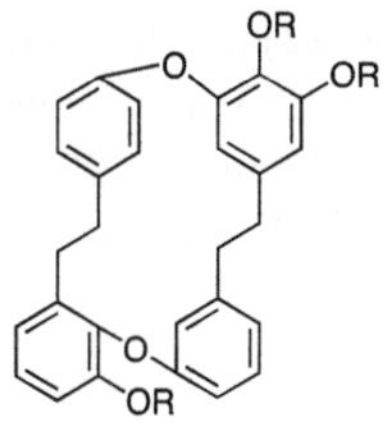

273 (marchantin A 13,1',6'-tripropanoyl ester, R = n-C_3H_7CO)
274 (marchantin A 13,1',6'-tributanoyl ester, R = n-C_4H_9CO)
275 (marchantin 13,1',6'-trihexanoyl ester, R = n-$C_6H_{13}CO$)

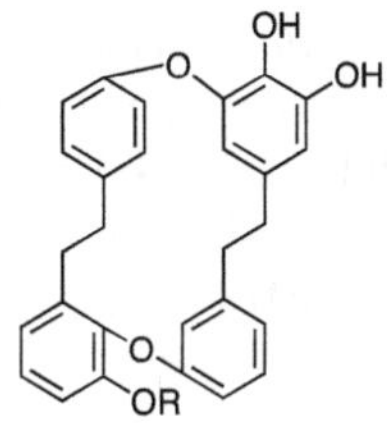

276 (marchantin A monopropanoyl ester, R = n-C_3H_7CO)
277 (marchantin A monobutanoyl ester, R = n-C_4H_9CO)
278 (marchantin A monohexanoyl ester, R = n-$C_6H_{13}CO$)

The antiproliferative activities of the products of marchantin A (**1**) were tested against the A549 lung and the MDA-MB-231 breast human cancer cell lines. All tested esters were less potently cytotoxic than the original **1**. However, they also indicated lower cytotoxicity against normal MRC-5 fibroblast cells. The monoesters showed improved cytotoxic activities over the corresponding peresterified products, presumably due to the presence of a free catechol moiety. The monohexanoyl ester **277** displayed the same IC_{50} value as **1** against MDA-MB-231 cells, although its cancer cell line selectivity was higher. In this way, regioselective enzymatic monoesterification was shown to enhance the selectivity of **1**. Compound **277** was also the most active among all derivatives produced against A549 cells, having slightly lower activity and selectivity in comparison to marchantin A [172].

4.4.2 Bibenzyls, Phthalides, and Phenanthrenes

Asakawa et al., summarized in 2020 and 2022 the distribution, structures and the biological activity of bibenzyls and bis-bibenzyls for 32 *Radula* identified species and four unidentified *Radula* species collected in Peru [173, 174]. These reviews covered the distribution of the psychoactive bibenzyls, *cis*-perrottetinene (**279**) and perrottetinenic acid (**280**), which are structurally very similar to the psychoactive (–)-*trans*-Δ^9-tetrahydrocannabinol (**281**).

A few liverworts such as species of *Frullania* are rich sources of bibenzyl and prenyl bibenzyls. Several phthalides and phenanthrenes, which might be biosynthesized from the bibenzyls, are also the chemical markers of some *Frullania* species.

On the basis of chemosystematic results of the Frullaniaceae, *Frullania* species may be divided into several major chemotypes, (1) sesquiterpene lactones, (2) bibenzyls, and (3) sesquiterpene lactones and bibenzyls, (4) diterpenes, (5) cyclocolorenones, and (6) monoterpenoids. Asakawa et al., [1] reported that *Frullania brittoniae* subsp. *truncatifolia* elaborated brittonins A (**282**) and B (**283**). The Chinese *F. inouei*, belonging to the second chemotype produced 3,3′,4,4′-tetramethoxybibenzyl (**284**), chrysotobibenzyl (**285**), and brittonins A (**282**) and B (**283**), which displayed cytotoxicity against KB, KB/VCR, K562, and K562/A02 cancer cells, with IC_{50} values of 1.1–29.8 μ*M* [175].

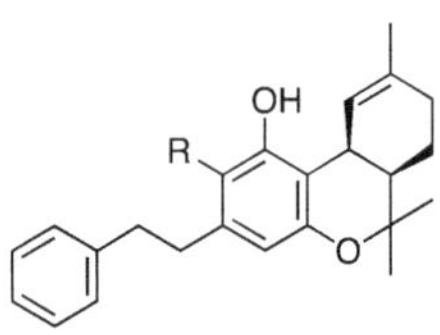

279 (*cis*-perrottetinene, R = H)
280 (perrottetinenic acid, R = COOH)

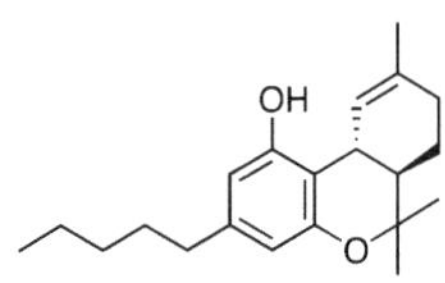

281 ((–)-*trans*-Δ⁹-tetrahydrocannabinol)

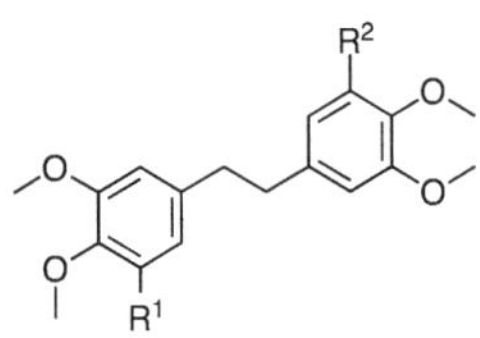

282 (brittonin A, $R^1 = R^2$ = OMe)
284 (3,3':4,4'-tetramethoxybibenzyl, $R^1 = R^2$ = H)
285 (chrysotobibenyl, R^1 = OMe, R^2 = OH)

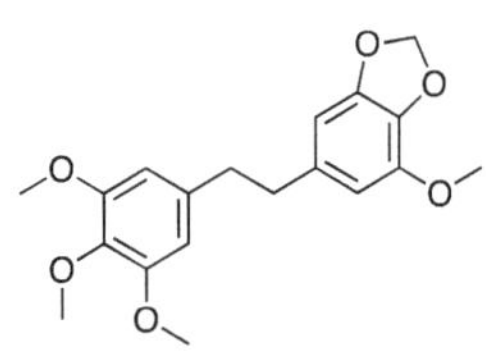

283 (brittonin B)

Lunularic acid (**196**), which is widely distributed in liverworts, showed cytotoxicity against HepG2 cells (IC_{50} = 7.4 μg/cm^3) [176].

The New Zealand *Plagiochila stephensoniana* elaborated the simple bibenzyl derivative, 3-methoxy-4′-hydroxybibenzyl (**286**) [160]. Compound **286** and the synthetic 3-methoxy-4-hydroxy-(*E*)-stilbene (**287**) and its (*Z*)-isomer (**288**) were tested against the P388 murine lymphocytic leukemia cell line. They showed IC_{50} values of > 25, 9.9, and 16.3 $\mu g/cm^3$, respectively [177].

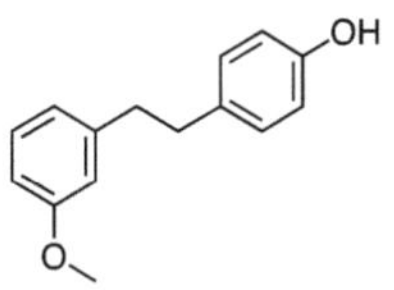

286 (3-methoxy-4'-hydroxy-bibenzyl)

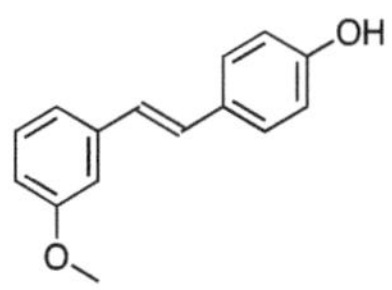

287 (3-methoxy-4'-hydroxy-(*E*)-stilbene)

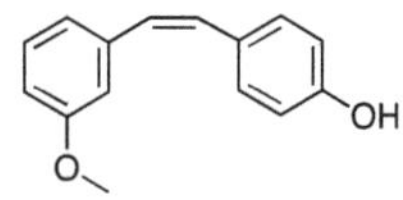

288 (3-methoxy-4'-hydroxy-(*Z*)-stilbene)

Almost of all *Radula* species belonging to the Radulaceae are distinct chemically from the other genera of liverworts, since they produce bibenzyls, prenyl bibenzyls, and/or bis-bibenzyls, and their related compounds as major components, of which many exhibit cytotoxic effects against various cancer cell lines [73, 174].

Bioactivity-guided fractionation of the ethanol extract of the Chinese *Radula apiculata* resulted in the isolation of eight dimeric prenyl bibenzyls named radulapins A–H (**289–296**) and four prenyl bibenzyls, 2-((2*S*)-hydroxymethyl)-2-methyl-5-phenylethyl-2*H*-chromen-7-ol (**297**), 2-((2*R*)-hydroxymethyl)-2-methyl-5-phenylethyl-2*H*-chromen-7-ol (**298**), (3*S*,7)-dihydroxy-2,2-dimethyl-5-phenylethylchroman-4-one (**299**), (3*R*,7)-dihydroxy-2,2-dimethyl-5-phenylethylchroman-4-one (**300**), 2-((2*S*)-hydroxy-3-methylbut-3-en-1-yl)-3-methoxy-5-hydroxybibenzyl (**301**), 2-((2*R*)-hydroxy-3-methylbut-3-en-1-yl)-3-methoxy-5-hydroxybibenzyl (**302**), and 2-(3-hydroxy-3-methylbutyl)-3,5-dihydroxybibenzyl (**303**), along with five known prenyl bibenzyls, 2-(3-hydroxy-3-methylbutyl)-3-hydroxy-5-methoxbibenzyl (**304**), 2-(2-hydroxy-3-methylbut-3-en-1-yl)-3,5-dihydroxybibenzyl (**305**), 2-(3-methyl-2-butenyl-3,5-dihydroxybibenzyl (**306**), 2-(3-methyl-2-butenyl)-3-hydroxy-5-methoxy-bibenzyl (**307**), 2-(2,3-epoxy-3-methylbutenyl)-3,5-dihydroxy-bibenzyl (**308**), one chromane, 2,2-dimethyl-5-(2-phenylethyl)-7-hydroxychromane (**309**) and one chromene, 2,2-dimethyl-5-(2-phenylethyl)-7-hydroxychromene (**310**) [2, 3]. The structures of the new products were determined using a combination of 2D-NMR spectroscopic, X-ray crystallography, optical rotation, and electronic circular dichroism measurements. Among the isolated compounds, radulapins A–H (**289–296**) showed noteworthy cytotoxic activity against the PC-3, A549, MCF-7, and NCI-H12199 cancer cell lines at IC_{50} 1.4–9.8 μM. Radulapin D (**292**) produced cell death in PC-3 prostate cancer cells through mitochondrial-derived apoptosis [178].

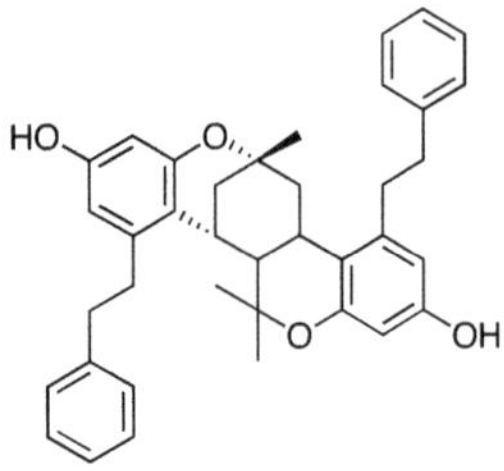

289 (radulapin A)

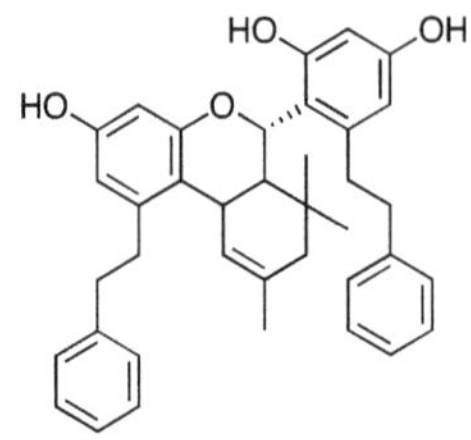

290 (radulapin B)

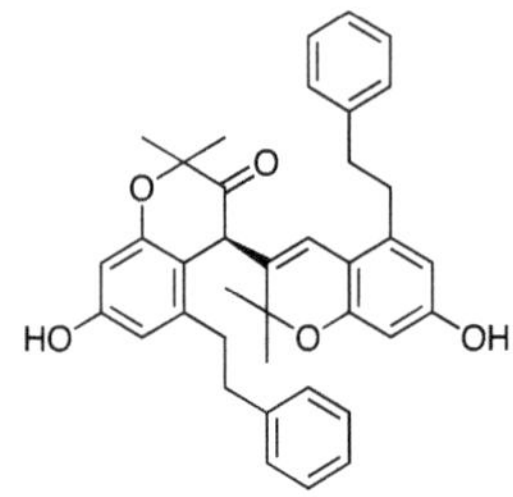

291 (radulapin C)

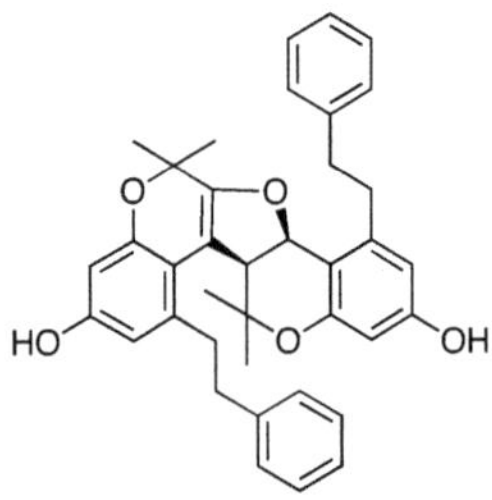

292 (radulapin D)

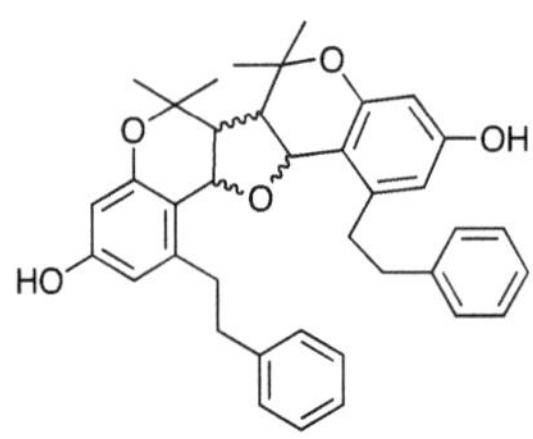

293 (radulapin E)

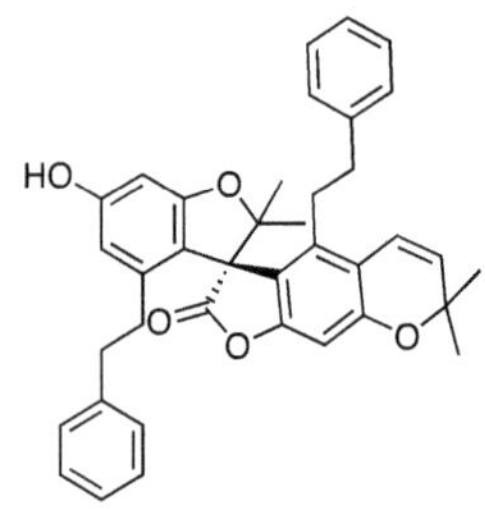

294 (radulapin F)

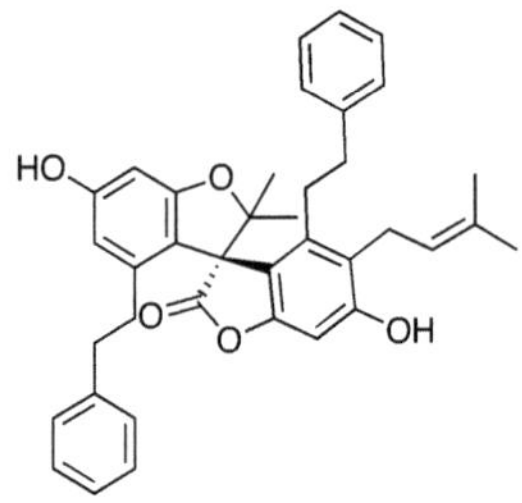

295 (radulapin G)

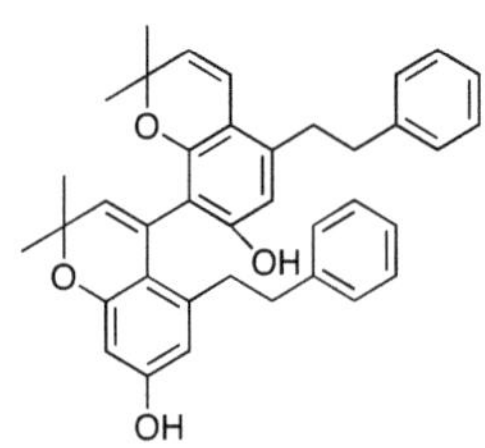

296 (radulapin H)

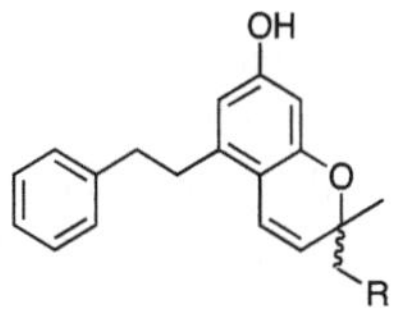

297 ((2*S*)-hydroxymethyl)-2-methyl-5-phenylethyl-2*H*-chromen-7-ol, R = OH)
298 ((2*R*)-hydroxymethyl)-2-methyl-5-phenylethyl-2*H*-chromen-7-ol, R = OH)
310 (2,2-dimethyl-7-hydroxy-5-phenylethylchromene, R = H)

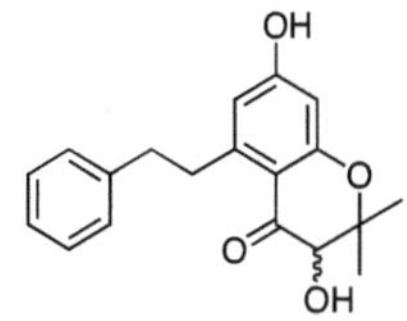

299 ((3*S*),7-dihydroxy-2,2-dimethyl-5-phenylethylchroman-4-one)
300 ((3*R*),7-dihydroxy-2,2-dimethyl-5-phenylethylchroman-4-one)

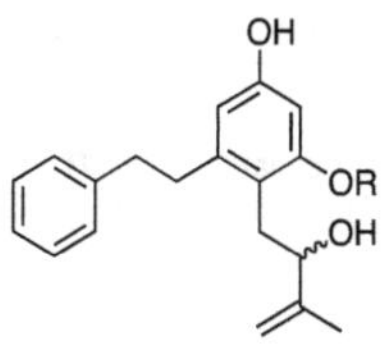

301 (2-(2*S*)-hydroxy-3-methylbut-3-en-1-yl)-3-methoxy-5-hydroxybibenzyl, R = Me)
302 (2-(2*R*)-hydroxy-3-methylbut-3-en-1-yl)-3-methoxy-5-hydroxybibenzyl, R = Me)
303 (2-(2-hydroxy-3-methylbut-3-en-1-yl)-3,5-dihydroxybibenzyl, R = H)

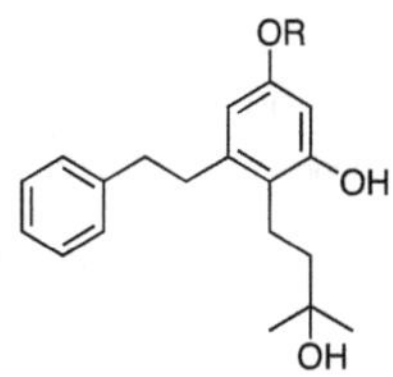

304 (2-(3-hydroxy-3-methylbutyl)-3,5-dihydroxybibenzyl, R = H)
305 (2-(3-hydroxy-3-methylbutyl)-3-hydroxy-5-methoxybibenzyl, R = Me)

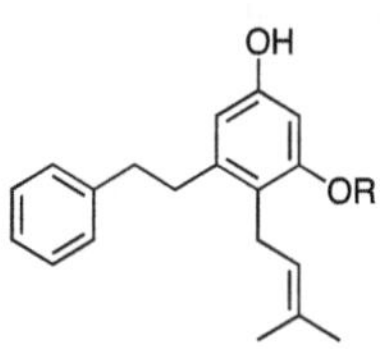

306 (2-(3-methyl-2-butenyl-3,5-dihydroxy-bibenzyl), R = H)
307 (2-(3-methyl-2-butenyl-3-methoxy-5-hydoxybibenzyl, R = Me)

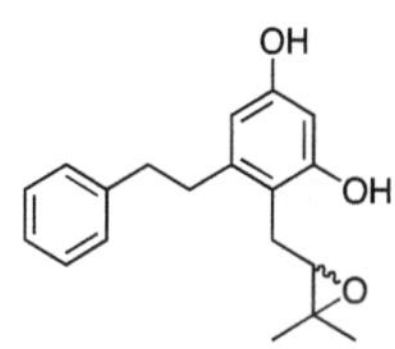

308 (2-(2,3-epoxy-3-methylbutenyl)-3,5-dihydroxy-bibenzyl)

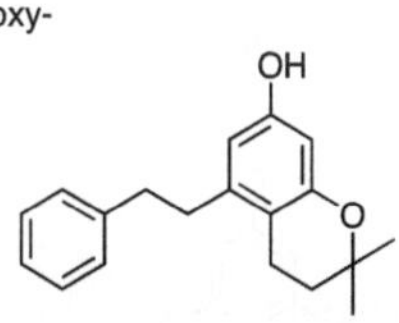

309 (2,2-dimethyl-5-(2-phenylethyl)-7-hydroxychroman)

From a 95% ethanol extract of the Chinese *Radula amoena*, six new prenyl bibenzyls, radulamoenins A–F (**311–316**), were isolated together with 3,5-dihydroxy-2-(3-methyl-2-butenyl)bibenzyl (**306**), 2,2-dimethyl-5-hydroxy-7-(2-phenylethyl)-chroman (**309**) and 2,2-dimethyl-7-hydroxy-5-(2-phenylethyl-chromene (**310**), which were described earlier [177, 179], and eight known analogues, 2-carbomethoxy-3,5-dihydroxystilbene (**317**), 3,5-dimethoxybibenzyl (**318**), 3,5-dihydroxybibenzyl (**319**), 3-hydroxy-5-methoxybibenzyl (**320**), 3-hydroxy-5-methoxy-2-(3-methyl-2-butenyl)bibenzyl (**321**), 3-hydroxy-5-methoxy-2-(3-hydroxy-3-methylbutyl)bibenzyl (**322**), 2,2-dimethyl-7-methoxy-5-(2-phenylethyl-chromene) (**323**), and 2,4-bis-(3-methyl-2-butenyl)-3,5-dihydroxybibenzyl (**324**).

The structures of the new compounds were elucidated by analysis of their spectroscopic data and by ECD calculations. Among the isolated bibenzyl derivatives, the two simple bibenzyls **317** and **318**, when evaluated against the HepG-2, SMMC-7721, and S549 human cancer cell lines, gave IC_{50} values of 18.0–27.6 and 16.6–19.1 μM, respectively [179].

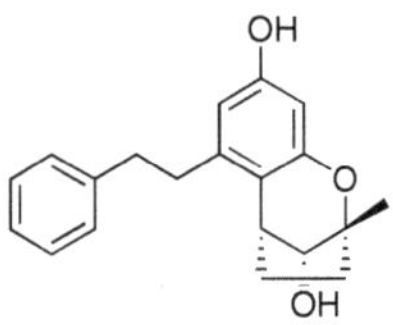

311 (radulamoenin A)

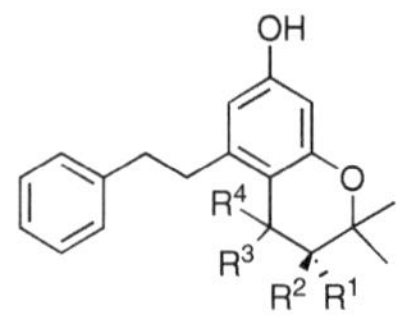

312 (radulamoenin B, $R^1 = R^2 = O$, $R^3 = R^4 = H$
313 radulamoenin C, $R^1 = OH$, $R^2 = H$, $R^3 = R^4 = O$)

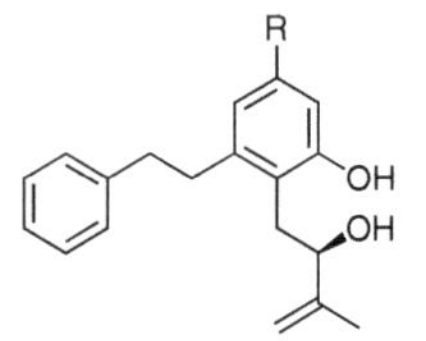

314 (radulamoenin D, R = OMe
315 (radulamoenin E, R = OH)

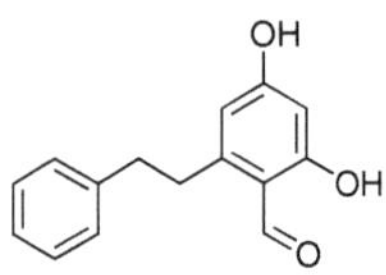

316 (radulamoenin F)

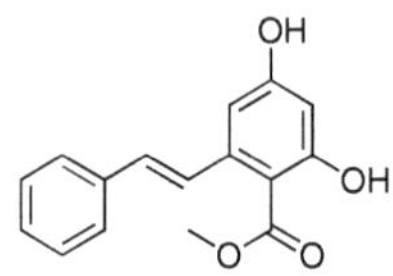

317 (2-carbomethoxy-3,5-dihydroxystilbene)

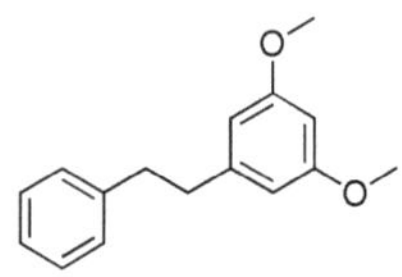

318 (3,5-dimethoxybibenzyl)

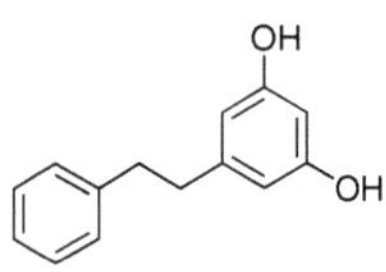

319 (3,5-dihydroxybibenzyl)

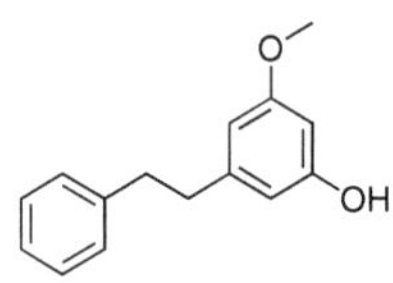

320 (3-hydrox-5-methoxybibenzyl)

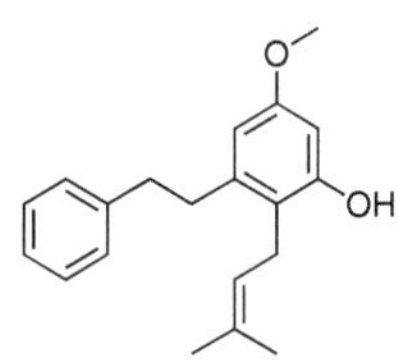

321 (3-hydroxy-5-methoxy-2-(3-methyl-2-butenyl)bibenzyl

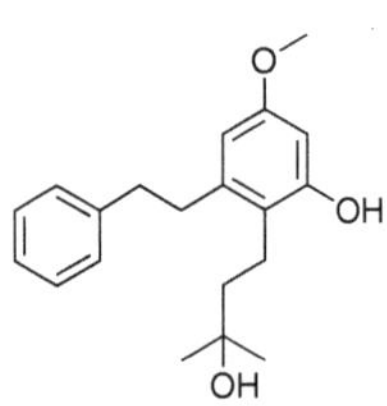

322 (3-hydroxy-5-methoxy-2-(3-hydroxy-3-methyl-butyl)bibenzyl))

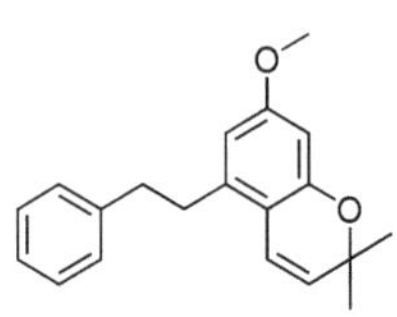

323 (2,2-dimethyl-7-methoxy-5-(2-phenylethyl-chromene))

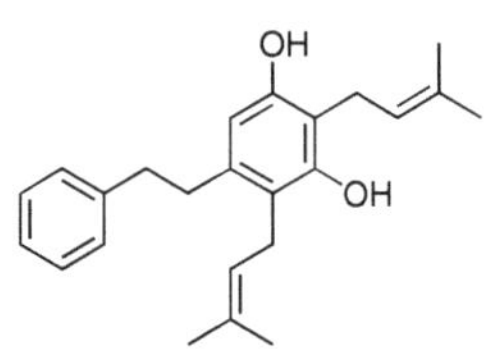

324 (2,4-bis-(3-methyl -2-butenyl)-3,5-dihydroxybibenzyl)

Nine new prenyl bibenzyls named radstrictins A–I (**325**–**333**) were isolated from the Chinese *R. constricta* together with 11 known analogues, 2-carbomethoxy-3,5-dihydroxy-4-(3-methyl-2-butenyl)-bibenzyl (**334**), radulanin A (**335**), radulanin L (**336**), 7-hydroxy-2,2-dimethyl-5-(2-phenyl-ethyl)chromane (**309**), 7-methoxy-2,2-dimethyl-5-(2-phenylethyl)chroman (**337**), (2*S*)-7-hydroxy-2-methyl-2-(4-methyl-3-pentenyl)-5-(2-phenyl-ethyl)chromene (**338**), 2-geranyl-3,5-dihydroxy-bibenzyl

(**339**), aglaiabbrevin C (**340**), bibenzyl/*o*-cannabicyclol hybrid (**341**), rasumatranin A (**342**), and rasumatranin B (**343**). Radstrictins A–F (**325**–**333**) were obtained as racemates or scalemic mixtures. Their stereostructures were elucidated by NMR, HRMS, and ECD methodology [179].

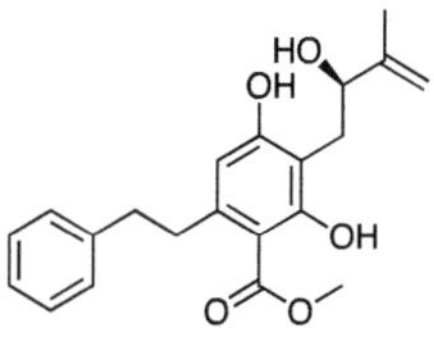

325 ((+)-radstrictin A)

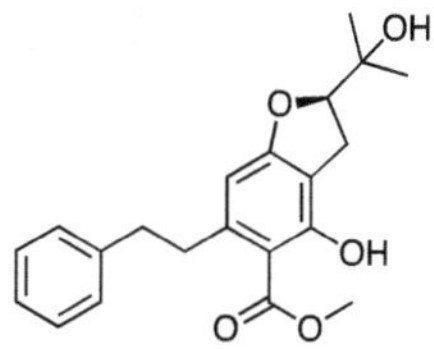

326 ((+)-radstrictin B)

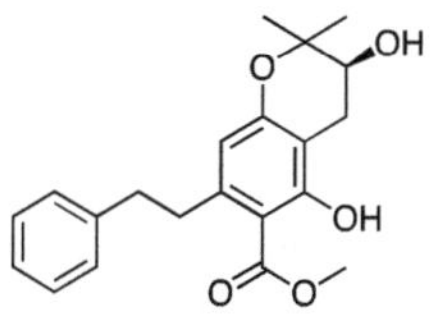

327 ((+)-radstrictin C)

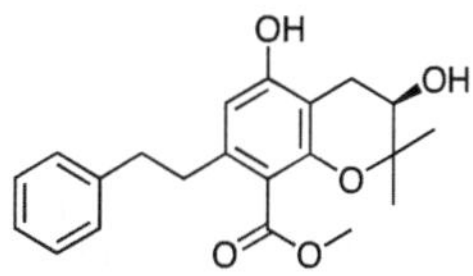

328 ((+)-radstrictin D)

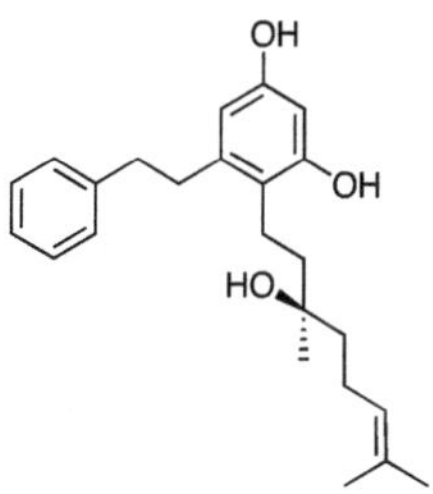

329 ((+)-radstrictin E)

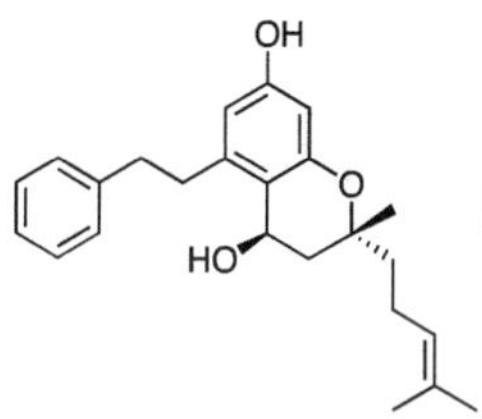

330 ((+)-radstrictin F)

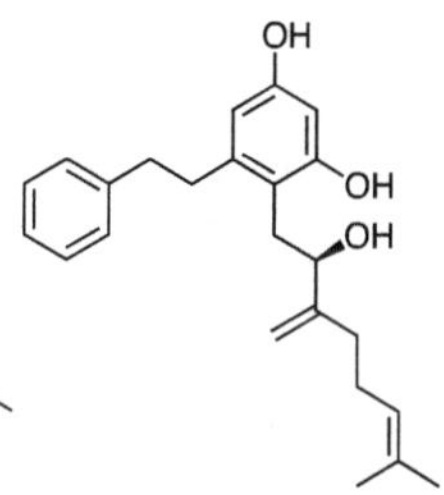

331 ((+)-radstrictin G)

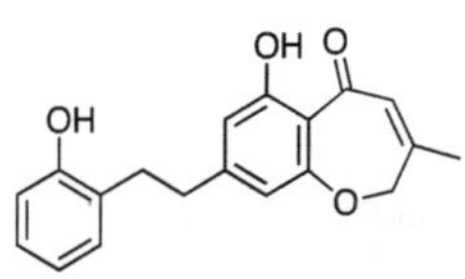

332 ((+)-radstrictin H)

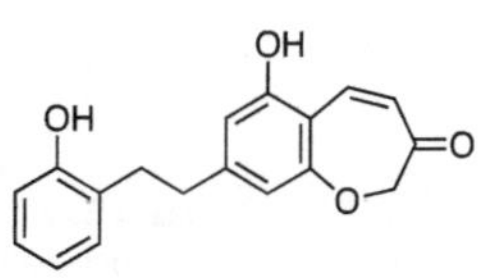

333 ((+)-radstrictin I)

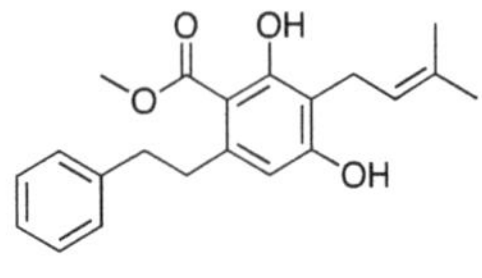

334 (2-carbomethoxy-3,5-dihydroxy-4-(3-methyl-2-butenyl)bibenzyl)

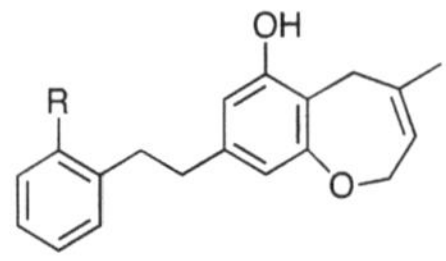

335 (radulanin A, R = H)
336 (radulanin L, R = OH)

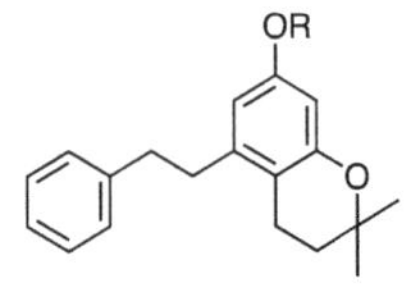

337 (7-methoxy-2,2-dimethyl-5-(2-phenylethyl)chromane, R = Me)

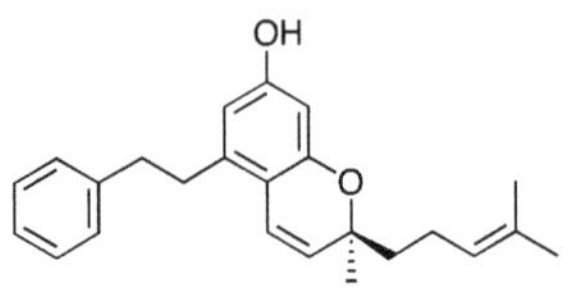

338 ((2*S*)-7-hydroxy-2-methyl-2-(4-methyl-3-pentenyl)-5-(2-phenylethyl)-chromene, = *o*-cannabichromene)

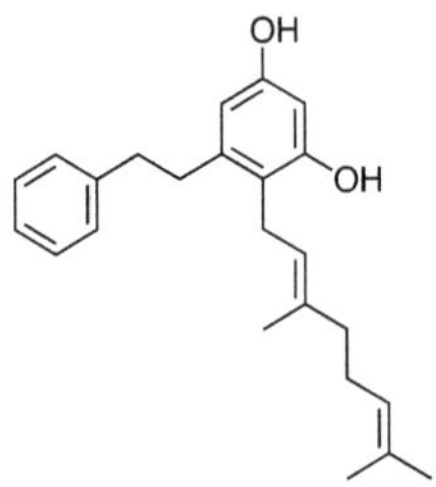

339 (2-geranyl-3,5-dihydroxy-bibenzyl)

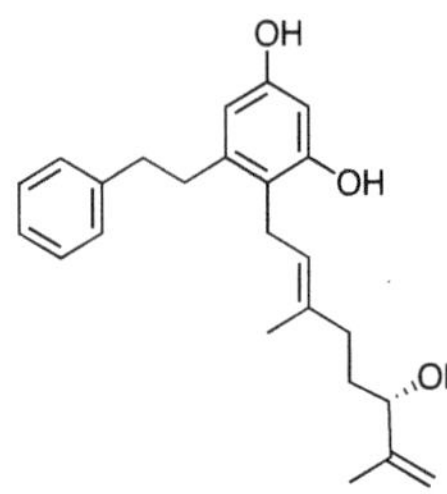

340 (2-(3,7-dimethyl-6α-hydroxy-oct-3,7-dienyl)-3,5-dihydroxy-bibenzyl, = aglaiabbrevin C)

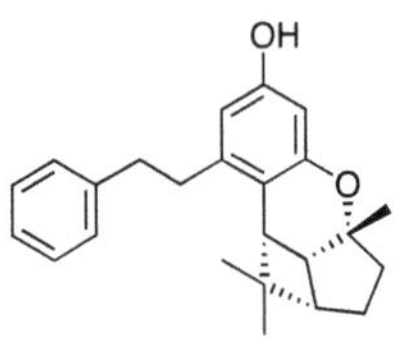

341 (bibenzyl/*o*-cannabicyclol hybrid)

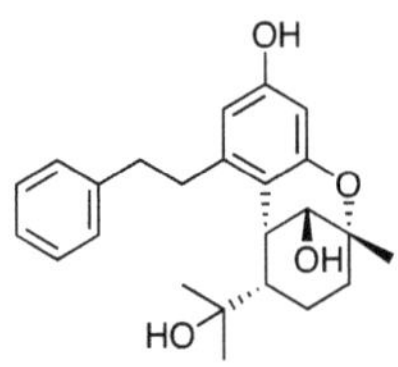

342 (rasmatranin A)

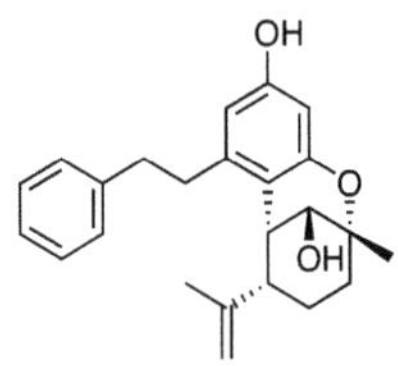

343 (rasmatranin B)

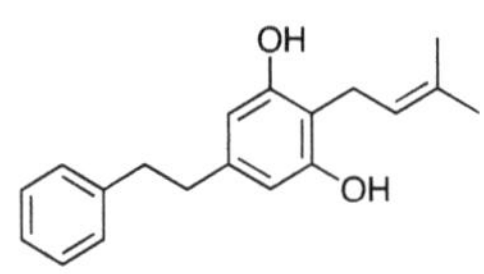

344 (3,5-dihydroxy-4-(3-methyl-2-butenyl)bibenzyl)

These new prenyl and previously known analogues might originate from the key precursors, 2-(3-methyl-2-butenyl)-3,5-dihydroxybibenzyl (**306**) and 3,5-dihydroxy-4-(3-methyl-2-butenyl)-bibenzyl (**344**), which were isolated as the major constituents in many *Radula* species, derived from 3,5-dihydroxybibenzyl (**320**) obtained by a coupling with 2,2-dimethylallyl diphosphate. Further dehydration, hydration, cyclization, oxidation and reduction reactions would afford prenyl bibenzyls. Bibenzyls **324–328** and **332–326** may be generated from precursor **344**, while compounds **309** and **337**, **329–331**, and **338–340**, could be formed by the coupling of 3,5-dihydroxybibenzyl (**320**) and 3-methyl-2-butenyl and geranyl diphosphates, respectively [179].

Among all of the prenyl bibenzyls, 2-carbomethoxy-3,5-dihydroxy-4(3-methyl-2-butenyl)-bibenzyl (**334**) possessed micromolar cytotoxicity against A549, NCI-H1299, HepG-2, HT-29, and KB cells with IC_{50} values of 6.0, 5.1, 6.3, 5.0, and 6.7 μ*M*, respectively. The cellular death triggered by compound **334** occurred via mitochondrial-derived paraptosis. Bibenzyl/*o*-cannabicyclol hybrid (**341**) also showed cytotoxicity, but only against the NCI-H1299 and A549 cancer cell lines, with IC_{50} values, in turn, of 7.5 and 9.8 μ*M* [180].

From a 95% ethanol extract of the Chinese *Radula sumatrana,* four new pairs of bibenzyl/meroterpenoid enantiomers, named rasumatrins A (**342**), B (**343**), C (**345**), and D (**346**), two new pairs of prenyl bibenzyl enantiomers named radulanins M (**347**) and N (**348**), and the known bibenzyl/o-cannabicyclol hybrid, 6-hydroxy-3-methyl-8-phenylethylbenzo[*b*]-oxcepin-5-one (**341**), 5-oxoradulanin A (**349**), radulanin A (**335**), tylimanthin B (**350**), radulanin I (**351**) and radulanin J (**352**) were obtained. The stereostructures of the new bibenzyls were determined by a combination of X-ray crystallographic and chiral phase HPLC-ECD analysis. This was the first isolation of bibenzyl *iso*-tetrahydrocannabinoids from a *Radula* species. Of those isolated, compound **349** showed the most potent cytotoxic activities against the MCF-7, PC-3, and SMMC-7721 human cancer cell lines, with IC_{50} values of 3.9, 6.6, and 3.9 μ*M*. Compounds **341, 343**, and **351** also possessed some growth inhibitory activities against these cell lines [181].

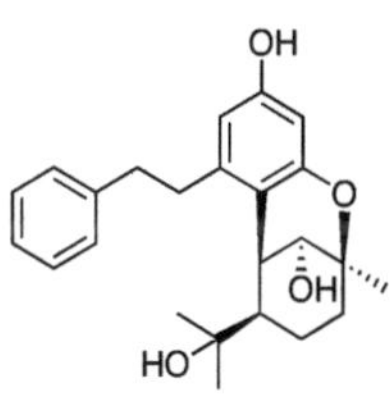

345 (rasumatramin C)

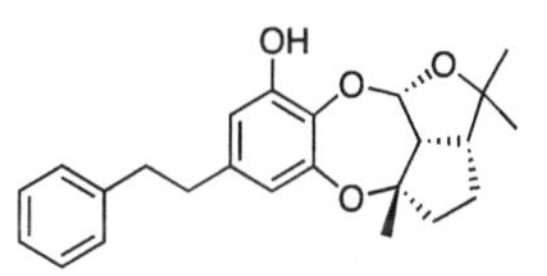

346 (rasumatramin D)

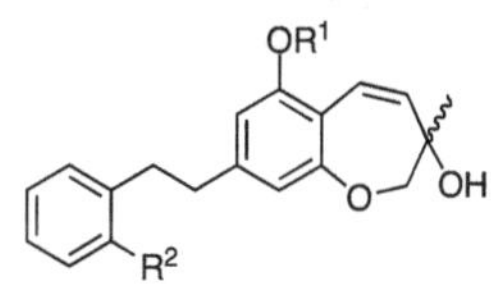

347 (radulanin M, $R^1 = R^2 = H$)
348 (radulanin N, $R^1 = H$, $R^2 = OH$)

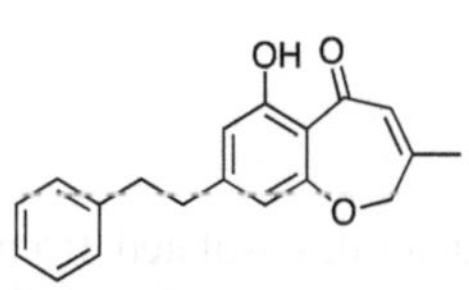

349 (5-oxoradulanin A)

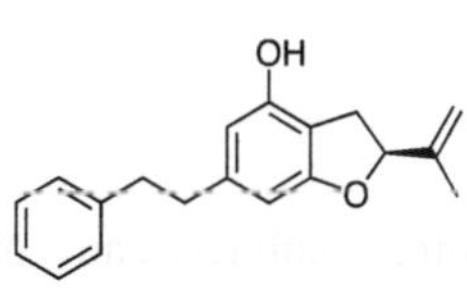

350 (tylimanthin B)

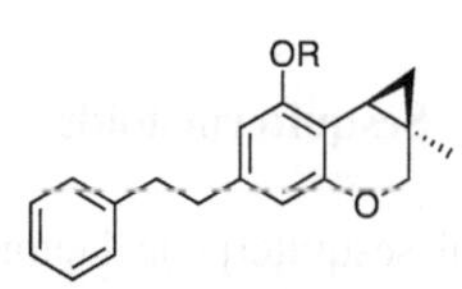

351 (radulanin I, R = H)
352 (radulanin J, R = Me)

The ether extracts of two unidentified *Frullania* species, one from Indonesia and the other from Tahiti, exhibited cytotoxic activity against the HL-60 (IC_{50} 6.7 and 1.6 μg/cm^3) and KB (3.3 and 11.2 μg/cm^3) cell lines [182]. Bioactivity-guided fractionation of the extract of the Indonesian sample led to the isolation of (+)-3*α*-(4′-methoxybenzyl)-5,7-dimethoxyphthalide (**353**), (–)-3*α*-(3′-methoxy-4′,5′-methylenedioxybenzyl)-5,7-dimethoxyphthalide (**354**), together with 3-methoxy-3′,4′-methylenedioxybibenzyl (**355**), 2,3,5-trimethoxy-9,10-dihydrophenanthrene (**357**), and atranorin (**358**), among which **353** and **354** possessed the most potent cytotoxic activities against HL-60 and KB cells, showing IC_{50} values of 0.92 and 6.3 μ*M*, and 0.96 and 5.5 μ*M*, respectively. Of the other compounds, **355** and its 2′-nitro derivative **356** were both found to be inactive against the HL-60 and KB cell lines [183].

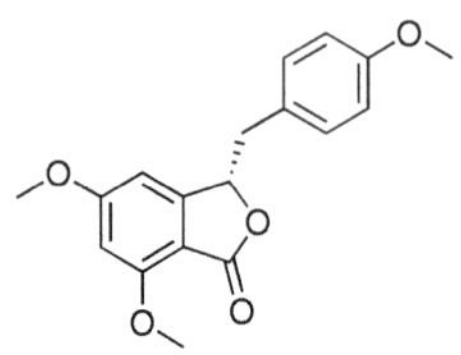

353 ((+)-3α-(4'-methoxybenzyl)-5,7-dimethoxyphthalide)

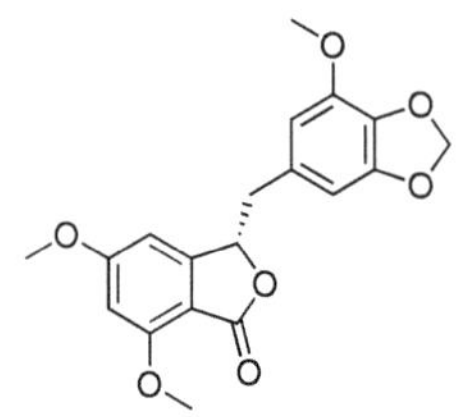

354 ((–)-3α-(3'-methoxy-4',5'-methylene-dioxybenzyl)-5,7-dimethoxyphthalide)

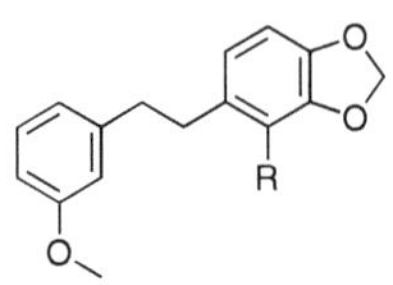

355 (3-methoxy-3',4'-methylenedioxybibenzyl, R = H)
356 (2-nitro-3,4-methylenedioxy-3'-methoxybibenzyl, R = NO_2)

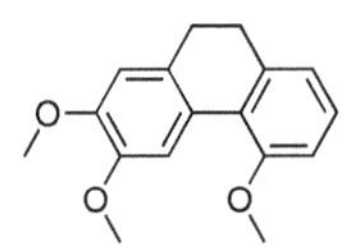

357 (2,3,5-trimethoxy-9,10-dihydrophenanthrene)

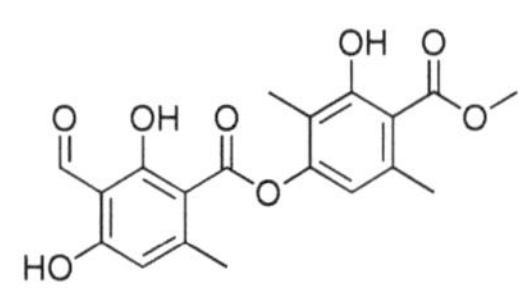

358 (atranorin)

4.4.3 Sesquiterpenoids

Several sesquiterpene lactones, germacranolides, and guaianolides isolated from liverworts exhibit cytotoxic activity against human KB nasopharyngeal and P-388 murine lymphocytic leukemia cells [2]. The crude ether extracts of the liverworts *Bazzania pompeana, Kurzia makinoana, Lophocolea heterophylla, Makinoa crispata, Marsupella emarginata, Pellia endiviifolia, Plagiochila fruticosa, P. ovalifolia, Porella caespitans, P. japonica, P. perrottetiana, P. vernicosa*, and *Radula perrottetii* showed cytotoxicity against P-388 cells (IC_{50} value range 4–20 μg/cm^3). In contrast, the crude extracts of *Frullania diversitexta, F. ericoides, F. muscicola, F. tamarisci* subsp. *obscura, Lepidozia vitrea, Pallavicinia subciliata, Plagiochila sciophila, Spruceanthus semirepandus*, and *Trocholejeunea sandvicensis* were inactive against this same cell line (IC_{50} values > 20 μg/cm^3) [3]. (–)-Polygodial (**90**), the major pungent sesquiterpene dialdehyde isolated from the liverworts *Porella* *vernicosa* complex, exhibited cytotoxic properties against a human melanoma cell line with an IC_{50} value of 2.0 μg/cm³ [3]. Glaucescenolide (**359**), from *Schistochila glaucescens*, displayed activity against P388 cells (IC_{50} value of 2.3 μg/cm³) [141].

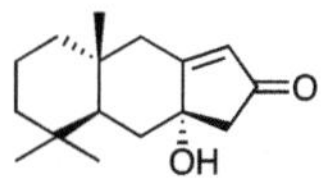

359 (glaucescenolide)

From the ether extract of an unidentified *Frullania* species collected in Tahiti [182], tulipinolide (**134**) and costunolide (**360**) were obtained, and the former germacranolide was cytotoxic against the HL-60 and KB cell lines (IC_{50} 4.6 and 2.7 μ*M*) [183].

Trocholejeunea sandvicensis is an abundant source of pinguisane-type sesquiterpenoids, among which dehydropinguisenol (**361**) exhibited weak cytotoxicity against KB cells (ED_{50} 12.6 μg/cm³) [38]. *Porella perrottetiana* is a large stem-leafy liverwort and a source of pinguisane sesqui- and sacculatane-type diterpenoids. Its ether and methanol extracts showed cytotoxic activity against HL-60 cells (IC_{50} 3.2 and 14.8 μg/cm³) and the former extract also possessed growth inhibitory activity against KB cells (15.1 μg/cm³). Activity-guided fractionation of the ether extract resulted in the isolation of acutifolone A (**362**) and 4*α*,5*β*-epoxy-8-*epi*-inunolide (**363**), which displayed cytotoxic activities against the HL-60 cell line (IC_{50} values of 2.75 and 2.8 μ*M*) but were less active against KB cells (IC_{50} values of 46.6 and 46.3 μ*M*).

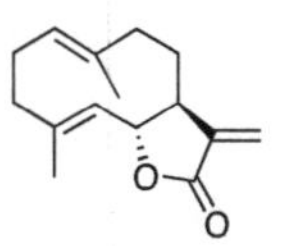

360 (costunolide)

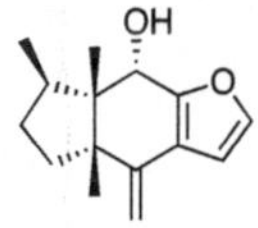

361 (dehydropinguisenol)

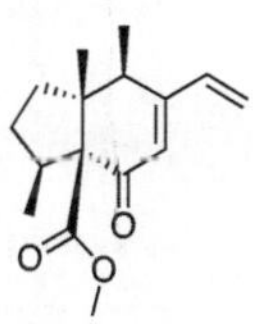

362 (acutifolone A)

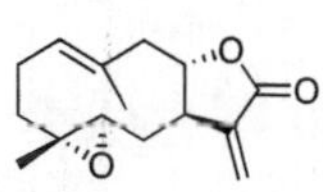

363 (4*α*,5*β*-epoxy-8-*epi*-inunolide)

7*α*-Hydroxypinguisenol-12-methyl ester (**365**), synthesized from 7-oxopingisenol-12-methyl ester (**364**) (Scheme 8), showed a decreased cytotoxic activity against the HL-60 cell line, and was inactive using KB cells (IC_{50} values of 83.1 and > 177 μ*M*) [183].

Scheme 8 Formation of 7α-hydroxypinguisenol 12-methyl ester (**365**) from 7-oxopinguisenol-12-methyl ester (**364**)

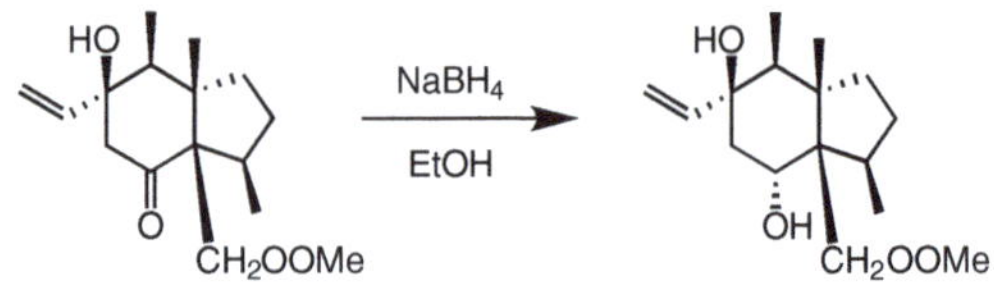

364 (7-oxopinguisenol-12-methyl ester

365 (7α-hydroxypinguisenol-12-methyl ester)

The French Polynesian liverwort *Chandonanthus hirtellus* (Plate 17) afforded a new sesquiterpene lactone, chandolide (**366**), which exhibited cytotoxicity against the HL-60 and KB cell lines (5.3 and 11.2 μg/cm^3), respectively [184].

Lepidozia species belonging to the Lepidoziaceae are abundant sources of sesquiterpenoids. Lepidozenolide (**367**) isolated from the Taiwanese *Lepidozia fauriana*, showed low micromolar cytotoxic potency when evaluated against P-388 cells (ED_{50} 2.1 μg/cm^3) [185].

Plate 17 *Chandonanthus hirtellus* (liverwort)

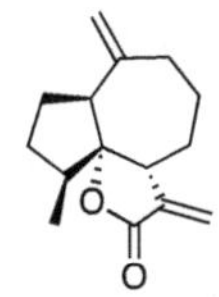

366 (chandolide)

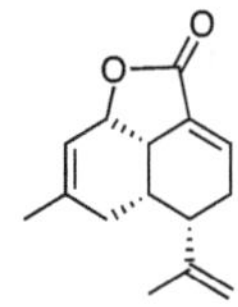

367 (lepidozenolide)

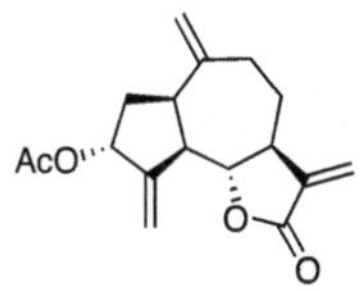

368 (zaluzanin D)

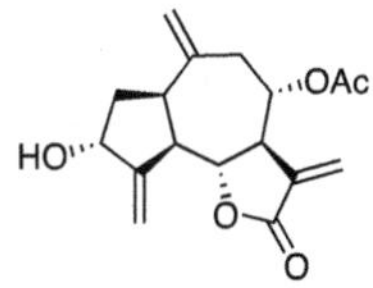

369 (8α-acetoxyzaluzanin C)

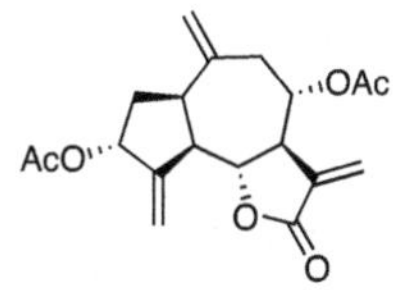

370 (8α-acetoxyzaluzanin D)

The major constituents of *Wiesnerella denudata* (Plate 3b) are the pungent tulipinolide (**134**) and the five- and seven-membered products, zaluzanin D (**368**), 8α-acetoxyzaluzanin C (**369**) and 8α-acetoxyzaluanin D (**370**), and all displayed cytotoxic activity against KB cells, with ED_{50} values in the range of 0.5–2.5 μg/cm^3 [38].

Species belonging to the genus *Chilosyphus* are rich sources of *ent*-eudesmane sesquiterpenoids, of which some possess pungency [1–3, 38]. 12-Hydroxychiloscyphone (**371**), obtained from *Chiloscyphus rivularis,* was tested against the RS322 (rad52), RS188N (RAD$^+$, and RS321 (rad52.top1) yeast strains. It gave IC_{12} values of 75 and 88 μg/cm^3 for strains RS321 and RS322, which is characteristic for a selective DNA-damaging agent that does not act as a topoisomerase-I or -II inhibitor. 12-Hydroxychiloscyphone (**371**) showed also cytotoxic activity against A549 human lung carcinoma cells (IC_{50} value 2.0 μg/cm^3) [186].

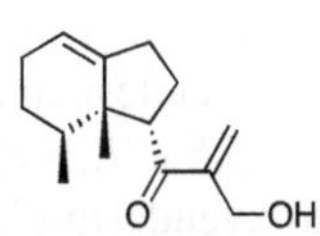

371 (13-hydroxychiloscyphone)

A Chinese collection of *Chiloscyphus polyanthus* var. *rivularis* produced seven new *ent*-eudesmanes, (3*S*,5S,7*R*,10*S*)-3-hydroperoxy-7-hydroxyeudesm-4(15)-ene (**372**), (3*S*,5*S*,7*R*,10*S*)-3,7-dihydroxyeudesm-4(15)-ene (**373**), (3*S*,5*S*,6*R*,7*R*,*10S*)-3,6,7-trihydroxyeudesm-4(15)-ene (**374**), (5*S*,6*R*,7*S*,10*R*)-6-acetoxy-7-hydroxy-2-oxoeudesm-(3*Z*)-ene (**375**), (6*S*,7*S*,10*R*)-6-acetoxy-7-hydroxy-2-oxoeudesm-(3*Z*)-ene (**376**), (7*S*,10*R*)-7-hydroxy-2-oxoeudesm-(3*Z*)-ene (**377**), (2*R*,5*S*,6*R*,7*S*,10*R*)-6,7-dihydroxy-2-methoxyeudesm-(3*Z*)-ene (**378**), along with six known analogues, *ent*-6β,7α-dihydroxy-eudesm-(3*Z*)-ene (**379**), eudesm-3-en-7α-ol (**380**), eudesm-3-en-6β,7α-diol (**381**), eudesm-3-ene-6α,7α-diol (**382**), eudesm-(3*Z*)-ene-6α-acetoxy-7α-ol (**383**), *ent*-6β,7β-dihydroxy-3-oxo-eudesm-(4E)-ene (**384**), and *ent*-7β-hydroxy-3-oxo-eudesm-(4E)-ene (**385**), among which the stereostructure of **372** was established by X-ray crystallographic analysis. Of these *ent*-eudesmanes, **372** showed weak cytotoxicity against A549 cells (IC_{50} 27.7 μ*M*) [187].

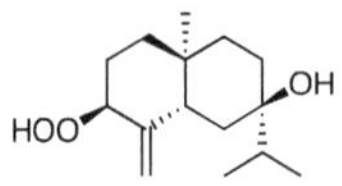

372 ((3*S*,5*S*,7*R*,10*S*)-3-hydroperoxy-7-hydroxyeudesm-4(15)-ene)

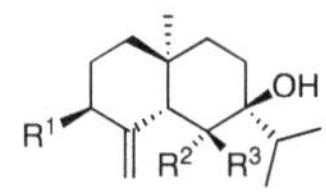

373 ((3*S*,5*S*,7*R*,10*S*)-3,7-dihydroxy-eudesm-4(15)-ene, R^1 = OH, R^2 = R^3 = H)
374 ((3*S*,5*S*,6*R*,7*R*,10*S*)-3,6,7-trihydroxy-eudesm-4(15)-ene, R^1 = R^2 = OH, R^3 = H)

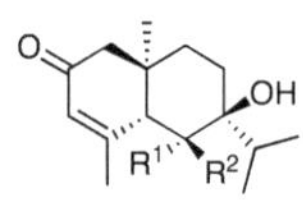

375 ((5*S*,6*R*,7*S*,10*R*)-6-acetoxy-7-hydroxy-2-oxoeudesm-(3*Z*)-ene, R^1 = H, R^2 = OAc)
376 ((6*S*,7*S*,10*R*)-6-acetoxy-7-hydroxy-2-oxoeudesm-(3*Z*)-ene, R^1 = OAc, R^2 = H)
377 ((7*S*,10*R*)-7-hydroxy-2-oxoeudesm-(3*Z*)-ene, R^1 = R^2 = H)

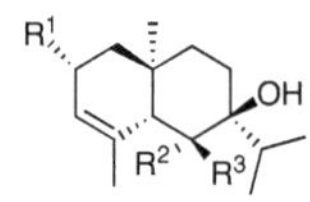

378 ((2*R*,5*S*,6*R*,7*S*,10*R*)-6,7-dihydroxy-2-methoxyeudesm-(3*Z*)-ene, R^1 = OMe, R^2 = OH, R^3 = H
379 (6β,7α-dihydroxy-eudesm-(3*Z*)-ene, R^1 = H, R^2 = OH, R^3 = H)
380 (eudesm-3-en-7β-ol, R^1 = R^2 = R^3 = H)
381 (eudesm-3-en-6α,7β-diol, R^1 = R^3 = H, R^2 = OH)
382 (eudesm-3-en-6β,7β-diol, R^1 = R^2 = H, R^3 = OH)
383 (eudesm-(3*Z*)-en-6α-acetoxy-7α-ol, R^1 = R^2 = OAc, R^3 = H)

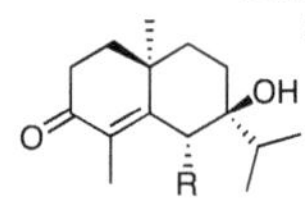

384 (*ent*-6β,7β-dihydroxy-3-oxo-eudesm-(4*E*)-ene, R = OH)
385 (*ent*-7β-hydroxy-3-oxo-eudesm-(4*E*)-ene, R = H)

A later reinvestigation of this liverwort by the same investigators led to the isolation of seven further new eudesmanes, (2*S*,5*S*,7*R*,10*R*)-7-hydroxy-2-methoxyeudesm-(3*Z*)-ene (**386**), (2*S*,5*S*,7*R*,10*R*)-2,7-dihydroxyeudesm-(3*Z*)-ene (**387**), (5*S*,7*R*,8*R*,10*S*)-7,8-dihydroxyeudesm-(3*Z*)-ene (**388**), 8-dihydroxy-3-oxoeudesm-(4*Z*)-ene (**389**), (3*S*,6*R*,7*S*,10*R*)-6,7-dihydroxy-3-methoxy-eudesm-(4*Z*)-ene (**390**), (6*R*,7*S*,10*R*)-6,7-dihydroxy-3-oxoeudesm-(1*Z*,4*Z*)-diene (**391**), and (4*S*,5*R*,7*R*,10*R*)-4,5,7-trihydroxy-3-oxoeudesmane (**392**), and seven known analogues, (3*S*,6*R*,7*S*,10*S*)-3,6,-trihydroxy-eudesm-(4*E*)-ene (**393**), (6*R*,7*S*,10*R*)-6.7-dihydroxy-3-oxo-eudesm-(4*E*)-ene (**394**), (5*S*,6*R*,7*S*,10*R*)-6,7-dihydroxy-2-oxo-eudesm-(3*Z*)-ene (**376**), (5*S*,7*R*,10*R*)-7-hydroxy-2-oxoeudesm-(3*Z*)-ene (**378**), (2*R*,5*S*,6*R*,7*S*,10*S*)-6,7-dihydroxy-2-methoxy-eudesm-(3*Z*)-ene (**379**), 7β-hydroxyeudesm-3-ene (**380**) and 7β-hydroxy-3-oxoeudesm-4-ene (**384**), with the absolute configurations of compounds **386–392** established by comparison of their NMR spectroscopic data with similar compounds isolated previously from liverworts and a consideration of their biogenesis. The cytotoxicity of these isolated compounds was evaluated against the HepG2 (human hepatocellular carcinoma) cancer cell line using an MTT assay and with doxorubicin as a positive control. Also, their antifungal activity was determined with *Candida albicans* wild-type strain

SC5314, having fluconazole as a positive control. However, none of the eudesmane sesquiterpenoids (**376–380**, **386–394**) evaluated showed either discernible cytotoxic or antifungal activities [188] (Plates 18, 19, and 20).

Plate 18 *Diplophyllum albicans* (liverwort)

Plate 19 *Mastigophora diclados* (liverwort)

Plate 20 *Pallavicinia ambigua* (liverwort)

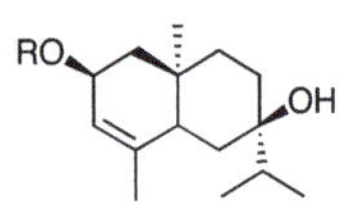

386 ((2*S*,5*S*,7*R*,10*R*)-7-hydroxy-2-methoxyeudesm-(3*Z*)-ene, R = Me)
387 ((2*S*,5*S*,7*R*,10*R*)-2,7-dihydroxy-eudesm-(3*Z*)-ene, R = H)

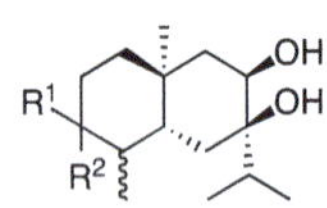

388 ((5*S*,7*R*,8*R*,10*S*)-7,8-dihydroxy-3-oxoeudesm-(4*Z*)-ene, R[1] = R[2] = H, Δ^3)
389 (8-dihydroxy-3-oxoeudesm-(4Z)-ene, R[1] = R[2] = O, Δ^4)

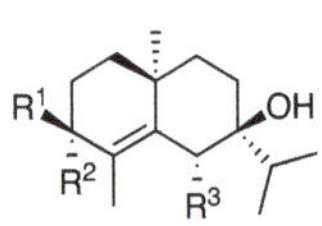

390 ((3*S*,6*R*,7S,10*R*)-6,7-dihydroxy-3-methoxy-eudesm-(4*Z*)-ene, R[1] = OMe, R[1] R[2] = H, R[3] = OH)
391 ((6*R*,7*S*,10*R*)-6,7-dihydroxy-3-oxoeudesm-(1*Z*,4*Z*)-diene, R[1] = R[2] = O, R[3] = OH, Δ^1)
393 ((3*S*,6*R*,7*S*,10*S*)-36,-trihydroxy-eudesm-(4*E*)-ene, R[1] = R[3] = OH, R[2] = H)
394 ((6*R*,7*S*,10*R*)-6,7-dihydroxy-3-oxo-eudesm-(4*E*)-ene, R[1] = R[2] = O, R[3] = OH)

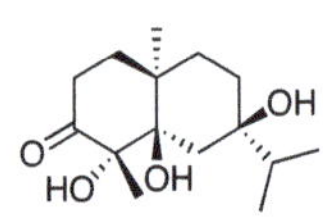

392 ((4*S*,5*R*,7*R*,10*R*)-4,5,7-trihydroxy-3-oxoeudesmane)

A major component of the Colombian *Clasmatocolea vermicularis* has been determined as the pungent eudesmane sesquiterpene lactone, *ent*-diplophyllolide (**132**). A bioactivity-directed study of the same New Zealand species led to the isolation of this lactone (**132**), which possesses cytotoxicity against P388 murine leukemia cells (IC_{50} 0.4 μg/cm^3) and BSC-1 African green monkey normal kidney epithelial cells, as well as antifungal activity against *Trichophyton mentagrophytes*. *Chiloscyphus polyanthos* is also pungent, with its taste also due to *ent*-diplophyllolide (**132**). The

Plate 21 *Diplophyllum albicans*, detail (liverwort)

New Zealand *C. subporosa* also elaborates this same eudesmane lactone, which is a chemical marker of *Clasmatocolea* and *Chiloscyphus* species belonging to the Lophocoleaceae. The presence of lactone **132** has been found in the liverworts *Diplophyllum albicans* (Scapaniaceae) (Plates 18 and 21), *Lophocolea bidentata* (Lophocoleaceae), *Plagiochila moritziana* (Plagiochilaceae), and *Tritomaria quinquendentata* (Lophoziaceae) [2]. *ent*-Diplophyllolide (**132**) isolated from the New Zealand *Chiloscyphus subporosa* and *Clasmatocolea vermicularis* [38] exhibited submicromolar cytotoxicity against the P388 cells (IC_{50} 0.4 μg/cm^3) [189].

(−)-*ent*-Arbusculin B (**395**) and (−)-*ent*-costunolide (**396**), purified from *Hepatostolonophora paucistipula*, exhibited cytotoxic effects against P388 cells, with IC_{50} values of 1.1 and 0.7 μg/cm^3, respectively [190].

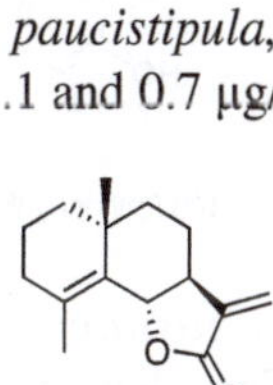

395 ((−)-*ent*-arbusculin B)

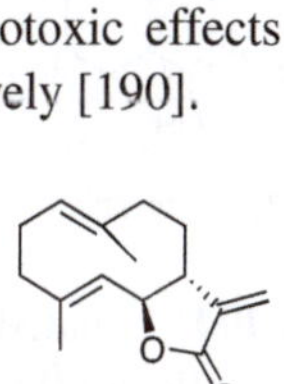

396 ((−)-*ent*-costunolide B)

Costunolide (**360**), obtained from *Frullania nisquallensis*, showed growth inhibitory activity against the A549 human lung carcinoma cell line with an IC_{50} value of 12 μg/cm^3, and moderate, but selective, DNA-damaging effects against the RS321N, RS322YK, and RS167K mutant yeast strains, with IC_{12} values of 50, 150, and 330 μg/cm^3, respectively [191].

Bazzanenyl caffeate (**397**) and naviculyl caffeate (**398**) from *Bazzania novae-zelandiae*, demonstrated growth inhibitory effects against P-388 cells with GI_{50} values in the range 0.8–1.1 μg/cm^3, although naviculol (**399**) was deemed as being inactive [192].

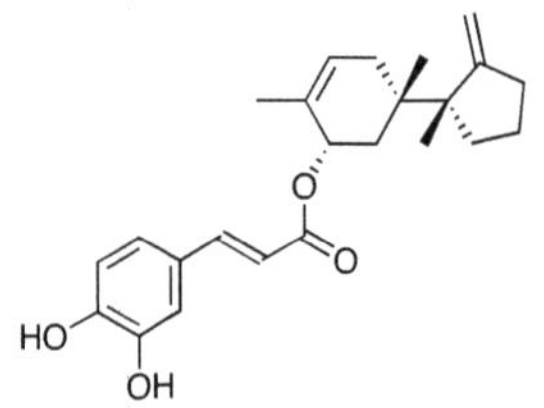

397 (bazzanenyl caffeate)

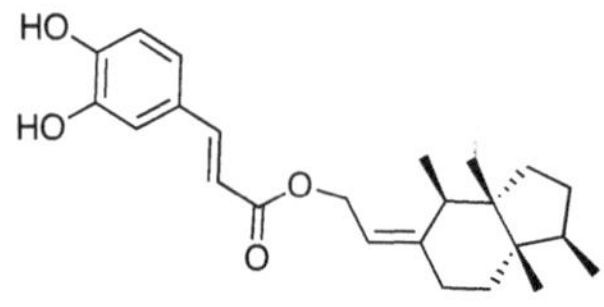

398 (naviculyl caffeate)

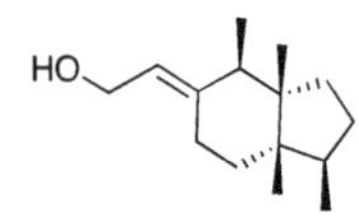

399 (naviculol)

The liverworts of the genus *Bazzania* speices belonging to the Lepidoziaceae, are sources of both sesquiterpenoids and cyclic bis-bibenzyls. A 95% EtOH extract of the Chinese *Bazzania albifolia* resulted in the isolation of a new compound, methyl 2-oxoaromadendra-1(10),3-dien-12-oate (**400**), along with three sesquiterpene alcohols, δ-cuparenol (**401**) and chiloscyphenols A (**402**) and B (**403**), among which compound **402** showed cytotoxicity against the MCF-7 and PC3 cell lines with IC_{50} values of 5.6 and 29.6 μM, respectively. The remaining compounds purified also showed weak growth inhibitory activities against these same tumor cell lines, having IC_{50} values of 14.8–64.8 and 38.5–57.9 μM, respectively [193].

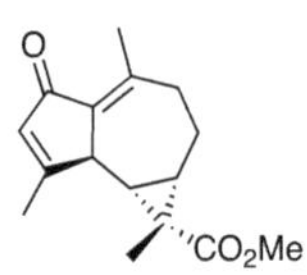

400 (methyl 2-oxo-aromadendra-1(10),3-dien-12-oate)

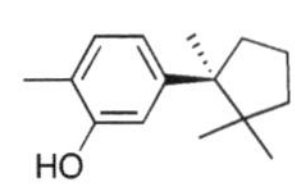

401 (δ-cuparenol)

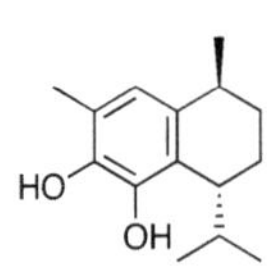

402 (chiloscyphenol A)

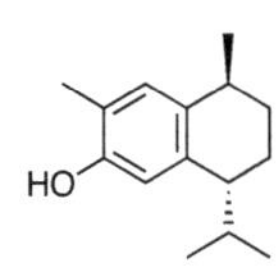

403 (chiloscyphenol B)

Riccardiphenol C (**404**), isolated from *Riccardia crassa*, exhibited weak inhibitory effects against BSC-1 cells at 60 μg/disk [194]. The ether and methanol extracts of the Tahitian *Mastigophora diclados* (Plate 19) displayed cytotoxic activity against HL-60 cells at IC_{50} 2.4 and 13.1 μg/cm^3 and KB cells at 14.6 and 32.5 μg/cm^3, respectively [195]. (–)-Diplophyllolide (**132**), α-herbertenol (**405**), (–)-herbertene-1,2-diol (**406**), mastigophorene C (**407**), and mastigophorene D (**408**), isolated from both extracts, were cytotoxic against HL-60 cells with IC_{50} values of 1.4, 12.8, 1.4, 2.4, and 2.5 μg/cm^3, respectively. In turn, they demonstrated antiproliferative effects against KB cells (IC_{50} values of 3.3, 12.5, 11.8, 14.8, and 14.2 μg/cm^3). 1-Hydroxy-2-methoxyherbertene (**409**) and herbertene-1,2-diacetate (**410**) showed less potent cytotoxicity than the parent compound herbertene-1,2-diol (**406**), against both HL-60 and KB cells. Moreover, the double bond isomer (–)-diplophyllin (**411**) did not produce any observable cytotoxicity against either of these cell lines [195].

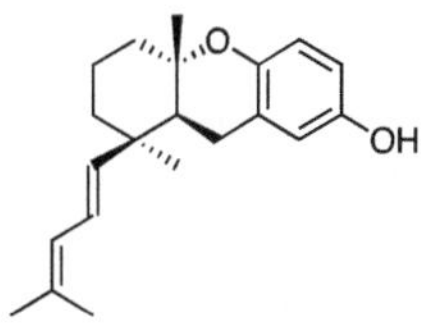

404 (riccardiphenol C)

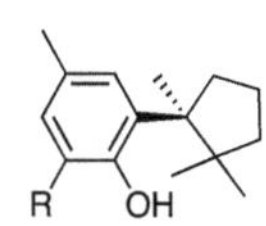

405 (α-herbertenol, R = H)
406 (herberten-1,2-diol, R = OH)

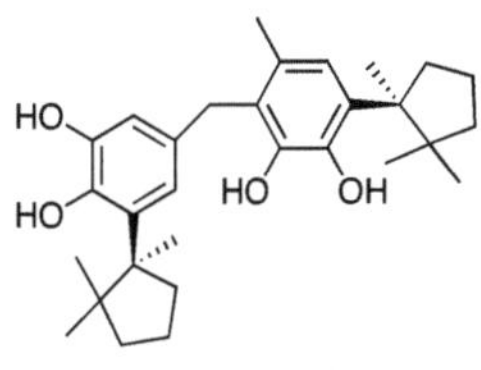

407 (mastigophorene C)

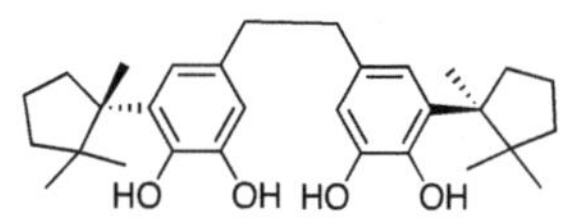

408 (mastigophorene D)

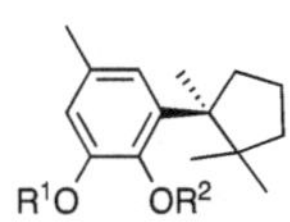

409 (1-hydroxy-2-methoxy-herbertene, R^1 = Me, R^2 = H
410 (herbertene-1,2-diacetate, $R^1 = R^2$ = Ac)

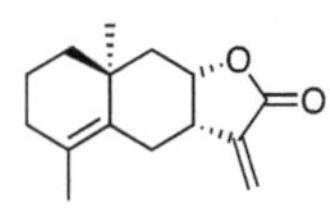

411 ((–)-diplophyllin)

Mastigophora diclados (Plate 19) is a source of herbertane sesquiterpene monomers and dimers that are useful chemical markers of this taxon. Komala et al., analyzed the Tahitian *M. diclados* and identified several herbertane sesquiterpenoids, and demonstrated that the cytotoxic effect of their crude extract of origin against a human breast cancer cell line (MCF-7) stemmed from the occurrence of herbertane sesquiterpenoids [196]. In 2000, Ng et al., reinvestigated the methanol extract of an east Malaysian specimen of *M. diclados* and isolated four already known compounds, namely, herbertene (**412**), α-herbertenol (**405**), herbertene-1,2-diol (**406**), and *ent*-7α-hydroxyeudesm-4(5)-en-6-one (**413**) [197]. Marsupellone (**414**) and acetoxymarsupellone (**415**) from the liverwort, *Marsupella emarginata*, showed cytotoxicity (ID_{50} = 1 μg/cm^3) against P-388 cells [40, 198].

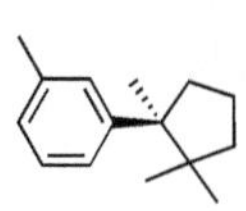

412 (herbertene)

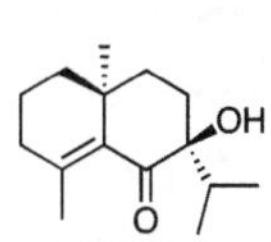

413 (*ent*-7α-hydroxyeudesm-4-en-6-one)

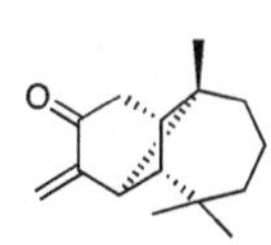

415 (marsupellone)

Asakawa and his group isolated the new skeletal *ent*-2,3-*seco*-aromadendrane sesquiterpene hemiacetal, plagiochiline A (**135**), from the liverwort *Plagiochila yokogurensis*, having a strong pungent taste, potent insect antifeedant activity against *Spodoptera exempta*, and cytotoxic activity against various human cancer cell lines [2, 38]. Subsequently, more than 50 compounds of the same type (e.g. **416–473**), with some possessing human cancer cell line antiproliferative effects, have been found mainly in Plagiochilaceae species grown in China, Europe, Japan, and South America [199]. Plagiochiline A (**135**), which is the major component in many *Plagiochila* species, was found to be cytotoxic against KB cells, with an ED_{50} value of 3.0 μg/cm^3 [38].

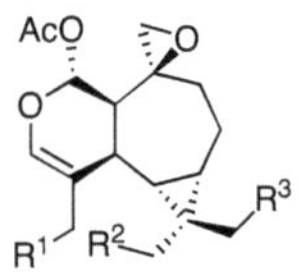

416 (plagiochiline B) R^1 = R^2 = OAc, R^3 = H
417 (plagiochiline D) R^1 = R^2 = R^3 = OAc
418 (plagiochiline G) R^1 = R^2 = OAc, R^3 = OH
419 (plagiochiline R) R^1 = R^3 = OAc, R^2 = H

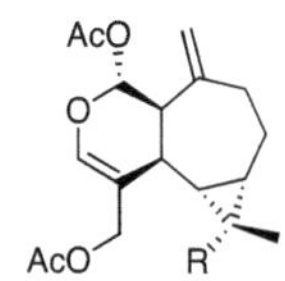

420 (plagiochiline C) R = Me
421 (plagiochiline T) R = CHO
422 (plagiochiline U) R = CO_2Me

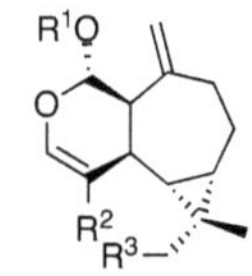

423 (plagiochiline H) R^1 = Ac, R^2 = Me, R^3 = H
424 (plagiochiline L) R^1 = Ac, R^2 = CO_2H, R^3 = H
425 (plagiochiline M) R^1 = Ac, R^2 = CO_2Me, R^3 = H
426 (plagiochiline S) R^1 = Ac, R^2 = Me, R^3 = OAc

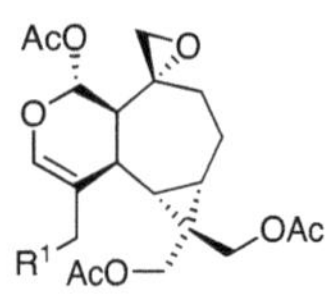

427 (plagiochiline E)

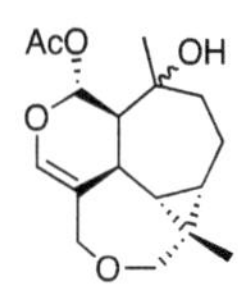

428 (plagiochiline F)

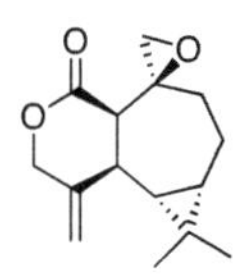

429 (plagiochiline J)

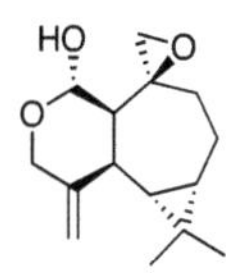

430 (plagiochiline K)

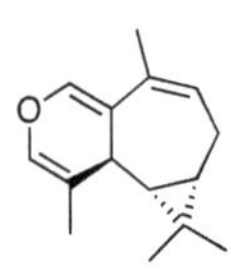

431 (plagiochiline N)

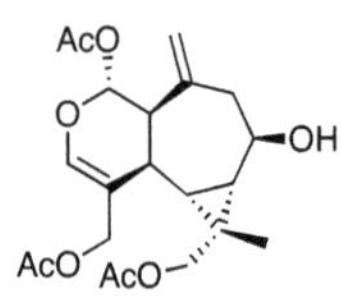

432 (plagiochiline O)

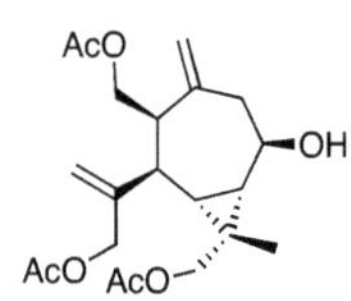

433 (plagiochiline P)

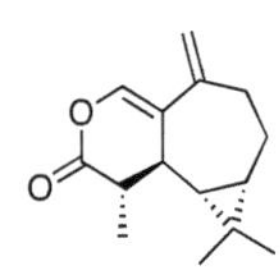

434 (plagiochiline Q)

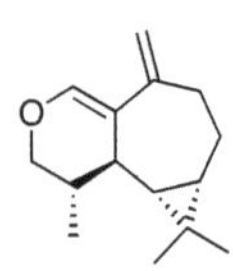

435 (plagiochiline W)

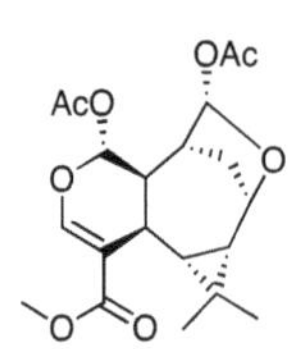

436 (plagiochiline V)

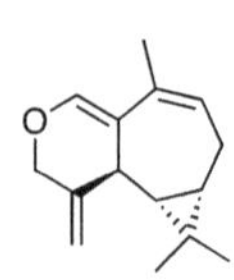

437 (plagiochiline X)

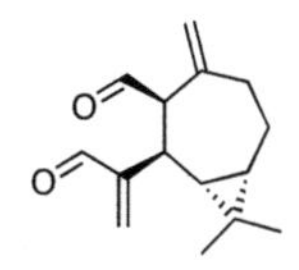

438 (plagiochilal A)

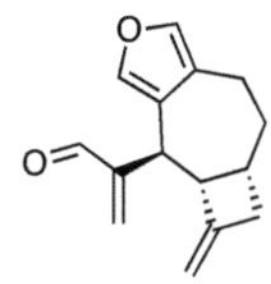

439 (neofurano-plagiochilal)

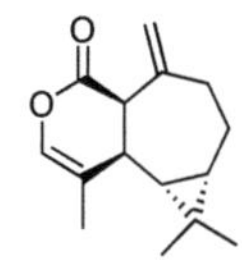

440 (plagiochide)

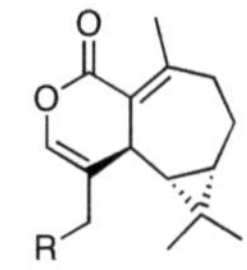

441 (isoplagiochilide, R = H)
442 (acetoxyisoplagiochilide, R = OAc)

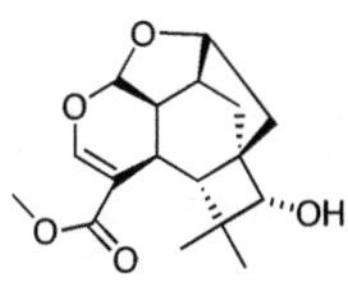

443 (plagiochianin A)

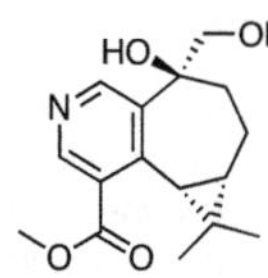

444 (plagiochianin B)

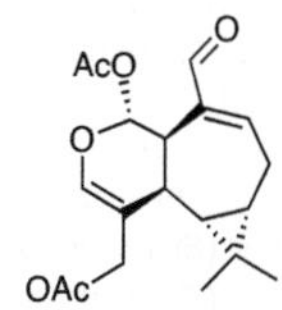

445 (ovalifolienal)

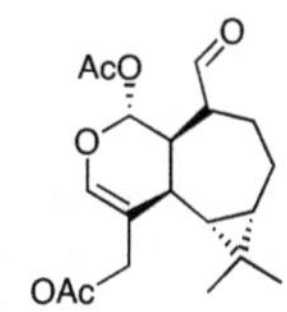

446 (10-dihydro-ovalifolienal)

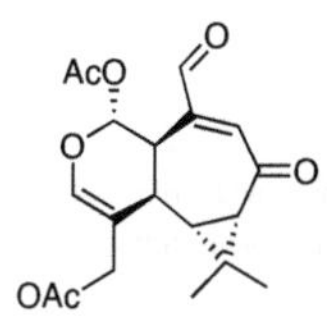

447 (ovalifolienalone)

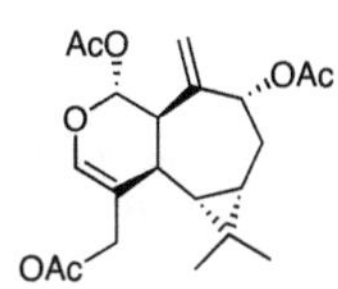

448 (9α-acetoxy-ovalifoliene)

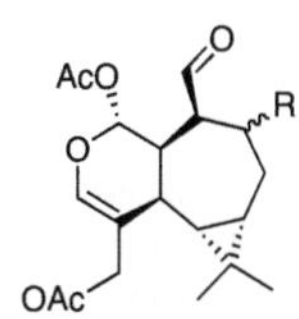

449 (9α-acetoxy-10β-ovalifolienal, R = α-OAc)
450 (9β-acetoxy-10β-ovalifolienal, R = β-OAc)

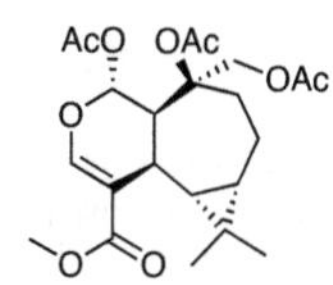

451 (9α-acetoxy-ovalifoliene)

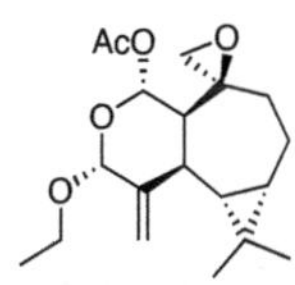

452 (3α-ethoxy-plagiochiline A)

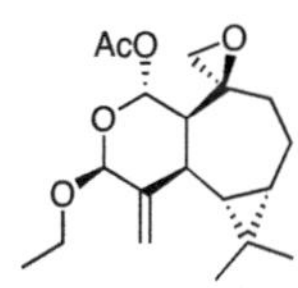

453 (3β-ethoxy-plagiochilne A2)

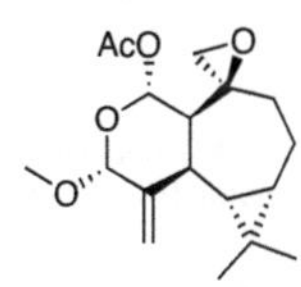

454 (3α-methoxy-plagiochiline A1)

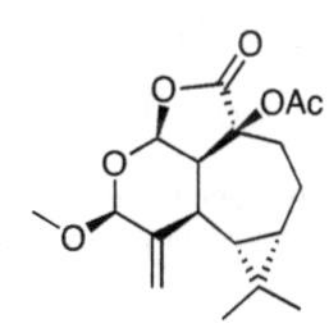

455 (3β-methoxy-plagiochiline A2)

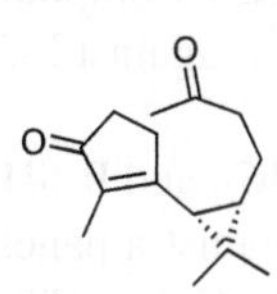

456 (3,10-dioxotaylori-4-ene)

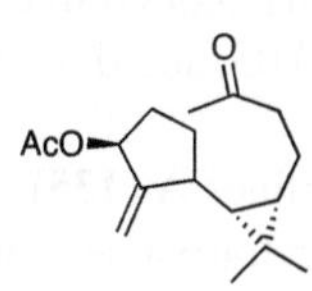

457 (3-acetoxytaylorione)

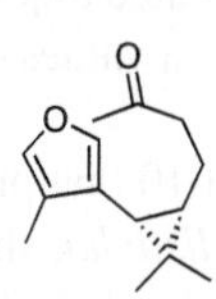

458 (2-*nor*-1,3-epoxy-1,10-*seco*-aromadendrane-1(5),3-dien-10-one)

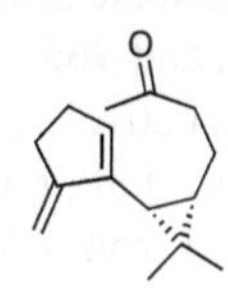

459 (taylorione-1(5),3-dien-10-one)

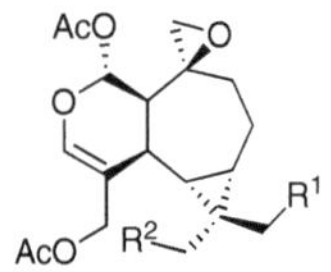

460 (plagiochiline A 15-yl octanoate) R^1 = octanoyl, R^2 = H
461 (plagiochiline A 15-yl decanoate) R^1 = decanyl, R^2 = H
462 (plagiochiline A 15-yl (4*Z*)-decenoate) R^1 = (4*Z*)-decenoyl, R^2 = H
463 (plagiochiline A 15-yl hexanoate) R^1 = hexanoyl, R^2 = H
464 (14-hydroxy-(plagiochiline A 15-yl (2*E*,4*E*,8*Z*)-tetradecatrienoate)
R^1 = (2*E*,4*E*,8*Z*)-tetradecatrienoyl, R^2 = OH
465 (14-hydroxy-(plagiochiline A 15-yl (2*E*,4*E*,8*Z*)-dodecadienoate) R^1 = (2*E*,4*E*)-dodecadienyl, R^2 = OH
466 (14-hydroxy-(plagiochiline A 15-yl (2*E*)-docenoate) R^1 = (2*E*)-dodecenoyl, R^2 = OH
467 (14-hydroxy-(plagiochiline A 15-yl (*E*)-4-hydroxycinnamate) R^1 = (*E*)-4-hydroxycinnamoyl, R^2 = OH
468 (14-hydroxy-(plagiochiline A 15-yl (*Z*)-4-hydroxycinnamate) R^1 = (*Z*)-4-hydroxycinnamoyl, R^2 = OH

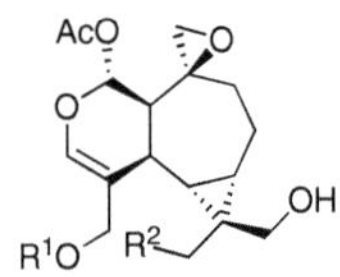

469 (15-hydroxy-(plagiochiline A 14-yl (*E*)-4-hydroxycinnamate) R^1 = Ac, R^2 = (*E*)-4-hydroxycinnamoyl
470 (15-hydroxy-(plagiochiline A 15-yl (*Z*)-4-hydroxycinnamate) R^1 = H, R^2 = (*Z*)-4-hydroxycinnamoyl
471 (plagicosin G) R^1 = H, R^2 = (*Z*)-cinnamoyl

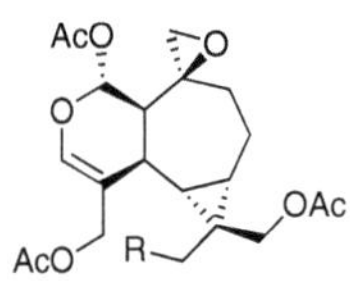

472 (plagiochiline-15-yl dodecanoate) R = dodecanoyl
473 (plagiochiline-15-yl (4*Z*)-decenoate) R = (4*Z*)-decenoyl

An ether extract of the Japanese *Plagiochila ovalifolia* showed inhibitory activity against P-388 murine lymphocytic leukemia cells, and the constituents, plagiochiline 15-yl octanoate (**460**) and 14-hydroxyplagiochiline A-15-yl-(2*E*,4*E*)-dodecadienoate (**465**), were isolated together with plagiochiline A (**125**). Compounds **135**, **460**, and **465** exhibited IC_{50} values of 3.0, 0.05, and 0.05 μg/cm^3 against P-388 cells, respectively [3, 199].

Aponte et al., reported in 2010 that plagiochilines A (**135**), I (**136**), and R (**419**), as isolated from *Plagiochila disticha*, showed cytotoxic activity against a panel of human tumor cell lines, namely, A-549, DU-145, H460, HCT116, HeLa, HT-29, K562, LNCaP, M-14, MCF-7, 3T3, and U937 cells, in comparison to VERO African green monkey normal kidney cells. Among them, compound **135** exhibited the most potent activities against the cancer cell lines, having GI_{50} values between 1.4 and 16.8 μ*M* [200].

The *Plagiochila* species have been divided into several chemotypes by Asakawa and his group [3]. The most characteristic one is that having 2,3-*seco*-aromadendrane-type sesquiterpenoids. From the Japanese *Plagiochila pulcherima,* which was classified to be of chemotype 1, plagiochiline A (**135**) was isolated together with several other compounds. Fractionation of the ethanolic extract of the same species of Chinese origin resulted in the isolation of 3*β*-ethoxyplagiochiline A2 (**453**) and the known 3*α*-methoxy-plagiochiline A1 (**454**), methoxyplagiochiline A2 (**455**), plagiochiline C (**420**), and plagiochiline J (**429**). Among the isolated sesquiterpenoids, compounds **453**–**455** derived from plagiochiline A (**135**) and plagiochiline C (**420**) [2] possessed cytotoxic activity against the HeLa, A172, and H450 cancer cell lines (IC_{50} value range 4.3–26.0 μM). Compound **455** displayed the greatest potencies against these three cell lines and induced the cytotoxicity of HeLa cells by promoting apoptosis [201].

The cytotoxic effects of the essential oil and ether extract of the Corsican *Plagiochila porelloide*s were evaluated against various different cancer and normal fibroblast cells. The ether extract showed cytotoxicity for both the J774 murine cancer macrophage and the WI38 human normal fibroblast cell lines with IC_{50} values of 1.3 and 3.0 $\mu g/cm^3$, respectively [202].

Although Asakawa and his associates [199] and Aponte et al., [200] reported plagiochiline A (**135**) from the liverwort genus *Plagiochila* as having cytotoxicity against several cancer cell lines, they did not investigate the mechanism for this cellular activity. Stivers et al., evaluated in 2018 the effects of plagiochiline A (**135**) on cell cycle progression in DU145 prostate cancer cells. On the basis of the experimental results obtained, they suggested that **135** inhibited cell division by preventing the completion of cytokinesis, particularly at the final abscission stage and also determined that this epoxy hemiacetal reduced DU145 cell survival in colonogenic assays and that it induced substantial cell death in these cells [203]. Bailly (2003) and Vergoten and Bailly (2023) summarized the distribution of the plagiochiline series in the *Plagiochila* genus and the mechanism of action of plagiochines with an epoxide in the molecule against the human cancer cell lines, especially prostate cancer [204, 205].

In 2008, Guo et al., evaluated the cancer cell growth inhibition properties of an ether extract of *Scapania verrucosa.* The extract, containing a mixture of sesquiterpenoids and fatty acids, when evaluated against small panel of four cell lines (A549, LoVo, HL-60 and QGY) was not generally cytotoxic (IC_{50} values of > 100, > 100, 43, and 90 $\mu g/cm^3$, respectively). In addition, the individual compounds present in the ether extract were not clarified [206].

Three solvent extracts of *Marchantia convoluta* were tested by Chen et al., in 2006 against two cancer cell lines. The ethyl acetate extract was inactive against H1299 cells and weakly cytotoxic for HepG2 cells, showing IC_{50} values of 100 and 50 $\mu g/cm^3$, respectively. In turn, the petroleum ether and *n*-butanol extracts of *M. convuluta* exhibited no observed cytotoxic effects [207]. This group suggested that the cytotoxic activity against HepG2 cells could be ascribed to the well-known sesquiterpene, caryophyllene oxide (**474**), as identified by GC/MS of the ethyl acetate extract [208].

From the New Zealand *Lepidolaena clavigera*, clavigerins A–D (**475–478**) were isolated and, when their cytotoxicity was evaluated, they showed weak activity (30 μg/disk) against BSC monkey kidney cells [209].

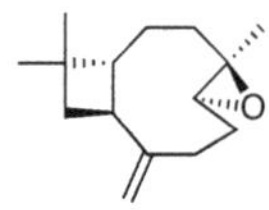

474 (β-caryophyllene oxide)

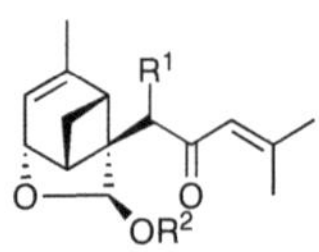

475 (clavigerin A, R^1 = OAc, R^2 = Ac
476 (clavigerin B, R^1 = H, R^2 = Ac)

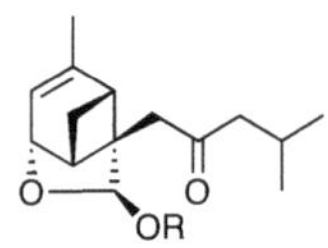

477 (clavigerin C, R = Ac)
478 (clavigerin D, R = Me)

The distribution of terpenoids and aromatic compounds of nine Ecuadorian liverworts, *Frullania brasiliensis*, *Herbertus acanthelius*, *H. juniperoideus*, *H. subdentatus*, *Plagiochila micropterys*, *Macrolejeunea pallescens*, *Marchantia plicata*, *Leptoscyphus hexagonus*, and *Syzygiella anomala*, was reported by Asakawa and his group [2, 3]. Altogether, 67 components from the essential oils of the Ecuadorian liverworts, *F. brasiliensis*, *H. juniperoideus*, *L. hexagonus*, and *S. anomala* were identified by GC/MS. The major components of each species were bicyclogermacrene (**101**), τ-muurolol (**479**), cabreuva oxide D (**480**), and shilphiperfola-5,7(14)-diene (**481**). However, bioactive compounds could not be isolated from any of the liverworts examined [210].

Riccia species belonging to the Ricciaceae are very tiny thalloid liverworts. These produce phytosterols including campesterol, stigmasterol and sitosterol, unsaturated fatty acids containing an acetylene moiety, and flavonoids (apigenin (**482**), luteolin (**483**), and their glycosides), which have been found in other liverworts and mosses [3]. Sharma et al., analyzed the *n*-hexane, ether, ethyl acetate, and methanol crude extracts of the Indian *Riccia billardieri* by GC/MS. The methanol extract gave evidence of weak cytotoxic activity against the HT-29 and HCT-116 cancer cell lines, but only when evaluated at a very high dose level [211].

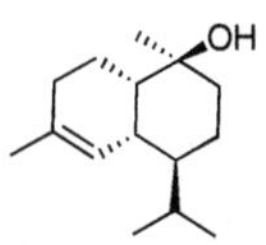

479 (τ-muurolol)

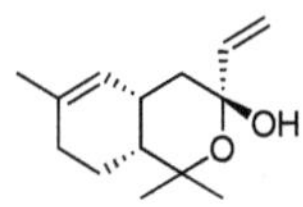

480 (cabreuva oxide)

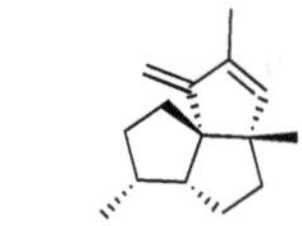

481 (shilphiperfola-5,7(14)-diene)

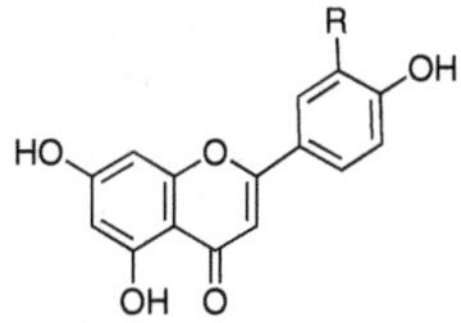

482 (apigenin, R = H)
483 (luteolin, R = OH)

There are some cryptic species of *Conocephalum conicum* in Europe and Japan. In 2020, Radulovic et al., studied the phytochemistry of Serbian *C. conicum* and isolated three new lepidozane, germacrane, and humulane sesquiterpenoids. Their stereostructures were elucidated as (1*Z*,4*E*)-lepidoza-1(10),4-dien-14-ol (**484**), *rel*-(1(10)*Z*,4*S*,5*E*,7*R*)-germacra-1(10),6-diene-11,14-diol (**485**) and *rel*-(1(10)*Z*,4*S*,5*E*,7*R*)-humula-1(10),5-diene-7,14-diol (**486**) by NMR data interpretation, and they co-occurred with conocephalenol (**487**). The previously reported bicyclogermacrene-14-al, which was isolated from the Japanese *C. conicum* [91], was revised structurally as isolepidozene-l4-al (**488**). The immunomodulatory effects of compounds **484**–**487** were evaluated against a model of non-stimulated and mitogen-stimulated rat splenocytes (SPC), The tested compounds showed varying degrees of cytotoxicity to non-stimulated splenocytes during the first 24 h: all four compounds caused cell death in the concentration range of 0.5–1.0 m*M*, while the highest concentration reduced viability by over 70%. Compounds **485** and **486** were found to exert immunosuppressive effects on concanavalin A-stimulated splenocytes, while not being cytotoxic at the same concentration. After 48 h of incubation, in the case of compound **487**, only the lowest concentration of 0.1 μ*M* was without any notable effects on the splenocytes [212].

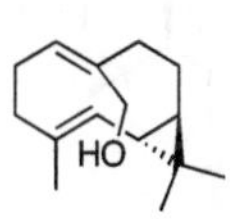

484 (1*Z*,4*E*)-lepidoza-1(10),4-dien-14-ol)

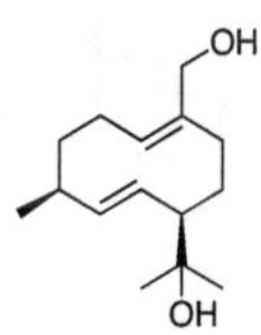

485 (*rel*-(1(10)*Z*,4*S*,5*E*,7*R*)-germacra-1(10),6-diene-7,14-diol)

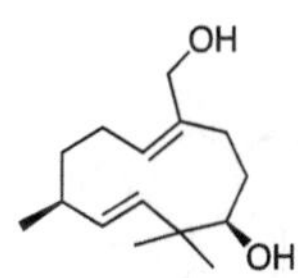

486 (*rel*-(1(10)*Z*,4*S*,5*E*,7*R*)-humula-1(10),5-diene-7,14-diol)

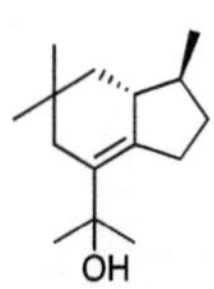

487 (conocepharenol)

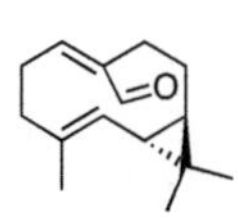

488 (isolepidozen-14-al)

4.4.4 Diterpenoids

Liverworts also produce a large number of different diterpenoids, inclusive of the abietane, cembrane, clerodane, cyathane, dolabellane, dolabrane, fusicoccane, harimane, kaurane, labdane, phytane, pimarane, rosane, sacculatane, sphenolobane, trachylobane, verticillane, vibrane, and viscidane types. Of these, *ent*-clerodanes, *ent*-kauranes and *ent*-labdanes are the major diterpenoids in liverworts [213]. Sacculatanes have not yet been isolated in any other types of bryophytes [1–3]. Many of the isolated diterpenoids have been evaluated for cytotoxicity against a number of cancer cell lines, as shown in Table 9.

The pungent diterpene dialdehyde, sacculatal (**128**), from *Pellia endiviifolia,* showed cytotoxic activity against a human melanoma cell line, with an IC_{50} value in the low micromolar range [3]. Compound **128** was also cytotoxic (IC_{50} values) for Lu1 (5.7), KB (3.2), LNCaP (7.6), and ZR-75-1 cells (7.6 μg/cm^3) [75].

Bazzania albifolia was found to produce not only cytotoxic sesquiterpenoids but also the fusicoccane diterpenoids, albifolione (**489**) and fusicoauritone (**490**), which also showed cytotoxicity against the MCF-7 and PC3 tumor cell lines with IC_{50} values of 37.7 and 24.7, and 59.1–43.1 μ*M*, respectively [193].

Ng et al., reinvestigated in 2000 a methanol extract of the east Malaysian *M. diclados* and isolated the new pimarane diterpene, pimara-8(14),15-dien-18-ol-19-oic acid (**491**) and the enantiomer of the known *ent*-chlorantene G (**492**), along with two known pimaranes, 4-*epi*-sandaracopimaric acid (**493**) and rosa-1(10),15-dien-18-oic acid (**494**). No biological evaluation work was performed on these isolated compounds [197].

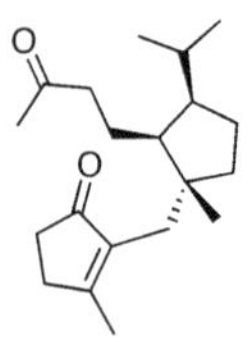

489 (albifolione)

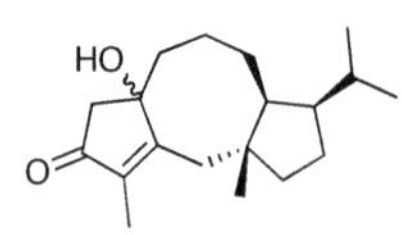

490 (fusicoauritone)

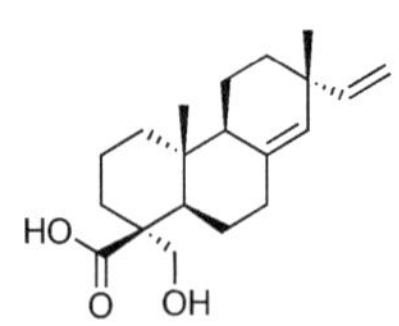

491 (pimara-8(14),15-dien-18-ol-19-oic acid)

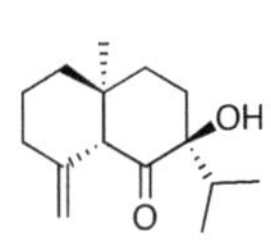

492 (*ent*-chloranten G)

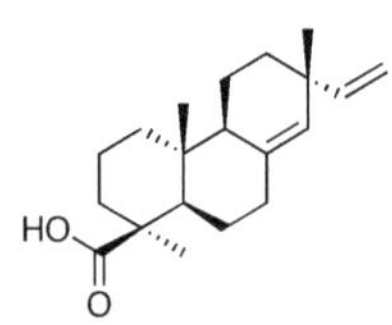

493 (4-*epi*-sandaracopimaric acid)

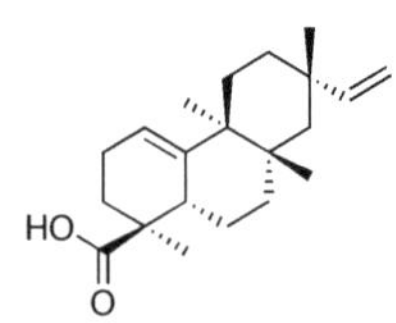

494 (rosa-1(10),15-dien-18-oic acid)

Fractionation of an ethanolic extract of the Chinese *Plagiochila pulcherrima* resulted in the isolation of three pimarane diterpenoids, 7*β*,11*α*-dihydroxypimara-8(14),15-diene (**495**), 1*β*,11*α*-dihydroxypimara-8(14),15-diene (**496**), and 11*β*-hydroxypimara-8(14),15-diene (**497**), which were newly isolated from the liverworts, and three known fusicoccane diterpenoids, fusicoauritone (**490**), 4*α*-hydroxyfusicocca-3(7)-en-6-one (**498**), and anadensin (**499**), along with several cytotoxic 2,3-*seco*-aromadendrane sesquiterpenoids that were described earlier. The presence of pimaranes is limited in the Plagiochilaceae, and only acanthoic acid (**500**) has been found in the New Zealand *Plagiochila deltoidea* [3]. All of the isolated pimarane diterpenoids from *P. pulcherrima* were inactive against the A172, HeLa, and H450 cancer cell lines [201].

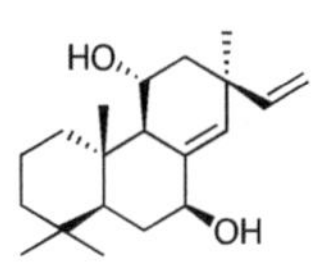

495 (7*β*,11*α*-dihydroxy-pimara-8(14),15-diene)

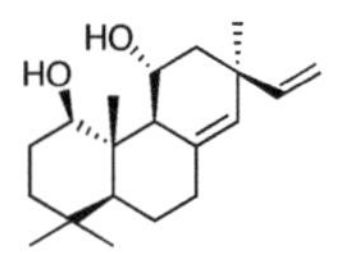

496 (1*β*,11*α*-dihydroxy-pimara-8(14),15-diene)

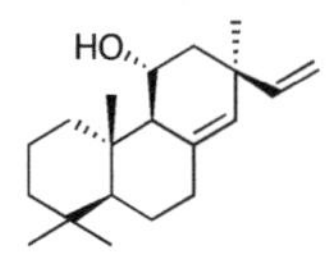

497 (11*α*-hydroxy-pimara-8(14),15-diene)

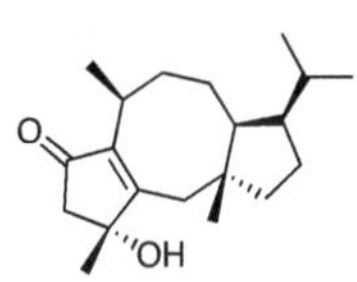

498 (4*α*-hydroxyfusicocca 3(7)-en-6-one)

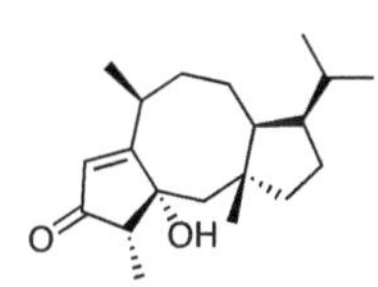

499 (anadensin)

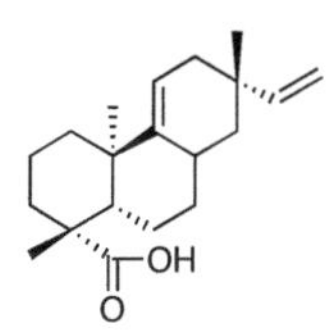

500 (acanthoic acid)

The Tahitian *Chandonanthus hirtellus* was found to produce three rare cembrane diterpenoids, 13,18,20-*epi-iso*-chandonanthone (**501**), (8*E*)-4*α*-acetoxy-12*α*,13*α*-epoxycembra-1(15),8-diene (**502**), and chandonanthine (**503**), in addition to seven fusicoccane-type diterpenes, fusicoauritone (**490**), fusicoauritone 6*α*-methyl ether (**504**), 6*β*,10*β*-epoxy-5*β*-hydroxyfusicocc-2-ene (**505**), fusicogigantenone A (**506**) and B (**507**), fusicogigantepoxide (**508**), and anadensin (**499**), and two verticillane diterpenoids, *ent*-verticillol (**509**) and *ent-epi*-verticillol (**510**). Also obtained were methyl salicylate (**511**), anastreptene (**512**), *allo*-aromadendrene (**513**), herbertene (**412**), spathulenol (**514**), fusicocca-3,5-diene (**515**), fusicocca-2,5-diene (**516**), thunbergene (**517**), and cembrene A (**518**) as volatile components, among which only compounds **499** and **501** demonstrated cytotoxicity against HL-60 cells (IC_{50}, 17.0 and 18.1 μg/cm^3, respectively) [214].

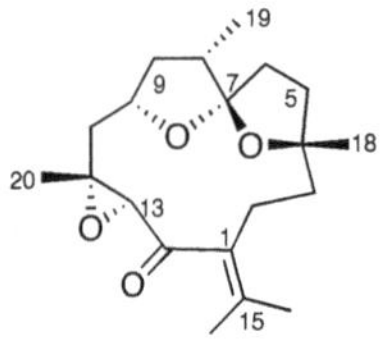

501 (13,18-20-*epi-iso*-chandonanthone)

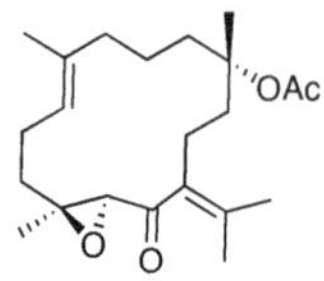

502 ((8*E*)-4α-acetoxy-12α,13α-epoxy-cembra-1(15),8-diene)

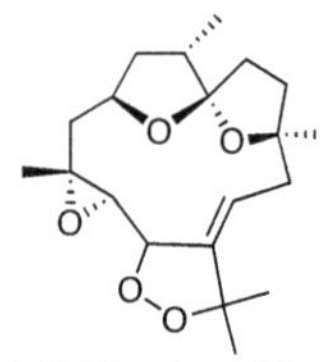

503 (chandonanthine)

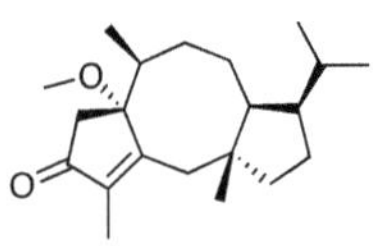

504 (fusicoauritone 6α-methyl ether)

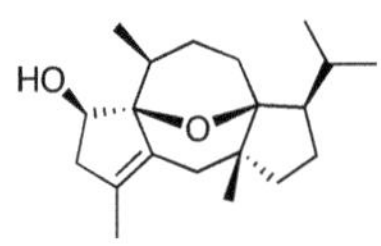

505 (6β,12β-epoxy-5β-hydroxyfusicocc-2-ene)

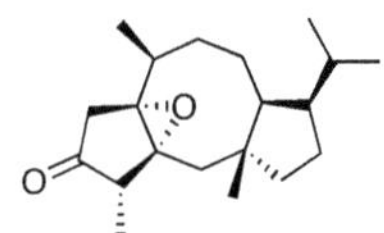

506 (fusicogigantone A)

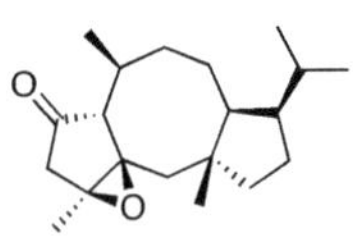

507 (fusicogigantone B)

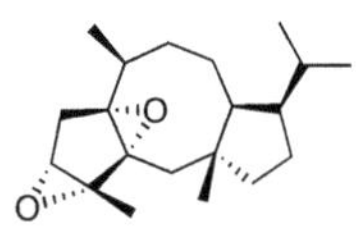

508 (fusicogigantepoxide)

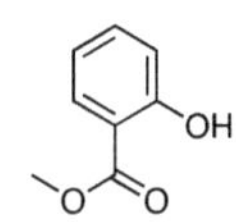

511 (methyl salicylate)

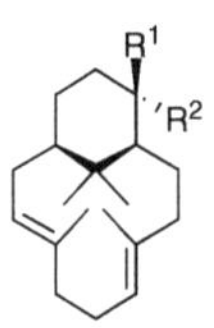

509 (*ent*-verticillol, R^1 = H, R^2 = OH)
510 (*ent-epi*-verticillol, R^1 = OH, R^2 = H)

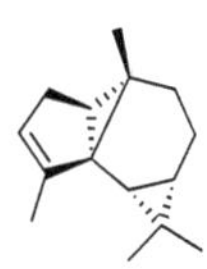

512 (anastreptene)

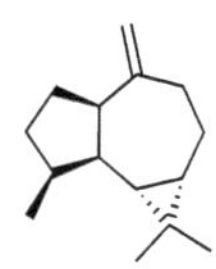

513 (*allo*-aromadendrene)

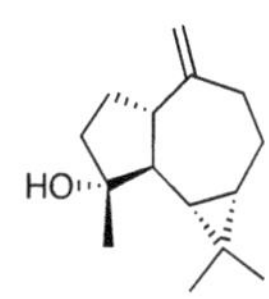

514 (spathulenol)

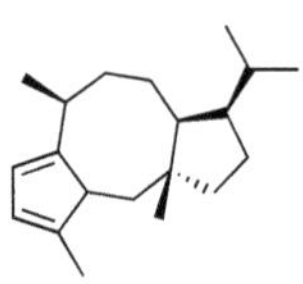

515 (fusicocca-3,5-diene)

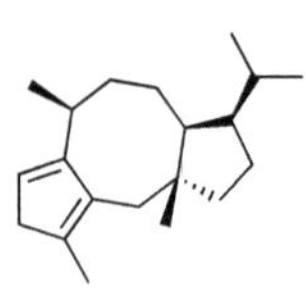

516 (fusicocca-2,5-diene)

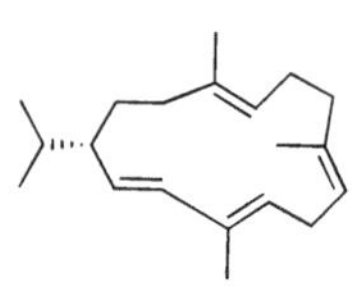

517 (thunbergene)

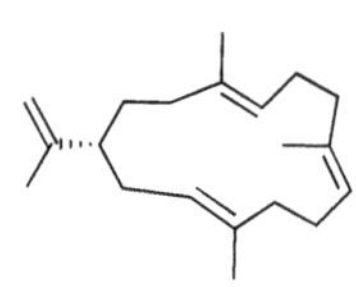

518 (cembrene A)

This same species collected in Malaysia also produced chandonanthone (**519**), isochandonanthone (**520**), and setiformenol (**521**) along with two new cembranes, 8,10-di-*epi*-chandonanthone (**522**) and 1β,15-dihydro-8,10-di-*epi*-chandonanthone (**523**) [215].

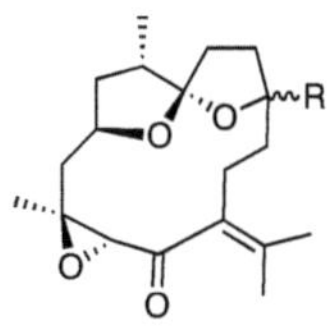

519 (chandonanthone, R = β-Me)
520 (*iso*-chandonanthone, R = α-Me)

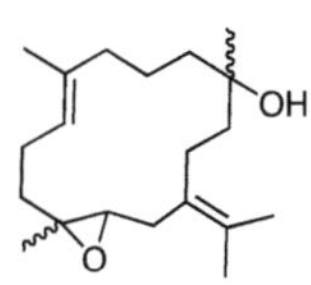

521 (setiformenol)

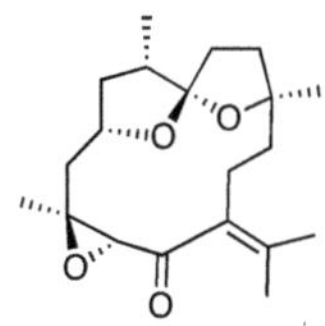

522 (8,10-di-*epi*-chandonanthone)

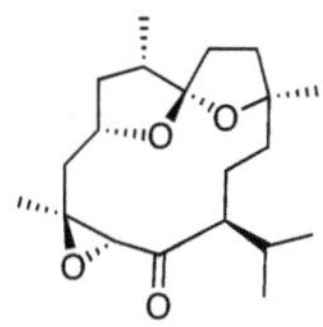

523 (1β,15-dihydro-8,10-di-*epi*-chandonanthone)

In 2014, Li et al., isolated several new cembrane diterpenoids, chandonanones A (**524**), B (**525**), and D–F (**527–529**) from the Chinese *Chandonanthus hirtellus*, and chandonanones B (**525**), C (**526**), E (**528**), and F (**529**) from *C. birmensis.* Also obtained were the known chandonanthone (**519**), isochandonanthone (**520**), and (8*E*)-4α-acetoxy-12α,13α-epoxycembra-1(15),8-diene (**530**), the fusicoccane diterpenoid anadensin (**499**), and 2,10–14-triacetoxy-7,8:18,19-diepoxy-dolabell-(3*E*)-ene (**531**). The absolute configurations of chandonanones A (**524**) and B (**525**) were established by single-crystal X-ray analysis with CuK_α radiation. The cytotoxic activity of all of the isolated compounds was tested against the DU145, PC3, A549, PC12, NCI-H292, NCI-H1299, and A172 cancer cell lines. Chandonanone A (**524**) displayed cytotoxic activity against PC12 and NCI-H1299 cells with IC_{50} values of 24.5 and 17.2 μ*M,* respectively. Chandonanone B (**525**) also showed weak activity against NCI-H1299 cells (IC_{50} value, 19.5 μ*M*). Anadensin (**499**) and the cembranes **526–529** were either determined as inactive (IC_{50} > 50 μ*M*) or showed only weak activities (IC_{50} values, 20–50 μ*M*) against these cell lines [216].

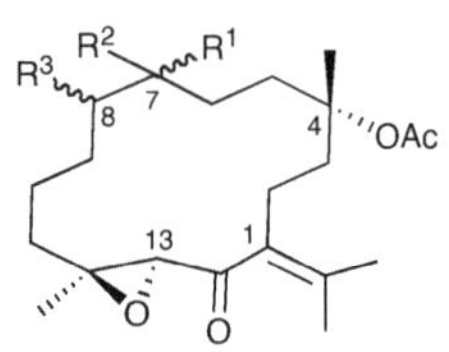

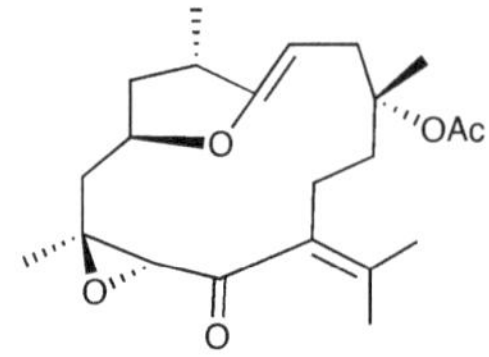

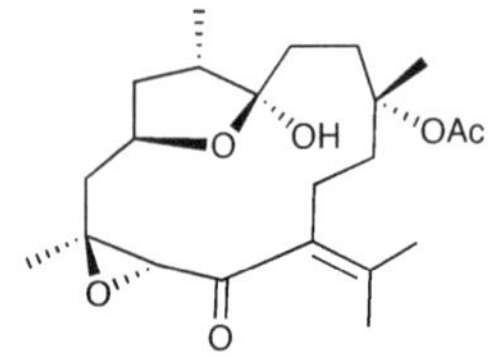

524 (chandonanone A,
$R^1 = R^2 = O$, $R^3 = \alpha$-Me)
525 (chandonanone B,
$R^1 = R^2 = O$, $R^3 = \beta$-Me)
526 (chandonanone C,
$R^1 = R^2 = O$, R^3 = Me, $\Delta^8(E)$)
527 (chandonanone D,
$R^1 = \alpha$-OH, $R^2 = \beta$-H, R^3 = Me, $\Delta^8(E)$)
530 ((8*E*)-4α-acetoxy-12α,13α-epoxycembra-1(15),8-diene,
$R^1 = R^2$ = H, R^3 = Me, $\Delta^8(E)$)

528 (chandonanone E)

529 (chandnanone F)

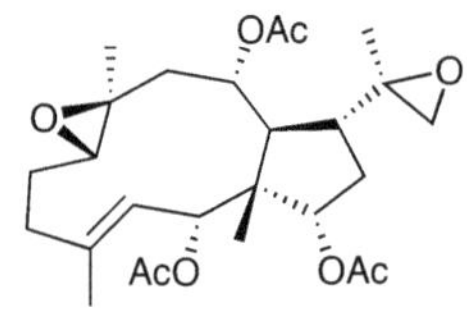

531 (2,10,14-triacetoxy-7,8:18,19-diepoxy-dolabell-(3*E*)-ene)

Four new labdanes, heteroscyphin A (= 6α,17:8β,13β:17,20-diepoxy-*ent*-labd-14-ene-17α-ol) (**532**), heteroscyphin B (= 6α,17:8β,13β:17,20-triepoxy-*ent*-labd-14-ene) (**533**), heteroscyphin C (= 8β,13β-epoxy-*ent*-labd-14-en-17,6β-olide) (**534**), and heteroscyphin D (= 6α-acetoxy-8β,13β-epoxy-*ent*-labd-14-en-17-oic acid) (**535**), along with the four known diterpenes, isomanool (**536**), and the three fusicoccanes, fusicoauritone (**490**), 4α-hydroxyfusicocca-3(7)-en-6-one (**498**), and anadensin (**499**), were isolated from the Chinese *Heteroscyphus tener*. The absolute configuration of **532** was established by single-crystal X-ray diffraction using CuK_α irradiation [217]. A cytotoxicity evaluation of the isolated compounds against eight human cancer cell lines (PC3, DU145, K562, A549, NCI-H252, NCI-H1299, NCI-H446, PWPE-1) and HBE human normal bronchial cells, showed compounds **499**, **534**, and **536** to exhibit IC_{50} values in the range 10.7–81.0 μ*M*. Isomanool (**536**) showed inhibitory effects on prostate cancer (PCa) cell proliferation, but a lesser inhibition of non-neoplastic prostate epithelial cells, and also induced cellular apoptosis in PCa cells through ROS-mediated DNA damage [217].

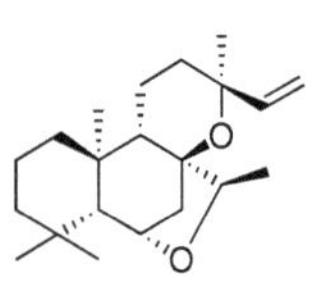

532 (heteroscyphin A)

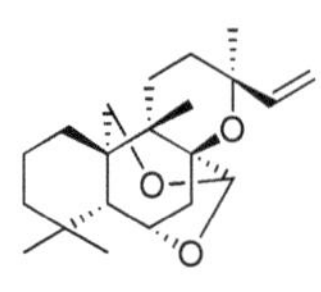

533 (heteroscyphin B)

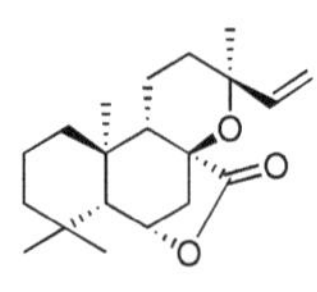

534 (heteroscyphin C)

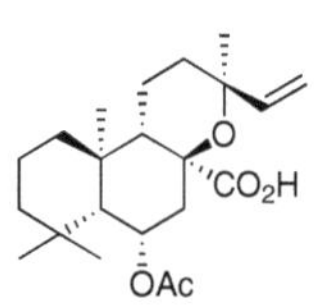

535 (heteroscyphin D)

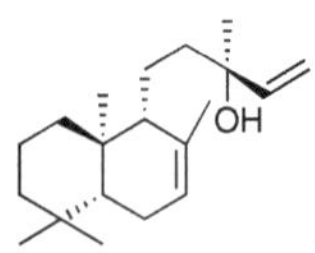

536 (isomanool)

From the Japanese *Jungermannia comata*, the acyclic bis-bibenzyl, perrottetin E (**183**), was isolated as the major component [3]. In 2016, Li et al., reinvestigated the phytochemical components of a 95% ethanol extract of a Chinese specimen of this species and isolated four new diterpenoids, kaur-16-ene-6β,9α-diol (**537**), kaur-16-ene-6β,9α,12β-triol (**538**), kaur-16-ene-5α,6β,9α-triol (**539**), and kaur-9α,19-diol (**540**), and five known analogs, kaur-16-en-19-oic acid (**541**), 9α-hydroxykaur-16-en-19-oic acid (**542**), kaur-16-en-19-al (**543**), kaur-16-en-19-ol (**544**), 17,19-dihydroxykaur-15-ene (**545**), along with the four trachylobane diterpenoids trachyloban-19-ol (**546**), trachyloban-19-oic acid (**547**), 13α-hydroxytrachyloban-19-oic acid (**548**), and trachyloban-9-en-19-oic acid (**549**) [218].

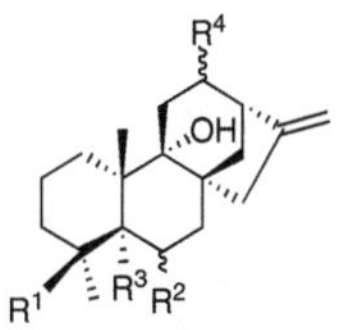

537 (kaur-16-ene-6β,9α-diol, R^1 = Me, R^2 = β-OH, R^3 = R^4 = H)
538 (kaur-16-ene-6β,9α,12β-triol, R^2 = R^4 = β-OH, R^3 = H)
539 (kaur-16-ene-5,6β,9α-triol, R^1 = Me, R^2 = β-OH, R^3 = OH, R^4 = H)
540 (kaur-16-ene-9α,19-diol, R^1 = CH_2OH, R^2 = R^3 = R^4 = H)

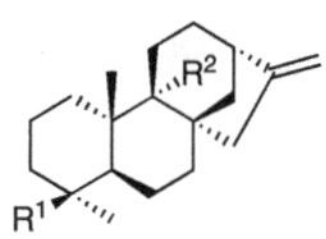

541 (kaur-16-en-19-oic acid, R^1 = CO_2H, R^2 = H)
542 (9α-hydroxy-kaur-16-en-19-oic acid, R^1 = CO_2H, R^2 = OH)
543 (kaur-16-en-19-al, R^1 = CHO, R^2 = H)
544 (kaur-16-en-19-ol, R^1 = CH_2OH, R^2 = H)

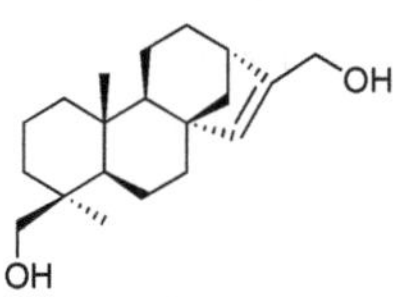

545 (17,19-dihydroxykaur-15-ene)

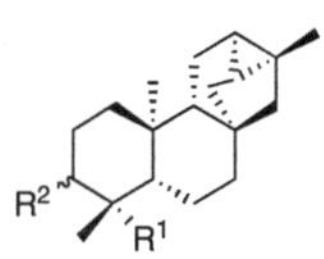

546 (trachyloban-19-ol, R^1 = CH_2OH, R^2 = H)
547 (trachyloban-19-oic acid, R^1 – CO_2H, R^2 – H)
548 (3α-hydroxytrachyloban-19-oic acid, R^1 = CO_2H, R^2 = α-OH)

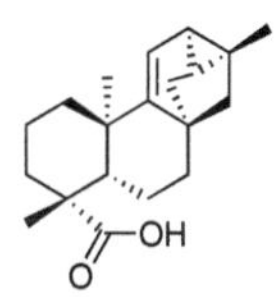

549 (trachyloban-9-en-19-oic acid)

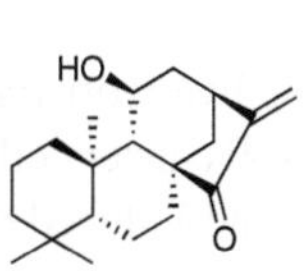

550 (*ent*-11α-hydroxy-kaur-16-en-15-one)

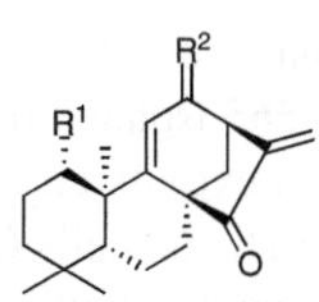

551 (*ent*-1β-hydroxy-9(11),16-kauradiene-15-one, R^1 = OH, R^2 = H_2)
552 (*ent*-9(11),16-kauradiene-12,15-dione, R^1 = H, R^2 = O)

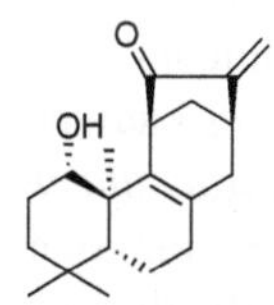

553 (jungermannenone A)

The kaurenes, *ent*-11α-hydroxykaur-16-en-15-one (**550**), *ent*-1β-hydroxy-9(11),16-kauradien-15-one (**551**), *ent*-9(11),16-kauradiene-12,15-dione (**552**), and jungermannenone A (**553**), isolated from an unidentified *Jungermannia* species from New Zealand, potently inhibited HL-60 cells with IC_{50} values, in turn, of 0.49, 7.0, 0.59, and 0.28 μ*M*. Treatment with the isolated *ent*-kaurenones (**550–553**) caused proteolysis of poly (ADP-ribose) polymerase, a sign of activation of the apoptotic process, whereas a feature of cell death induced by treatment with compounds **531** and *ent*-9(11),16-kauradiene-12,15-dione (**552**) was necrosis. Treatment with jungermannenone A (**553**) induced apoptosis in HL-60 cells [219].

The similar *ent*-kaurane and kaurene diterpenoids, *ent*-11α-hydroxykauran-15-one (**554**), *ent*-6β-hydroxykaur-16-en-15-one (**555**), and jungermannenones B–D (**556–558**) from an indentified *Jungermannia* species were evaluated for their cytotoxicity against HL-60 cells. While compound **554** was determined as being inactive (IC_{50} > 100 μ*M*), diterpenoids **555–558** showed potent growth inhibitory activities with IC_{50} values of 0.4, 1.2, 1.3, and 0.8 μ*M*, respectively [220].

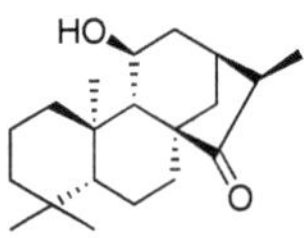

554 (*ent*-11α-hydroxy-kauran-15-one)

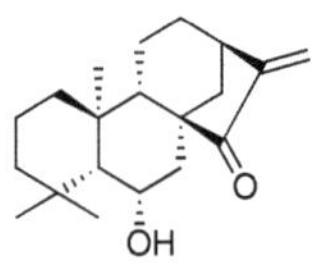

555 (*ent*-6β-hydroxy-kaur-16-en-15-one)

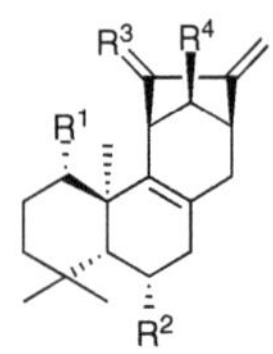

556 (jungemannenone B, $R^1 = R^2 = R^4 = H, R^3 = O$)
557 (jungemannenone C, $R^1 = R^2 = H, R^3 = O, R^4 = OH$)
558 (jungemannenone D, $R^1 = R^4 = OH, R^2 = H, R^3 = O$)

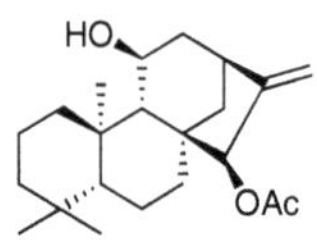

559 (*ent*-11α-hydroxy-15α-acetoxykaur-16-ene)

ent-11α-Hydroxykaur-16-en-15-one (**550**) from a New Zealand *Jungermannia* species showed cytotoxic activity against P-388 cells with an IC_{50} value of 0.5 μg/cm^3, while its acetate (**559**) was inactive [221].

The *ent*-kaurenes **550** and **560–565** isolated from the Japanese liverwort *Jungermannia truncata*, were evaluated for cytotoxicity against the HL-60 cancer cell line. Of these, compound **550** inhibited HL-60 cells (IC_{50} 0.82 μ*M*) and induced apoptosis, partly through a caspase-8-dependent pathway. The presence of an enone group in this class of metabolites appears to be essential for the induction of apoptosis and the activation of caspases in human leukemia cell lines [222, 223].

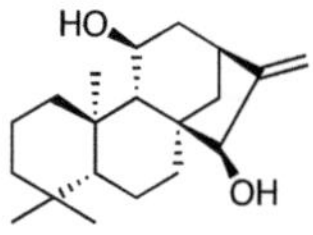

560 (*ent*-11α,15α-dihydroxy-kaur-16-ene)

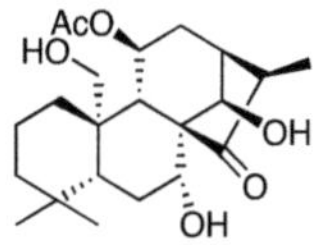

561 (*ent*-11α-acetoxy-7β,14α,20-trihydroxy-kauran-15-one)

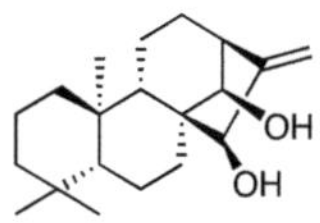

562 (*ent*-14α,15α-dihydroxykaur-16-ene)

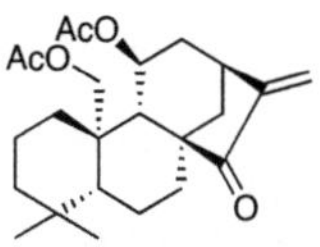

563 (*ent*-11α,20-di-acetoxy-16-kauren-15-one)

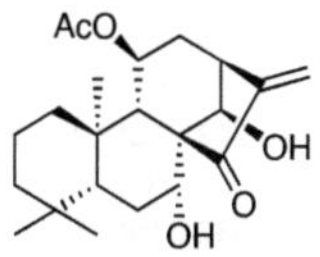

564 (*ent*-11α-acetoxy-7β,14α-dihydroxy-16-kaurene)

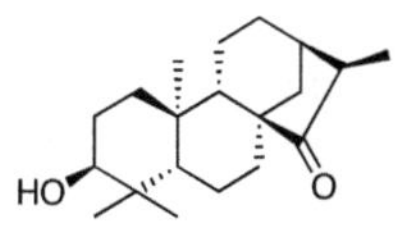

565 (*ent*-3α-hydroxy-kauran-15-one)

The *ent*-kaurenes, **550**, **552**, **564**, and **565**, as well as the rearranged *ent*-kaurene, jungermannenone A (**553**), selectively inhibited nuclear factor-κB (NF-κB)-dependent gene expression following treatment with TNF-α. Compound **550**, in combination with TNF-α, caused a dramatic increase in apoptosis in HL-60 and K562 human leukemia cells, accompanied by activation of caspases, and when combined with camptothecin, also caused an increase in apoptosis [224, 225].

Jungermannenones A–D (**553**, **556–558**), obtained from *Jungermannia fauriana*, induced DNA fragmentation and nuclear condensation. Both are biochemical markers of apoptosis induction, which occurred through a caspase-independent pathway. Compounds **553** and **558** showed inhibitory activity for NF-κB, which is a transcriptional factor of antiapoptotic factors. Thus, *ent*-kaurene diterpenoids from liverworts may represent promising antitumor lead compound candidates [226, 227].

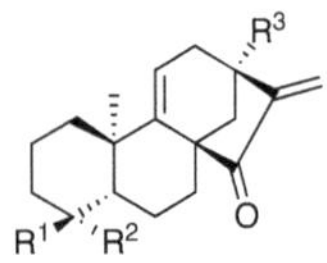

567 (13,18-dihydroxy-9(11),16-kauradien-15-one, $R^1 = CH_2OH$, $R^2 = Me$, $R^3 = OH$)
568 (13-hydroxy-*ent*-9(11),16-kauradien-18-al-15-one, $R^1 = CHO$, $R^2 = Me$, $R^3 = OH$)
569 (13,19-dihydroxy-*ent*-9(11),16-kauradien-15-one, $R^1 = Me$, $R^2 = CH_2OH$, $R^3 = OH$)
570 (18-hydroxy-*ent*-9(11),16-kauradien-15-one, $R^1 = CH_2OH$, $R^2 = Me$, $R^3 = H$)
571 (13-hydroxy-*ent*-9(11),16-kauradien-15-one, $R^1 = R^2 = Me$, $R^3 = OH$)

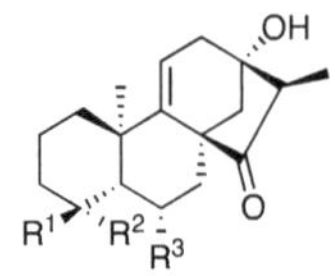

575 ((16*R*)-13-hydroxy-*ent*-9(11)-kauren-15-one, $R^1 = R^2 = Me$, $R^3 = H$)
576 ((16*R*)-13,18-dihydroxy-*ent*-9(11)-kauren-15-one, $R^1 = CH_2OH$, $R^2 = Me$, $R^3 = H$)
578 ((16*R*)-13-hydroxy-*ent*-9(11)-kauren-18-al-15-one, $R^1 = CHO$, $R^2 = Me$, $R^3 = H$)
579 ((16*R*)-13,19-dihydroxy-*ent*-9(11)-kauren-15-one, $R^1 = Me$, $R^2 = CH_2OH$, $R^3 = H$)
580 ((16*R*)-6α,13,18-trihydroxy-*ent*-9(11)-kauren-15-one, $R^1 = CH_2OH$, $R^2 = Me$, $R^3 = OH$)

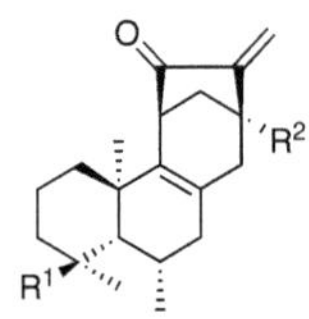

572 (13-hydroxyjungermannenone B, $R^1 = Me$, $R^2 = OH$)
573 (13,18-dihydroxyjungermannenone B, $R^1 = CH_2OH$, $R^2 = OH$)
574 (18-hydroxyjungermannenone B, $R^1 = CH_2OH$, $R^2 = H$)

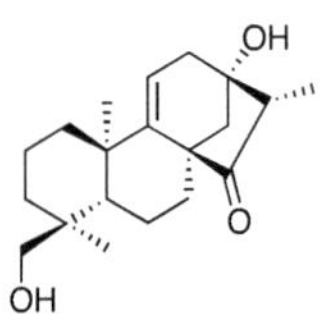

577 ((16*S*)-13,18-dihydroxy-*ent*-9(11)-kauren-15-one)

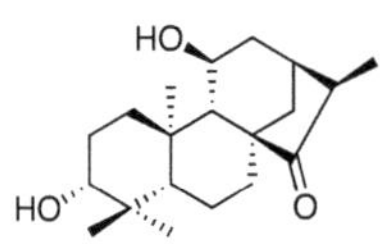

581 ((16*R*)-3α,11β-dihydroxy-*ent*-kauran-15-one)

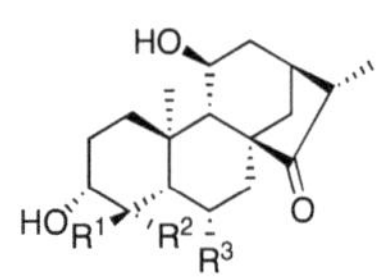

582 ((16*S*)-3α,11β-dihydroxy-*ent*-kauran-15-one)

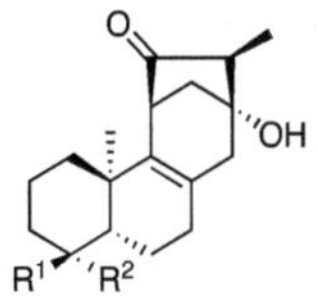

583 (16α,17-dihydro-13,18-dihydroxyjungermanenone B, $R^1 = CH_2OH$, $R^2 = Me$)
584 (16α,17-dihydro-13-hydroxyjungermanenone B, $R^1 = R^2 = Me$)
585 (18-formyl-16α,17-dihydro-13-hydroxyjungermanenone B, $R^1 = CHO$, $R^2 = Me$)
586 (16α,17-dihydro-13,19-dihydroxyjungermanenone B, $R^1 = Me$, $R^2 = CH_2OH$)

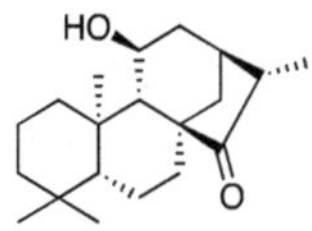

587 ((16*S*)-11β-hydroxy-*ent*-kauran-15-one)

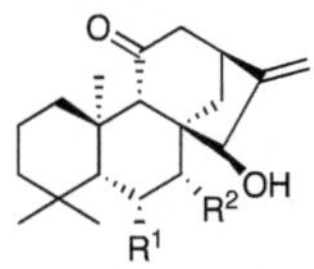

588 (6α,15β-dihydroxy-*ent*-kauren-11-one, $R^1 = OH$, $R^2 = H$)
589 (6α,7α,15β-trihydroxy-*ent*-16-kauren-11-one, $R^1 = R^2 = OH$)
590 (7α,15β-dihydroxy-*ent*-16-kauren-11-one, $R^1 = H$, $R^2 = OH$)
591 (15β-hydroxy-*ent*-16-kauren-11-one, $R^1 = R^2 = H$)

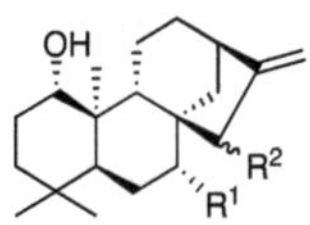

592 (1α,15β-dihydroxy-*ent*-16-kaurene, $R^1 = H$, $R^2 = \beta$-OH)
593 (1α,15α-dihydroxy-*ent*-16-kaurene, $R^1 = H$, $R^2 = \alpha$-OH)
596 (1α,7α-dihydroxy-*ent*--16-kaurene, $R^1 = OH$, $R^2 = H$
597 (7α-acetoxy-1α-hydroxy-*ent*-16-kaurene, $R^1 = OAc$, $R^2 = H$)

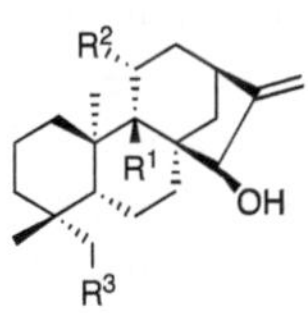

594 (9β,15β-dihydroxy-*ent*-16-kaurene, $R^1 = OH$, $R^2 = R^3 = H$)
595 (11α,15β-dihydroxy-*ent*-16-kaurene, $R^1 = R^3 = H$, $R^2 = OH$)
603 (19-hydroxy-*ent*-16-kaurene, $R^1 = R^2 = H$, $R^3 = OH$)

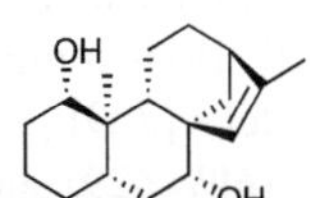

598 (1α,7α-dihydroxy-*ent*-15-kaurene)

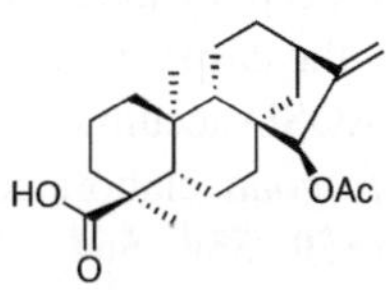

599 (15β-acetoxy-*ent*-kaur-16-en-18-oic acid)

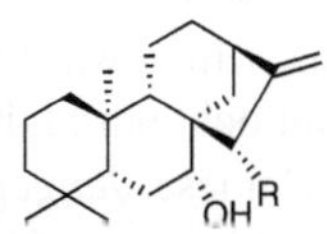

601 (7α,15α-dihydroxy-*ent*-16-kaurene, R = OH)
602 (7α-hydroxy-*ent*-16-kaurene, R = H)

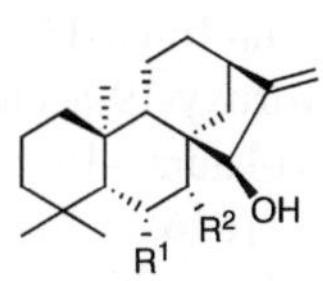

600 (7α,15β-dihydroxy-*ent*-16-kaurene, $R^1 = H$, $R^2 = OH$)
604 (6α,15β-dihydroxy-*ent*-16-kaurene, $R^1 = OH$, $R^2 = H$)
605 (15β-hydroxy-*ent*-16-kaurene, $R^1 = R^2 = H$)

In 2015, Lin et al., isolated 20 new (**567–586**) and four known kaurane diterpenes (**550**, **554**, **560**, and **587**) from the Chinese *Lepidozia fauriana*. The same group also obtained 20 new (**609–629**) and ten known kaurenoids (**550**, **554**, **562**, **586**, **592**, **604**, **612**, **613**, **630**, and **631**) from the Chinese *L. hyalina* and prepared the synthetic products **606–608** from steviol (**566**), as shown in Scheme 9. The isolated new *ent*-kauranes (**550**, and **567–574**) and the synthetic products **606–608**) were evaluated against a panel of 18 cancer and normal cell lines (PC3, DU145, LNCaP, K562, A549, NCI-H1299, NCI-H446, MCF-7, HepG2, MDA-MB-231, SKOV3, LOVO, T24, HL-60, SH-SY5Y, A172, Saos-2, and RWPE-1). Among the tested kauranes, with the exception of **574**, *ent*-kaurenes **550**, **567–573** showed cytotoxicity against all cancer cell lines, with *ent*-11α-hydroxy-16-kauren-15-one (**550**), 13-hydroxy-9(11),16-kauradien-15-one (**571**), 13-hydroxy-jungermannenone B (**572**), and 13,18-dihydroxyjungermannenone B (**573**) exhibiting potent cytotoxicity with IC_{50} values of 1.6–6.3 μM. Steviol-15-one (**606**), its methyl ester (**607**) and piperazine ethyl ester (**608**), prepared from steviol (**566**) as shown in Scheme 9, were less active than compounds **550**, **572**, and **573** [228].

The liverwort genus *Jungermannia* species (Jungermanniaceae) are abundant sources of *ent*-kaurane diterpenoids. Nagashima et al., in 2015, purified the rearranged *ent*-kaurenones jungermannenones A–D (**553**, and **556–558**) from an unidentified New Zealand *Jungermannia* species [220].

In 2016, Guo et al., reported the cytotoxicity of "jungermannenones A and B" against a cancer cell line panel. However, the structures of these two compounds are not identical to those of "jungermanenones A and B", which were isolated from the New Zealand *Jungermannia* species by Asakawa's group. Thus, the names of these two compounds should be revised to 13-hydroxyjungermanenone B (**572**) and 13,18-dihydroxy-jungermannenone B (**573**), respectively. Both compounds, **572** and **573**, showed strong antiproliferative effects against PC3, DU145, LNCaP, K562, A549, NCI-H1299, NCI-H446, MCF-7, HepG2, and RWPE-1 cells, with IC_{50} values of 1.2–18.3 μM. Compound **572** (IC_{50} 1.5 μM) and **573** (IC_{50} 5 μM) induced PC3 cell apoptosis, which was attenuated by the caspase inhibitor Z-VAD. Both compounds caused mitochondrial damage and ROSD accumulation in PC3 cells [229].

The Chinese *Jungermannia tetragona* elaborated 20 new (**609–629**) and ten known *ent*-kaurane diterpenoids (**550**, **554**, **564**, **586**, **592**, **604**, **612**, **613**, **630**, and **631**). Their absolute structures were elucidated by a combination of analysis of NMR spectoscopy, high-resolution MS, ECD, and single-crystal X-ray analysis as *ent*-11α-hydroxy-19,20-diacetoxykaur-16-en-15-one, (**609**), *ent*-11α,20-dihydroxy-19-acetoxykaur-16-en-15-one (**610**), *ent*-20-hydroxykaur-16-en-15-one (**611**), *ent*-11α-hydroxy-19-acetoxy-(16*S*)-kauran-15-one (**614**), 11α,20-dihydroxy-19-acetoxy-(16*S*)-kauran-15-one (**615**), *ent*-11α,20-dihydroxy-19-acetoxy-(16*R*)-kauran-15-one (**616**), *ent*-11α-hydroxy-20-acetoxy-(15*S*)-kauran-15-one (**617**), *ent*-11α-hydroxy-20-acetoxy-(16*R*)-kauran-15-one (**618**), *ent*-11α-hydroxy-19,20-diacetoxy-(16*S*)-kauran-15-one (**619**), *ent*-11α-hydroxy-19,20-diacetoxy-(16*R*)-kauran-15-one (**620**), *ent*-11α,20-dihydroxy-(16*S*)-kauran-15-one (**621**), *ent*-11α,19-dihydroxy-20-acetoxy-(16*S*)-kauran-15-one (**622**),

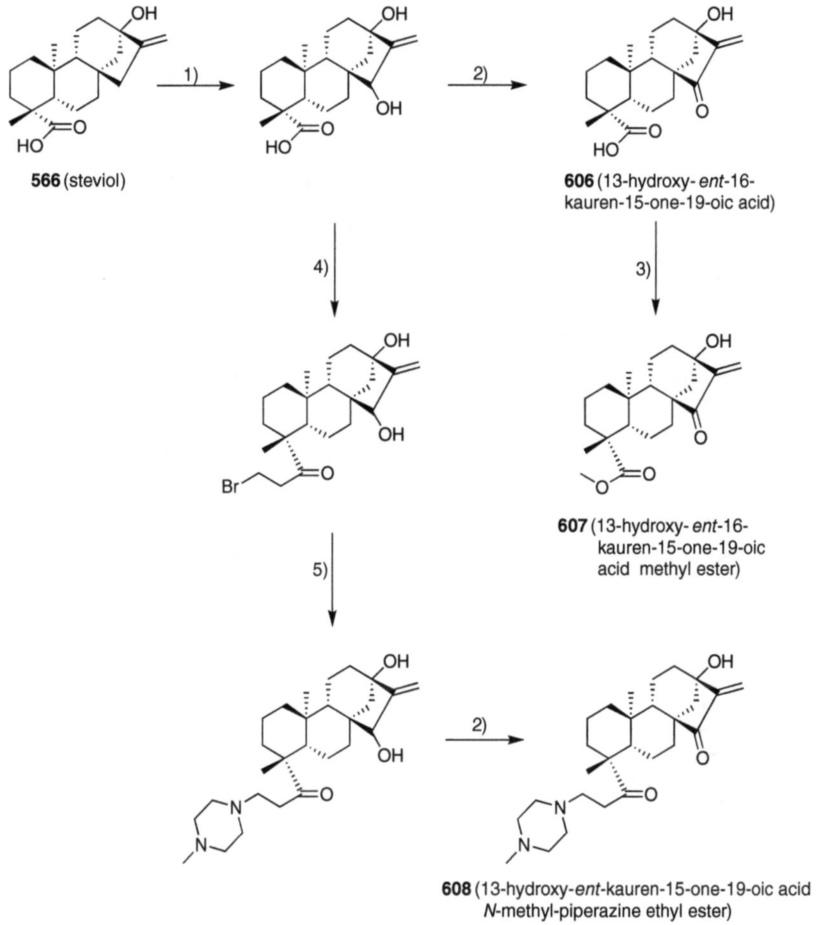

Scheme 9 Synthetic products **606–608** from steviol (**566**). (1) SeO_2, *t*-BuO_2H,THF (2) PDC, DMF (3) MeI, DMF (4) $Br(CH_2)_2B$, K_2CO_3, DMF (5) *N*-methylpiperazine, DMF

ent-11*α*,20-epoxy-19-acetoxy-(16*R*)-kauran-15-one (**623**), *ent*-11*α*,16*α*-epoxy-15*α*,20-dihydroxy-(16*S*)-kaurane (**624**), *ent*-11*α*-hydroxykaur-16-en-15-one (**560**), *ent*-11*α*,15*α*-dihydroxy-20-acetoxykaur-16-ene (**625**), *ent*-11*α*,15*α*-dihydroxy-19,20-diacetoxykaur-16-ene (**626**), *ent*-6*β*,12*α*,15*α*-trihydroxy-20-acetoxykaur-16 ene (**627**), *ent*-11*β*,20-epoxy-15*α*-hydroxykaur-16-ene (**628**), *ent*-11*β*,20-epoxy-15*α*,19-diacetoxykaur-16-ene (**629**), *ent*-11*α*-hydroxy-19-acetoxykaur-16-en-15-one (**612**), *ent*-11*α*-hydroxy-20-acetoxykaur-16-en-15-one (**613**), *ent*-kaur-11*α*-hydroxy-16-en-15-one (**550**), *ent*-12*α*-hydroxy-(16*S*)-kauran-15-one (**586**), *ent*-(16*R*)-11*α*-hydroxy-20-acetoxykauran-15-one (**630**), *ent*-(16*R*)-11*α* hydroxykauran-15-one (**554**), *ent*-(16*S*)-3*α*,11*α*-dihydroxykauran-15-one (**631**), *ent*-5*β*,11*α*,15*α*-trihydroxykaur-16-ene (**604**), *ent*-1*β*,15*α*-dihydroxykaur-16-ene (**592**), and *ent*-14*α*,16*α*-dihydroxykaur-16-ene (**562**) [230].

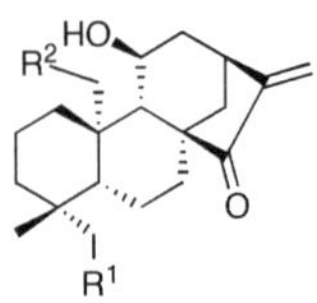

609 (*ent*-11α-hydroxy-19,20-diacetoxy-kaur-16-en-15-one, R1 = R2 = OAc)
610 (*ent*-11α,20-dihydroxy-19-acetoxy-kaur-16-en-15-one, R1 = OAc, R2 = OH)
611 (*ent*-11α,20-dihydroxykaur-16-en-15-one, R1 = H, R2 = OH)
612 (*ent*-11α-hydroxy-19-acetoxykaur-16-*ent*-15-one, R1 = OAc, R2 = H)
613 (*ent*-11α-hydroxy-20-acetoxykaur-16-en-15-one, R1 = H, R2 = OAc)

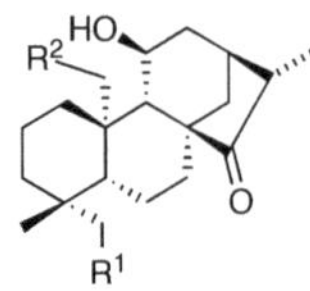

614 (*ent*-11α-hydroxy-19-acetoxy-(16*S*)-kauran-15-one, R1 = OAc, R2 = H)
615 (*ent*-11α,20-dihydroxy-19-acetoxy-(16*S*)-kauran-15-one, R1 = OAc, R2 = OH)
617 (*ent*-11α-hydroxy-20-acetoxy-(16*S*)-kauran-15-one, R1 = H, R2 =OAc)
619 (*ent*-11α-hydroxy-19,20-diacetoxy-(16*S*)-kauran-15-one, R1 = R2 = OAc)
621 (*ent*-11α,20-dihydroxy-(16*S*)-kauran-15-one, R1 = H, R2 = OH)
622 (*ent*-11α,19-dihydroxy-20-acetoy-(16*S*)-kauran-15-one, R1 = OH, R2 = OAc)

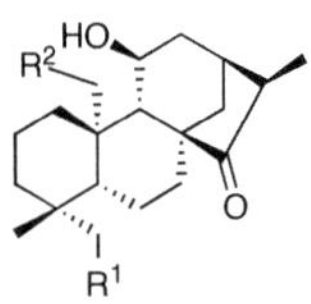

616 (*ent*-11α,20-dihydroxy-19-acetoxy-(16*R*)-kauran-15-one, R1 = OAc, R2 = OH)
618 (*ent*-11α-hydroxy-20-acetoxy-(16*R*)-kauran-15-one, R1 = H, R2 = OAc)
620 (*ent*-11α-hydroxy-19,20-diacetoxy-(16*R*)-kauran-15-one, R1 = R2 = OAc)

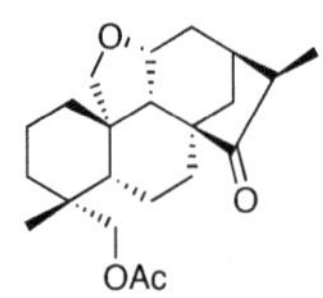

623 (*ent*-11β,20-epoxy-19-acetoxy-(16*R*)-kauran-15-one)

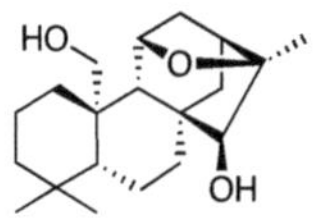

624 (*ent*-15α,20-dihydroxy-11α,16α-epoxy-(16*S*)-kaurane)

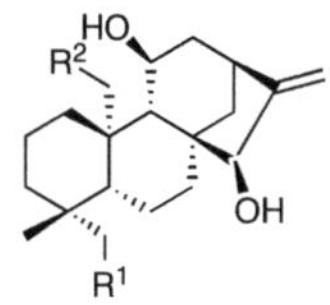

625 (*ent*-11α,15α-dihydroxy-19-acetoxy-kaur-16-ene, R^1 = H, R^2 = OAc)
626 (*ent*-11α,15α-dihydroxy-19,20-diacetoxy-kaur-16-ene, $R^1 = R^2$ = OAc)

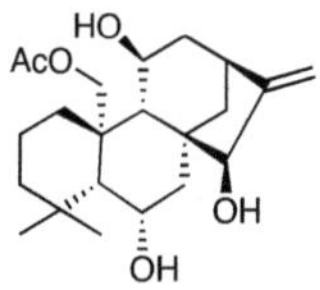

627 (*ent*-6β,11α,15α-trihydroxy-20-acetoxy-kaur-16-ene)

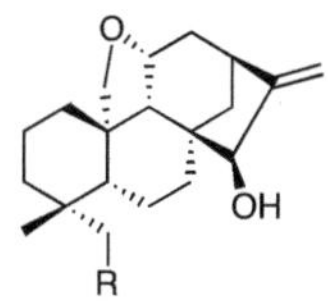

628 (*ent*-11β,20-epoxy-15α-hydroxy-kaur-16-ene, R = H)

629 (*ent*-11β,20-epoxy-15α-hydroxy-19-acetoxykaur-16 ene, R = OAc)

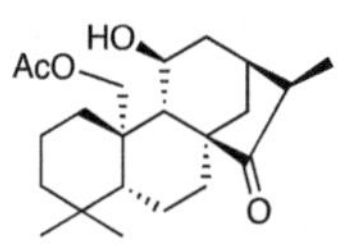

630 (*ent*-11α-hydroxy-20-acetoxy-(16*R*)-kauran-15-one)

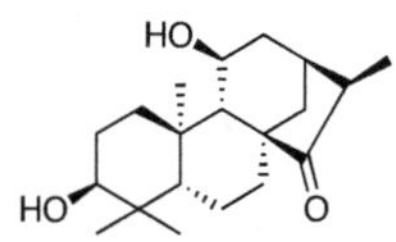

631 (*ent*-3α,11α-dihydroxy-(16*R*)-kauran-15-one)

The cytotoxic effects of the newly isolated compounds **609–611**, **614**, **615**, **620**, and **622** and the known *ent*-kaurenes **612**, **613**, and **550** were evaluated against the A549 lung, A2780 ovarian, HepG2 liver, 7860 renal human cancer cell lines, and also HBE (normal human bronchial epithelial) cells. Compounds **609–611** and **612**, **613**, and **550**, all possessing an exomethylene cyclopentane ring, showed potent activity against most of these cancer cell lines. Among the tested compounds, **550** displayed the most significant activity against each cancer cell line listed above (IC_{50} values, in turn, of 3.8, 0.9, 4.2, and 4.9 μ*M*). The cytotoxicity of **550** to several normal cells, such as HBE, HUVEC (human umbilical endothelial cells), and HL-7702 cells (normal human liver cell line), was significantly lower than that against the A549 cancer cell line [230].

ent-1β-Hydroxykauran-12-one (**632**), isolated from *Paraschistochila pinnatifolia*, and 1α-hydroxy-*ent*-sandaracopimara-8(14),15-diene (**633**), from *Trichocolea mollissima*, were only weakly active (IC_{50} values of 15 and > 25 μg/cm^3), when evaluated against P-388 cells [231].

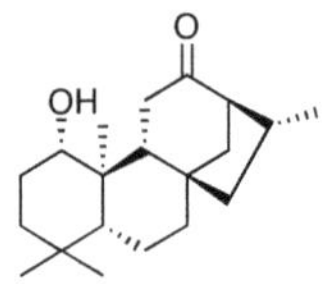

632 (*ent*-1β-hydroxy-kauran-12-one)

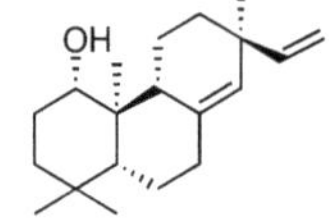

633 (1α-hydroxy-*ent*-sandara-copimara-8(14),15-diene)

An ethanol extract of *Lepidolaena taylorii*, which showed cytotoxicity against the P-388 cell line with an IC_{50} value of 1.3 μg/cm^3, was purified to give the *ent*-8,9-*seco*-kaurane diterpenoids, rabdoumbrosanin (**634**), 16,17-dihydrorabdoumbrosanin (**635**), and 8,14-epoxyrabdoumbrosanin (**636**), and their related compounds **637–640**, and the *ent*-kaur-16-en-15-ones **641–645**.

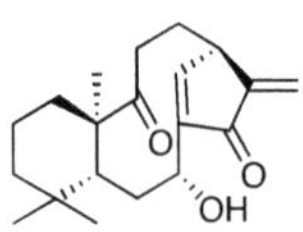

634 (rabdoumbrosanin)

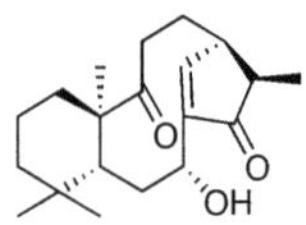

635 (16,17-dihydro-rabdoumbrosanin)

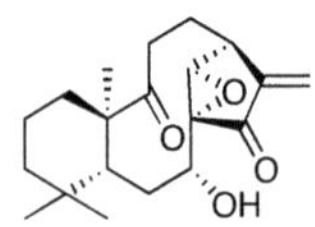

636 (8,14-epoxyrabdoumbrosanin)

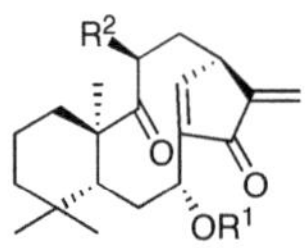

637 (7-acetoxyrabdoumbrosanin, R[1] = Ac, R[2] = H)
638 (11β-hydroxy-7-acetoxy-rabdoumbrosanin, R[1] = Ac, R[2] = OH)
639 (11β-acetoxyrabdoumbrosanin, R[1] = H, R[2] = OAc)
640 (11β-hydroxyrabdounbrosanin, R[1] = H, R[2] = OH)

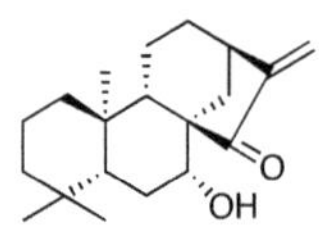

641 (*ent*-7β-hydroxy-16-kauren-15-one)

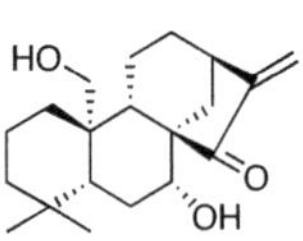

642 (*ent*-7β,20-dihydroxy-16-kauren-15-one)

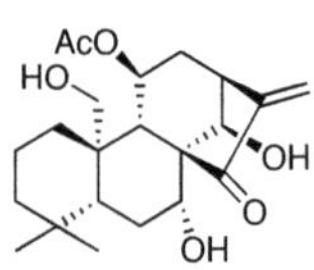

643 (*ent*-7β,15α,20-trihydroxy-11α-acetoxy-16-kauren-15-one)

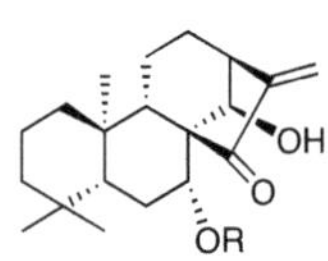

644 (*ent*-7β,14α-dihydroxy-16-kauren-15-one, R = H)
645 (*ent*-7β-acetoxy-14α-hydroxy-16-kauren-15-one, R = Ac)

Also, *L. palpebrifolia* elaborated the 8,9-*seco*-kauranes **634** and **636**. Their cytotoxicity was tested against the P-388 murine lymphocytic leukemia and several leukemia and solid tumor human cancer cell lines. Compounds **634** and **636** showed the most potent cytotoxic activities (mean IC_{50} values of 0.006 and 0.27 μg/cm^3; GI_{50} values of 0.1 and 1.2 μ*M*, respectively). Compound **635** also showed cytotoxicity against P-388 cells at 0.8 μ*M*. Compounds **634** (containing 10% of **635**) and **636** showed differential cytotoxicity in vitro when tested against five leukemia cell lines.

For compound **634**, an average IC_{50} value of 0.4 μM was obtained but, however, leukemia cell growth was not inhibited by **636** ($IC_{50} > 50\ \mu M$). The growth of seven colon cancer cell lines was also inhibited by **634** (mean IC_{50} value, 6 μM) [232].

Compounds **634** and **636** were tested in an in vivo hollow fiber model system, in which neither compound was active at the doses tested (18 and 12 mg/kg for **634**, and 150 and 100 mg/kg for **636**). Compound **640** showed the highest activity against several leukemia cell lines (mean GI_{50} 0.3 μM), and was least active against various central nervous system cancer cell lines (mean GI_{50} 6 μM) [232]. Among the isolated compounds, *ent*-8,9-*seco*-kaurenes **634**, **636**, and **640** showed selective activity. The mode of action for the cytotoxicity of the *ent*-8,9-*seco*-kaur-16-en-15-one and *ent*-kaur-16-en-15-one series was supported by a Michael addition of a thiol to the C-16–C-17 double bond of **634**, but the C-8–C-14 double bond of **635** was relatively unreactive [232, 233].

The *Porella* species are divided chemically into several chemotypes. A typical group is type I, belonging to the *Porella vernicosa* complex, which contains pungent drimane, non-pungent pinguisane-type sesquiterpenoids, and type III compounds (pinguisane-sacculatane) [1–3, 108]. The Japanese *Porella perrottetiana* and New Caledonian *P. viridissima* produced the sacculatane-type diterpene dialdehyde, perrottetianal A (**158**), which showed potent cytotoxic activity against the A2780 ovarian cancer cell line, with an IC_{50} value of 1.6 $\mu g/cm^3$) [234]. The latter species also contained santalane-type sesquiterpenoids, such as α-photosantalol (**646**), (*Z*)-*epi*-β-santalol (**647**), α- (**224**) and β-santalol (**225**), which are very rare in liverworts. *P. viridissima* is similar chemically to the Japanese *P. caespitans* var. *setigera* since the latter species also elaborates α-santalane-(12*R*),13-diol (**648**) [2].

In 2023, Chien et al., studied the Vietnamese *P. perrottetiana* and isolated a new sacculatane, perrottetianal E (**649**), along with the known perrottetianals A (**158**) and B (**650**), and a new oplopanone sesquiterpene, oplopanone C (**651**). Their cytotoxicity was tested aginst the KB, HepG2, and A549 cell lines, and gave IC_{50} values ranging from 6.6 to 103 μM [235].

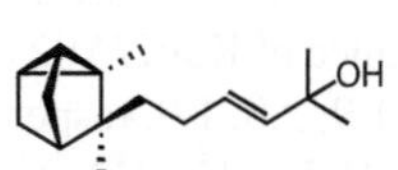

646 (α-photosantalol)

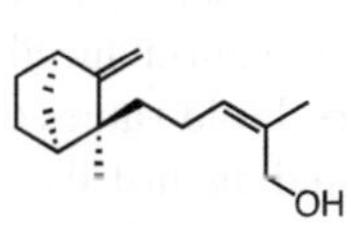

647 (*epi*-β-santalol)

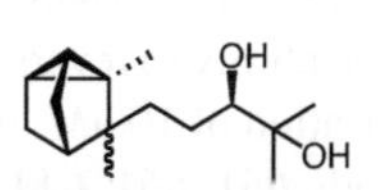

648 (α-santalene-(12*R*),13-diol)

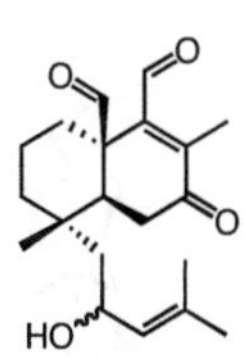

649 (perottetianal E = 7-oxoperrotetianal B)

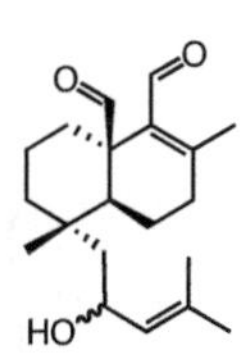

650 (perottetianal B)

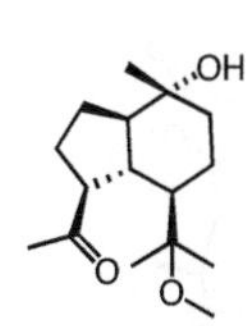

651 (oplopanone C)

The fractionation of the Chinese *Plagiochila nitens* led to the isolation of seven sacculatane diterpenoids that were named plagiochilarins A–G (**652–658**), along with a new *seco*-aromadendrane sesquiterpene ester, plagiochilarin H (**459**), together with anadensin (**499**), cinnamolide (**659**), and loliolide (**660**). The structures of the new compounds were determined by a combination of single-crystal X-ray diffraction and by analysis of their NMR and MS spectroscopic data. However, none of the isolated compounds showed significant cytotoxicity against the Hepa-1c1c7 mouse hepatoma cell line at the concentrations used (IC_{50} > 50 μM) [236].

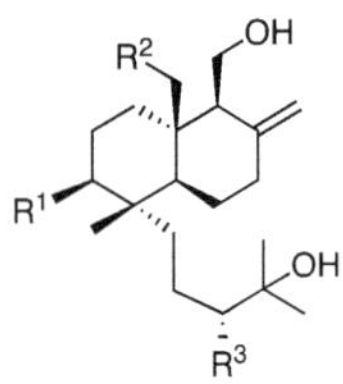

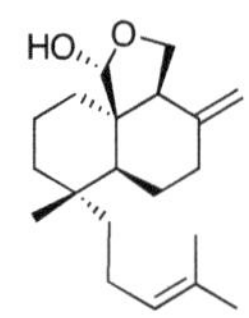

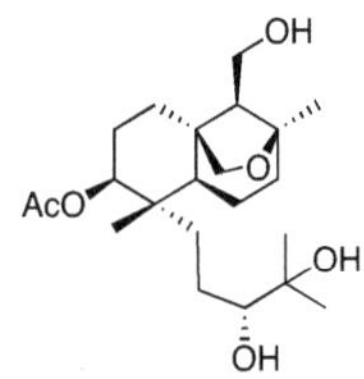

652 (plagiochilarin A, $R^1 = R^3$ = OAc, R^2 = OH)
653 (plagiochilarin B, R^1 = OAc, $R^2 = R^3$ = OH)
654 (plagiochilarin C, R^1 = H, R^2 = OH, R^3 = OAc)
655 (plagiochilarin D, R^1 = H, $R^2 = R^3$ = OH)
656 (plagiochilarin E, R^1 = H, R^2 = OAc, R^2 = OH)

657 (plagiochilarin F)

658 (plagiochilarin G)

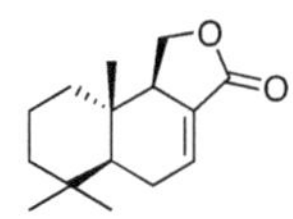

659 (cinnamolide)

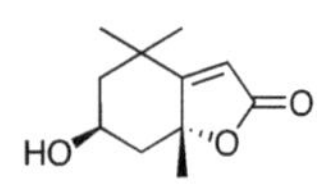

660 (loliolide)

Pallavicinia subciliata belonging to the Metzgeriales is a very distinct liverwort since it biosynthesizes stereochemically complex rearranged labdanes [237] for which their stereostructures were elucidated by the research term of the present author and by Chinese groups [238]. In 2012, Wang et al., [239] also isolated labdanes, comprised by the new pallambins A and B (**661–662)** and the previously known pallambins C and D (**663–664**) from *P. ambigua* (Plate 20), and showed that they possess the ability to reverse the adriamycin-induced resistance of K562/A02 cells at a concentration of 10 μM, with reversal fold values of 4.3, 1.9, 2.0, 1.9, respectively. Compounds **661–661–644** were tested against the A172, HeLa, HepG2, and U87 human cancer cell lines using a MTT assay. However, all compounds were deemed inactive (IC_{50} > 10 μM) against all four cell lines [239].

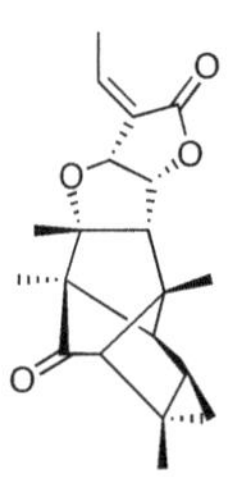

661 (pallambin A)

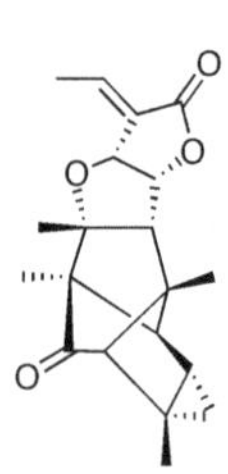

662 (pallambin B)

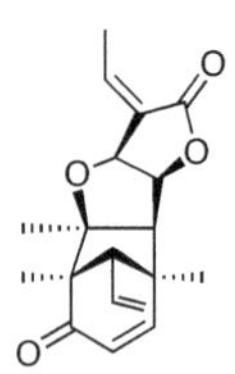

663 (pallambin C)

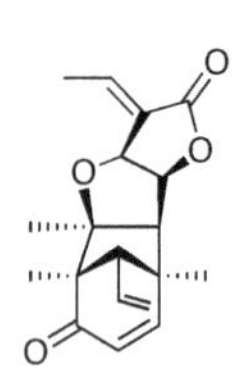

664 (pallambin D)

Scapania is a source of labdane diterpenoids that are the significant chemical markers of this genus. In 2015, Zhang et al., studied the phytochemical components of the Chinese *Scapania irrigua* and isolated 17 new labdanes **665–681** from the 95% ethanol extract, namely, scapairrin A ((5*R*,8*R*,9*R*,10*R*,12*S*,14*S*)-8,12-epoxy-5,14-dihydroxy-13(16)-labdene) (**665**), scapairrin B ((5*R*,8*R*,9*R*,10*R*,12*S*,14*R*)-8,12-epoxy-5,14-dihydroxy-13(16)-labdene) (**666**), scapairrin C ((5*R*,8*R*,9*R*,10*R*, 12*S*,14*S*)-8,12-epoxy-5,14-dihydroxy-13(16)-labden-1-one) (**667**), scapairrin D ((5*R*,8*R*,9*R*,10*R*,12*S*,14*R*)-8,12-epoxy-5,14-dihydroxy-13(16)-labden-1-one) (**668**), scapairrin E ((1*S*,5*R*,8*R*,9*R*,10*R*,13*S*,14*S*)-8,12-epoxy-1,5,14-trihydroxy-13(16)-labdene) (**669**), scapairrin F ((1*S*,5*R*,8*R*,9*R*,10*R*,13*S*,14*R*)-8,12-epoxy-1,5,14-trihydroxy-13(16)-labdene) (**670**), scapairrin G (8*α*,12*β*-epoxy-1*α*,5*α*-dihydroxy-13(16)-labden-14-one) (**671**), scapairrin H (8*α*,12*α*-epoxy-1*α*,5*α*-dihydroxy-13(16)-labden-14-one) (**672**), scapairrin I (8*α*,12*β*-epoxy-5*α*-hydroxy-13(16)-labden-14-one) (**673**), scapairrin J (8*α*,12*β*-epoxy-5*α*-hydroxy-13(16)-labdene-14-dione) (**674**), scapairrin K (8*α*,12*β*-epoxy-5*α*-hydroxy-(13*E*)-labdene) (**675**), scapairrin L (8*α*,12*β*-epoxy-5*α*-hydroxy-13(16)-labden-1-one) (**676**), scapairrin M ((1*S*,5*R*,8*R*,9*R*,10*S*,13*S*,14*R*) 13,14-epoxy-1,5,8-trihydroxylabdan-12-one) (**677**), scapairrin N ((1*S*,5*R*,8*R*,9*R*,10*S*,13*R*,14*S*)-13,14-epoxy-1,5,8-trihydroxy-labdan-12-one) (**678**), scapairrin O ((1*S*,5*R*,8*R*,9*R*,10*R*,13*R*,14*R*)-13,14-epoxy-5,8-dihydroxylabdan-12-one) (**679**), scapairrin P ((1*S*,5*R*,8*R*,9*R*,10*R*,13*R*,14*S*)-13,14-epoxy-5,8-dihydroxy-labdan-12-one) (**680**), and scapairrin Q (13-*nor*-5*α*- hydroxy-9(11)-labden-8,12-olide) (**681**), including six pairs of diastereoisomers. Also obtained were the three known analogues, 5*α*,8*α*,9*α*-trihydroxy-(13*E*)-labda-12-one (**682**),1*α*,5*α*,8*α*-trihydroxy-(13*E*)-labda-12-one (**683**), and 5*α*,8*α*-dihydroxy-(13*E*)-labda-12-one (**684**) [1, 2]. The structures of **665** and **666** were elucidated by X-ray crystallographic analysis with CuK_{α} irradiation. All compounds were evaluated for their cytotoxicity against five cancer cell lines and the non-cancerous HUVEC line. Among them, scapairrins G–J (**671–674**) gave IC_{50} values in the range 4.1–8.7 μ*M* for the A2780, HeLa, HT-29, and MDA-MB231 cell lines, but were inactive (IC_{50} > 10 μ*M*) against A549 cancer cells and the HUVEC line The remaining compounds were inactive. A conjugated double bond functionality in the labdane molecule is necessary for the exhibition of cancer cell line antiproliferative activity [240].

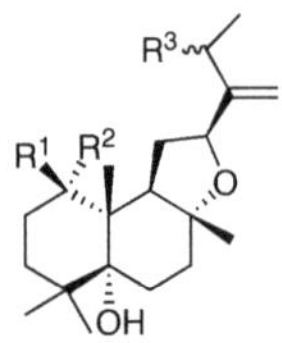

665 (scapairrin A, $R^1 = H$, $R^2 = H$, $R^3 = \beta$-OH)
666 (scapairrin B, $R^1 = H$, $R^2 = H$, $R^3 = \alpha$-OH)
667 (scapairrin C, $R^1 = R^2 = O$, $R^3 = \beta$-OH)
668 (scapairrin D, $R^1 = R^2 = O$, $R^3 = \alpha$-OH)
669 (scapairrin E, $R^1 = H$, $R^2 = \alpha$-OH, $R^3 = \beta$-OH)
670 (scapairrin F, $R^1 = R^2 = R^3 = \alpha$-OH)

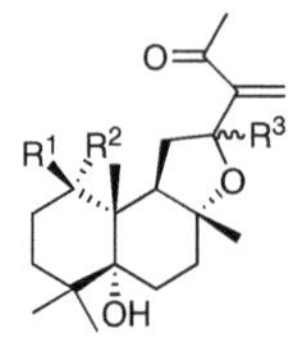

671 (scapairrin G, $R^1 = H$, $R^2 = \alpha$-OH, $R^3 = \alpha$-H)
672 (scapairrin H, $R^1 = H$, $R^2 = \alpha$-OH, $R^3 = \beta$-H)
673 (scapairrin I, $R^1 = R^2 = O$, $R^3 = \alpha$-H)
674 (scapairrin J, $R^1 = R^2 = H$, $R^3 = \alpha$-H)

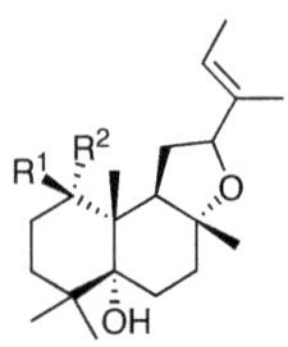

675 (scapairrin K, $R^1 = R^2 = H$)
676 (scapairrin L, $R^1 = R^2 = O$)

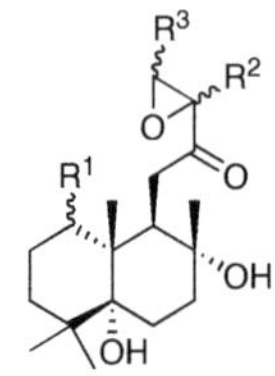

677 (scapairrin M, $R^1 = \alpha$-OH, $R^2 = R^3 = \alpha$-Me)
678 (scapairrin N, $R^1 = \alpha$-OH, $R^2 = R^3 = \beta$-Me)
679 (scapairrin O, $R^1 = H$, $R^2 = R^3 = \beta$-Me)
680 (scapairrin P, $R^1 = H$, $R^2 = R^3 = \alpha$-Me)

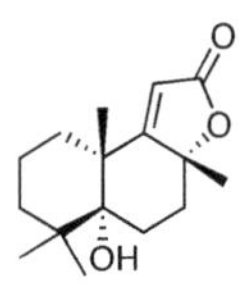

681 (scapairrin Q)

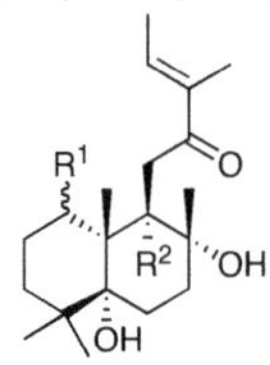

682 ($5\alpha,8\alpha,9\alpha$-trihydroxy-(13*E*)-labda-12-one, $R^1 = H$, $R^2 = \alpha$-OH)
683 ($1\alpha,5\alpha,8\alpha$-trihydroxy-(13*E*)-labda-12-one, $R^1 = \alpha$-OH, $R^2 = H$)
684 ($5\alpha,8\alpha$-dihydroxy-(13*E*)-labda-12-one, $R^1 = R^2 = H$)

Scapania undulata contains labdane monomers and dimers, of which some have a potently bitter taste [3]. The Japanese *S. undulata* produces two unusual labdane dimers, scapaundulins A (**685**) and B (**686**), together with three labdane monomers [3]. This same Chinese species was studied by Kang et al., in 2015, who isolated a new labdane monomer named scapanin C (**687**) (= scapaundulin C), together with the four known labdanes, scapaundulin A (**685**), $5\alpha,8\alpha,9\alpha$-trihydroxy-(13*E*)-labden-12-one (**688**), $5\alpha,8\alpha$-dihydroxy-(13*E*)-labden-12-one (**689**), and (13*S*)-15-hydroxylabd-8(17)-en-19-oic acid (**690**). Compounds **688** and **689** showed weak cytotoxicity against A2780 human ovarian carcinoma cells with IC_{50} values of 19.5 and 16.9 μM, respectively [241].

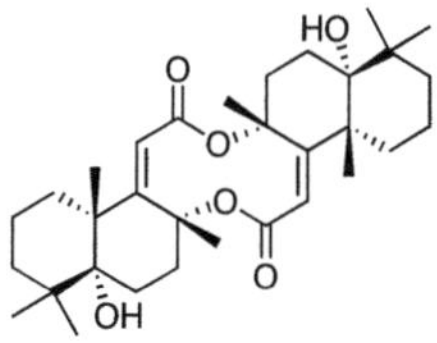

685 (scapaundulin A)

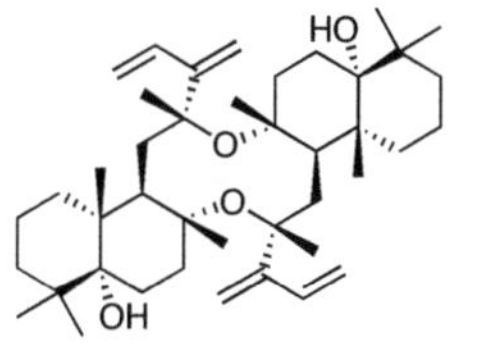

686 (scapaundulin B)

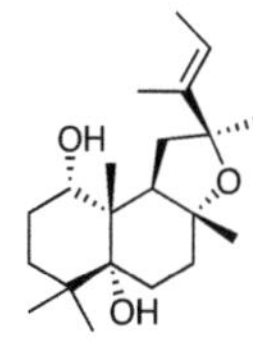

687 (scapaundulin C)

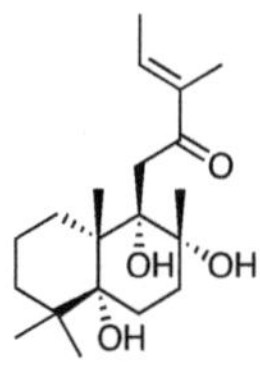

688 (5α,8α,9α-trihydroxy-(13*E*)-labden-12-one)

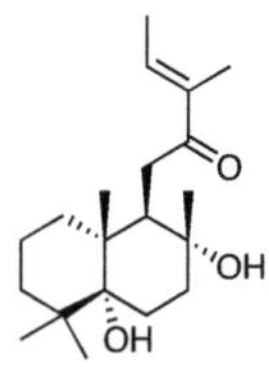

689 (5α,8α-dihydroxy-(13*E*)-labden-12-one)

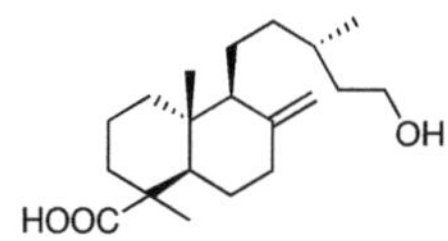

690 ((13*S*)-15-hydroxy-labd-8(17)-en-19-oic acid)

Five caged *cis*-clerodane diterpenoids named scaparvins A–E (**691**–**695**) were isolated from the Chinese *Scapania parva*, along with two known related clerodanes, parvitexins B (**696**) and C (**697**) [242]. Their absolute stereostructures were elucidated by a combination of 2D-NMR spectroscopic and ECD data interpretation. However, when evaluated for their cytotoxic activity against four human cancer cell lines (HeLa, KB, MCF-7, and PC-3) all these compounds were regarded as inactive ($IC_{50} > 10\ \mu M$).

Frullania muscicola belongs to chemotype II of the *Frullania* genus. It afforded a new diterpene named muscicolin (**698**), which exhibited weak cytotoxicity against BEL-7402, HT-29, KB, and PG cells (IC_{50} 20–50 μg/cm^3) [243].

Two new clerodane diterpenoids, *cis*-cleroda-3,13-dien-16,18-dihydroxy-15-oic acid-15,16-olide (**699**), and 15,16-epoxy-*cis*-cleroda-3,13(16),14-trien-18-ol (**700**), and three known analogues, 15,16-epoxy-*cis*-cleroda-3,13(16),14-trien-18-al (**701**), (–)-5-*epi*-hardwickiic acid (**702**), and *cis*-cleroda-3,13-diene-16-hydroxy-15,18-dioic acid-15,16-olide (**703**) were isolated from *Gottschelia schizopleura*, belonging to the Jungermanniaceae. Their cytotoxicity was examined using the Hep-G2, A-549, HEP-G2, LoVo, and MDA-MB-435 human cancer cell lines. Only compound **699** showed cytotoxicity against LoVo and MDA-MB-435 cells, with IC_{50} values of 7.9 and 8.3 μg/cm^3, respectively. It is noteworthy that compound **702** possessing a furan ring was inactive against all four cancer cell lines ($IC_{50} > 10$ μg/cm^3), while its *trans*-isomer, hardwickiic acid, has been obtained earlier from a higher plant source and found to possess potent activity against MDA-MB-435 cells ($IC_{50} < 1$ μg/cm^3) [244].

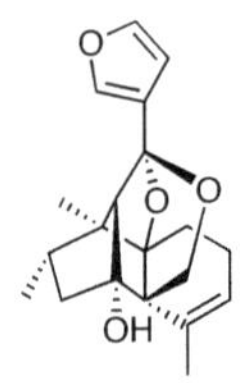

691 (scaparvin A)

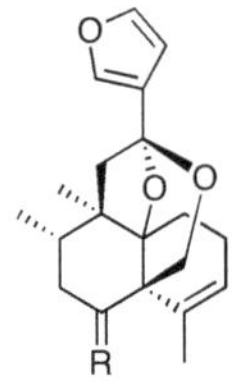

692 (scaparvin B, R = O)
693 (scaparvin C, R = α-OH, β-H)
694 (scaparvin D, R = α-OAc, β-H)
696 (parvitexin B, R = α-H, β-OAc)
697 (parvitexin C, R = α-H, β-OH)

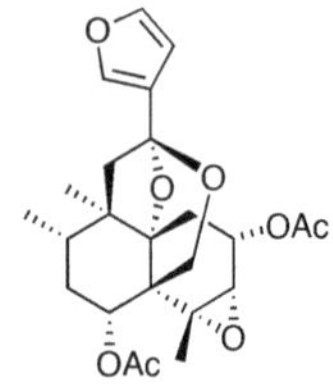

695 (scaparvin E = scaparin C 2-acetate)

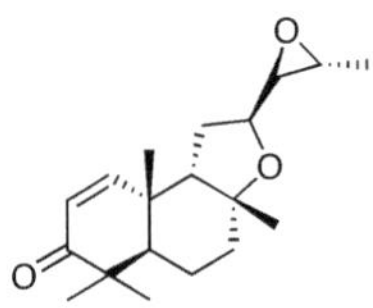

698 (muscicolin)

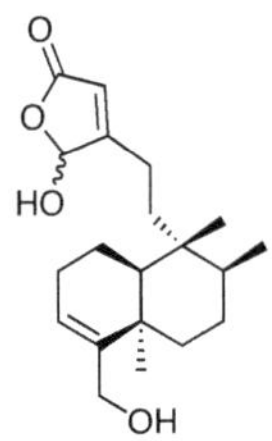

699 (*cis*-cleroda-3,13-dien-16,18-dihydroxy-15-oic acid-15,16-olide)

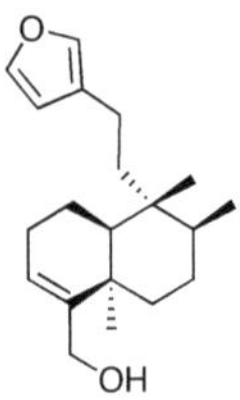

700 (15,16-epoxy-*cis*-cleroda-3,13(16),14-trien-18-ol)

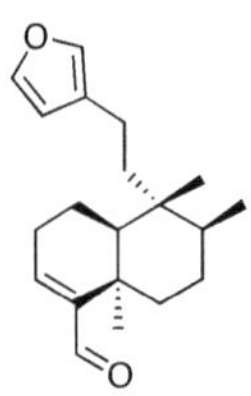

701 (15,16-epoxy-*cis*-cleroda-3,13(16),14-trien-18-al)

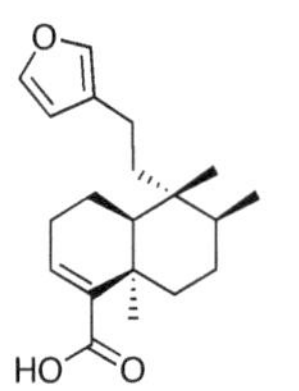

702 ((–)-5-*epi*-hardwickic acid)

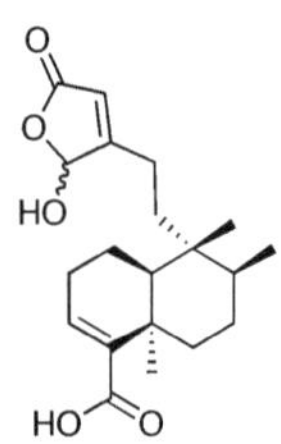

703 ((–)-*cis*-cleroda-3,13-diene-16-hydroxy-15,18-dioic acid-15,16-olide)

In 2018, Ng et al., purified a methanol extract of the West Malaysian *Gottschelia schizopleura* to give two new *cis*-clerodane diterpenoids, schizopleurolides A (**704**) and B (**705**), which were obtained along with four known analogues, 15,16-epoxy-*cis*-cleroda-3,13(16),14-trien-18-ol (**700**), 15,16-epoxy-*cis*-cleroda-3,13(16),14-trien-18-al (**701**), 5-*epi*-hardwickiic acid (**702**), and (13*E*)-*cis*-*ent*-cleroda-3,13-dien-15-ol (**706**). Their cytotoxic activity against HL-60 (human promyelocytic leukemia) and B16-F10 (murine melanoma) cells was investigated. Compounds **704** and **706** showed weak cytotoxicity against HL-60 and B16-F10 cells, with IC_{50} values of 38.5 and 47.3, and 36.1 and 44.3 μM, respectively [245].

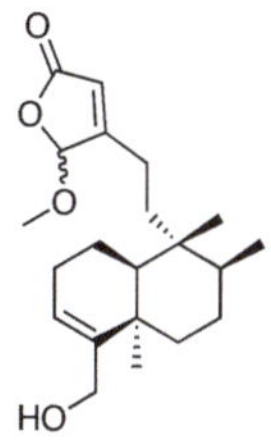

704 (schizopleurolide A)

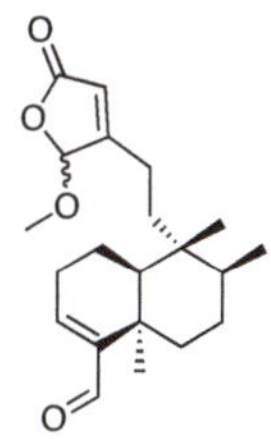

705 (schizopleurolide B)

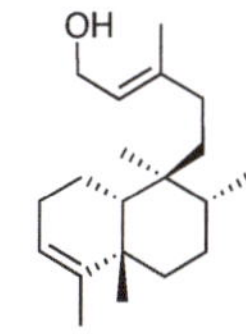

706 ((13*E*)-*cis*-*ent*-cleroda-3,13-dien-15-ol)

Notoscyphus lutescens (Plate 22) is a known source of dolabrane-type diterpenoids. In 2014, Wang et al., described the isolation of notolutesins A–J (**707**–**716**), together with *ent-trans*-coumminol and the corresponding acid, *ent-trans*-coummunic acid (**717**), 15,16-bis-*nor*-13-oxo-8(17),(11*E*)-labdadiene-18-ol (**718**), and (–)-pimara-9(11),15-dien-19-ol (**719**), with the latter compound isolated earlier from the Malaysian liverwort *Jungermannia truncata* [3]. Among the isolated dolabranes, notolutesin A (**707**) showed cytotoxic activity against PC3 human prostate cancer cells (IC_{50} 6.2 μ*M*), although it was found to be inactive against the A549, DU145, and H1688 cell lines (IC_{50} > 10 μ*M*) [246].

Plate 22 *Notoscyphus lutescens* (liverwort)

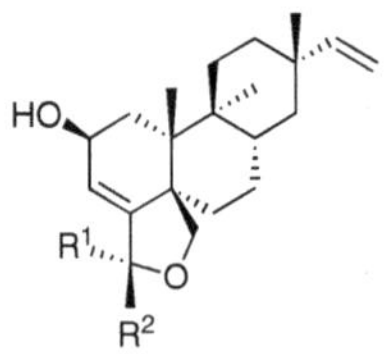

707 (notolutesin A, $R^1 = R^2 = O$)
708 (notolutesin B, $R^1 = \alpha$-H, $R^2 = \beta$-OMe)
709 (notolutesin C, $R^1 = \alpha$-H, $R^2 = \beta$-OEt)

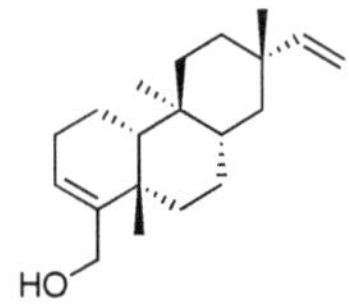

710 (notolutesin D)

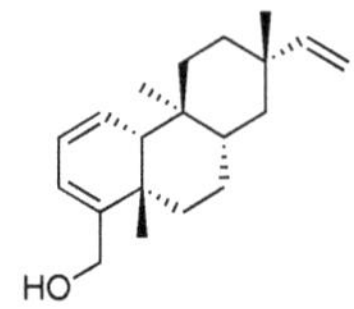

711 (notolutesin E)

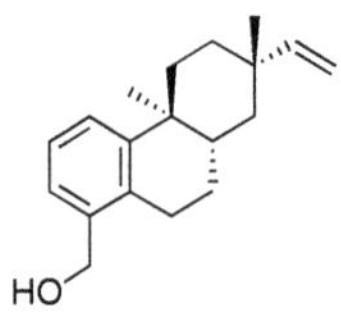

712 (notolutesin F)

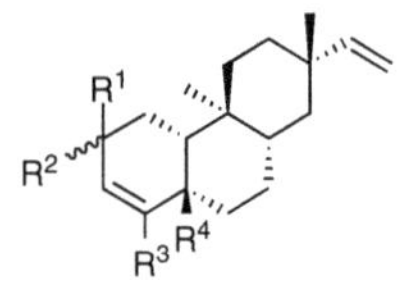

713 (notolutesin G, $R^1 = \alpha$H, $R^2 = \beta$-OH, $R^3 = R^4 = CH_2OH$)
714 (notolutesin H, $R^1 = \alpha$H, $R^2 = \beta$-OH, R^3 = CHO, R^4 = Me)
715 (notolutesin I, $R^1 = R^2 = O$, $R^3 = CH_2OH$, R^4 = Me)
716 (notolutesin J, $R^1 = R^2 = O$, $R^3 = R^4$ = Me)

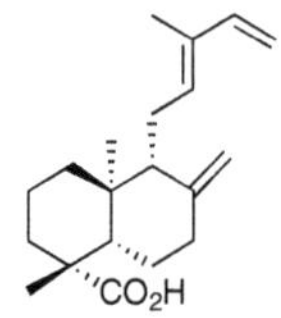

717 (*ent-trans*-communic acid)

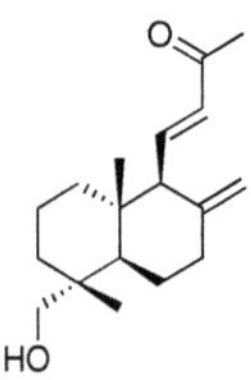

718 (15,16-bis-*nor*-13-oxo-8(17),(11*E*)-labdadien-18-ol)

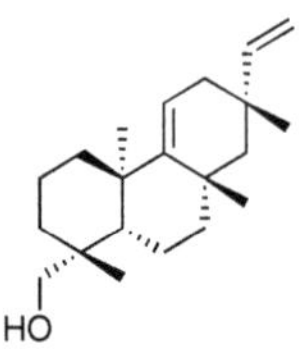

719 ((–)-pimara-9(11),15-dien-19-ol)

Further phytochemical investigation of the Chinese *Notoscyphus collenchymatosus* resulted in the isolation of six new dolabranes, notolutesins K–P (**720**–**725**), together with the previously known dolabranes, notolutesins A (**707**), C (**709**), E (**711**), H (**714**), and J (**716**), and the pimarane diterpenoid, (–)-pimara-9(11),15-dien-19-ol (**719**). Compounds **725** and **708** might be artefacts resulting from the extraction and isolation procedures since ethanol and methanol were used as solvents. Notolutesin P (**725**) has an oxygen bridge between C-2 and C-19, which is the first such example among the dolabrane diterpenoids. All isolated compounds were evaluated for their cytotoxicity against the MCF-7 and HCC-1428 human breast cancer and HT-29 human colorectal carcinoma cells. Notolutesin P (**725**) was the most potent component among all the compounds isolated, with IC_{50} values, in turn of 5.2 (MCF-7), 4.8 (HCC-1428), and 3.5 (HT-29) μ*M*, respectively [247].

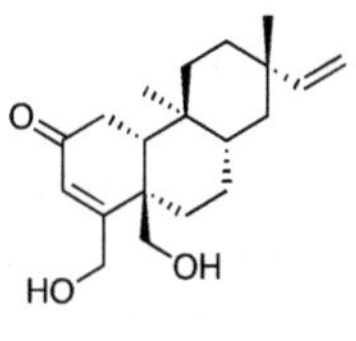

720 (notolutesin K)

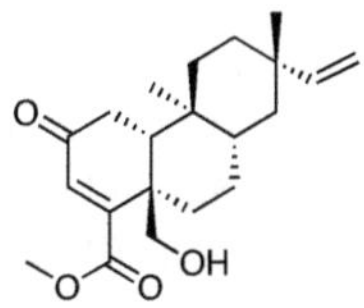

721 (notolutesin L)

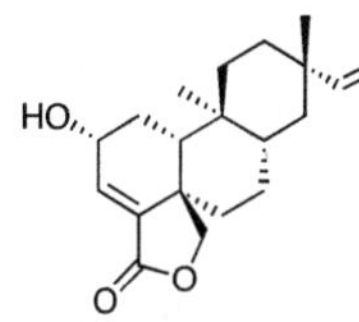

722 (notolutesin M)

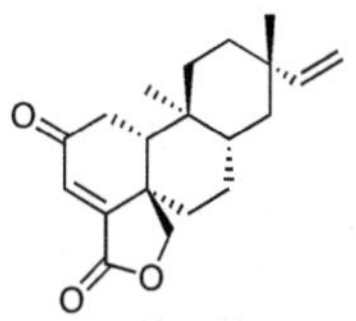

723 (notolutesin N)

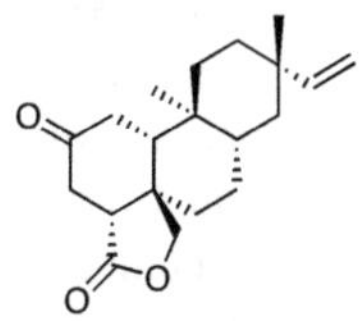

724 (notolutesin O)

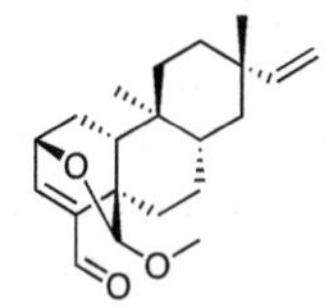

725 (notolutesin P)

The Japanese *Plagiochila sciophila* elaborates sweet-mossy odorous humulane-type sesquiterpenes, as described earlier. This liverwort produces not only these rare sesquiterpenes but also bis-bibenzyls and various fusicoccane diterpenoids [2, 3]. Kenmoku et al., isolated in 2014 seven fusicoccanes named fusicosciophins A–E (**726**–**730**) and 8- (**731**) and 9-deacetylfusicosciophin E (**732**) from *P. sciophila*. These compounds showed seed dormancy breaking activity (see later in Sect. 4.3.4) as well as cytotoxicity against HL-60 and KB cells, respectively. Compounds **727** and **729** exhibited moderate differentiation-inducing activity (ED_{50} 31.2 and 59.1 μM, respectively) against HL-60 cells, while the same two compounds each showed a higher selectivity index (IC_{50}/EC_{50}, > 3.4 and 6.4, respectively) than fusicoccin A (**726**). All fusicoccanes were determinded as being inactive against KB cells [248].

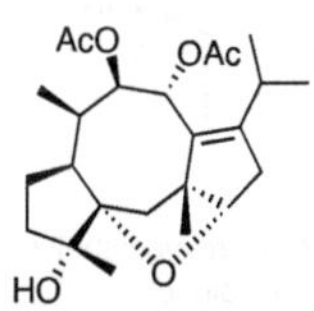

726 (fusicosciophin A)

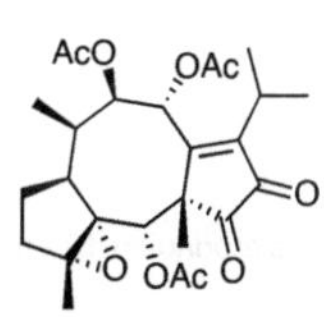

727 (fusicosciophin B)

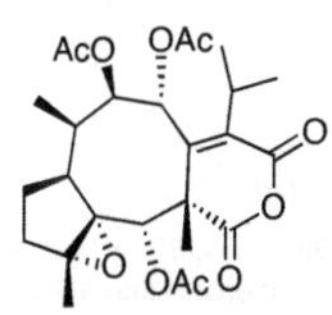

728 (fusicosciophin C)

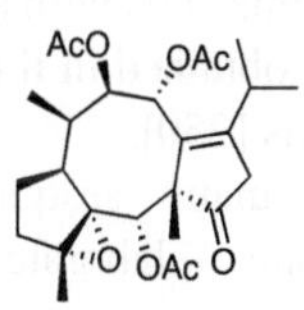

729 (fusicosciophin D)

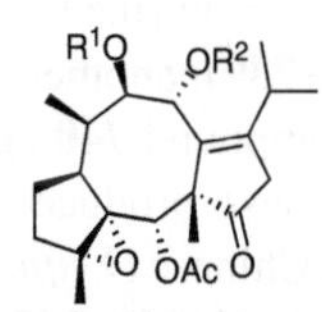

730 (fusicosciophin E, $R^1 = R^2 = H$)
731 (8-deacetylfusicosciophin E, $R^1 = H$, $R^2 = Ac$)
732 (9-deacetylfusicosciophin E, $R^1 = Ac$, $R^2 = H$)

Clerodane diterpenoids are distributed in many liverworts [2, 3]. However, cancer cell line cytotoxicity studies on these compounds are limited.

Liverwort species in the genus *Schistochila* are distributed in Taiwan and in the southern hemisphere, especially in New Zealand, and their chemical diversity is very high since they biosynthesize not only alkylphenols but also diterpenoids and bis-bibenzyls [3]. Ng et al. reported in 2016 the isolation of a new 5,10-*seco*-clerodane named schistochilic acid D (**733**) and three known clerodane diterpenoids, (–)-3,13(16),14-*cis*-clerodadien-18-oic acid (**735**), (–)-3,(12*E*),14-*cis*-clerodadien-18-oic acid (**736**), and *cis*-3,14-clerodadien-13-ol (**737**) and a known dolabellane diterpene alcohol, 12*β*-hydroxydolabella-(3*E*,7*E*)-diene (**738**), from the Bornean *Schistochila acuminata.* The *cis*-clerodanes **735** and **736** have already been isolated from this same Taiwanese species together with six analogues and compound **737** from the Japanese liverwort, *Jungermannia infusca* [2]. Compound **735** was also found in *Schitochila aligera* collected in Borneo [2]. Schistochilic acid A (**734**), a 5,10-*seco*-clerodane, for which the structure is very similar to that of **733**, was isolated from the New Zealand *S. nobilis* [2]. Compounds **737** and **738** showed weak cytotoxic effects (IC_{50} values of 40 and 62 μg/cm^3) against the B16-F10 cell line [249].

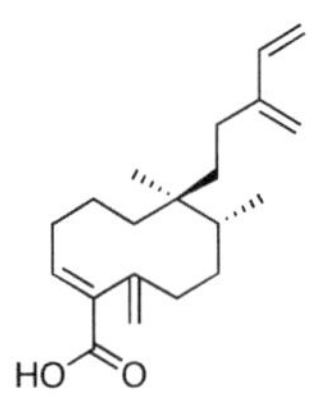

733 (schistochilic acid D)

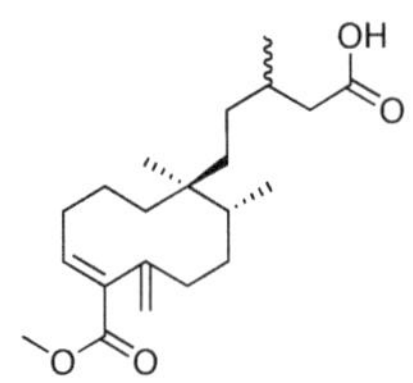

734 (schistochilic acid A)

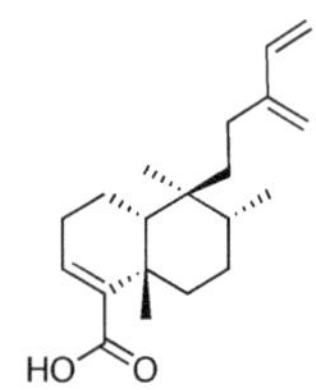

735 ((–)-3,13(16),14-*cis*-clerodadien-18-oic acid)

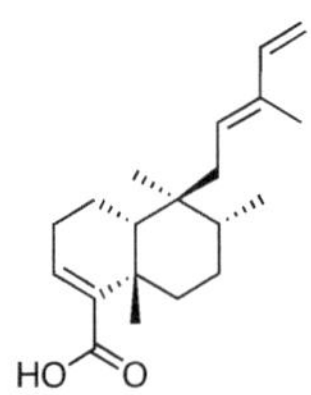

736 ((–)-3,(12*E*),14-*cis*-clerodadien-18-oic acid)

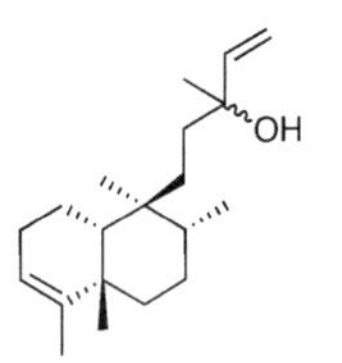

737 (*cis*-3,14-clerodadien-13-ol)

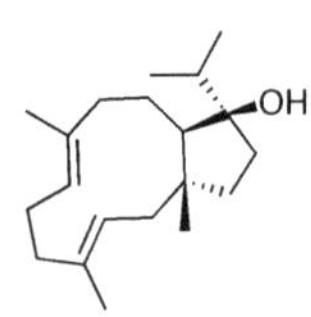

738 (12*β*-hydroxydolabella-(3*E*,7*E*)-diene)

A new atisane diterpene, atisane-2 (**740**) from *Lepidolaena clavigera* showed cytotoxicity against P-388 murine lymphocytic leukemia cells with an IC_{50} value of 16 μg/cm^3. Atisane-1 (**739**), another new compound obtained in this investigation, was less active than compound **740** against P-388 cells [250].

Diplophyllum species produce pungent eudesmane sesquiterpenoids, as mentioned earlier. The Chinese *D. apiculatum* was found to elaborate three new diterpenoids, diplapiculins A–C (**741**–**743**). Their absolute structures were elucidated by a combination of ECD, X-ray crystallographic, and 2D NMR spectroscopic analyses. In addition, drimenol (**744**), albicanol (**745**), chiloscyphenol A (**402**), fusicoauritone (**490**), and methyl 2,6-dihydroxy-3,4-dimethylbenzoate (**746**) were obtained. Their cytotoxic effects was evaluated against three human cancer cell lines (A549, DU145, and PC-3), but no significant activity was observed in a MTT assay using adriamycin as a positive control [251].

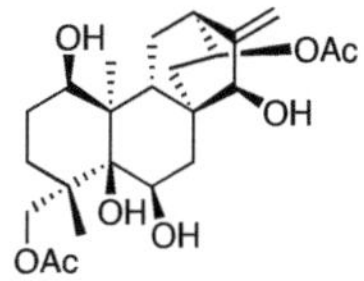

739 (atisane-1)

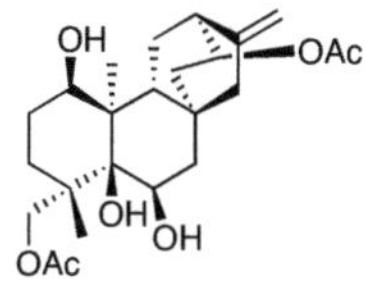

740 (atisane-2)

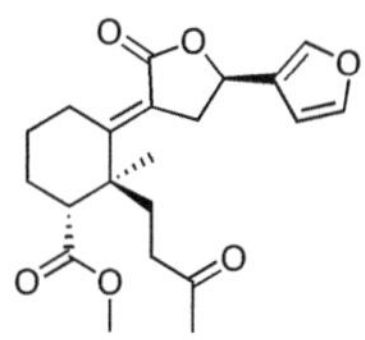

741 (diplapiculin A)

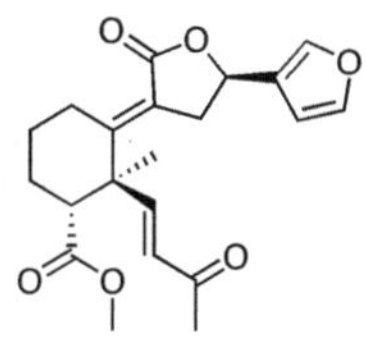

742 (diplapiculin B)

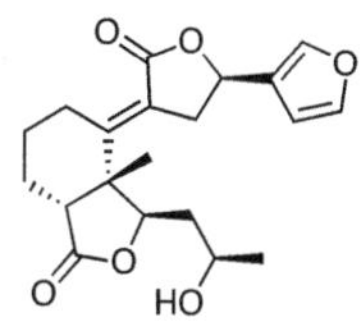

743 (diplapiculin C)

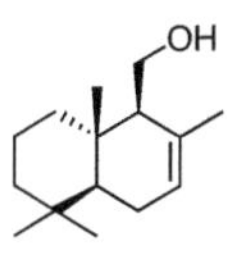

744 (drimenol)

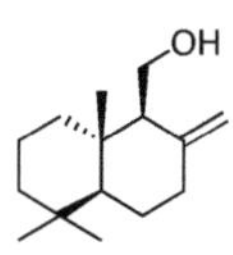

745 (albicanol)

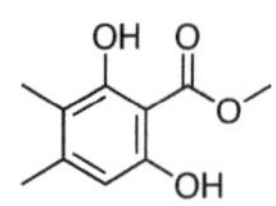

746 (methyl 2,6-dihydroxy-3,4-dimethylbenzoate)

4.4.5 Triterpenoids

The occurrence of triterpenoids in bryophytes is rare. The Chinese *Lepidozia reptans* produces nine 9,10-cycloartanes, lepidozin A (3*β*,10*β*,16*α*-trihydroxy-9,10-*seco*-9,19-cyclolanost-9(11),24(25)-diene) (**747**), lepidozin B ((23*R*)-16*α*,23-epoxy-3*β*,10*β*-dihydroxy-9,10-*seco*-9,19-cyclolanost-9(11),24(25)-diene) (**748**), lepidozin C ((23*S*)-16*α*,23-epoxy-3*β*,10*β*-dihydroxy-9,10-*seco*-9,19-cyclolanost-9(11),24(25)-diene) (**749**), lepidozin D (3*β*,10*β*,16*α*,25-tetrahydroxy-9,10-*seco*-9,19-cyclolanost-9(11),23(24)-diene) (**750**), lepidozin E (3*β*,10*β*,16*α*-trihydroxy-25-methoxy-9,10-*seco*-9,19-cyclolanost-9(11),23(24)-diene) (**751**), lepidozin F (3*β*,10*β*,16*α*-trihydroxy-24-methyl-9,10-*seco*-9,19-cyclolanost-9(11),23(24′)-diene) (**752**), lepidozin G (3*β*,10*β*,16*α*-trihydroxy-24-oxo-9,10-*seco*-9,19-cyclolanost-9(11),25(26)-diene) (**753**), lepidozin H ((24*R*)-3*β*,10*β*,16*α*,24-pentahydroxy-9,10-*seco*-9,19-cyclolanost-9(11),25(26)-diene) (**754**), lepidozin I ((24*S*)-3*β*,10*β*,16*α*,24-pentahydroxy-9,10-*seco*-9,19-cyclolanost-9(11),25(26)-diene) (**755**), and lepidozin J ((3*R*,5*R*,8*S*,9*S*,10*R*,13*R*,13*S*,17*R*,20*R*,24*R*)-16,24-epoxy-3*β*,7*β*,25-trihydroxycyclolanostane) (**756**). Also obtained were a new sesquiterpene, lepidoza-2,7-dione (**757**), and four known compounds, 6*α*-hydroxyconfertifolin (**758**), loliolide (**660**), 4*α*,7*β*-dihydroxy-eudesmane (**759**), and *ent*-7*α*-hydroxy-2-oxo-eudesma-(3*Z*)-ene (**760**). The stereostructures of these new derivatives were elucidated using a combination of spectroscopic data interpretation and X-ray crystallographic analysis. Among the isolated products, lepidozin A (**747**), F (**752**), and G (**753**) possessed cytotoxic activity against PC-3, A549, H466, and H3255 cells, with IC_{50} values ranging from 4.2 to 9.6 μ*M* [252].

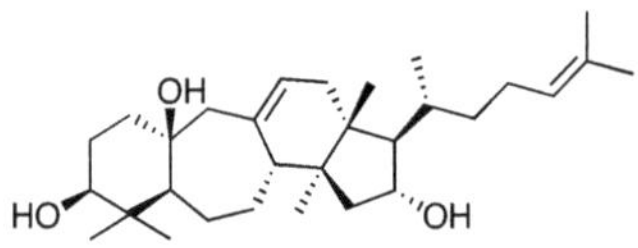

747 (lepidozin A)

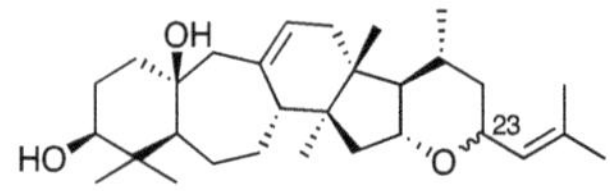

748 (lepidozin B (23**R**))
749 (lepidozin C (23**S**))

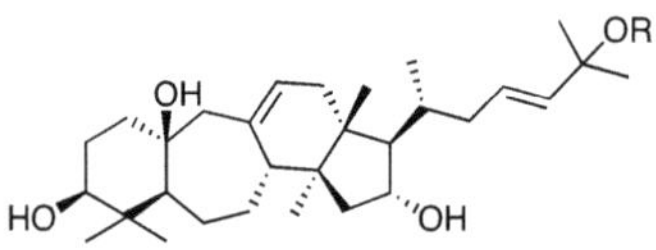

750 (lepidozin D, R = H)
751 (lepidozin E, R = Me)

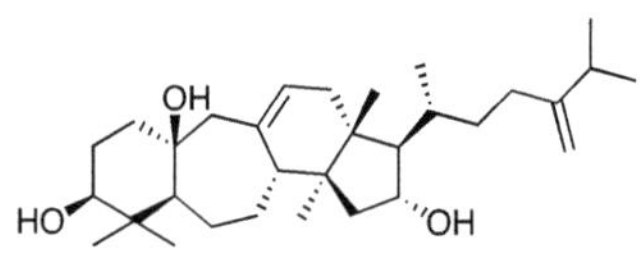

752 (lepidozin F)

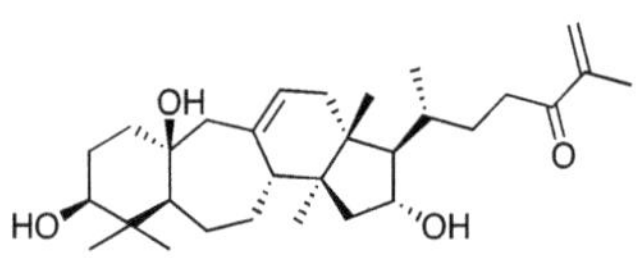

753 (lepidozin G)

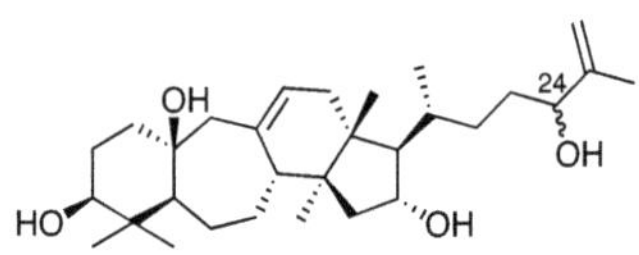

754 (lepidozin H (24**S**))
755 (lepidozin I (24**R**))

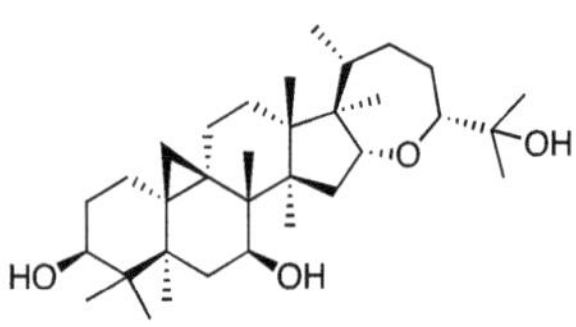

756 (lepidozin J)

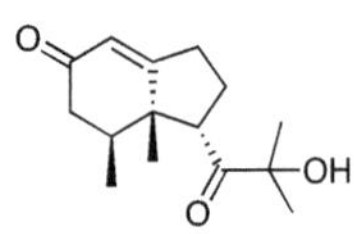

757 (lepidoza-2,7-dione)

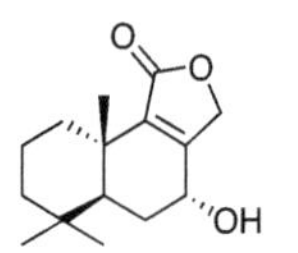

758 (6α-hydroxyconfertifoline)

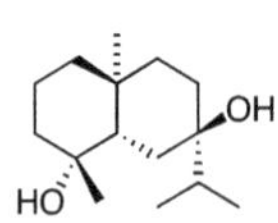

759 (4α,7β-dihydroxy-eudesmane)

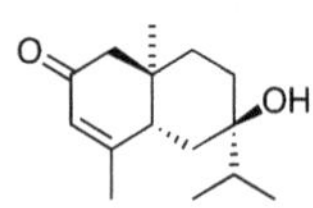

760 (*ent*-7α-hydroxy-2-oxo-eudesma-(3*Z*)-ene)

A methanol extract of the Bornean *Lepidozia borneensis* showed antiproliferative activity against several human cancer cells, although the triterpenoids suspected to be active were not isolated [253].

α-Zeorin (**761**) has been isolated from several liverworts, such as *Reboulia hemisphaerica* and *Targionia* species, and displayed cytotoxic activity against P-388 cells with an IC_{50} value of 1.1 μg/cm^3 [254, 255]. Three ursane triterpenoids from the liverwort *Ptilidium pulcherrimum*, namely, ursolic acid (**762**), acetylursolic acid (**763**) and 2α,3β-dihydroxyurs-12-en-28-oic acid (**764**), showed inhibition of the growth of PC3 human prostate cancer cells, with IC_{50} values obtained between 10.1 and 39.7 μ*M* [256].

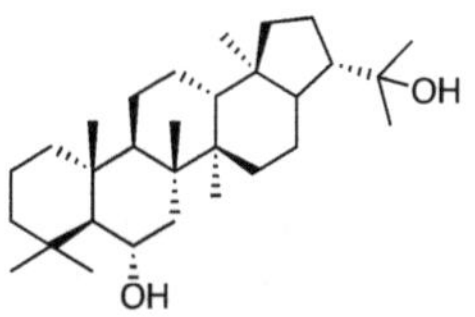

761 (α-zeorin)

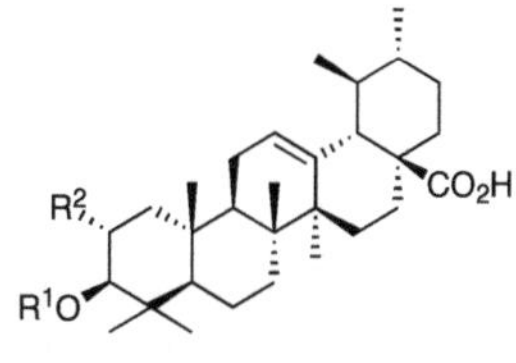

762 (ursolic acid, R¹ = R² = H)
763 (acetylursolic acid, R¹ = Ac, R² = H)
764 (2α,3β-dihydroxyurs-12-en-28-oic acid) R¹ = H , R² = OH

4.4.6 Monoterpenoids

2α,5β-Dihydroxybornane-2-cinnamate (**765**) from *Conocephalum conicum* exhibited weak cytotoxic activity against human HepG2 cells, with an IC_{50} value of 4.5 μg/cm^3 [176]. Some monoterpenoids found in liverworts, such as (+)-bornyl acetate (**56**), demonstrated potent apoptosis-inducing activities in cultured cells of *Marchantia polymorpha.* Apoptosis induced by monoterpenoids occurs through the production of active oxygen species, such as H_2O_2 [257].

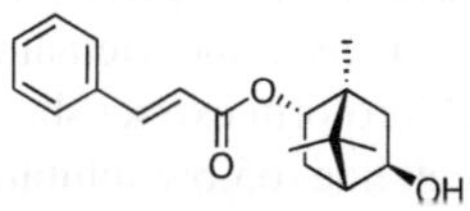

765 (2α,5β-dihydroxybornane-2-cinnamate)

4.4.7 Acetophenones

Acetophenones are rare phytochemicals in bryophytes. 2-Hydroxy-3,4,6-trimethoxyacetophenone (**766**) and 2-hydroxy-4,6-dimethoxyacetophenone (**767**) from *Plagiochila fasciculata,* were inactive against P-388 murine lymphocytic leukemia cells with IC_{50} values of > 50 μg/cm^3 [258].

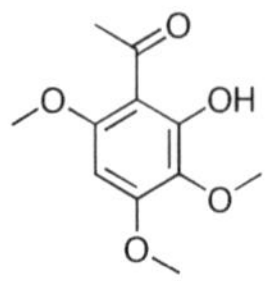

766 (2-hydroxy-3,4,6-trimethoxy-acetophenone)

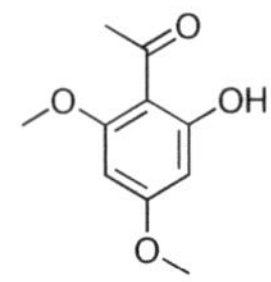

767 (2-hydroxy-4,6-dimethoxy-acetophenone)

4.4.8 Benzoates

Trichocolea lanata and *T. tomentella* produce tomentellin (**768**), which showed inhibitory activity against African green monkey kidney epithelial (BSC-1) cells at 15 μg/cm^3, with no antiviral effects against herpes simplex or polio viruses. Demethyl-tomentellin (**769**) from *T. tomentella* showed a similar cell growth inhibitory effect, indicating that both an allylic ether and a conjugated enone substructure are required for such activity [259]. Methyl 4-((2*E*)-3,7-dimethyl-2,6-octadienyl)oxy-3-hydroxybenzoate (**770**), isolated from *T. hatcheri*, was inactive (IC_{50} > 100 μ*M*) against both KB and SK-MEL-3 human cancer cells, as well as NIH 3T3 fibroblasts [260].

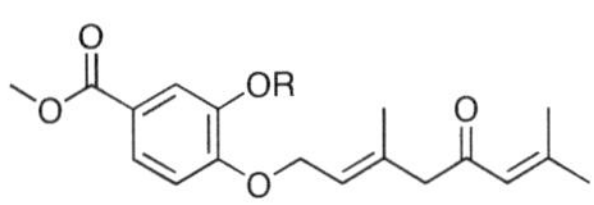

768 (tomentellin, R = Me)
769 (demethyltomentellin, R = H)

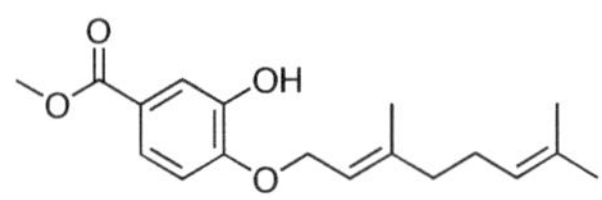

770 (methyl 4-((2*E*)-3,7-dimethyl-2,6-octadienyl)oxy-3-hydroxy benzoate)

In 2018, Tan et al., investigated the biological effects of the *n*-hexane, chloroform, ethyl acetate, and ethanol extracts of the Vietnamese *Marchantia polymorpha*. Among these four extracts, the chloroform extract showed the the best antiproliferative activity against the MCF-7 cell line (65.6% inhibition at the concentration used) [261].

The Vietnamese leafy liverwort, *Denotarisia linguifolia*, belonging to the Jamesoniellaceae, was extracted with *n*-hexane, chloroform, ethyl acetate, and ethanol. The first three of these extracts showed weak cytotoxic effects against MCF-7 cells, with IC_{50} values between 43 and 79 μg/cm^3 [262].

The European liverwort, *Porella platyphylla* mainly produces pinguisane sesquiterpenoids [3]. The *n*-hexane and chloroform extracts of this species were evaluated biologically. Both extracts showed evidence of cytotoxic effects against A2780, HeLa, and T47D cells, when evaluated at 10 and 30 μg/cm^3 [263].

4.5 Cytotoxic Activity (Mosses)

About 100 mosses that were extracted with organic solvents inclusive of *n*-hexane, ethanol, methanol, chloroform, ethyl acetate, acetone and DMSO, as well as water, with their solvent extracts then evaluated for cytotoxic activity against several human cancer cell lines. Among them, 60 mosses possessing such cytotoxic effects are listed in Table 9.

The ether extract of *Isothecium subdiversiforme* indicated potent cytotoxic activity against P-388 cells. From this fresh moss (20.3 kg), four maytansinoids (**771–774**) were obtained in yields of 2.0×10^{-5} to 8.0×10^{-8} g. These were obtained also from the fermentation broth of *Noccardia* species, the African plant *Maytenus buchananii,* and the Indian plant *Trewia nudiflora.* 15-Methoxyansamitocine P-3 (**771**) showed the most potent cytotoxic effects (IC_{50} 0.002 μg/cm^3) against P-388 cells [38]. The alcoholic or acidic extracts of *Polytrichum juniperinum* exhibited growth inhibitory activity against tumors caused by Sarcoma 37 introduced by intramuscular injection in CAF_1 mice. However, no active compound was isolated from this moss [38].

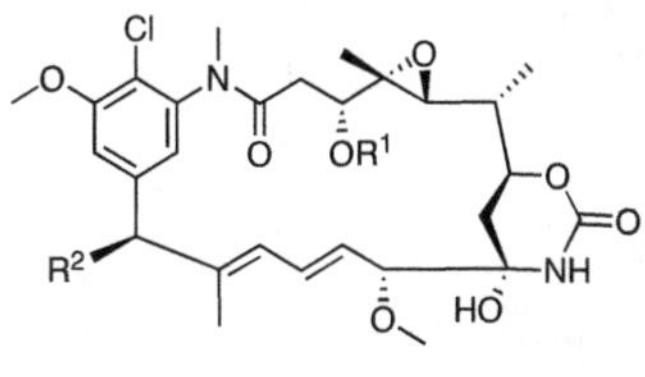

771 (15-methoxyansamitocine P-3, R^1 = COCH(Me)$_2$, R^2 = OMe)
772 (maytansinoid derivative, R^1 = COCH(Me)$_2$, R^2 = H)
773 (maytansinoid derivative, R^1 = CCHMe-NMe-COCH(Me)$_2$, R^2 = H)
774 (maytansinoid derivative, R^1 = CCHMe-NMe-COCH(Me)$_2$, R^2 = OMe)

Altogether, 168 organic solvents and aqueous extracts of 41 Hungarian mosses (*Atrichum, Bryum, Ceratodon, Funaria, Neckera, Paraleucobryum, Plagiomnium, Polytrichum, Rhytidiadelphus, Schistidium,* and *Thuidium* species) and one liverwort (*Porella platyphylla,* which was mentioned above), were evaluated for their antiproliferative activities against three human cancer cell lines (A2780 ovarian, HeLa cervical, and T47D breast) using the MTT assay. A total of 99 extracts derived from 41 species exerted > 20% inhibition of proliferation of at least one of these cancer cell lines at 10 μg/cm^3. The highest antiproliferative activity was observed in the case of *Paraleucobryum longifolium.* For *Brachythecium rutabulum, Climaciuum dendroides, Encalypta streptocarpa, Neckera besseri, Pleurozium schreberi,* and *Pseudolenerbskeella nervosa,* more than one extract was active in the cellular growth inhibition assay used. Species in the families Amblystegiaceae and Brachytheciaceae gave the highest numbers of antiproliferative extracts [263, 264].

In 2015, Klavina et al., tested the ethanol extracts of 13 mosses against six mammalian cancer cell lines (A431, A549, B16-F10, C6, CaCo-2, and MCF-7). Of these, *Sphagnum magellanicum* (Plate 23), *Dicranum polysetum* (Plate 24) and

Pleurozium schreberi showed growth inhibitory activities (0.9, 3, and 5 μg/cm^3, respectively) against C6 rat glioma cells [265].

Two pimarane diterpenoids, momilactones A (**775**) and B (**776**), which were identified as phytoalexins in rice, were isolated from the moss *Hypnum plumaeforme* (Hypnaceae) [266]. Momilactone B (**776**) was shown to reduce the viability of HT-29 and SW620 human colon cancer cells when evaluated in the dose range of 0.5–5 μ*M* [267].

Plate 23 *Sphagnum magellanicum* (moss)

Plate 24 *Dicranum polysetum* (moss)

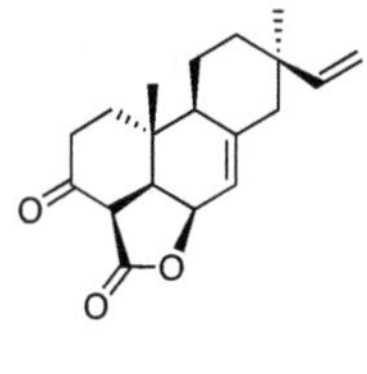

775 (momilactone A)

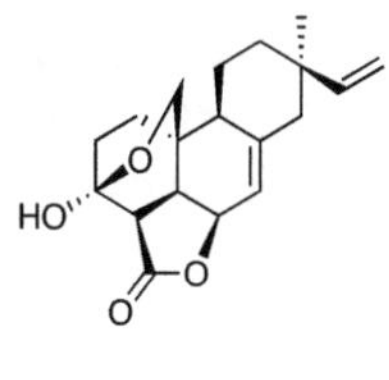

776 (momilactone B)

Zheng et al., isolated five novel benzonaphthoxanthenones, ohioensins A–E (**777**–**781**), from *Polytrichum ohioense.* They were evaluated for their cytotoxic effects against four human cancer cell lines (A549, HT-29, KB, and MCF-7) and one murine cancer cell line (P388). All of compounds **777**–**781** showed an IC_{50} value in the range 1.0–9.7 μg/cm^3 for at least one of these cell lines [268].

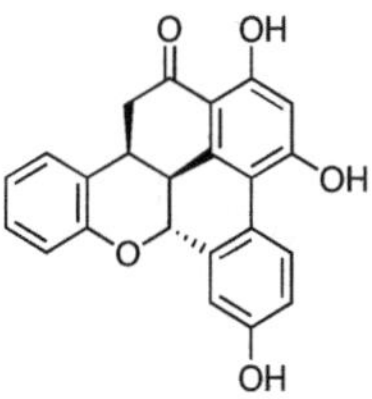

777 (ohioensin A)

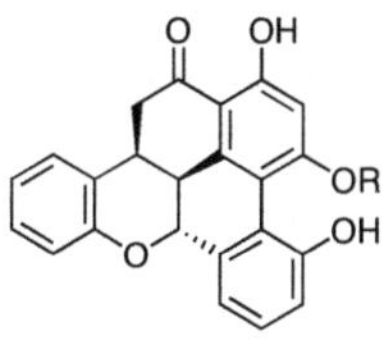

778 (ohioensin B, R = Me)
779 (ohioensin C, R = H)

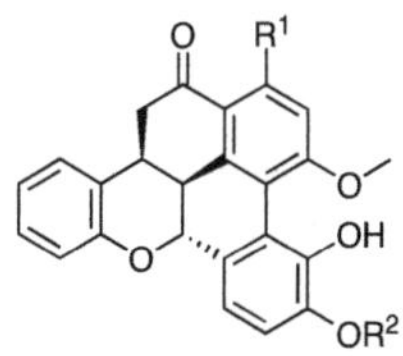

780 (ohioensin D, R^1 = OH, R^2 = H)
781 (ohioensin E, R^1 = OH, R^2 = Me)

Pallidisetin A (**782**) and pallidisetin B (**783**), isolated from the moss *Polytrichum pallidiscetum*, exhibited cytotoxicity against human melanoma (RPMI-7951) and human glioblastoma multiforme (U-251) cells, with ED_{50} values of 1.0 and 1.0 μg/cm^3 and 2.0 and 2.0 μg/cm^3, respectively [269].

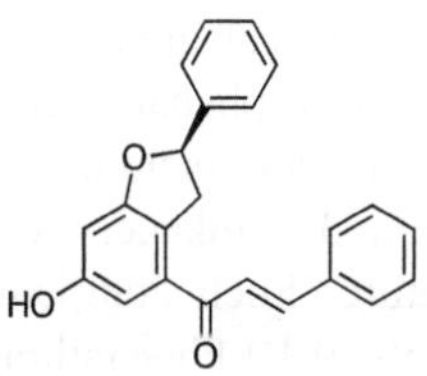

782 (pallidisetin A)

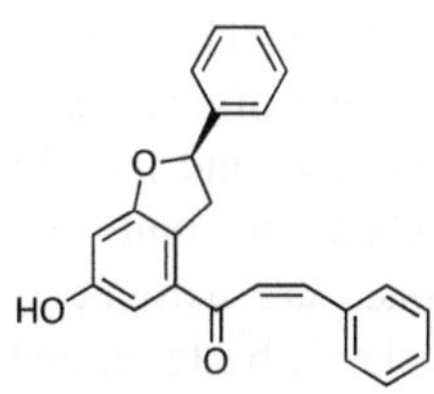

783 (pallidisetin B)

The moss *Polyrichium commune* (Plate 2) was found to contain three cytotoxic compounds, 1-*O*-methylohioensin B (**784**), 1-*O*-methyldihydroohioensin B (**785**) and 1,14-di-*O*-methyldihydroohioensin B (**786**). Compound **784** proved to be cytotoxic for HT-29 human colon adenocarcinoma, RPMI-7951 human melanoma, and U-251 human glioblastoma multiforme cells, with ED_{50} values of 1.0, 1.0, and 2.0 μg/cm^3, respectively. Compound **785** showed inhibitory activity only against U-251 cells (ED_{50} 0.8 μg/cm^3), while **786** inhibited the growth of the A549 (ED_{50} 1.0 μg/cm^3) and RPMI-7951(ED_{50} 1.0 μg/cm^3) cell lines, respectively [269]. Fu et al., isolated two unusual styryl flavanones, communins A (**787**) and B (**788**) together with ohioensin H (**789**) from *P. commune*. When evaluated for cytotoxicity against the A549, HepG2, LOVO, MDA-MB-435, and 6 T-CEM human cancer cell lines, none of these isolated compounds gave an IC_{50} value of < 5 μg/cm^3) [270].

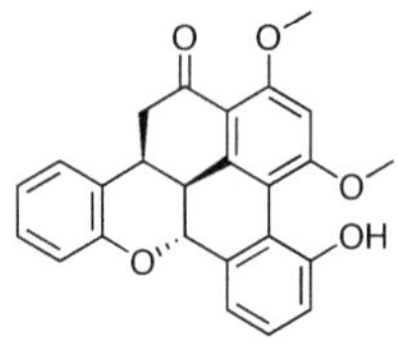

784 (1-*O*-methylohioensin B)

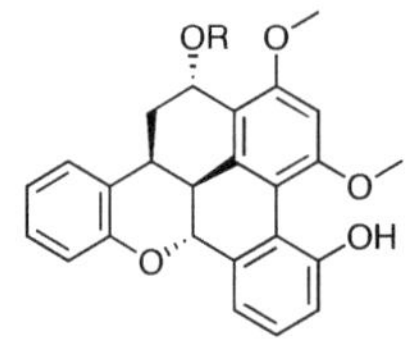

785 (1-*O*-methyldihydro-ohioensin B, R = H)
786 (1,14-di-*O*-methyldihydro-ohioensin B, R = Me)

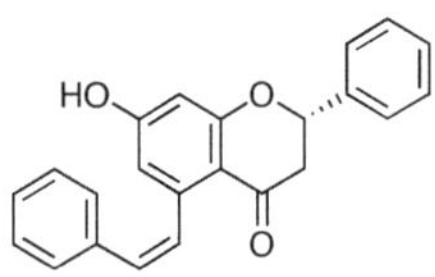

787 (communin A)

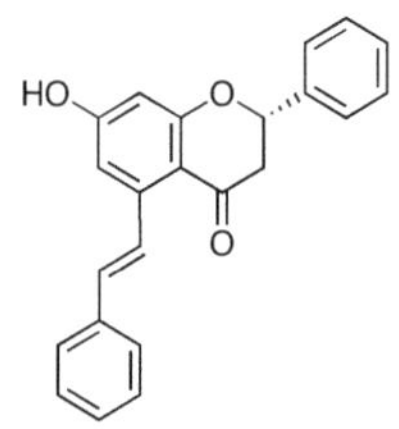

788 (communin B)

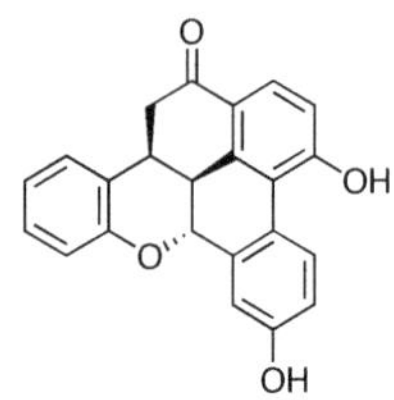

789 (ohioensin H)

Paraleucobryum longifolium, belonging to the Dicranaceae, produces pigments with a violet color. Cuspor et al., isolated in 2020 five new dark-violet compounds, which were determined as 9,10-phenanthrene dimers and named leucobryns A–E (**790**–**794**), together with diosmetin triglycoside (**795**), from an 80% methanol extract of this moss. The stereostructures of the new compounds were established by ECD measurements and TDA-ECD and TDDFT-SOR calculations [271].

Leucobryns are the first representatives of dimeric phenanthrenes with a 9,10-orthoquinone structure, and leucobryns B–E (**791**–**794**) include a monoterpenoid side chain in which the units are joined by a 3,4-linkage. Monomeric 9,10-phenanthraquinones are extremely rare in Nature. Such compounds have only been found in lichens, bacteria, and higher plants. 9,10-Phenanthraquinones from microorganisms have strong antibiotic properties and antiviral activity [271].

The occurrence of bibenzyl-phenanthrenes and bibenzyl-dihydrophenanthrenes in some bryophytes, although they have been isolated only from liverworts, gives further evidence for the suggestion that leucobryns are biosynthesized by dimerization of bibenzyls with a C–C linkage, followed by oxidative steps yielding a 9,10-dihydro functionality. As seen in leucobryns A–E (**790**–**794**), certain bibenzyls are substituted with a prenyl, geranyl, or farnesyl group [1–3].

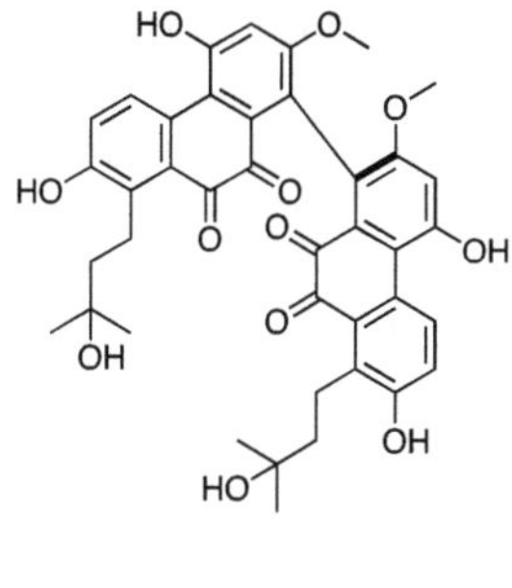

790 (leucobryn A)

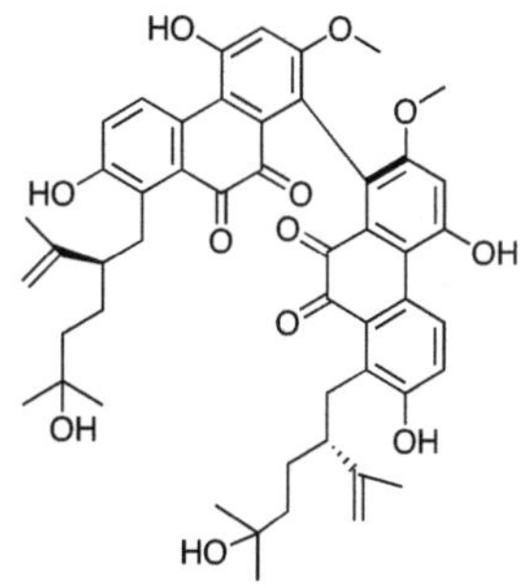

791 (leucobryn B)

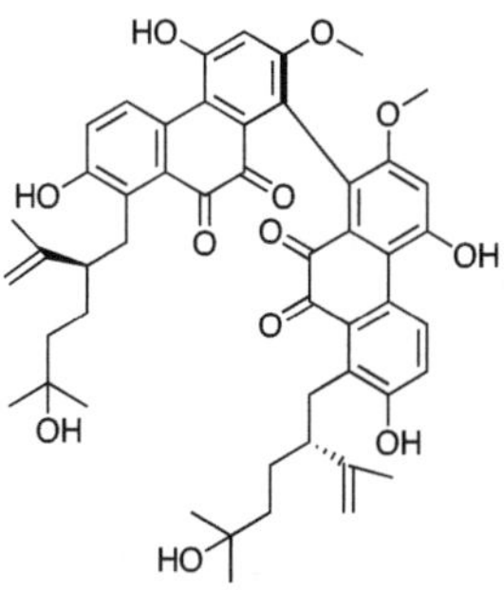

792 (leucobryn C)

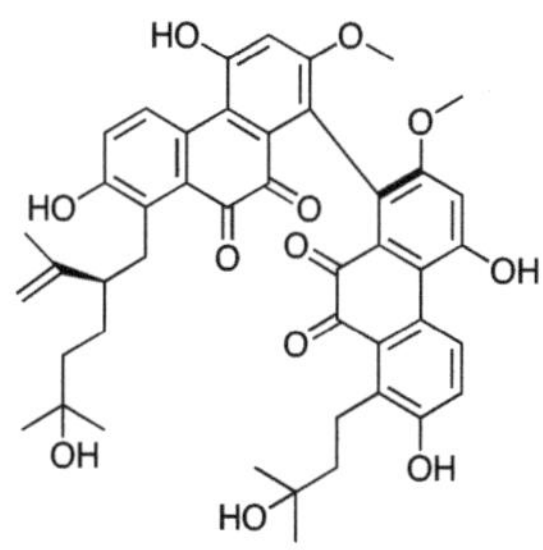

793 (leucobryn D)

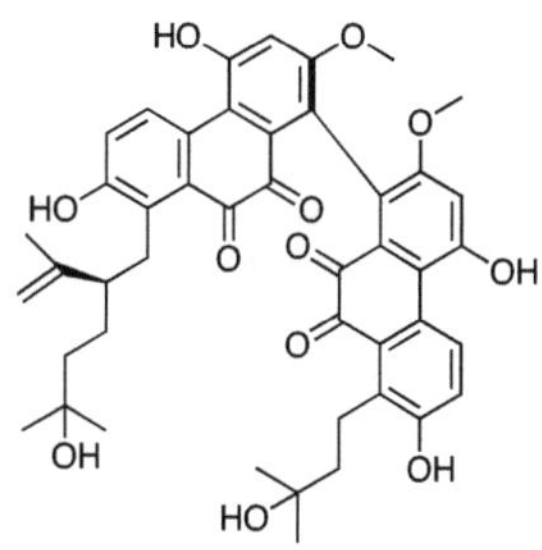

794 (leucobryn E)

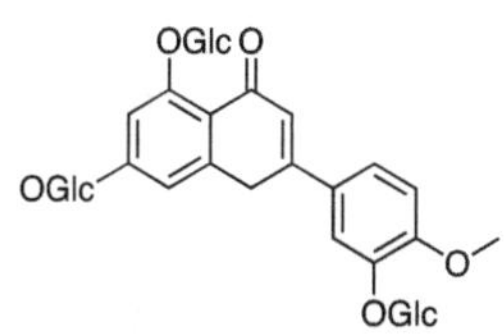

795 (diosmetin triglycoside)

The five leucobryns possessed weak antiproliferative activities against four human cancer cell lines (A-2780, HeLa, MDA-MB-231, and SiHa) cells with IC_{50} values for leucobryn A (**790**) in the range 40.4–50.8 μ*M*. Leucobryns B (**791**) and C (**792**) were less active than **790** [271].

Liquid chromatography/MS analysis indicated that the extracts of 96% ethanol, ethanol/water (1/1), ethyl acetate, and water of *Hypnum cupressiforme* showed the presence of 14 secondary metabolites, namely, six phenolic acids, gallic (**796**), protocatechuic (**797**), 5-*O*-caffeoyl-quinic (**798**), *p*-hydroxybenzoic (**799**), caffeic (**800**), and *p*-coumaric acid (**801**), and eight flavonoids, apigenin (**482**), quercetin 3-*O*-rutinoside (**802**), quercetin 3-*O*-glucoside (**803**), isorhamnetin-3-*O*-glucoside (**804**), eriodictyol (**805**), naringenin (**806**), kaempferol (**807**), and acacetin (**808**). However, the percentages of these compounds in each extract were different. The predominant components of the ethanol extract were **807**, **799**, **797**, and **801**. The chemical profile of the ethanol/water extract was very similar to that of the ethanol extract. The ethyl acetate and water extracts contained **799** and **801**, respectively. The cytotoxicity of the extracts was evaluated against the HCT-16 and MDA-MB-231 human cancer cell lines. The ethanol/water (1/1), ethyl acetate, and water extracts, which contained simple aromatic acids and the flavonoid **807** as major components, showed antiproliferative effects against the MDA-MB-231 breast cancer cell line at a concentration of 10 μg/cm^3. However, these three extracts did not show any inhibitory activity against the HCT-16 human colon cancer cell line [272]

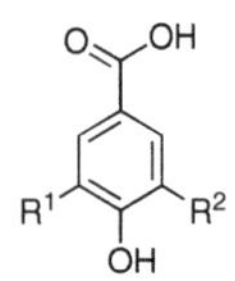

796 (gallic acid, R^1 = R^2 = OH)
797 (protocatechuic acid, R^1 = OH, R^2 = H)
799 (*p*-hydroxybenzoic acid, R^1 = R^2 = H)

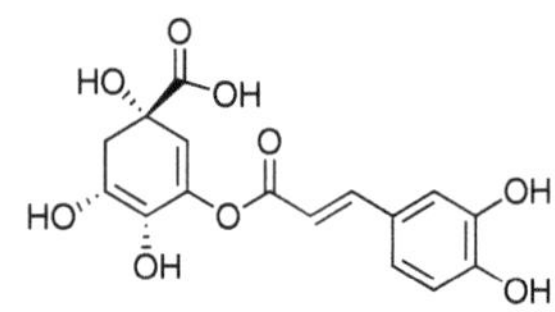

798 (5-*O*-caffeoyl quinic acid)

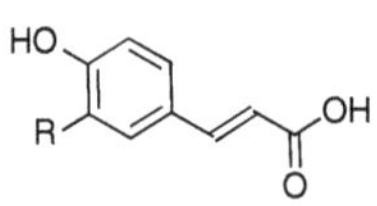

800 (caffeic acid, R = OH)
801 (*p*-coumaric acid, R = H)

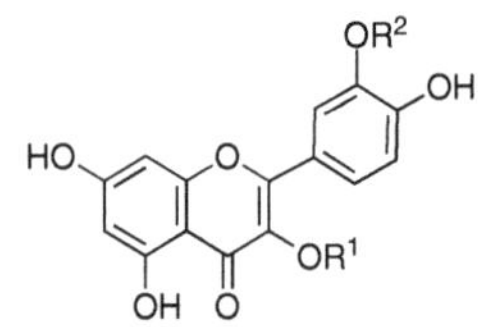

802 (quercetin-3-*O*-rutinoside, R^1 = rutinose, R^2 = H)
803 (quercetin-3-*O*-glucoside, R^1 = glucose, R^2 = H)
804 (isorhamnetin-3-*O*-glucoside, R^1= glucose, R^2 = Me)

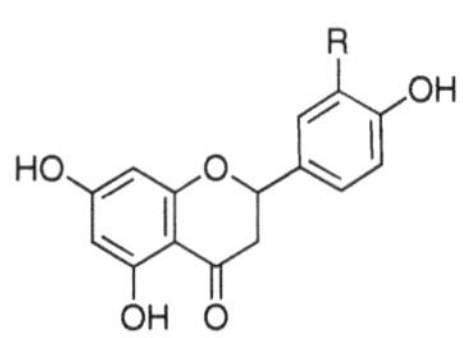

805 (eriodictyol, R = OH)
806 (naringenin, R = H)

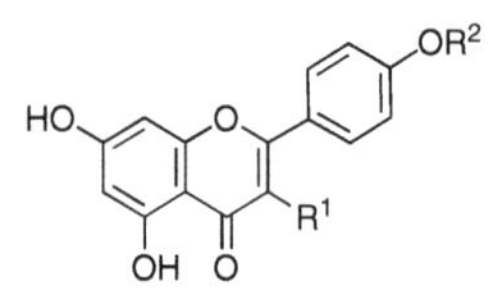

807 (kaempferol, R^1 = OH, R^2 = H)
808 (acacetin, R^1 = H, R^2 = Me)

In 2022, a chemical analysis of the essential oil of the Chinese moss species *Plagiomnium acutum* (Plate 25) was carried out by Li et al., using GC/MS. They identified dolabella-3,7-dien-18-ol (**809**) as the major component, along with a number of monoterpenes inclusive of α- (**71**) and β-pinene (**72**), camphene (**73**), myrcene (**810**), Δ^4-carene (**811**), limonene (**104**), 1,8-cineole (**105**), (*E*)-β-ocimene (**812**), γ-terpinene (**74**), linalool (**46**), 2-bornanone (= camphor) (**813**), and α-terpinene (**68**), and several sesquiterpenoids (β-elemene (**37**), α-cedrene (**36**), *ent*-β-cedrene (**814**), γ-elemene (**815**), aromadendrene (**816**), β-copaene (**817**), α-caryophyllene (= α-humulene) (**818**), α-acoradiene (**819**), 4-*epi*-acoradiene (**820**), β-selinene (**42**), δ-selinene (**821**), ledene (**822**), α-calacorene (**823**), nerolidol (**40**), spathulenol (**514**), butanoic acid ethyl ester (**824**), globulol (**825**), selina-6-en-4-ol (**826**), α-cadinol (**827**), and drimenol (**744**). In addition, they identified the diterpenoid bioformene (**828**) and the polyketides *n*-hexanal (**112**), heptanal (**829**), 3-octanone (**830**), and octanal (**831**), and the aromatic compound phenylacetaldehyde (**832**). The authors

Plate 25 *Plagiomnium acutum* (moss)

indicated that they had detected dolabella-3,7-dien-18-ol (**809**) for the first time in a moss. However, in 1998 the group of the present volume author isolated this same dolabellane diterpenoid from an ether extract of *P. acutum*, when collected in Japan, and established its absolute configuration as the enantiomer to that found in marine organisms [274]. In the Japanese *P. acutum*, the *ent*-sesquiterpenoid, *ent*-β-cedrene (**814**) was also found. This ether extract also contained α-cedrene (**36**) and α-acoradiene (**819**), which might be enantiomers of the same compounds found in higher plants. This is the first example of the isolation of *ent*-sesqui- and diterpenoids from the Bryophyta. However, the same *ent*-sesqui- and the other diterpenoids have been isolated from, or detected, in many liverworts (Marchantiophyta). The essential oils and dolabella-3,7-dien-18-ol (**809**) showed weak growth inhibitory activites against the A549 and HepG2 cancer cell lines with IC_{50} values of 23.6 and 25.8 μg/cm^3, respectively [273].

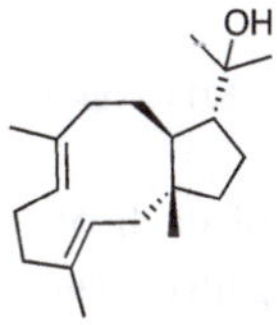

809 (dolabella-3,7-dien-18-ol)

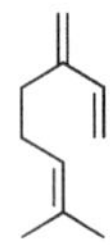
810 (myrcene)

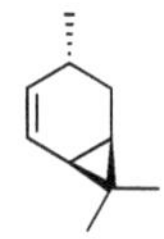
811 (Δ^4-carene)

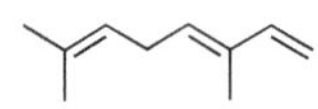
812 ((*E*)-β-ocimene)

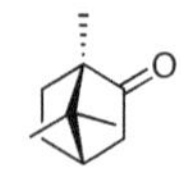
813 (2-bornanone)

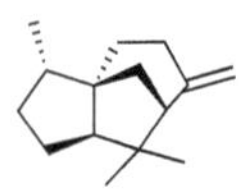
814 (*ent*-β-cedrene)

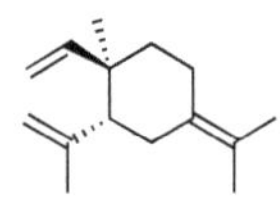
815 (γ-elemene)

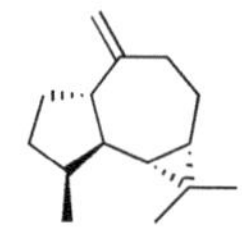
816 (aromadendrene)

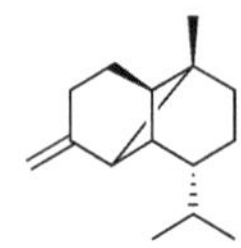
817 (β-copaene)

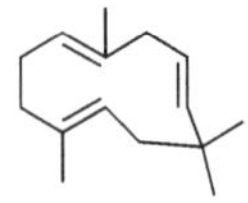
818 (α-caryophyllene)

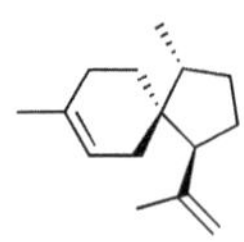
819 (α-acoradiene)

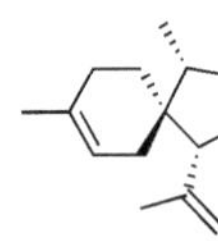
820 (4-*epi*-acoradiene)

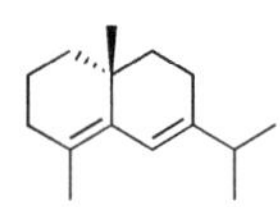
821 (δ-selinene)

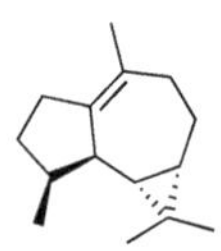
822 (ledene)

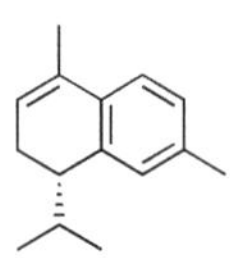
823 (α-calacorene)

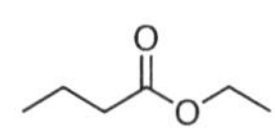
824 (butanoic acid ethyl ester)

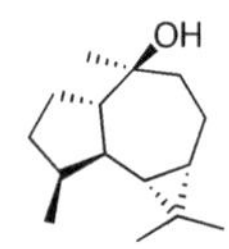
825 (globulol)

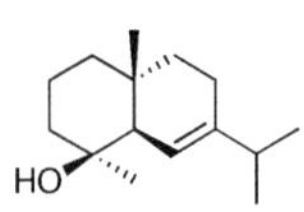
826 (selin-6-en-4-ol)

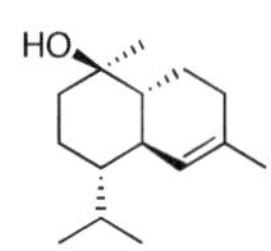
827 (α-cadinol)

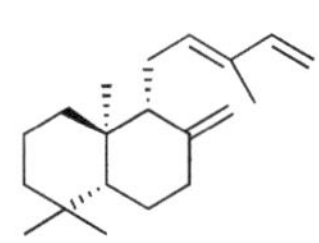
828 (biformene)

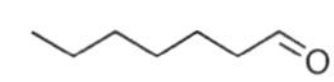
829 (heptanal)

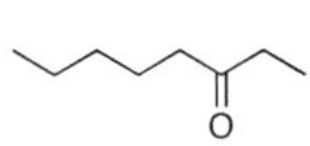
830 (3-octanone)

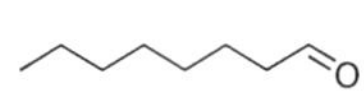
831 (octanal)

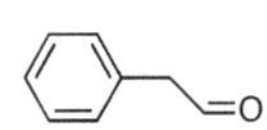
832 (phenylacetaldehyde)

The total phenolic constituents of each of five mosses, *Ceratodon purpureus*, *Dyprodon purvinatus*, *Hypnum cupressiforme*, *Rhytidiadelphus squarossus*, and *Tortulla muralis* were estimated by the Folin-Ciocalteu method, using gallic acid (**796**) as a standard. The individual phenolic components of each species were identified using an ultra-performance liquid chromatographic (UPLC) system coupled with quadruple-time of flight mass spectrometry (Q-TOF-MS).

Ceratodon purpureus was found to contain coumaroylquinic acid (**833**), kaempferol-3-galactoside (**834**), eriodictyol hexoside (**835**), apigenin hexoside (**836**), quercetin-3-*O*-rutinoside (**802**), kaempferol-3-glucoside (**837**), a procyanidin dimer (**838**), unknown kaempferol derivatives I (**839**) and II (**840**), and synapoyl hexoside (**841**), among which **837** was the major flavonoid.

Rhytidiadelphus squarrosus produced catechin (**842**), chlorogenic acid (**843**), caffeic acid (**800**), quercetin 3-rutinoside (**802**), kaempferol 3-rutinoside (**844**), apigenin hexoside (**836**), 3,5-dicaffeoylquinic acid (**843a**), and kaempferol derivative II (**840**), among which apigenin hexoside (**836**) occurred as the most abundant flavonoid.

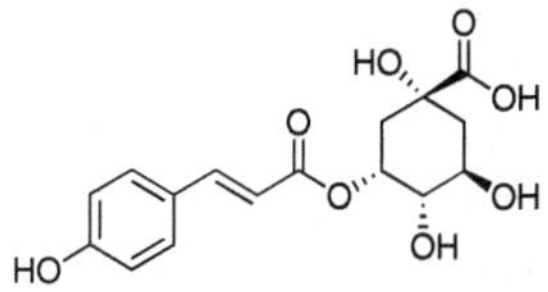

833 (coumaroylquinic acid)

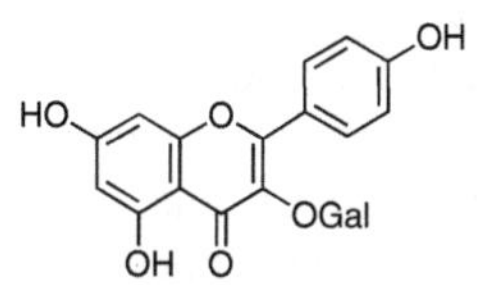

834 (kaempferol-3-*O*-galactoside)

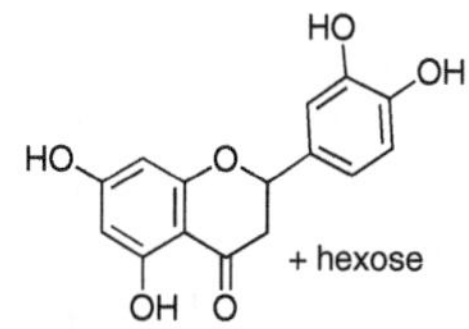

835 (eriodictyol hexoside)

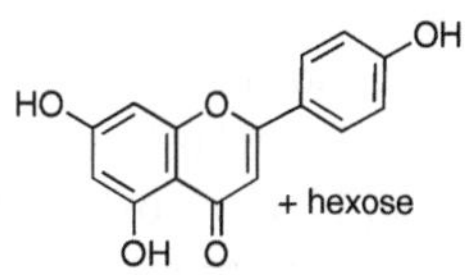

836 (apigenin hexoside)

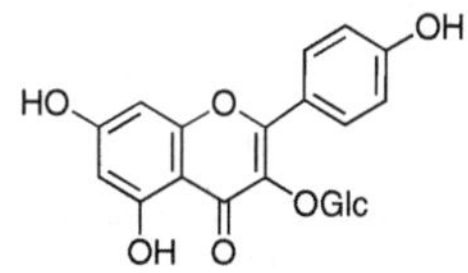

837 (kaempferol-3-*O*-glucoside)

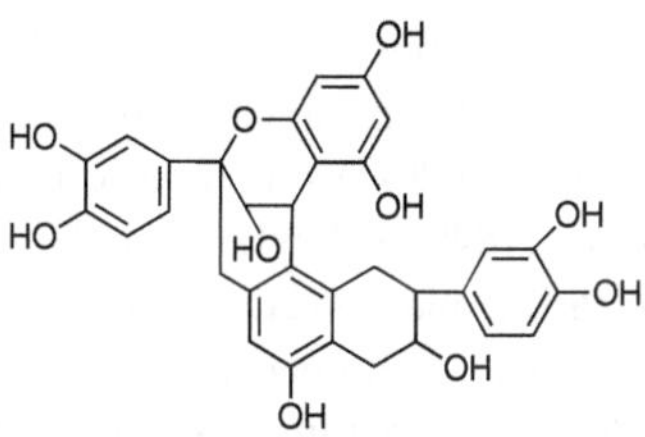

838 (procyanidin dimer)

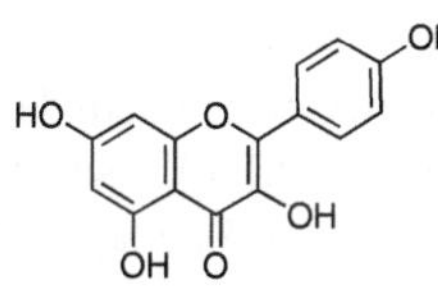

839 (kaempferol derivative I)
840 (kaempferol derivative II; both with unknwon substituents)

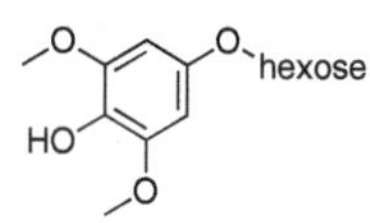

841 (synapoyl hexoside)

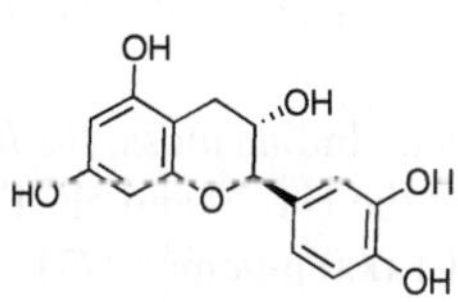

842 ((+)-catechin)

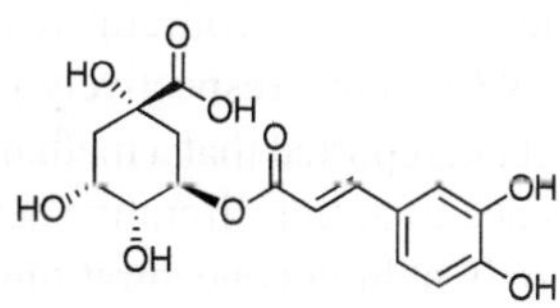

843 (chlorogenic acid)

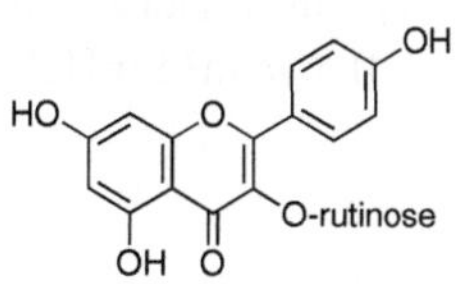

844 (kaempferol rutinoside)

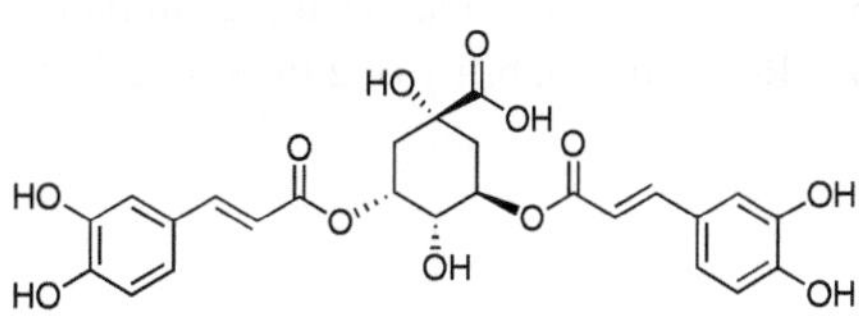

843a (3,5-dicaffeoylquinic acid)

The crude extracts of each moss were evaluated for their cytotoxic activity against the L929 mouse fibroblast cell line using a MTT assay. All of the extracts tested did not affect the viability of these cells at a concentration up to 0.125 mg/cm^3 for the *T. muralis* extract and up to 0.5 mg/cm^3 for the *H. cupressiforme* extract. Of all the extracts evaluated, the most potent effect was noted for the *R. squarrosus* extract that gave a cell viability level of 68.6% when applied at 0.13 mg/cm^3 [275].

Two mosses, *Rhytidiadelphus triquetrus* and *Tortella tortuosa*, were extracted with *n*-hexane, chloroform, ethyl acetate, methanol, water, water/ethyl acetate, and water/butanol. Their crude extracts were analyzed by GC/MS and their antiproliferative and cytotoxic effects on C6 and HeLa cancer cell lines evaluated by Yagliaglu et al., [276]. The *n*-hexane extract was methylated to give methyl esters that were also analyzed by GC/MS. The major fatty acids from these mosses were found to be palmitic, arachidic, arachidonic acids, linolenic, and linoleic acids, which occurred along with 4,7,7-trimethylbi-cyclo[3.3.0]octan-2-one. The moss *R. triquetrus* also contained isopropyl palmitate, stearic acid, and 11-octadecenoic acid as minor constituents. In turn, *T. tortuosa* produced isopropyl palmitate, stearic acid, oleic acid, 7-octadecenoic acid, 11-octadecenoic acid, 11,13-eicosadienic acid, ethyl arachidonyl, 5,8,11,14,17-eicocapentaenoic acid, 1-octadecene, and 1-heptadecene as minor compounds, and also *n*-alkenes.

The phenolic compounds of each solvent extract of both mosses were analyzed by HPLC-TOF-MS to identify several known compounds. These comprised gallic acid (**796**), gentisic acid (**845**), *p*-hydroxybenzoic acid (**799**), protocatechuic acid (**797**), caffeic acid (**800**), *p*-hydroxybenzaldehyde (**846**), rutin (**847**), *p*-coumaric acid (**801**), ferulic acid (**848**), naringenin (**806**), rosmarinic acid (**849**), salicylic acid (**850**), quercetin (**851**), chlorogenic acid (**843**), ellagic acid (**852**), and resveratrol (**853**) in *Rhytidiadelphus triquetrus*. In turn, *Tortella tortuosa* also contained all of these phenolic compounds, except for rosmarinic acid (**849**) and quercetin (**851**). Among the phenolic products detected, *p*-hydroxybenzoic acid (**846**) was the predominant constituent of both mosses. The *n*-hexane and ethyl acetate extracts of the mosses showed the highest antiproliferative activities against HeLa cells with IC_{50} values of 17.9 and 33.3 μg/cm^3, respectively [276].

Singh et al., reported that a methanol extract of an Indian moss, *Barbula javanica*, gave evidence of antiproliferative activity against the HT-29 and HCT-6 cancer cell lines, but only at high concentrations of 100 and 1,000 μg/cm^3 [277].

An ethanol extract of the Turkish moss *Bryum capillare* (Plate 26) was evaluated for its cytotoxicity against three human cancer cell lines, HeLa, MCF-12A, and SKBR 3, and a viability of 24% was determined at 1.0 mg/cm^3 for HeLa cells [278].

Plate 26 *Bryum capillare* (moss)

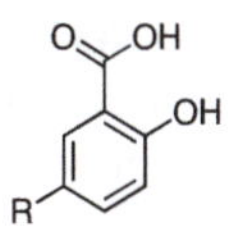

845 (gentisic acid, R = OH)
850 (salicylic acid, R = H)

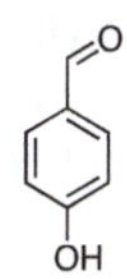

846 (*p*-hydroxy-benzaldehyde)

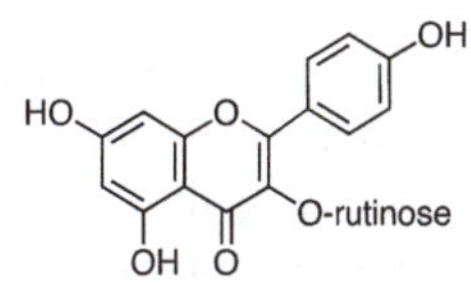

847 (rutin)

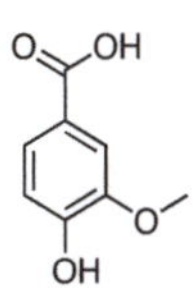

848 (ferulic acid)

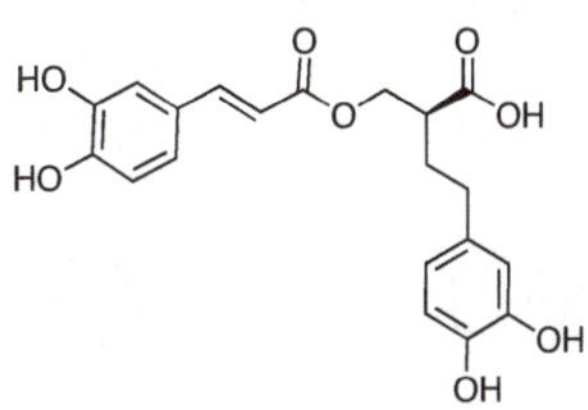

849 (rosmarinic acid)

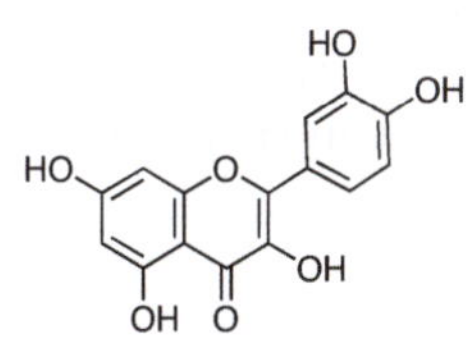

851 (quercetin)

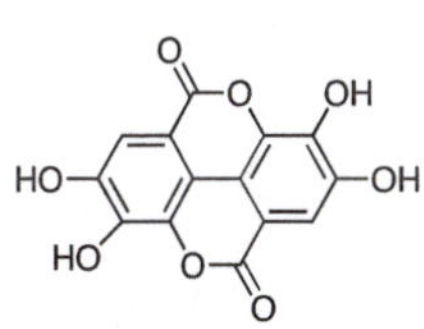

852 (ellagic acid)

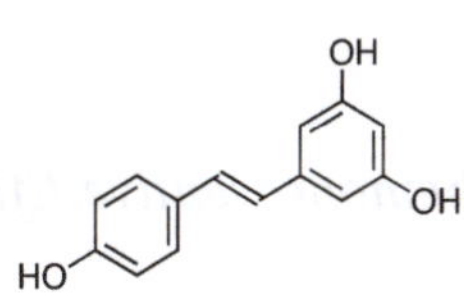

853 (resveratrol)

Using a MTT assay, the cytotoxic effects of three solvent extracts ((ethanol, ethanol/water (1/1), and ethyl acetate)) of *Hedwigia ciliata* (Plate 27) were evaluated against the HCT-16 and MDA-MB-231 cancer cell lines. A cytotoxic effect was not seen for HCT-16 cells, while the three extracts showed growth inhibitory effects against MDA-MB-231 cells. Of these, the ethanol/water (1/1) and ethyl acetate extracts possessed the best activities (higher than 50% inhibition) [279].

Plate 27 *Hedwigia ciliata* (moss)

The essential oil containing β-caryophyllene (**35**), β-bazzanene (**854**) and β-chamigrene (**855**) obtained from the moss, *Phyllogonium viride*, was evaluated for cytotoxicity against MCF-7, HCT-116, and HaCaT cells, although no significant activities were observed [280].

The ethanol and ethanol/water (1/1) extracts of four *Sphagnum* species, namely, *S. girgensohnii*, *S. magellanicum*, *S. palustre*, and *S. squarrosum*, were evaluated for cytotoxicity against NHDF normal human fibroblast cells. It was found that *S. magellanicum* exhibited a weak cytotoxicity with 70% inhibition of growth at a concentration of 500 μg/cm^3. The other *Sphagnum* species were inactive [281].

854 (β-bazzanene)

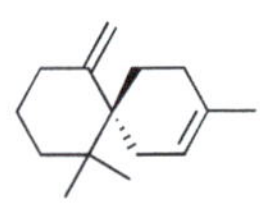

855 (β-chamigrene)

4.6 Cancer Chemopreventive Effects

Li et al., reported in 2014 an efficient total synthesis of plagiochin G (**856**), an unnatural bis-bibenzyl, together with its ester derivatives **857–860**. The cancer chemopreventive effects of these compounds and the three synthetic intermediates **861–863** of **856** against an Epstein-Barr virus early antigen (EBV-EA) model were investigated, using glycyrrhetic acid as a positive control. All eight compounds that were tested showed inhibitory effects toward EBV-EA activation induced by 12-*O*-tetradecanoylphorbol-13-acetate (TPA), without also showing cytotoxicity to Raji cells. Plagiochin G (**856**) demonstrated the highest potency with 88, 45, and 19% inhibition at 1000, 500, and 100 mol ratio/TPA, respectively, and an IC_{50} value 481 μ*M*, and preserved the viability of Raji cells. The four ester derivatives (**857**–**863**) exhibited similar inhibitory activities to those of **856**, while the three synthetic intermediates **860**–**863** of **856** were comparably or slightly less active (Table 10) [282].

Table 10 Cancer chemopreventive effects (relative ratio[a] of EBV-ET activation with respect to a positive control (100%) in the presence of plagiochin G (**856**) and related compounds)

Compound	Compound concentration (mol ratio[c]/TPA[b])				Activity value IC_{50} (μ*M*)	Refs.
	1000	500	100	10		
Plagiochin G 2-amino-11,5′,13′-trimethyl ether (**857**)	14.9 (70)[d]	58.2	83.1	100	491	[282]
Plagiochin G 2-amino-11,5′,13′-trimethyl ether (**858**)	15.3 (70)	59.6	84.6	100	500	
Plagiochin G trimethyl ether (**861**)	13.0 (70)	56.8	81.1	100	490	
Plagiochin G (**856**)	11.5 (70)	54.3	80.1	100	481	
Plagiochin G triacetate (**859**)	13.9 (70)	57.9	82.4	100	495	
Plagiochin G tri-[3-(1H-imidazol-1-yl)]-propanoate (**860**)	13.0 (60)	55.4	79.1	100	479	
Plagiochin G tri-[3-(1H-1,2,4-triazol-1-yl)]-propanoate (**862**)	13.8 (60)	56.0	80.0	100	482	
Plagiochin G tri-[3-(1H-tetrazol-1-yl)]-propanoate (**863**)	14.0 (60)	57.6	81.6	100	488	
Glycyrrhetic acid[e]	7.4 (60)	35.7	83.2	100	413	

[a] Values represent percentages relative to the positive control value (100%)
[b] TPA concentration was 20 ng/cm^3
[c] The molar ratio of compound relative to TPA required to inhibit 50% of the positive activated with 32 pmol control
[d] Values in parentheses of Raji Cells. In all other experiments, viability was over 80%
[e] Positive control

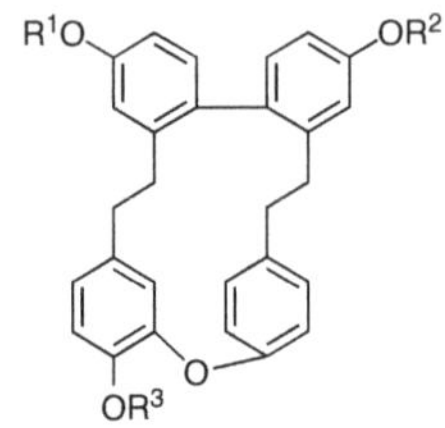

856 (plagiochin G, $R^1 = R^2 = R^3 = H$)
857 (plagiochin G triacetate, $R^1 = R^2 = R^3 = COCH_3$)
358 (ester derivative, $R^1 = R^2 = R^3 = COCH_2CH_2$-1-imidazolyl)
359 (ester derivative, $R^1 = R^2 = R^3 = COCH_2CH_2$-1-(1,2,4-triazolyl))
360 (ester derivative, $R^1 = R^2 = R^3 = COCH_2CH_2$-1-(1,2,3,4-tetrazolyl))

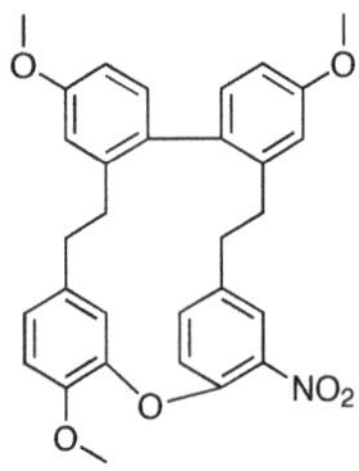

861

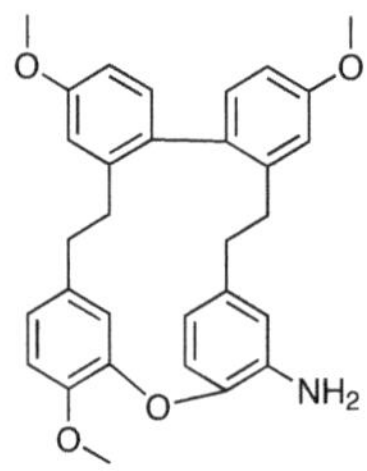

862

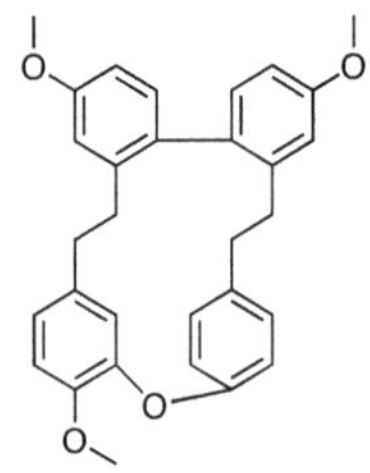

863

4.7 Genotoxic Activity

The European and Japanese specimens of *Pellia endiviifolia* elaborate the pungent diterpene dialdehyde, sacculatal (**128**) as the major metabolite. This species also contains the bis-bibenzyls, perrottetin E (**183**), 10′-hydoxyperrottetin E (**247**), and 10,10′-dihydroxyperrottetin E (**248**). In 2021, Ivkovic et al., studied the biological effects of a methanol extract of *P. endiviifolia* and these three bis-bibenzyls using human peripheral blood cells as a model system. The genotoxicity testing conducted indicated that 25 and 100 μ*M* concentrations of the three bis-bibenzyls displayed genotoxic and antiproliferative effects in the above-mentioned cells, as revealed by significant, concentration-dependent enhancement of the micronuclei incidence, and a decrease in the cytokinesis-block proliferation index, when compared to a control. An apoptosis assessment indicated their substantial proapoptotic effects, which were not concentration-dependent. The methanol extract and the three bis-bibenzyls did not induce oxidative stress in human peripheral blood cells, as shown by their insignificant effects on erythrocyte catalase activity and lymphocyte malondialdehyde production. On the basis of the above results, the authors suggested that the bis-bibenzyls evaluated possessed genotoxic, antiproliferative, and proapoptotic effects, which are desired properties of anticancer drugs [283].

In 2019, Onbasli et al., investigated the genotoxic properties of an ethanol extract of the Turkish moss, *Hypnum andoi*. Concentrations of 0.5 and 1 mg/cm^3 of the extract showed 35 and 53% genotoxic inhibition, respectively, against human lymphocyte DNA [284]. The same group reported on the genotoxicity and antigenotoxicity of an ethanol extract of the moss, *Bryum capillare*, using human lymphocytes. The ethanol extract did not show any DNA damage to human lymphocytes at concentrations of 0.5 and 1 mg/cm^3, whereas the positive control (H_2O_2) caused 54% DNA

Table 11 Genotoxic and antigenotoxic activities of bryophyte extracts

Species name[a]	Extract	Cells	Activity values	Refs.
[M] *Bryum capillare*		Human lymphocytes	Genotoxic activity Tail length (μm/tail moment)	[278]
	EtOH		28.96/0.00 (500 μg/cm^3)	
	EtOH		29.46/0.26 (1000)	
	H_2O_2 (positive control)		74.0/18.00 (50)	
			Tail DNA damage (%)	
	EtOH		0.00 (500 μg/cm^3)	
	EtOH		0.05 (1000)	
	H_2O_2		54 (50 μ*M*)	
			Antigenotoxic activity Tail length (μm)/tail moment	
	EtOH + H_2O_2		52.07 (500 μg/cm^3)/11.00	
	EtOH + H_2O_2		41.03 (1000)/8.26	
	H_2O_2		74.0 (50 μ*M*)/18.00	
			Tail DNA (damage %)/inhibition %)	
	EtOH + H_2O_2		33.00 /39 (500 μg/cm^3)	
	EtOH + H_2O_2		25.05/ 55 (1000 μg/cm^3)	
	H_2O_2		54 (50 μ*M*)	
[M] *Hypnum andoi*	EtOH	Human lymphocytes	% inhibition (μg/cm^3) 35 (500) 53 (1000)	[281]
[L] *Pellia endiviifolia*	Perrottetin E (**183**) 10′-Hydroxyperrotin E (**247**), 10,10′-dihydroxyperrotin E (**248**)	Human peripheral blood cells	25–100 m*M* genotoxic, antiproliferative, and proapoptotic effects	[284]

[a] [M] Moss, [L] Liverwort

damage. At these same concentrations, the ethanol extract inhibited the genotoxic activity of 50 μ*M* of H_2O_2, by 39 and 55%, respectively (Table 11) [278].

4.8 Antimicrobial Activity

In 1982 and 1990, antibacterial compounds from several liverworts, including a *Bazzania* species, *Conocephalum conicum, Diplophyllum albicans, Dumortiera hirsuta, Marchantia polymorpha, Metzgeria furcata, Lunularia cruciata, Pellia endiviifolia, Plagiochila platyphylla,* and a *Radula* species were reported. It was found that marchantin A (**1**), isolated from the Japanese *Marchantia polymorpha,* showed both antibacterial and antifungal activity [1–3, 38]. Since then, the antibacterial activity of the solvent (*n*-hexane, diethyl ether, ethyl acetate, chloroform, ethanol, methanol, dimethysulfoxide, acetone, and water) extracts and essential oils of many bryophytes have been evaluated and number of technical reports have appeared [60, 263, 286, 307, 314, 316, 321], as summarized in Tables 12 and 13.

The Japanese and German *Marchantia polymorpha* and other related *Marchantia* species produce the cyclic bis-bibenzyl, marchantin A (**1**) and related compounds. Marchantin A (**1**) exhibited antibacterial activity against *Acinetobacter calcoaceticus* (*MIC* 6.3 μg/cm^3), *Alcaligenes faecalis* (100 μg/cm^3), *Bacillus cereus* (12.5 μg/cm^3), *B. megaterium* (25 μg/cm^3), *B. subtilis* (25 μg/cm^3), *Cryptococcus neoformans* (12.5 μg/cm^3), *Enterobacter cloacae, Escherichia coli, Proteus mirabilis, Pseudomonas aeruginosa, Salmonella typhimurium* (all 100 μg/cm^3), and *Staphylococcus aureus* (3.1–25 μg/cm^3) [38]. Kamaory et al., reported in 1995 that **1** isolated from the Hungarian *M. polymorpha* showed antimicrobial activity against the Gram-negative bacteria, *Pasteurella multococida, Pseudomonas aeruginosa, Haemophylus infulensae,* and *Neissera meningitidis* with *MIC* values of 4.5, 84.5, 72.7, 72.7 n*M*, respectively, and against the Gram-positive *S. aureus, Streptococcus pyogenes,* and *S. viridans,* with *MIC* values of 6.9, 9.1, and 18.1 n*M*, respectively [285].

The antibacterial activities of the methanol extracts of three Serbian liverworts, *Conocephalum conicum, Marchantia polymorpha,* and *Pellia endiviifolia* and of marchantin A (**1**), the major component of *M. polymorpha,* were evaluated against four Gram-positive bacteria, *Bacillus subtilis, Clavibacter michiganensis, Listeria monocytogenes,* and *Staphylococcus aureus,* and against four Gram-negative bacterial strains, *Escherichia coli, Pseudomonas aeruginosa, P. syringae,* and *Xanthomonas arboricola.* The Gram-positive bacteria, with the exception of *C. michiganensis,* were sensitive to the methanol extract of each liverwort species and to marchantin A (**1**). An *M. polymorpha* extract showed the most effective antibacterial activity against *B. subtilis* (inhibition zone 14 mm). The methanol extracts of all liverwort species and marchantin A (**1**) used in the *MIC* assay had no observed antibacterial activity against Gram-negative microorganisms, while the growth of Gram-positive bacteria was inhibited with *MIC* values ranging from 0.06 to 5 mg/

Table 12 Antimicrobial compounds and solvent extracts from bryophytes

Species name[a]	Compound or type of solvent extract	Bacterial name	Activity value minimal inhibitory concentration (*MIC* μg/cm^3, μ*M*) (Inhibition zone: mm)	Refs.
[L] *Acrobolbus saccatus*	5-Oxoradulanin (**349**) Saccatene A (**908**) Saccatene B (**909**) Saccatene C (**910**)	*Listeria monocytogenes*	*MIC* (μg/cm^3) 25	[308]
		Salmonella enteritidis	25	
		Yesinia enterocolitica	25	
[L] *Chandonanthus hirtellus*	Chandonanthone (**519**) Isochandonanthone (**520**) Chandonanol (**912**)	*Staphylococcus aureus*	*MIC*/*MBC* (μg/cm^3) 275/1250	[309]
		Escherichia coli	250/1250	
		S. aureus	250/1250	
		E. coli	250/1250	
		S. aureus	100/180	
		E. coli	120/200	

(continued)

Table 12 (continued)

Species name[a]	Compound or type of solvent extract	Bacterial name	Activity value minimal inhibitory concentration (*MIC* μg/cm³, μ*M*) (Inhibition zone: mm)	Refs.
[L] *Conocephalum conicum* [L] *Marchantia polymorpha* [L] *Pellia endiviifolia*	MeOH MeOH Marchantin A (**1**) MeOH		*MIC*/*MBC*(μg/cm³)	[306]
		Bacillus subtilis *Clavibacter michiganensis* *Listeria monocytogenes* *Staphylococcus aureus* *L. monocytogenes* *Staphylococcus aureus* *Bacillus subtilis* *L. monocytogenes* *Staphylococcus aureus* *L. monocytogenes* *Staphylococcus aureus*	Inactive Inactive Inactive Inactive 2500 156 1000 500–1000/1000 62 5000 315/5000	
[M] Dicranium polysetum	EtOAc subfractions of *D. polysetum* (DPE)	*P. larvae* strains: *PB3.2B, PB6A, SV27B, PB31B, PB35, ERICI:*	*MIC* μg/cm³ 1.4–30.7	[321]
	DPE-1–6	*DMS27,030*		
		ERICII:DMSZ25,430,		
		ERICIII LMG16,252		
		P. dendritiformis strains: *PB22F, PB31A*	5.6–90.0	
		P. alvei strains: *M6, PB33*	5.6–92.7	
		Melissococcus plutonius II clone: Mp1	2.8–22.5	

(continued)

Table 12 (continued)

Species name[a]	Compound or type of solvent extract	Bacterial name	Activity value minimal inhibitory concentration (*MIC* μg/cm^3, μ*M*) (Inhibition zone: mm)	Refs.
	Glycer-2-yl-hexadeca-4-yne-(7Z,10Z,13Z)-trienoate (**921**)	The above four strains	8.12–65.0	
	Poriferasterol (**922**)	The above four strains	2.09–33.4	
	γ-Taraxasterol (**923**)	The above four strains	1.8–28.8	
[M] *Dicranum scoparium*	Apigenin-7-tri-glycoside (**917**)	*Enterobacter aerogenes* *Escherichia coli* *Proteus mirabilis*	256 μg/cm^3 256 256	[310]
[L] *Dumortiera hirsuta*	Lunularin (**103**)	*Pseudomonas aeruginosa*	64 μg/cm^3	[176]
[L] *Frullania dilatata*	EtOH (frullanolide (**3**), 2,3-dimethylanisole (**1133**), linoleic, palmitic (**1132**), and valerenic (**1134**) acid containing)		Inhibition zone (mm)	[317]
			50 100 200 mm^3	
		Enterococcus faecalis	7 8 9	
		E. faecium	7 8 8	
		Listeria monocytogenes	8 9 9	
		Providencia rustigianii	7 8 9	
		Staphylococcus aureus	9 10 11	
		S. hominis	7 8 –	
	EtOH	*E. faecalis*	*MIC* (mg/cm^3) 0.1285	
		E. faecalis	1.028	
		L. monocytogenes	0.257	
		P. rustigianii	1.028	

(continued)

Table 12 (continued)

Species name[a]	Compound or type of solvent extract	Bacterial name	Activity value minimal inhibitory concentration (*MIC* µg/cm^3, µ*M*) (Inhibition zone: mm)	Refs.
		S. aureus	0.257	
	H_2O	*Listeria monocytogenes*	21.44	
		Staphylococcus aureus	21.44	
[L] *Frullania dilatata*	3,4′-Dimethoxy-4-hydroxybibenzyl (**901**) 3-Hydroxy-4′-methoxy-bibenzyl (**902**)	*Sarcina lutea*	*MIC* (µg/cm^3) 400	[243]
		Bacillus subtilis	100	
		B. sporecereus	50	
		Sarcina lutea	200	
[M] *Hedwigia ciliata*	Lucenin-2 (**914**)	*Enterobacter aerogenes* *E. cloacae* *Escherichia coli* *Klebsiella pneumoniae* *Pseudomonas aeruginosa*	8 µg/cm^3 64 16 64 8	[279]
[L] *Jungermannia exsertifolia* subsp. *cordifolia*	*ent*-Trachyloban-17-al (**894**)	*Mycobacterium tuberculosis*	MIC_{90} (µg/cm^3) 24	[294]
	ent-3*β*-Acetoxy-19-hydroxytrachylobane (**895**)		59	
	ent-Trachylobane-3-one (**896**)		50	
	ent-3*β*-Hydroxy-trachylobane (**897**)		51	
	ent-3*β*-Acetoxy-trachylobane (**898**)		111	
[L] *Lepidozia fauriana*	Lepidozenolide (**367**)	*Staphylococcus aureus*	100 µg/cm^3	[185]
[L] *Lunularia cruciata*	Perrottetin E (**183**)	*Pseudomonas aeruginosa*	150 µ*M*	[152]

(continued)

Table 12 (continued)

Species name[a]	Compound or type of solvent extract	Bacterial name	Activity value minimal inhibitory concentration (*MIC* μg/cm^3, μ*M*) (Inhibition zone: mm)	Refs.
[L] *Marchantia polymorpha* (Japanese specimen)	Marchantin A (**1**)	*Acinetobacter calcoaceticus*	*MIC* (μg/cm^3) 6.25	[1–3, 38]
		Alcaligenes faecalis	100	
		Bacillus cereus	12.5	
		B. megaterium	25	
		B. subtilis	25	
		Cryptococcus neoformans	12.5	
		Enterobacter cloacae	1	
		Escherichia coli	00	
		Proteus mirabilis	100	
		Pseudomonas aeruginosa	100	
		Salmonella typhimurium	100	
		Staphylococcus aureus	3.13–25	
[L] *Marchantia polymorpha* (Hungarian specimen)	Marchantin A (**1**)	*Escherichia coli*	*MIC* (n*M*) 108.4	[285]
		Haemophilus influenzae	72.7	
		Neisseria meningitidis	72.7	
		Pasteurella multocida	4.5	
		Proteus mirabilis	121.1	

(continued)

Table 12 (continued)

Species name[a]	Compound or type of solvent extract	Bacterial name	Activity value minimal inhibitory concentration (*MIC* μg/cm^3, μ*M*) (Inhibition zone: mm)	Refs.
		Pseudomonas aeruginosa	84.5	
		Staphylococcus aureus	6.8	
		S. faecalis	145.4	
		S. pyogenes	9.1	
		S. viridans	18.1	
[L] *Mastigophora diclados* (Malagasy sp.)	EtOH	*Staphylococcus aureus*	(Disk diffusion) Inhibition zone (mm)	[292]
	α-Formyl-herbertenol (**888**)		16	
	α-Herbertenol (**405**)		15	
	Herbertene-1,2-diol (**406**)		13	
	Mastigophorene C (**407**)		17	
	β-Herbertenol (**886**)		16	
[L] *Mastigophora diclados* (Tahitian sp.)			(Disk diffusion)	[182, 195]
	Et_2O	*Bacillus subtilis*	16 μg/cm^3	
	MeOH	*Staphylococcus aureus*	64	
	(–)-Herbertene-1,2-diol (**406**)	*B. subtilis*	2–8	
	Mastigophorene C (**407**) Mastigophorene D (**408**)		2–8	
	1-Hydroxy-2-methoxy-herbertene (**409**)		32	
	Herbertene-1,2-diacetate (**410**)		16	

(continued)

Table 12 (continued)

Species name[a]	Compound or type of solvent extract	Bacterial name	Activity value minimal inhibitory concentration (*MIC* μg/cm^3, μ*M*) (Inhibition zone: mm)	Refs.
[L] *Pellia endiviifolia*	Et_2O	*Bacillus subtilis*	*MIC* (μg/cm^3) 16	[38, 49]
		Staphylococcus aureus	16	
	Sacculatal (**128**)	*Cryptococcus neoformans*	50	
		Streptococcus mutans	8	
[L] *Pellia endiviifolia*			MIC_{90} (μg/cm^3)	[288]
	Lunularin (**103**)	*Pseudomonas aeruginosa*	64 *MIC* (μ*M*)	
	Perrottetin F (**183**)	*Staphylococcus aureus*	100	
[L] *Plagiochila banksiana* [L] *Plagiochila deltoidea* [L] *Plagiochila fasciculata* [L] *Plagiochila stephensoniana* [L] *P. stephensoniana*			Inhibition zone (mm)	[177]
	EtOH (0.05 mg of bibenzyl/g dry liverwort)	*Bacillus subtilis*	5	
	EtOH (0.54)	*B. subtilis*	Not detected	
	EtOH (trace, unable to resolve)	*B. subtilis*	1 mm	
	EtOH (10.5)	*B. subtilis*	4 mm	
	3-Methoxy-4′-hydroxy-bibenzyl (**286**)	*Escherichia coli*	Inactive	
		B. subtilis	3 mm (12 μg/cm^{3}dose) 5 (500) 5 (60)	
	(*E*)-3-Methoxy-4′-hydroxystilbene (**287**)[b]	*E. coli*	Inactive	

(continued)

Table 12 (continued)

Species name[a]	Compound or type of solvent extract	Bacterial name	Activity value minimal inhibitory concentration (*MIC* μg/cm^3, μ*M*) (Inhibition zone: mm)	Refs.
		B. subtilis	2 (12) 4 (60) 6 (300)	
[L] *P. suborbiculata*	(*Z*)-3-Methoxy-4′-hydroxystilbene (**288**)[b]	*E. coli*	1 (60)	
		B. subtilis	4 (12) 9 (60)	
		B. subtilis	9 (300) 2 mm	
[L] *Plagiochila fasciculata*	2-Hydroxy-4,6-di-methoxyacetophenone (**767**)	*Escherichia coli* *Proteus mirabilis* *Staphylococcus aureus*	100 μg/cm^3	[258]
[L] *Plagiochila ovalifolia*	Chlorophyll-metabolites: phaeophytins **890–893**	*Bacillus subtilis* *Escherichia coli*	No values given	[293]
[M] *Dicranum scoparium* *Plagiomnium affine*	Apigenin (**482**)	*Enterobacter aerogenes* *E. cloacae* *Escherichia coli* *Klebsiella pneumoniae* *Proteus mirabilis* *Pseudomonas aeruginosa* *Salmonella typhi*	4 μg/cm^3 128 4 128 16 8 128	[310]
[M] *Plagiomnium affine*	Vitexin (**916**)	*Enterobacter cloacae*	128 μg/cm^3	[310]
		Escherichia coli	256	
		Proteus mirabilis	128	

(continued)

Table 12 (continued)

Species name[a]	Compound or type of solvent extract	Bacterial name	Activity value minimal inhibitory concentration (*MIC* μg/cm³, μ*M*) (Inhibition zone: mm)	Refs.
[M] *Plagiomnium cuspidatum*	Saponarin (**915**)	*Enterobacter aerogenes* *E. cloacae* *E. faecalis* *Escherichia coli* *Klebsiella pneumoniae* *Proteus mirabilis* *Pseudomonas aeruginosa* *Salmonella typhi*	4 μg/cm³ 128 2048 4 54 8 8 64	[310]
Porella chiliensis		*Pseudomonas aeruginosa*	% Inhibition of biofilm formation at 5 μg/cm³	[239]
	Anadensin (**499**)		47	
	ent-Spathulenol (**931**)		41	
	ent-4*β*,10*α*-Dihydro-aromadendrene		47	
[L] *Porella vernicosa* complex	Polygodial (**90**)	*Cryptococcus neoformans*	*MIC* (μg/cm³) 100	[38]
		Staphylococcus aureus	50–100	
[L] *Radula javanica* [L] *R. kojana* [L] *R. perrottetii* and many other *Radula* species [L] *Radula oyamensis*	2-Geranyl-3,5-dihydroxybibenzyl (**899**)	*Staphylococcus aureus*	*MIC* (μg/cm³) 20	[38]
	3-Hydroxy-5-methoxy-4-(3-methyl-2-butenyl)-bibenzyl (**900**)	*S. aureus*	3	

(continued)

Table 12 (continued)

Species name[a]	Compound or type of solvent extract	Bacterial name	Activity value minimal inhibitory concentration (*MIC* μg/cm^3, μ*M*) (Inhibition zone: mm)	Refs.
[L] *Reboulia hemisphaerica*	Riccardin C (RC) (**120**)	*Staphylococcus aureus* OM 481/ OM 584	*MIC* (IC_{50}) μg/cm^3 13.2/3.2	[289]
	2-Hydroxy-RC (**864**)[b]		16/16	
	3-Hydroxy-RC (**865**)[b]		8/8	
	RC-1′-methyl ether (**866**)[b]		> 100 / > 100	
	RC-11-methyl ether* (**867**)[b]		> 100 / > 100	
	RC-13′-methyl ether (**868**)[b]		> 100 / > 100	
	11,1′-Didehydroxy-RC (**869**)[b]		> 100 / > 100	
	1′,13′-Didehydroxy-RC (**870**)[b]		1.6/ > 100	
	11,13′-Hydroxy-RC (**871**)[b]		3.2/ > 100	
	13′-Dehydroxy-RC (**872**)[b]		3.2/3.2	
	11-Dehydroxy-RC (**873**)[b]		3.2/3.2	
	1-Hydroxybiphenyl ether (**874**)[b]		6.3/6.3	
	1,4′-Dihydroxy-biphenyl (**875**)[b]		> 100 / > 100	
	3,4′-Dihydroxy-bibenzyl (**876**)[b]		> 100 / > 100	
[L] *Reboulia hemisphaerica* [L] *Dumortiera hirsuta* [L] *Marchantia polymorpha*	Riccardin C (RC) (**120**)	*Staphylococcus aureus* OM 481/ OM 584/N 315	*MIC* (IC_{50}) μg/cm^3 2/2/2	[290]
	Riccardin D (**253**)		8/4/4	

(continued)

Table 12 (continued)

Species name[a]	Compound or type of solvent extract	Bacterial name	Activity value minimal inhibitory concentration (*MIC* μg/cm^3, μ*M*) (Inhibition zone: mm)	Refs.
	Isoriccardin D (**881**)		4/4/4	
	2-Fluoro-RC (**877**)[b]		8/8/4	
	1′-Deoxy-1′-fluoro-RC (**878**)[b]		32/32/32	
	1′-Demethyl-*neo-iso*-marchantin C (**879**)[b]		32/32/32	
	1′-Dehydroxy-5′-hydroxy-isoplagiochin D (**880**)[b]		32/32/32	
	1-Hydroxybiphenyl ether (**874**)[b]		8/8/8/	
[L] *Riccardia crassa*	Riccardiphenol C (**404**)	*Bacillus subtilis*	60 μg/disk	[194]
[L] *Riccardia marginata*	2,4,6-Trichloro-3-hydroxy–bibenzyl (**877**), 2,4-dichloro-3-hydroxy-bibenzyl (**978**), 2-chloro-3-hydroxy-bibenzyl (**979**)	*Bacillus subtilis*	30 μg/disk	[339]
[L] *Schistochila glaucescens*		*Bacillus subtilis*	60 μg/disk Inhibition zone (mm)	[141]
	Marchantin C (**197**)		2.0	
	Neomarchantin A (**212**)		1.5	
	Neomarchantin B (**213**)		2.0	
[M] *Sphagnum fimbriatum*	Isoplagiochin D-1 (**882**)[b]	Methicillin-resistant *Staphylococcus aureus*	*MIC* (μg/cm^3) > 128	[291]

(continued)

Table 12 (continued)

Species name[a]	Compound or type of solvent extract	Bacterial name	Activity value minimal inhibitory concentration (*MIC* μg/cm^3, μ*M*) (Inhibition zone: mm)	Refs.
	Isoplagiochin D-2 (**883**)[v]		0.5	
	Isoplagiochin D-3 (**884**)[b]		2	
[M] *Sphagnum magellanicum*	EtOH (containing caffeic (**800**), chlorogenic (**843**), *p*-coumaric (**801**), 3,4-dihydroxybenzoic (**903**), gallic (**796**), salicylic (**850**), and vanillic (**905**) acid)	*Escherichia coli*	*MIC* (μg/cm^3) 1162.3	[305]
		Erwinia carotovora subsp. *carotovora*	581.3	
		Salmonella typhi	1162.3	
		Streptococcus type β	1162.3	
		Vibrio cholerae	581.3	
[L] *Trichocolea hatcheri*	Methyl 4-[(2*E*)-3,7-dimethyl-2,6-octadienyl]oxy]3-hydroxybenzoate (**770**)	*Staphylococcus epidermidis*	*MIC* 1 (mg/cm^3)	[260]

[a] [L] liverwort, [M] moss
[b] Synthetic

cm^3, with the minimum bacterial concentrations (MBC) being from 1 to 5 mg/ cm^3. The most potent antibacterial activity was shown towards *S. aureus* by the bis-benzyl **1** with a *MIC* value of 0.06 mg/cm^3. A *MIC* value of 0.003–0.025 mg/cm^3 of compound **1** isolated from the Japanese *M. polymorpha* was reported in 1990 by Asakawa and colleagues, as shown in Table 12. The methanol extracts of both *M. polymorpha* and *Pellia endiviifolia* each showed an antibacterial effect against *L. monocytogenes*, with *MIC* values of 2.5 and 5.0 mg/cm^3, respectively, and against *S. aureus* (*MIC* 0.16 and 0.32 mg/cm^3, respectively).

Gram-negative bacteria were resistant to all the extracts tested and to marchantin A (**1**). The methanol extract of *C. conicum* did not show an antibacterial effect against any of the bacteria tested, although several different researchers have reported that some solvent extracts of *C. conicum* do exhibit antibacterial activity, as indicated in Table 12. The use of ^{1}H NMR spectra showed the crude extracts of *M. polymorpha* and *Pellia endiviifolia* to contain bis-bibenzyls, while these characteristic compounds were absent in *C. conicum* [286].

The antibacterial and antifungal activities of the *n*-hexane, chloroform, methanol, and water extracts of the Indian *Marchantia polymorpha* were evaluated against Gram-positive (*Bacillus subtilis* and *Streptococcus mutans*) and and Gram-negative (*Klebsiella pneumoniae*) bacteria and two fungal pathogens (*Candida albicans* and *Rhizopus oryzae*), using an agar well diffusion method. A concentration of 200 mg/ cm^3 of each extract was prepared and their inhibition zones (mm) were evaluated. Of these, the methanol extract of *M. polymorpha* showed the best antifungal activity against *Candida albicans* (18 mm), followed by *Rhizopus oryzae* (17 mm). Antibacterial activity was found also against *Klebsiella pneumoniae* (15 mm), in addition to *Bacillus subtilis* (14 mm) [287].

In order to obtain further antimicrobial compounds from liverwort bis-bibenzyls, the acyclic bis-bibenzyl, perrottetin F (**184**), was incubated with *Aspergillus niger* to give three new products, 8-hydroxyperrottetin F (**270**), 3,4-dihydroxy-5-benzyloxybibenzyl (**271**) and perrottetin F-6′-sulfate (**272**), although these showed weaker antimicrobial effects against *Pseudomonas aeruginosa, Staphylococcus aureus*, and *S. marcescens* than the parent compound, perrottetin F (**184**). The synthesis of short-chain acyl homoserine lactone bacterial-quorum-sensing molecules was inhibited by these three new metabolites [288].

Sawada et al., reported in 2012 that riccardin C (**182**) isolated from the Japanese liverwort *Reboulia hemisphaerica* exhibited potent antibacterial activity, with *MIC* values of 3.2 and 3.2 μg/cm^3 against methicillin-resistant *Staphylococcus aureus* (MRSA) strains OM481 and OM584, comparable to those of the clinically used drugs, vancomycin, and linezolid. Furthermore, ten analogues (**864–873**) and three fragments (**874–876**) of riccardin C (**182**) were synthesized and their antibacterial activities were evaluated (see Scheme 10 for the synthesis of 2- (**872**) and 3-hydroxyriccardin C (**873**)). Bis-bibenzyls with a single methoxy and one hydroxy group did not show anti-MRSA activity except for compounds **868** and **869**, which gave activity only towards strain OM481. However, bis-bibenzyls with two and three phenolic hydroxy groups (**870** and **871**, and **182**) showed anti-MRSA activity against two strains, with *MIC*s of 3.2 μg/cm^3, each. Compounds **872** and **873**, which possess one more hydroxy group than riccardin C, exhibited decreased levels of anti-MRSA activity. It is noteworthy that 1-hydroxybiphenyl ether (**874**), which constitutes the A and C ring of riccardin C (**182**), showed high anti-MRSA activity (*MIC* 6.3 μg/cm^3) against two strains. Neither 1,4′-dihydroxybiphenyl (**875**) nor 3,4′-dihydroxybibenzyl (**876**) showed anti-MRSA effects [289].

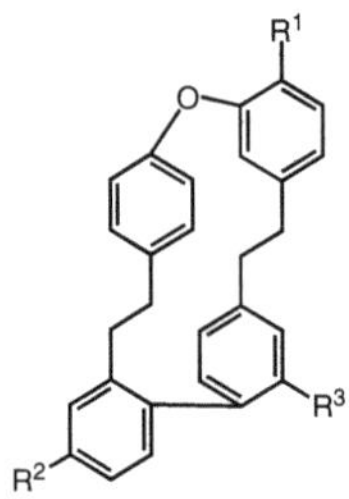

864 (R^1 = OMe, R^2 = R^3 = H)
865 (R^1 = H, R^2 = OMe, R^3 = H)
866 (R^1 = H, R^2 = H, R^3 = OMe)
867 (R^1 = OH, R^2 = H, R^3 = H)
868 (R^1 = H, R^2 = OH, R^3 = H)
869 (R^1 = H, R^2 = H, R^3 = OH)
870 (R^1 = OH, R^2 = OH, R^3 = H)
871 (R^1 = OH, R^2 = H, R^3 = OH)

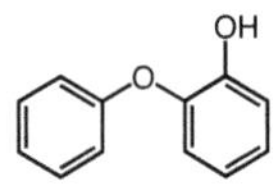

874

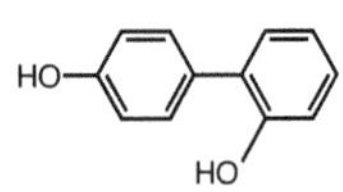

875

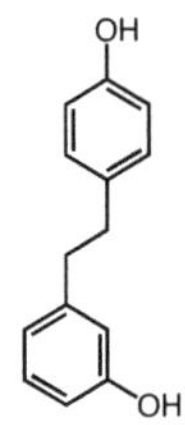

876

Sawida et al., also synthesized two fluororiccardin C derivatives, 2-fluororiccardin C (**877**) (Scheme 11) and 1′-deoxy-1′-fluororiccardin C (**878**), as well as demethyl-*neo-iso*-marchantin C (**879**) and 1′-dehydroxy-5′-hydroxy-isoplagiochin D (**880**). Their anti-MRSA activities were tested along with those of natural riccardin C (**182**), riccardin D (**253**), and isoriccardin D (**881**) against the OM481, OM 584, and N315 strains by means of a liquid microdilution method. The natural riccardin series (**182**, **253**, and **881**) and the synthetic 2-fluoro- (**877**) and 1′-deoxy/1′-fluororiccardin C (**878**) exhibited potent anti-MRSA activities (IC_{50} 1–8 μg/cm^3) against the three strains used. However, synthetic demethyl-dehydroxy-neoisomarchantin C (**879**) did not show anti-MRSA activity against these three strains, and 1′-dehydroxy-5′-hydroxyisoplagiochin D (**880**) exhibited only weak anti-MRSA activity (IC_{50} 32 μg/cm^3). 1-Hydroxybiphenyl ether (**874**) was re-evaluated with respect to its antimicrobial activity. It showed potent anti-MRSA activity with an IC_{50} value of 8 μg/cm^3 against the three strains used [290].

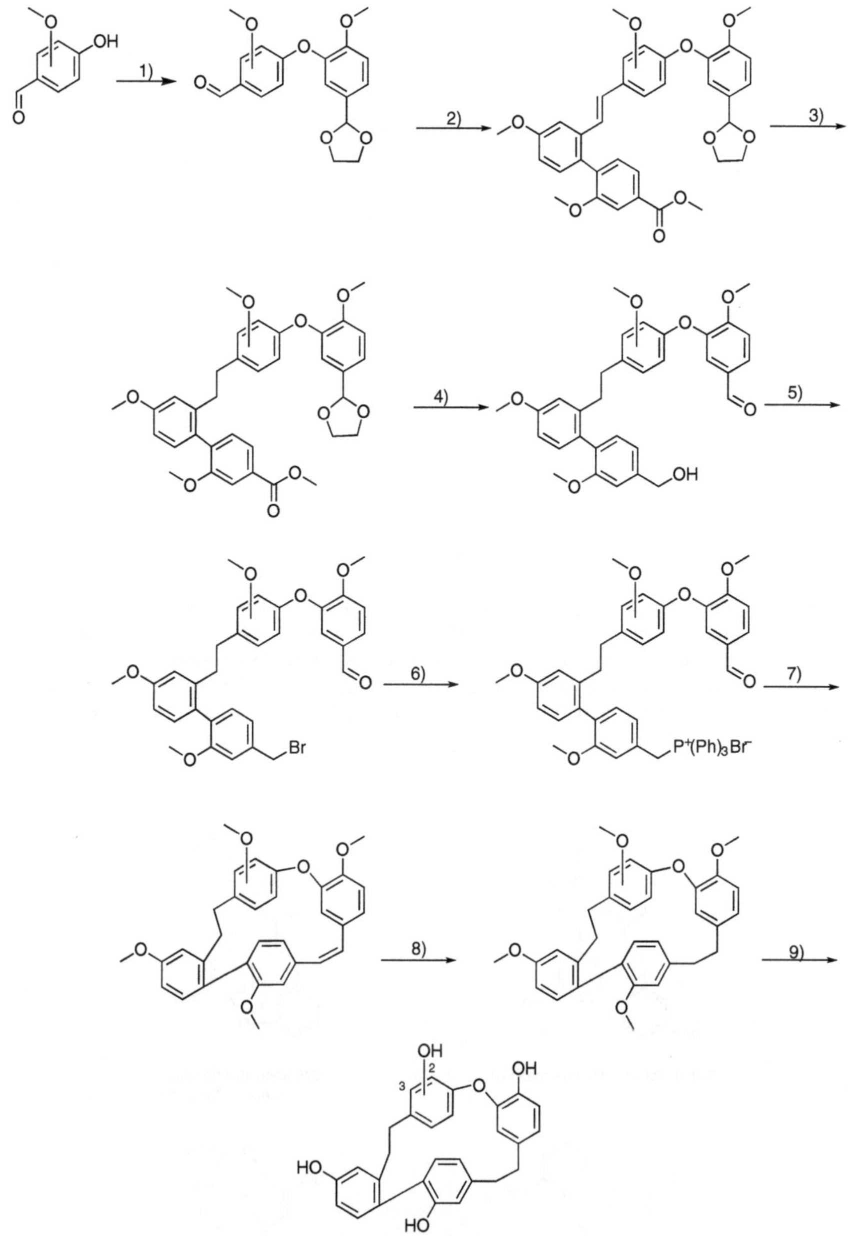

872 (11'-dehydroxy-2-hydroxyriccardin C)
873 (11'-dehydroxy-3-hydroxyriccardin C)

Scheme 10 Preparation of 11′-dehydroxy-2-hydroxyriccardin C (**872**) and 11′-dehydroxy-3-hydroxyriccardin C (**873**). (1) 4-fluoro-3(or 2)-methoxybenzaldehyde, K_2CO_3, DMF, (2) 5-methoxy-2(2-methoxy-4-methoxy)carbonylphenylmethyltriphenylphosphonium bromide, K_2CO_3, CH_2Cl_2, (3) 10% Pd-C, H_2, (4) a: $LiAlH_4$, Et_2O, b: aq HCl (5) CBr_4, PPh_3, benzene, (6) PPh_3, MeCN, (7) NaOMe, CH_2Cl_2, phosphonium salt, CH_2Cl_2, (8) 10% Pd-C, H_2, AcOEt, (9) BBr_3, CH_2Cl_2

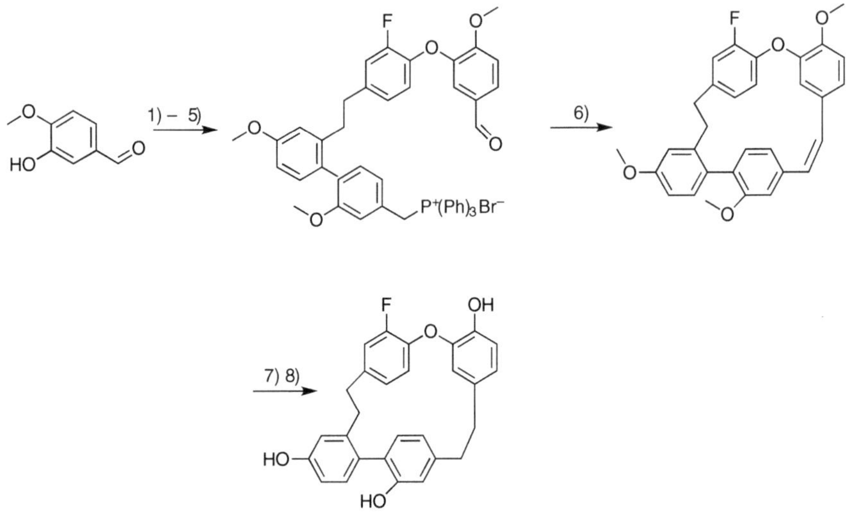

877 (2-fluororiccardin C)

Scheme 11 Preparation of fluororiccardin C (**877**). (1) a: ethylene glycol, PPTS, benzene, b: 3,4-difluorobenzaldehyde, K_2CO_3, DMF, (2) 5-methoxy-2-(2-methoxy-4-methoxycarbonylphenylmethyltriphenylphosphonium bromide, K_2CO_3, 18-crown-6, CH_2Cl_2, (3) a: 10% Pd-C, H_2, 0.4 Mpa, TEA, AcOEt, b: $LiAlH_4$, THF, c: aq HCl, (4) CBr_4, PPh_3, benzene, (5) PPh_3, MeCN, (6) NaOMe, CH_2Cl_2 phosphonium salt, CH_2Cl_2, (7) 10% Pd-C, H_2, 0.4 Mpa, TEA, AcOEt, (8) BBr_3, CH_2Cl_2

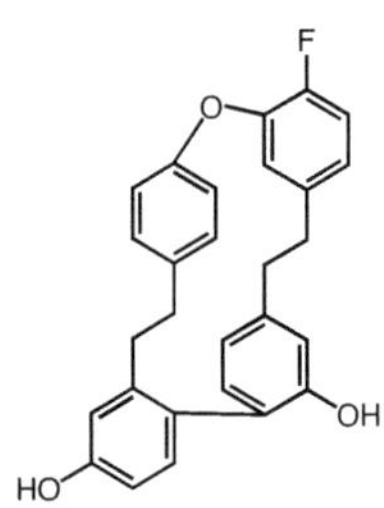

878 (1'-deoxy-1'-fluororiccardin C)

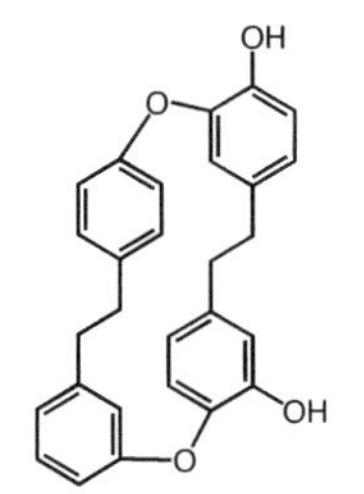

879 (demethyl-dehydroxy-neoisomarchantin C)

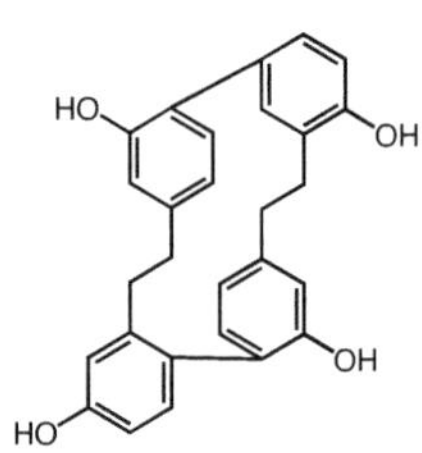

880 (1'-dehydroxy-5'-hydroxy-isoplagiochin D)

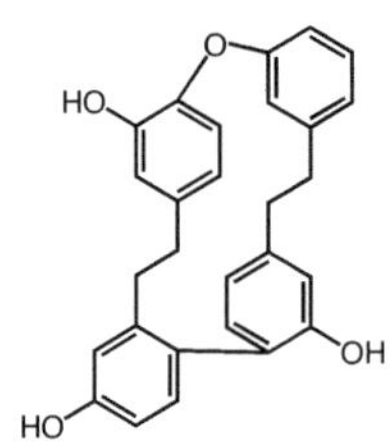

881 (isororiccardin D)

In order to understand further the structure-activity relationships of the riccardin C class of bis-bibenzyls, Onada et al., synthesized the isoplagiochins **882–884** and evaluated their anti-MRSA effects. Isoplagiochin D-1 (**882**) gave an IC_{50} value of > 125 μg/cm^3, while the latter two compounds (**883** and **884**) possessed potent anti-MRSA activities (0.5 and 2 μg/cm^3). The potency shown for compounds **883** and **884** is due to their more rigid conformational structures than occurs in compound **882**.

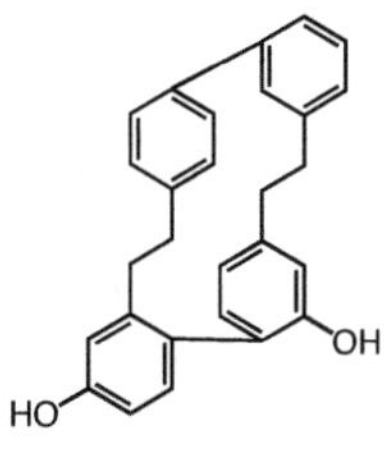

882 (isoplagiochin D-1)

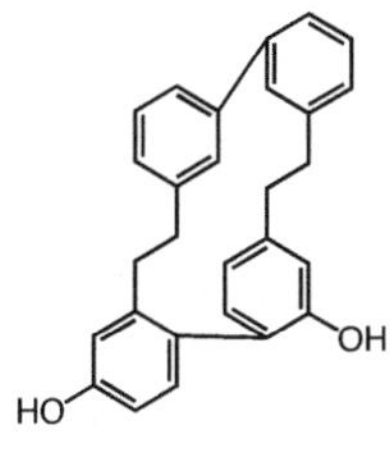

883 (isoplagiochin D-2)

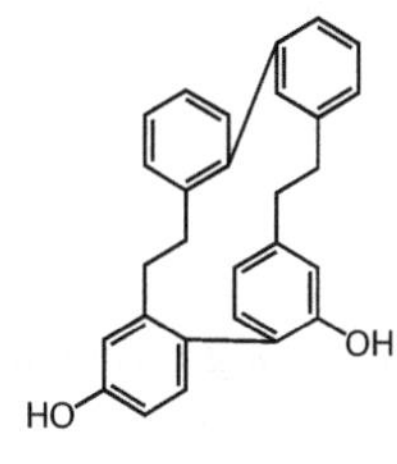

884 (isoplagiochin D-3)

The effects of the above-mentioned synthetic bis-bibenzyls on ethidium bromide inflow and outflow from cells and on intercellular Na^+ and K^+ concentrations were investigated. It was found that the synthetic bis-bibenzyls elicited their anti-MRSA activities by damaging the cell membrane of *S. aureus*, thereby increasing its permeability [291].

The crude ether and methanol extracts of the Tahitian *Mastigophora diclados* showed antimicrobial activities against *Bacillus subtilis* and *Staphylococcus aureus* (*MIC* 16 and 64 μg/cm^3) [182]. Bioactivity-guided fractionation of both extracts gave (–)-diplophyllolide (**132**), (–)-α-herbertenol (**405**), (–)-herbertene-1,2-diol (**406**), (–)-mastigophorene C (**407**), (–)-mastigophorene D (**408**), diplophyllin (**411**), and mastigophorene A (**885**), among which **406**, **407**, and **408** showed moderate antimicrobial effects against *B. subtilis* at *MIC* values of 2–8 μg/cm^3. 1-Hydroxy-2-methoxyherbertene (**409**) and herbertene-1,2-diacetate (**410**), derived from herbertene-1,2-diol (**406**), exhibited *MIC* values of 16 and 32 μg/cm^3, respectively, against *B. subtilis* [195]. The herbertane sesquiterpenoids, α-herbertenol (**405**), herbertene-1,2-diol (**406**), mastigophorene C (**407**), β-herbertenol (**886**), mastigophoric acid methyl ester (**887**), α-formyl-herbertenol (**888**), and 1,2-dihydroxyherberten-12-al (**889**), isolated from the Madagascan *M. diclados*, were tested against *Staphylococcus aureus*, using an agar diffusion method. These sesquiterpenoids showed weaker activity than the standard antibiotics chloramphenicol (22 mm) and kanamycin (23 mm). Of the compounds tested, mastigophorene C (**407**), a dimer of herbertene-1,2-diol (**406**), showed the most potent antibacterial activity at 17 mm while **406** also displayed significant activity at 13 mm [292].

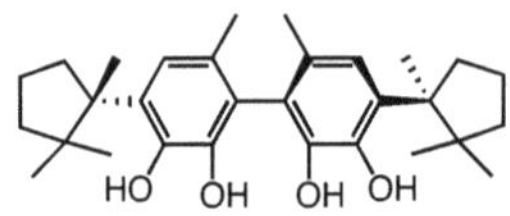

885 (mastigophorene A)

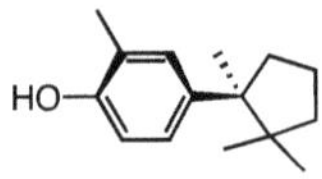

886 (β-herbertenol)

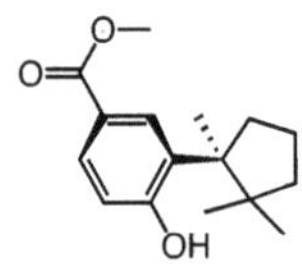

887 (mastigophoric methyl ester)

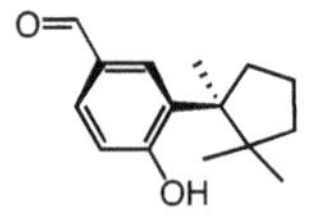

888 (formyl-α-herbertenol)

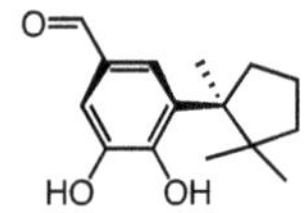

889 (1,2-dihydroxy-herberten-12-al)

2-Hydroxy-4,6-dimethoxyacetophenone (**767**) from *Plagiochila fasciculata* showed weak growth inhibitory activity against *Escherichia coli*, *Proteus mirabilis* and *Staphylococcus aureus* at 100 μg/cm^3 [258].

The antimicrobial activities of the ethanol extracts of five mosses, *Ceratodon purpureus*, *Dryptodon pulvinatus*, *Hypnum cupressiforme*, *Rhytidiadelphus squarrosus*, and *Tortula muralis* were evaluated against *Enterococcus faecalis*, *Escherichia coli*, *Pseudomonas aeruginosa*, *Staphylococcus aureus*, and *S. epidermidis* [275]. The moss extracts exhibited no evident antimicrobial activity at a concentration range tested up to 1 μg/cm^3.

The chlorophyll metabolites, phaeophytin a (**890**), 13^2-hydroxy-(13^2-*S*)-phaeophytin a (**891**), 13^2-hydroxy-(13^2*R*)- phaeophytin a (**892**), and 13^2-(MeO$_2$)-(13^2*R*)-phaeophytin a (= phaeophytin a hydroperoxide) (**893**), isolated from the methanol-soluble extract of a cell suspension culture of *Plagiochila ovalifolia*, showed antimicrobial activity against *Bacillus subtilis* and *Escherichia coli* [293].

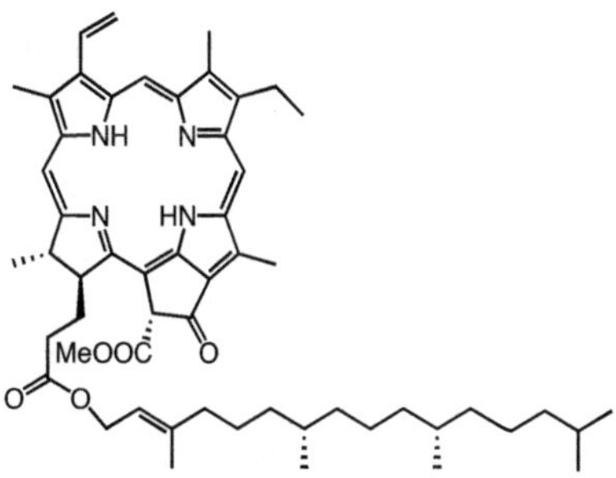

890 (phaeophytin a)

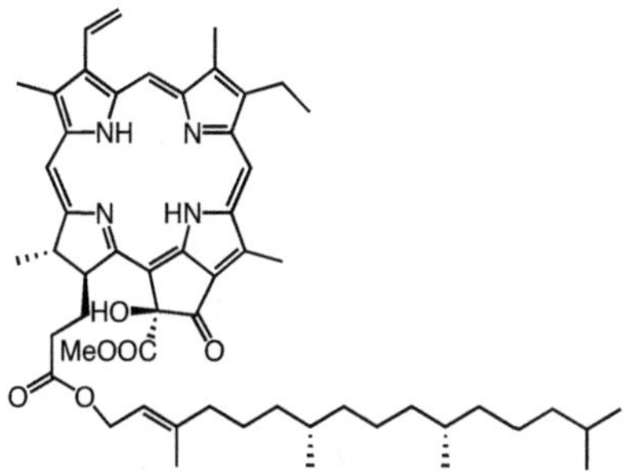

891 (13^2-hydroxy-(13^2S)-phaeophytin a)

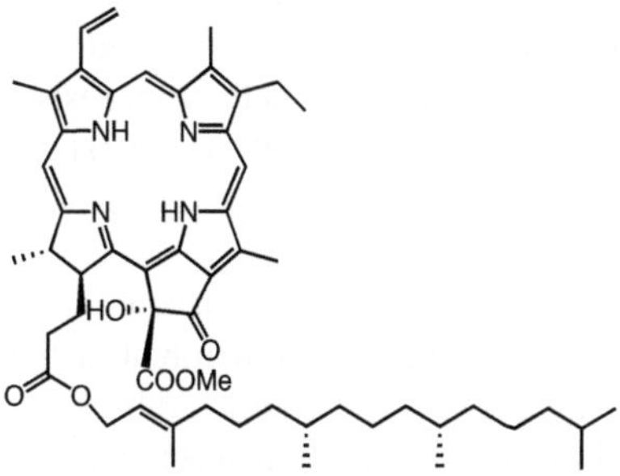

892 (13^2-hydroxy-(13^2R)-phaeophytin a)

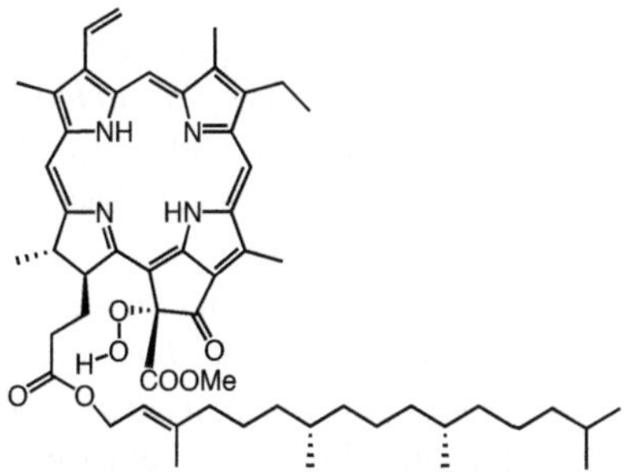

893 (13^2-peroxy-(13^2R)-phaeophytin a)

In the search for new antituberculosis lead compounds from bryophytes, Scher et al., isolated 14 trachylobane diterpenoids from the liverwort *Jungermannia exsertifolia* subsp. *cordifolia,* among which *ent*-trachyloban-17-al (**894**) showed the most significant activity against the virulent *Mycobacterium tuberculosis* H37Rv strain, with a MIC_{90} value of 24 µg/cm^3. *ent-3β*-Acetoxy-19-hydroxytrachylobane (**895**), *ent*-trachyloban-3-one (**896**), *ent-3β*-hydroxytrachylobane (**897**), and *ent-3β*-acetoxytrachylobane (**898**) demonstrated weak inhibitory activities against the same microbe (MIC_{90} values: 59, 50, 51, and 111 µg/cm^3, respectively) [294].

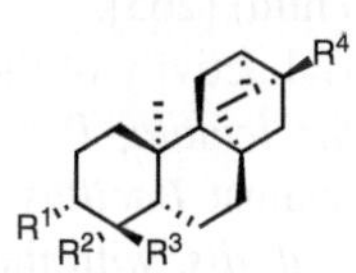

894 (*ent*-trachyloban-17-al, R^1 = H, R^2 = R^3 = Me, R^4 = CHO)
895 (*ent-3β*-acetoxy-19-hydroxytrachylobane, R^1 = OAc, R^2 = CH_2OH, R^3 = R^4 = Me)
897 (*ent-3β*-hydroxytrachylobane, R^1 = OH, R^2 = R^3 = R^4 = Me)
898 (*ent-3β*-acetoxytrachylobane, R^1 = OAc, R^2 = R^3 = R^4 = Me)

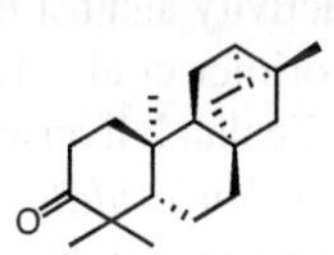

896 (*ent*-trachyloban-3-one)

The essential oil of *Marchesinia mackaii* showed antibacterial activity against *Bacillus subtilis, Escherichia coli, Salmonella pullorum, Staphylococcus aureus,* and *Yersinia enterocolitica* [295].

The ether extract of *Pellia endiviifolia* led to the growth inhibition of *Staphylococcus aureus* with an *MIC* value of 16 μg/cm^3. The pungent diterpene dialdehyde, sacculatal (**128**), isolated from this species, showed potent antibacterial activity against *Streptococcus mutans* (a causative organism of dental caries), exhibiting a LD_{50} value of 8 μg/cm^3. Compound **128** also exhibited the same level of activity against *Bacillus subtilis* and *Cryptococcus neoformans* with *MIC* values, in turn, of 16 and 50 μg/cm^3. Interestingly, (–)-polygodial (**90**) possessing a 1,2-dialdehyde group, was inactive (LD_{50} 100 μg/cm^3) in this particular bioassay [38, 49].

Lunularin (**103**) from *Dumortiera hirsuta* also showed antimicrobial activity against *Pseudomonas aeruginosa* at a concentration of 64 μg/cm^3 [176]. Lepidozenolide (**367**), isolated from *Lepidozia fauriana*, was active against methicillin-resistant *Staphylococcus aureus* at 100 μg/cm^3 [185]. Riccardiphenol C (**404**), a constituent of *Riccardia crassa*, showed antibacterial activity against *Bacillus subtilis* at 60 μg/disk [194].

Methyl 4-[(2*E*)-3,7-dimethyl-2,6-octadienyl)oxy]-3-hydroxybenzoate (**770**), obtained from *Trichocolea hatcheri*, showed very weak antimicrobial activity against *Staphylococcus epidermidis* (*MIC* 1 mg/cm^3), while a structurally similar methyl benzoate with a prenyl ether group exhibited weak antifungal activity [260].

In 2015, Klavina et al., scrutinized the antimicrobial effects of the ethanol extract of 13 mosses against four bacterial species, *Bacillus cereus*, *Escherichia coli*, *Pseudomonas aeruginosa*, and *Staphylococcus aureus*. All extracts showed antimicrobial activity against *Bacillus cereus*. The highest activities against this organism*s* were found for the extracts from *Climacium dendroides* and *Polytrichum commune* (inhibition zone diameter for each, 12 mm) and the lowest activities for the extracts of *Hylocomium splendens* and *Sphagnum magellanicum* (9 mm). The most potent inhibitory activity against *Staphylococcus aureus* was observed for the extract prepared from *Polytrichum commune* (15 mm). The extracts of *Climacium dendroides*, *Ptilium crista-castrensis*, *Rhytidiadelphus triquetrus*, and *Sphagnum magellanicum* showed antibacterial activity against *Escherichia coli* (10 mm) [265].

In 1993, Lorimer et al., studied the antibacterial activity of the ethanol extracts of five New Zealand liverworts, *Plagiochila fasciculata*, *P. stephensoniana*, *P. banksiana*, *P. suborbiculata*, and *P. deltoidea*, against *Bacillus subtilis*. The first four of these species inhibited the growth of *B. subtilis*, with the inhibition zones of each extract being 1, 4, 5, and 2 mm, respectively, at the concentration level tested. The bioactivity-guided fractionation of the ethanol extract of *P. stephensoniana* resulted in the isolation of 3-methoxy-4′-hydroxybibenzyl (**286**), which exhibited antibacterial activity against *B. subtilis* with a *MIC* of 3–5 μg/cm^3. Two dehydro products of compound **286**, 3-methoxy-4′-hydroxy-(*E*)-stilbene (**287**) and 3-methoxy-4′-hydroxy-(*Z*)-stilbene (**288**), were synthesized in order to compare the antimicrobial activities between the natural (**286**) and synthetic **287** and **288**. The former stilbene synthesized showed growth inhibition of *B. subtilis* and *Escherichia* *coli* with *MIC* 2–6 μg/cm^3, while the latter had the same activity against *B. subtilis* and *E. coli* with *MIC* values of 4–9 and 1 μg/cm^3, respectively [177].

Japanese *Radula* species are well-known sources of bibenzyls and prenyl bibenzyls. 2-Geranyl-3,5-dihydroxybibenzyl (**899**), from *R. perrottetii, R. kojana,* and *R. javanica*, and 3-hydroxy-5-methoxy-4-(3-methyl-2-butenyl)-bibenzyl (**900**), from *R. oyamensis*, showed antimicrobial activities against *Staphylococcus aureus* at *MIC* values of 20 and 3 μg/cm^3, respectively [38].

Frullania species are divided into several chemotypes, and *F. muscicola*, belonging to the bibenzyl-type, was found to contain two known bibenzyls, 3,4′-dimethoxy-4-hydroxybibenzyl (**901**) and 3-hydroxy-4′-methoxybibenzyl (**902**), together with a new labdane, muscicolin (**698**), and four flavones. Compound **901** showed weak antimicrobial activity against *Sarcina lutea* with a *MIC* value of 400 μg/cm^3 and compound **902** gave growth inhibitory activities against *Bacillus subtilis, B. spore-cereus*, and *Sarcina lutea* with *MIC* values of 100, 50, and 200 μg/cm^3, respectively [243].

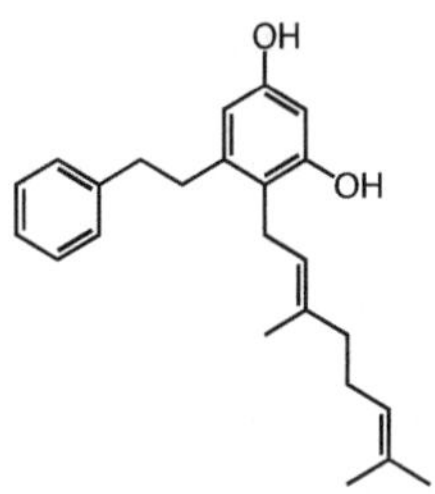

899 (2-geranyl-3,5-dihydroxybibenzyl)

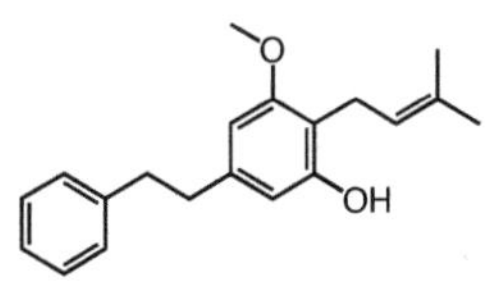

900 (3-hydroxy-5-methoxy-4-(3-methyl-2-butenyl)bibenzyl)

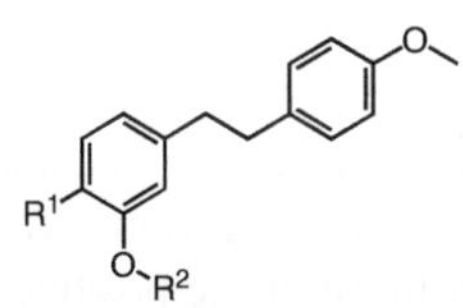

901 (3,4′-dimethoxy-4-hydroxybibenzyl, R^1 = OH, R^2 = Me)
902 (3-hydroxy-4′-methoxybibenzyl, R^1 = R^2 = H)

Xiao et al., confirmed in 2005 that an 80% ethanol extract of *Marchantia convoluta* containing the flavonoids, apigenin (**482**) and luteolin (**483**), their glucuronides, and quercetin (**851**), displayed weak growth inhibitory effects against *Bacillus enteritidis, Escherichia coli, Diplococcus pneumoniae, Salmonella typhi, Staphylococcus aureus*, and *Streptococcus pyogenes* at *MIC* values of 0.6–2.5 mg/cm^3, but not against *Bacillus dysenteriae, Pseudomonas aeruginosa*, and the fungus *Candida albicans* [296].

The antibacterial activity of the crude methanol and flavonoid-free extracts of *Marchantia polymorpha* were examined against three bacterial strains, *Escherichia coli, Proteus mirabilis*, and *Staphylococcus aureus*. Both extracts showed their best activity against *S. aureus* (inhibition zone 20.6 and 19.6 mm, and *MIC* values of 0.28 and 0.31 μg/cm^3, and *MBC* values of 1.13 and 0.31 μg/cm^3, respectively [297].

Gahtori et al., tested in 2011 the methanol and chloroform extracts of *Marchantia polymorpha* against three Gram-negative bacteria strains, *Pasteurella multocida, Salmonella enterica,* and *Xanthomonas oryzae* pv. *oryzae.* Both extracts showed antimicrobial activity against *P. multocida* and *X. oryzae* at *MIC/MBC* 2.50/1.25 and 2.75/1.25 μg/cm^3, respectively [298].

Joshi et al., evaluated in 2022 the antibacterial activity of an 80% methanol extract of the liverwort *Plagiochasma appendiculata* and the moss *Sphagnum fimbriatum* against the two bacterial strains, *Bacillus subtilis* and *Escherichia coli.* Both extracts showed zones of inhibition of 0.9 and 11 mm, and 0.9 and 10 mm, against *B. subtilis* and *E. coli,* respectively [299].

Ilhan et al., described in 2006 the antibacterial activity of acetone and methanol extracts of the Turkish moss *Palustriella commutata* against eleven bacterial strains, *Bacillus mycoides, B. cereus, B. subtilis, Enterobacter aerogenes, Enterococcus faecium, Escherichia coli, Micrococcus luteus, Klebsiella pneumoniae, Pseudomonas aeruginosa, Staphylococcus aureus,* and *Yersinia enterocolitica.* With the exception of *E. faecium* and *S. aureus,* the acetone extract inhibited of all the above-mentioned bacteria (concentration range 7–12 μ*M*), while the methanol extract showed weaker activities than those of the acetone extract. *Bacillus mycoides, B. subtilis, E. aerogenes, E, faecium, P. aeruginosa,* and *S. aureus* were insensitive towards the moss methanol extracts [300].

The methanol extract of the Serbian *Ptilidium pulcherrimum* gave evidence of antimicrobial activity against *Enterobacter cloacae, Escherichia coli, Micrococcus flavus,* and *Staphylococcus aureusi,* with *MIC* values of 10–20 μg/cm^3 and an *MBC* value of 20 μg/cm^3, respectively [301]. The DMSO extract of another Serbian moss, *Rhodobryum ontariense,* exhibited antibacterial activity against *Bacillus cereus, Escherichia coli, Enterobacter cloacae, Listeria monocytogenes, Micrococcus flavus, Pseudomonas aeruginosa, Salmonella typhimurium,* and *Staphylococcus aureus* at *MIC* and *MBC* values in the range of 1 to 3 μg/cm^3. These activities were found to be 10 to 30 times less potent than those of the positive controls, streptomycin and ampicillin [302].

Veljic et al., confirmed in 2010 that the methanol extracts of the three mosses *Ctenidium molluscum, Fontinalis antipyretica* var. *antipyretica,* and *Hypnum cupressiforme* displayed moderate antimicrobial activities against the Gram-positive bacteria *Bacillus subtilis, Micrococcus flavus,* and *Staphylococcus epidermidis,* and the Gram-negative bacteria *Escherichia coli* and *Salmonella enteritidis,* having *MIC/MBC* values in the range of 10–20 mg/cm^3. The antibacterial activities of the methanol extracts of the species tested were greater against the Gram-negative bacteria than against the Gram-positive bacteria used [303].

Antimicrobial bioassays of the *n*-hexane, ethyl acetate, and methanol extracts of four Sri Lankan mosses, *Calymperes motoleyi*, a *Fissidens* species, *Hypnum cupressiforme*, and *Sematophyllum demissum*, and two liverworts in the genera *Marchantia* and *Plagiochila* were carried out in 2020 by Kirisanth et al., against *Bacillus subtilis, Pseudomonas aeruginosa* and *Staphylococcus aureus* at a 500 µg/disk concentration. The *n*-hexane and ethyl acetate extracts of the *Fissidens* species inhibited the growth of *B. subtilis* (inhibition zone 10.5 mm), *P. aeruginosa* (7.3 mm), *S. aureus* (10.5 mm), *B. subtilis* (9.2 mm), and *S. aureus* (8.3 mm), respectively. The *n*-hexane extract of the *Marchantia* species showed an antibacterial effect only against *S. aureus* (8.3 mm). The methanol extracts of *C. motoley* and *S. demissum* exhibited zones of inhibition of 7.3 and 8.3 mm, and 7.3 and 12.3 mm, respectively, for *B. subtilis* and *S. aureus.* However, the methanol extract of *H. cupressiforme* showed an antibacterial effect only against *S. aureus* (8.5 mm) [304].

An investigation of the antibacterial compounds from the Chilean moss, *Sphagnum magellanicum*, has resulted in the identification of eight simple aromatic compounds, caffeic (**800**), chlorogenic (**843**), salicylic (**850**), 3,4-dihydroxybenzoic (**903**), syringic (**904**), and vanillic (**905**) acids, sculetin (**906**), as well as scopoletin (**907**). In 2009, Montenegro et al., tested an ethanol extract containing caffeic (**800**), chlorogenic (**843**), *p*-coumaric (**801**), gallic (**796**), salicylic (**850**), and vanillic (**905**) acids, of *S. magellanicum,* which is an economically important moss in Chile, against the Gram-negative *Azotobacter inelandii, Erwinia carotovora,* subsp. *carotovora, Enterobacter aerogenes, Escherichia coli, Pseudomonas aeruginosa, Salmonella typhi,* and *Vibrio cholerae,* the Gram-positive bacterium *Staphylococcus aureus* subsp. *aureus,* and *Streptococcus* type β. From this investigation, *E. carotovora* subsp. *carotovora* and *V. cholerae* were inhibited by the extract with *MIC* 581 µg/cm^3, while this ethanol extract also inhibited the growth of *E. coli* and *Streptococcus* type β with *MIC* 1162 µg/cm^3 [305].

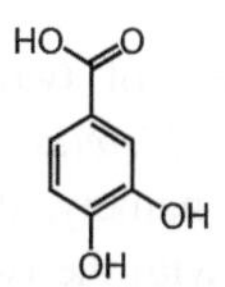

903 (3,4-dihydroxybenzoic acid)

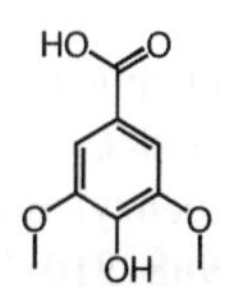

904 (syringic acid)

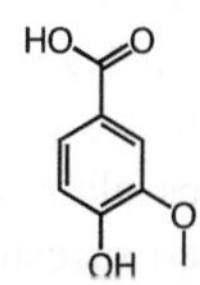

905 (vanillic acid)

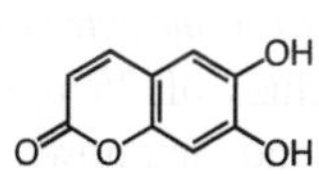

906 (sculetin)

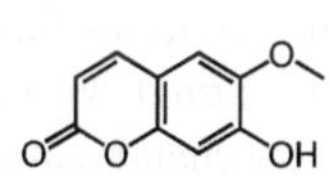

907 (scopoletin)

Antibacterial activity of the solvent extracts (*n*-hexane, chloroform, *n*-butanol, and methanol) of two liverworts, *Conocephalum conicum* and *Plagiochasma appendiculatum*, and two mosses, *Bryum argenteum* and *Mnium marginatum* were evaluated against the Gram-positive bacterial *Bacillus cereus*, *B. subtilis*, *Staphylococcus aureus*, and *Streptococcus pyogenes*, and the Gram-negative bacteria, *Enterobacter aerogenes*, *Enterococcus faecalis*, *Escherichia coli*, *Klebsiella pneumoniae*, *Proteus mirabilis*, and *Pseudomonas aeruginosa*, using disk diffusion and microdilution assays [306]. *Staphylococcus aureus* was the most sensitive strain of those tested against the *n*-butanol partition of *M. marginatum*, with the largest inhibition zone of 102.9% in comparison to the standard drug erythromycin. The *n*-butanol extract also showed antibacterial activity against *B. cereus* (92.5%), *B. subtilis* (94.7%), *E. coli* (106.8%), and *S. pyogenes* (92%). *Pseudomonas aeruginosa* was the most sensitive among the Gram-negative bacteria tested, with inhibition zones, in turn, of 92.5 and 120.3% for the chloroform extracts of *P. appendiculatum* and *C. conicum*. With the exceptions of *P. appendiculatum* (*MIC* 1.87 mg/cm^3) when evaluated against *B. argenteum*, and *M. marginatum* (*MIC* 1.25 mg/cm^3) against *B. subtilis*, *E. faecalis*, and *P. aeruginosa*, all tested bryophyte extracts displayed discrete antibacterial activities [306].

The antimicrobial activity of aqueous and ethanol extracts of eleven mosses and nine liverworts were evaluated against *Bacillus cereus*, *Escherichia coli*, and *Staphylococcus aureus*. The 73% ethanol extracts of the mosses, *Atrichum undulatum* (Plate 28), *Dicranum scoparium*, *Eurhynchium angustirete*, *Hylocomium splendens*, *Polytrichum commune*, *Rhodobryum roseum*, *Rhytidiadelphus squarrosus*, and the liverworts, *Frullania dilatata* and *Lophocolea heterophylla* showed moderate antibacterial activity against *S. aureus* (MIC_{80} 3–33% for all plant extracts). The aqueous extracts of *D. scoparium*, *H. splendens*, *R. roseum*, *P. commune*, *F. dilatata*, *L. heterophylla*, and the liverwort, *Ptilidium pulcherrimum* also inhibited the growth of *S. aureus* (*MIC* 13–33%) [307].

Ng et al., analyzed the phytochemical constituents of two Bornean liverworts, *Acrobolbus saccatus* [308] and *Chandonanthus hirtellus* [309]. The first of these species elaborated five bibenzyl derivatives, namely, three new prenyl bibenzyls named saccatenes A–C (**908**–**910**), together with the two known prenyl bibenzyls, radulanin A 5-one (**349**) and 2,2-dimethoxy-5-hydroxy-6-carboxy-7-(2-phenylethyl)-chromane (**911**) [173]. Among these compounds **349**, **908**, and **909** exhibited antimicrobial activity against *Listeria monocytogenes*, *Salmonella enteritidis*, and *Yersinia enterolitica* with *MIC* values of 25 μg/cm^3 [308]. The latter liverwort afforded the new chandonanol (**912**) and the known chandonanthone (**519**) and isochandonanthone (**520**) [216], which were shown to possess weak antibacterial activity against *Escherichia coli* and *Staphylococcus aureus* with *MIC* values of 100–275 μg/cm^3 [309].

Plate 28 *Atrichum undulatum* (moss)

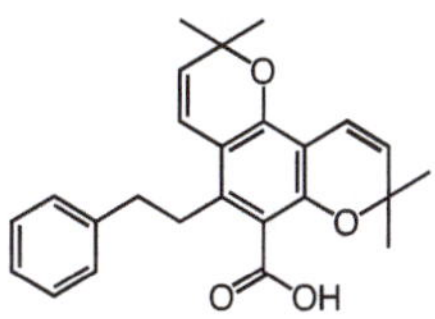

908 (saccatene A)

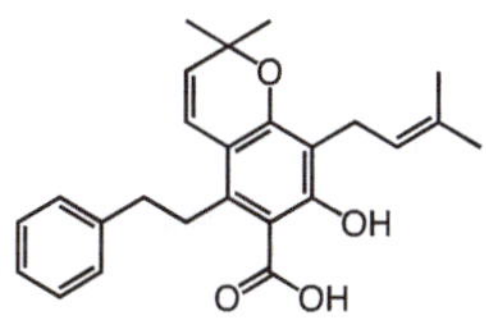

909 (saccatene B, R = H)

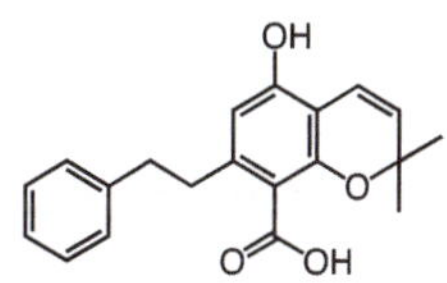

910 (saccatene C, R = Me)

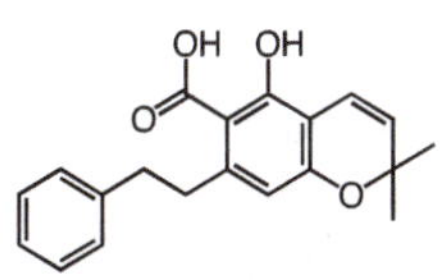

911 (2,2-dimethyl-5-hydroxy-6-carboxy-7-(β-phenethyl)chromene)

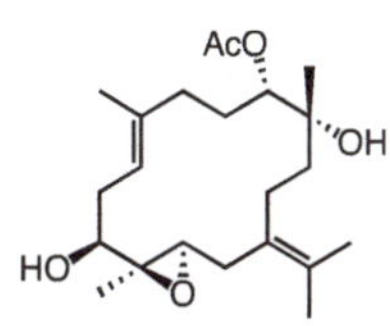

912 (chandonanol)

Basile et al., (1999) reported seven flavonoids, bartramiaflavone (**913**), lucenin 2 (**914**), saponarin (**915**), apigenin (**482**), vitexin (**916**) apigenin-7-triglycoside (**917**), and luteolin-7-*O*-neohesperioside (**918**), isolated from five mosses, *Bartramia pomiformis*, *Dicranum scoparium*, *Hedwigia ciliata*, *Plagiomnium affine*, and *P. cuspidatum*, to possess antimicrobial activity specifically against the Gram-negative bacteria *Enterobacter aerogenes*, *Enterobacter cloacae*, *Escherichia coli*, *Klebsiella pneumoniae*, *Proteus mirabilis*, *Pseudomonas aeruginosa*, and *Salmonella typhi*, and the Gram-positive bacterium, *Enterococcus faecalis*. The flavones mentioned above strongly inhibited the growth of *E. aerogenes* and *E. coli* (*MIC* 4–8 μg/cm^3). However, these flavonoids did not show antibacterial activity against either *Proteus* *vulgaris* or *Staphylococcus aureus* [310]. Eryodictiol (**919**) and hesperitin (**920**) were not encountered in any of the mosses investigated.

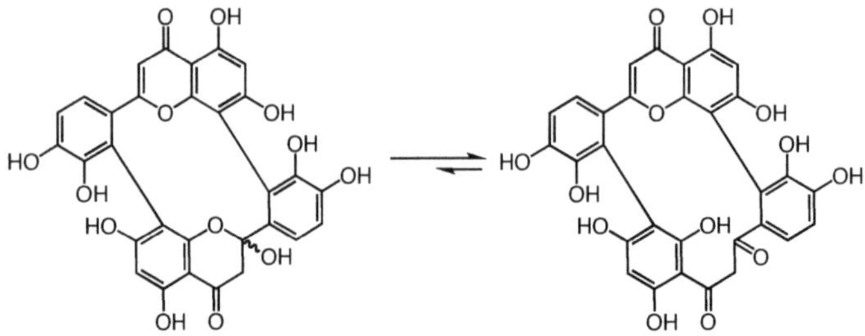

913 (bartamiaflavone)

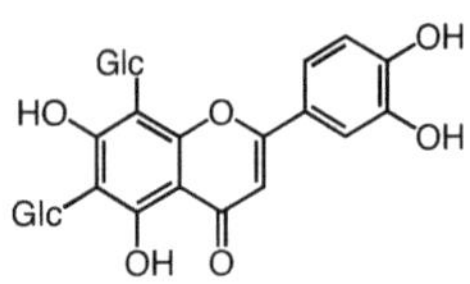

914 (lucenin 2)

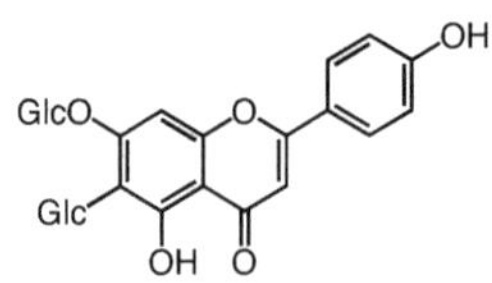

915 (saponarin)

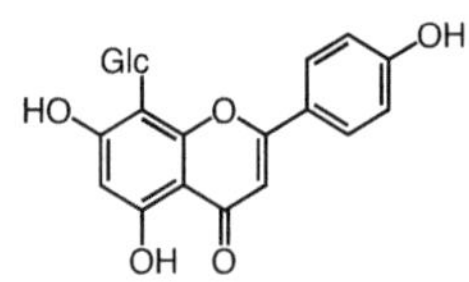

916 (vitexin)

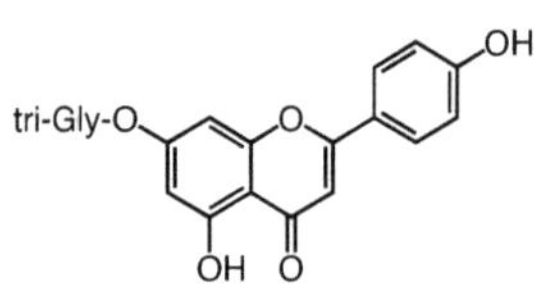

917 (apigenin triglycoside)

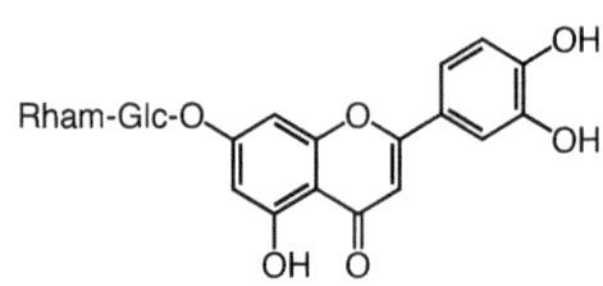

918 (luteolin-7-*O*-neohesperoside)

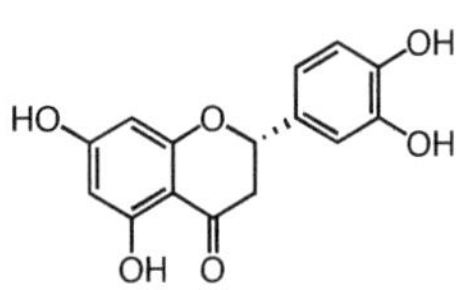

919 (eryodictiol)

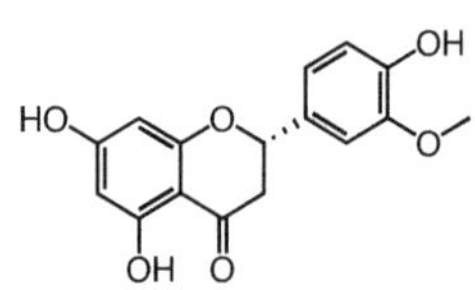

920 (hesperitin)

The methanol extracts of the liverwort, *Plagiochila beddomei* and two mosses *Leucobryum bowringii* and *Octoblepahrum albidum* were analyzed by reversed-phase HPLC to detect gallic (**796**), protcatechuic (**797**), *p*-hydroxybenzoic (**799**), caffeic (**800**), coumaric (**801**), ferulic (**848**), vanillic (**905**), cinnamic, and chlorogenic (**843**) acids. The extracts containing aromatic acids of this type showed antibacterial activity against *Bacillus cereus*, *B. subtilis*, *Cryptococcus neoformans*, *Escherichia coli*, *Klebsiella pneumoniae*, *Proteus vulgaris*, *Pseudomonas aeruginosa*, *Salmonella typhimurium*, and *Staphylococcus aureus* at *MIC/MBC* concentration ranges of 0.0625–2.0/0.25–3.0 mg/cm^3) [311].

The antimicrobial activities of the aqueous, methanol and ethanol extracts of an Indian moss, *Entodon nepalensis* were tested against three bacteria, *Bacillus subtilis*, *Escherichia coli*, and *Salmonella typhimurium.* The ethanol extract at a concentration of 0.04–0.06 g/cm^3 showed more pronounced antibacterial effects against the three bacteria than the methanol extract. For *E. coli* and *S. typhimurium*, the extract concentration mentioned above showed greater levels of inhibition than against *B. subtilis*. The ethanol extract of the moss also exhibited larger zones of inhibition than those of the aqueous and methanol extracts [312].

The antimicrobial activity of an ethanol extract of the Turkish moss *Hypnum andoi* was evaluated against 17 bacteria. The disk diffusion testing of this extract showed inhibition zones for *Enterobacter aerogenes, Escherichia coli, Klebsiella pneumoniae, Salmonella Kentucky*, and *Staphylococcus carnosus* of between 7 and 10 mm, at a concentration range of 40–100 μg [313].

Two Turkish mosses, *Cinclidotus fontinaloides* and *Palustriella commutata,* were extracted with ethanol and water and the extracts tested for their antibacterial activity. The ethanol extract of both mosses demonstrated weak inhibitory activities against *Escherichia coli, Bacillus subtilis, Staphylococcus aureus*, and *Pseudomonas aeruginosa* (*MIC* and *MIB*: > 100 and > 200 mg/10 cm^3, respectively) [314].

Vollor et al., (2018) evaluated the antimicrobial activity of *n*-hexane and chloroform extracts of 14 Hungarian mosses against 11 bacterial strains using the disk-diffusion method. Altogether, 19 samples of 15 taxa showed moderate antimicrobial activities including the most active *Plagiomnium cuspidatum*, which was active against eight of the test strains, with MRSA (*Staphylococcus aureus*) being the most susceptible to the species tested [263].

An Indian moss, *Atrichum undulatum*, was extracted with ethanol, methanol and water, and its crude extracts were evaluated for antibacterial activity against *B. subtilis*, *E. coli*, and *S. typhimurium*. However, their activities were only very weak (*MIC* 0.8–1.2 mg) [315].

Six Turkish mosses, *Bryum capillare*, *Grimmia anodon*, *Orthotrichum rupestre*, *Pleurochaete squorrosa*, *Syntrichia ralis*, and *Tortella* sp. were extracted with methanol, ethanol, acetone, and chloroform, respectively, and the antibacterial effects of these extracts were evaluated against *Bacillus subtilis*, *B. cereus*, *Salmonella* sp., *Staphylococcus aureus*, *Pseudomonas aeruginosa*, and *Escherichia coli*. With the exception of *Bacillus subtilis* and *Escherichia coli*, the growth of *B. cereus*, *Salmonella* sp., *S. aureus*, and *Pseudomonas aeruginosa* were inhibited by *G. anodon* (inhibition zone: 6–8 mm). Almost all of the solvent extracts of *O. rupestre*, *P. squorrosa*, and *Syntrichia ruralis* showed weak antimicrobial activities against some of these bacteria, but extracts of *T. tortuosa* were deemed inactive. While crude extracts of bryophytes have tended to be inactive against Gram-negative bacteria, it is noteworthy that the methanol and ethanol extracts of *B. cappillare* and *G. anodon* showed moderate antibacterial activities against the Gram-negative bacterium, *P. aeruginosa* (inhibition zone: 6 mm). The growth of *E. coli* was also inhibited by the methanol and ethanol extracts of *G. anodon* and *P. squarrosa* (inhibition zones of 7 and 8 mm, respectively). Furthermore, the ethanol extract of *O. rupestre* and the methanol extract of *S. ruralis* both showed the same level of inhibitory activity against *E. coli* (7 mm) [316]. The ethanol extract of the Turkish *Bryum capillare* exhibited weak antibacterial activity against *S. aureus* (*MIC* > 125 μg/cm^3) and biofilm inhibition (3 and 5%) at 50 and 100 mm^3/cm^3, respectively [278].

An ethanol extract of *Frullania dilatata* containing the sesquiterpene lactone frullanolide (**3**) showed weak antibacterial activity against *Enterococcus faecalis*, *E. faecium*, *Listeria monocytogenes*, *Staphylococcus aureus*, and *S. hominis* (inhibition zones 7–10 mm) (50–200 mm^3; *MIC* values from 0.13 to 21.4 mg/cm^3) [317].

Five liverworts, *Marchantia emarginata*, *M. paleacea*, *Pellia epiphylla*, *Plagiochasma rupestre*, and *Reboulia hemisphaerica* were extracted with ethanol, methanol, and water, respectively, and the antimicrobial activity of each extract was evaluated against a Gram-negative and a Gram-positive bacterial strain, *Achromobacter xylosoxidans* and *Bacillus lichenformis*, respectively The growth of both bacteria was inhibited by the methanol and ethanol extracts of four liverworts, *Marchantia emarginata*, *P. paleacea*, *Pellia epiphylla*, *Plagiochasma rupestre*, and *Reboula hemisphaerica* (zones of inhibition from 7 to 18 mm). The aqueous extract of *M. paleacea* also showed some antibacterial activity [318].

The chloroform, acetone, ethanol, and water extracts of the Indian liverwort, *Plagiochasma appendiculatum* were evaluated for antimicrobial activity against eleven microorganisms. The ethanol extract strongly inhibited the growth of *Escherichia coli*, *Proteus mirabilis*, and *Salmonella typhimurium*, with a *MIC* value in each case of 2.5 μg/cm^3. *Micrococcus luteus* was inhibited by the ethanol and aqueous extracts of *P. appendiculatum* with an *MIC* value of 10 μg/cm^3. The growth of *P. mirabilis* was also inhibited by the aqueous extracts of this same liverwort, with an *MIC* value of 5 μg/cm^3. The other extracts showed weak antimicrobial effects against *Micrococcus luteus*, *Bacillus subtilis*, *B. cereus*, *Enterobacter aerogenes*, *Escherichia coli*, *Klebsiella pneumoniae*, *Proteus mirabilis*, *Staphylococcus aureus*, *Salmonella typhimurium*, *Streptococcus pneumoniae*, and *Pseudomonas aeruginosa* with an overall MIC_{50} range from 100 to 200 μg/cm^3 [319].

The methanol extract of the Turkish liverwort *Plagiochila asplenioides* showed weak antibacterial activity against *Escherichia coli* (inhibition zone 4.4 mm) and *Salmonella typhimurium* (4.1 mm) [320].

A methanol extract of the Turkish moss, *Dicranum polysetum* was fractionated to give glycer-2-yl-hexadeca-4-yne-(7*Z*,10*Z*,13*Z*)-trienoate (**921**), poriferasterol (**922**), and γ-taraxasterol (**923**). These compounds were isolated and identified for the first time from *D. polysetum.* This initial extract and the various solvent partitions, and isolated compounds obtained were tested against the honeybee bacteria pathogen, *Paenibacillus larvae.* The lowest concentration exhibiting antimicrobial activity of the methanol extract was 10.4 μg/cm^3, while the highest one was 19.0 μg/cm^3 for a *n*-hexane fraction. This *MIC* value also represented the *P. larvae* sporicidal dose as well. The methanol extract and *n*-hexane/ethyl acetate fractions, and several ethyl acetate subfractions had high antimicrobial effects against American Foulbrood- and European Foulbrood-causing bacteria causing honey bee infections. The ethyl acetate fractions and subfractions showed antimicrobial activities against the honeybee larvae pathogens evaluated with the *MIC* range of 1.8–130.0 μg/cm^3. Also, compounds **922** and **923** were the most effective in giving a *MIC* range of 1.8 to 33.4 μg/cm^3. The extractives of *D. polysetum* were postulated for use an as alternative approach to protect honey bee larvae from infectious bacterial diseases [321].

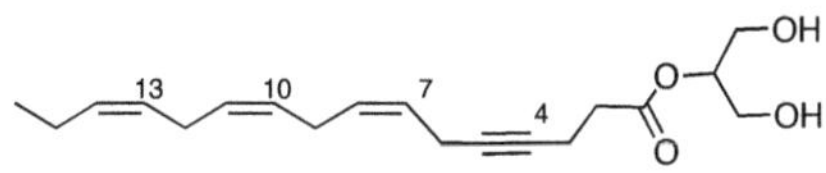

921 (glycer-2-yl-hexadeca-4-yne-(7*Z*,10*Z*,13*Z*)-trienoate)

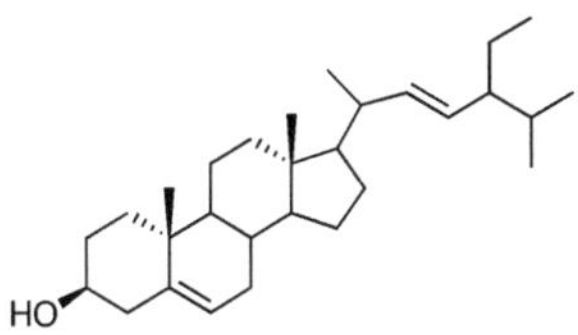

922 (poliferasterol)

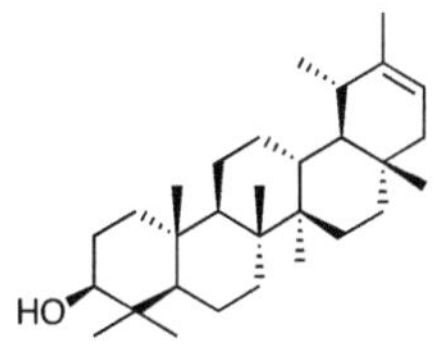

923 (γ-taraxasterol)

The antimicrobial activity of the ethanol, methanol, acetone, and chloroform extracts of five Turkish pleurocarpic mosses, namely, *Anomodon viculosus*, *Homalothecium sericeum*, *Hypnum cupressiforme*, *Leucodon sciuroides*, and *Platyhypnidium riparioides*, were examined against the six bacteria, *Bacillus subtilis*, *B. cereus*, *Salmonella* sp., *Pseudomonas aeruginosa*, and *Escherichia coli.* The methanol extracts of *H. sericeum* and *P. riparioides* and the ethanol extract of *L. sciuroides* were each found to possess antimicrobial activity against the Gram-negative bacterium *Pseudomonas aeruginosa.* The growth of the Gram-negative bacterium *E. coli* was also inhibited by the methanol extracts of *A. viculsus, H. sericeum*, and *P. riparioides* (inhibition zones 7.4 to 7.8 mm). The acetone and chloroform extracts of *H. sericeum* also inhibited the growth of *E. coli* (8.0–8.2 mm). All extracts of *A. viculosus* and *P. riparioides* showed antimicrobial effects against *B. subtilis* (6.7–7.0 mm). *Salmonella* sp. were inhibited by all solvent extracts of *P. riparioides*, and by the methanol and chloroform extracts of *H. cupressiforme* (6.6–8.0 mm). The methanol extract of *H. cericeum*, *H. cupressiforme*, and *P. riparioide*s showed growth inhibition against *B. cereus* (6.4–7.3 mm) [322].

Porella species are divided into nine chemotypes, such as those containing drimane-pinguisane, pinguisane, and sacculatane, and Gilabert et al., reported in 2011 a number of secondary metabolites from the Argentine liverwort, *Porella chilensis.* They isolated two new fusicoccane diterpenoids, 3*α*-hydroxy-6-oxofusicocc-4-ene (**924**) and 4*α*-hydroxy-6-oxofusicocc-3(7)-ene (**925**), and two known analogues, anadensin (**499**) and fusicoauritone (**490**), along with four pinguisane sesquiterpenoids, pinguisenol (**926**), norpinguisone (**927**), norpinguisane acetate (**928**), and norpinguisone methyl ester (**929**), as well as two aromadendrane sesquiterpenoids, *ent*-spathulenol (**931**) and *ent*-4*β*,10*α*-dihydroxyaroma-dendrane (**932**). Configurational details of the new fusicoccane **925** were determined by comparison with africanes, inclusive of compound **930**. This was the first example of the presence of fusicoccanes in the genus *Porella.* Thus, *P. chiliensis* is representative of a new chemotype, the pinguisane-aromadendrane-fusicoccane type in the Porellaceae. Compounds **499**, **931**, and **932** inhibited biofilm formation of the human pathogen *Pseudomonas aeruginosa*, at 53 and 47%, 45 and 41%, and 48 and 37%, at 50 and

5 μg/cm^3, respectively. These compounds also showed slight decreases in bacterial growth and interfered with the processes of quorum sensing at the same doses [323].

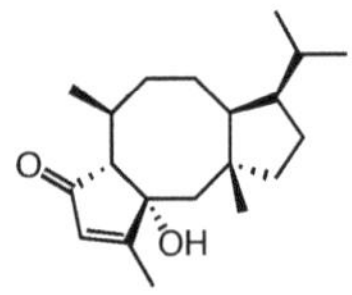

924 (3α-hydroxy-6-oxo-fusicocc-4-ene)

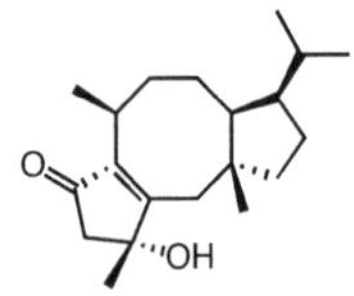

925 (4α-hydroxy-6-oxo-fusicocc-3(7)-ene)

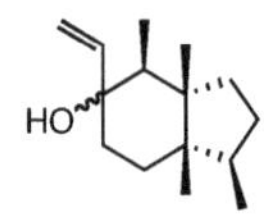

926 (pinguisenol)

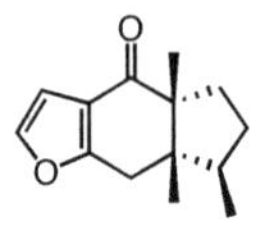

927 (norpinguisone)

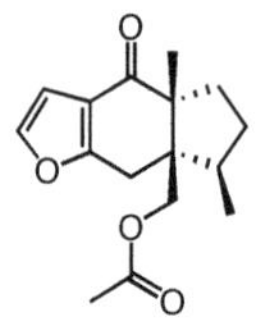

928 (norpinguisane acetate)

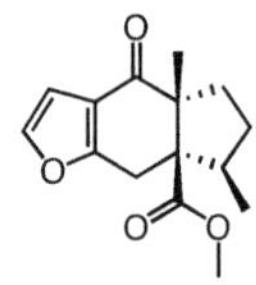

929 (norpinguisone methyl ester)

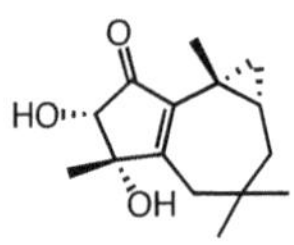

930 ((1a*S*,5*S*,6*S*,7b*R*)-5,6-dihydroxy-3,3,5,7b-tetramethyl-1,1a,2,3,4,5,6,7b-octahydro-7*H*-cyclopropa[*e*]azulen-7-one)

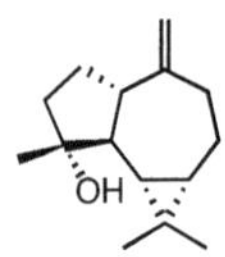

931 (*ent*-spathulenol)

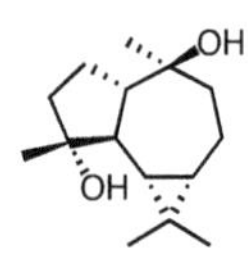

932 (*ent*-4β,10α-dihydroxy-aromadendrane)

4.9 Antifungal Activity

Bryophytes possess not only antibacterial properties but also antifungal effects. Many species of liverworts and mosses collected in China, India, Japan, New Zealand, Sri Lanka, Turkey, Europe, and Madagascar have been extracted using non-polar and polar solvents and their antifungal activities were evaluated, followed by isolation of the active compounds from each crude extract (Table 13).

Glaucescenolide (**359**) from *Schistochila glaucescens*, exhibited antifungal activity against *Trichophyton mentagrophytes* [141] whereas *ent*-1β-hydroxykauran-12-one (**632**) from *Paraschistochila pinnatifolia*, demonstrated only weak antifungal activity at a IC_{50} value of 15 μg/cm^3 against *Candida albicans* [231].

3-Methoxy-4′-hydroxybibenzyl (**286**) from the New Zealand *Plagiochila stephensoniana* showed *MIC* and *MFC* (minimum fungicidal activity) against *Trichophyton mentagrophytes* and *Candida albicans* at 125 and 62.5 μg/cm^3, respectively [177]. The synthetic analogues, 3-methoxy-4′-hydroxy-(*E*)-stilbene (**287**) and its (*Z*)-isomer (**288**) possessed improved fungicidal activities against these two fungi with

Table 13 Compounds, solvent extracts, and essential oils from bryophytes with antifungal activity

Species name[a]	Compound or solvent extract	Fungus name	Activity value	Refs.
[L] *Asterella angusta*	Perrottetin E (**183**) Riccardin B (**216**) Marchantin M (**217**) Plagiochin E (**219**) Dihydroptychantol (**228**) Asterelin A (**937**) Asterelin B (**938**) 11-*O*-Demethyl marchantin I (**939**) Marchantin H (**940**) Marchantin P (**941**)	*Candida albicans*	*MIQ/MIC* (μg/cm^3) 2.0/128 0.5/32 10.0/512 0.3/16 0.8/64 2.0/128 10.0/517 0.4/32 4.0/256 15.0–512	[329]
[M] *Atrichum undulatum*	DMSO extract (axenic culture) DMSO (grown in Nature)	*Aspergillus fumigatus*	*MIC/MFC* (mg/cm^3) 0.5–1.0/1.0–2.0	[250]
		A. versicolor *Penicillium*	0.5/1.0	
		*funiculosum*lePara>	0.5/1.0	
		P. ochrochloron	0.5/1.0	
		Trichoderma viride	0.1/1.0	
		Aspergillus fumigatus	0.5/1.0	
		A. versicolor	0.5/2.0	
		Penicillium funiculosum	0.5/1.0	
		P. ochrochloron	1.0/2.0	

(continued)

Table 13 (continued)

Species name[a]	Compound or solvent extract	Fungus name	Activity value	Refs.
		Trichoderma viride	1.0/1.0	
[L] *Balantiopsis cancellata*	CH_2Cl_2 Isotachin B (**14**) β-Phenylethyl benzoate (**971**) (2*R*)-Hydroxy-β-phenylethyl benzoate (**972**) β-Phenylethyl (*E*)-cinnamate (**973**) β-Phenethyl (*Z*)-cinnamate (**974**)	*Cladosporium herbarum*	0.001 mg/spot 0.006 0.1 0.05 (not determined) (not determined)	[337]
[L] *Bazzania albifolia*	Chiloscyphenol A (**402**)	*Candida albicans* (SC5314)	*MIC* (μg/cm^3) 16	[334]
		C. albicans (11D)	8	
		C. albicans (23E)	15	
		C. albicans (CA1)	16	
		C. albicans (148)	16	
		C. albicans (162)	16	
		C. albicans (28A)	16	
		C. albicans (28D)	16	
		C. albicans (281)	16	
		C. krusei (CK1)	32	
		C. glabrata (CG1)	16	
		C. tropicalis (CT1)	8	
		C. tropicalis (CT3)	32	
		C. parapsilosis (CP1)	0.5–1	

(continued)

Table 13 (continued)

Species name[a]	Compound or solvent extract	Fungus name	Activity value	Refs.
	AMB (amphotericin B) Fluconazole		2– > 128	
[L] *Bazzania harpago*	Cneorubin X (**1003**), gymnomitr-3(15)-en-4*β*-ol (**1004**)		IC_{50} (μg/cm^3)	[334]
		Lagenidium thermophylum	50/25	
		Haliphthoros sabahensis	100/100	
		Haliphthoros milfordensis	100/100	
[L] *Bazzania trilobata*	6′,8′-Dichloro-isoplagiochin C (**943**)	*Botrytis cinerea*	IC_{50} (μg/cm^3) 18.9	[331]
		Cladosporium cucumerinum	17.5	
		Pyricularia oryzae	3.9	
		Septoria tritici	23.5	
	Isoplagiochin D (**462**)	*B. cinerea*	7.6	
		C. cucumerinum	13.0	
		P. oryzae	4.0	
		S. tritici	15.0	
	6′-Chloro-*iso*-plagiochin D (**945**)	*B. cinerea*	50.6	
		C. cucumerinum	30.8	
		P. infestans	29.2	
		P. oryzae	2.6	
		S. tritici	4.5	

(continued)

Table 13 (continued)

Species name[a]	Compound or solvent extract	Fungus name	Activity value	Refs.
	Viridiflorol (**946**) Gymnomitrol (**947**)	*B. cinerea*	> 125	
		C. cucumerinum	> 125	
		P. oryzae	105.2	
		S. tritici	> 125	
	5-Hydroxycalamenene (**948**)	*B. cinerea*	103.2	
		C. cucumerinum	80.5	
		P. infestans	0.1	
		P. oryzae	59.4	
		S. tritici	29.0	
	7-Hydroxycalamenene (**949**)	*B. cinerea*	> 126	
		C. cucumerinum	97.0	
		P. oryzae	1.7	
		S. tritici	53.0	
	Drimenol (**744**)	*B. cinerea*	14.2	
		C. cucumerinum	11.8	
		P. infestans	0.9	
		P. oryzae	4.1	
		S. tritici	10.0	
	Drimenal (**950**)	*B. cinerea*	> 125	
		C. cucumerinum	6.6	
		P. oryzae	> 125	
		S. tritici	80.1	

(continued)

Table 13 (continued)

Species name[a]	Compound or solvent extract	Fungus name	Activity value	Refs.
	7-Hydroxycalamenene (**949**)	*B. cinerea*	81.8	
		C. cucumerinum	> 125	
		P. infestans	< 0.03	
		P. oryzae	61.6	
		S. tritici	17.6	
		Plasmopara viticola	250 ppm	
[M] *Bryum capillare*	EtOH/MeOH/Me_2CO/ $CHCl_3$		Inhibition zone (mm)	[316]
		Saccharomyces cerevisiae	–/8/8/8	
		Candida albicans	–/–/–/–	
[M] *Grimmia anodon*		*Saccharomyces cerevisiae*	–/8/8/9	
		Candida albicans	–/–/–/–	
[M] *Orthotrichum rupestre*		*Saccharomyces cerevisiae*	7/9/–/–	
		Candida albicans	–/–/–/–	
[M] *Pleurochaete squarrosa*		*Saccharomyces cerevisiae*	7/7/7/–	
		Candida albicans	–/–/–/–	
[M] *Syntrichia ruralis*		*Saccharomyces cerevisiae*	–/–/–/–	
		Candida albicans	7/7/–/–	
[M] *Tortella tortuosa*		*Saccharomyces cerevisiae*	–/–/–/–	
		Candida albicans	–/–/7/–	
	Ampicillin/erythromycin/vancomycin/ketoconazole*	*Saccharomyces cerevisiae*	–/–/–/24	
	*Positive controls [147]	*Candida albicans*	–/–/–/21	

(continued)

Table 13 (continued)

Species name[a]	Compound or solvent extract	Fungus name	Activity value	Refs.
[M] *Bryum capillare*	EtOH		*MIC* (μg/cm^3)	[278]
		Candida albicans	> 125	
[M] *Cinclidotus fontinaloides* [M] *Palustriella commutate*	EtOH		*MIC*/*MIB* (μg/cm^3)	[314]
		Candida albicans	> 100	
			> 200	
[M] *Ctenidium molluscum* [M] *Fontinalis antipyretica* [M] *Hypnum cupressiforme*	MeOH		*MIC*/*MFC* (mg/cm^3)	[303]
		Aspergillus flavus	2.5–5/5	
		A. *fumigatus*	5/5	
		A. *niger*	5/5	
		Penicillium funiculosum	2.5–5/5	
		P. ochrochloron	2.5–5/5	
		Trichoderma viride	2.5–5/5	
[L] *Chiloscyphus subporosa* [L] *Clasmatocolea vermicularis*	Diplophyllolide (**965**)	*Trichophyton mentagrophytes*	Inhibition zone (mm): 4	[189]
[M] *Dicranum japonicum* [M] *Dicranum scoparium*	4-Cyclopentenone-(2*S*),(2′*Z*)-pentenyl-(3*S*)-2′′-octynoic acid (**1018**) 2-Cyclopentenone-2-[(2′*Z*)-pentenyl]-3-(1′′,2′′)-octadienoic acid (**1019**)	*Botrytis cinerea*	IC_{50} (μg/cm^3) 60	[38]
[M] *Dryptodon pulvinatus*	EtOH	*Candida glabrata*	0.5 mg/cm^3	[275]
[L] *Frullania dilatata*	EtOH (containing frullanolide (**3**), 2,3-dimethylanisole (**1133**) linoleic, palmitic (**1132**), valerenic acid (**1134**))	*Candida albicans* *C. tropicalis*	*MIC* (μg/cm^3) 0.26 0.51	[317]

(continued)

Table 13 (continued)

Species name[a]	Compound or solvent extract	Fungus name	Activity value	Refs.
[L] *Frullania muscicola*	3,4′-Dimethoxy-4-hydroxybibenzyl (**901**)	*Epidermophyton floccosum*	*MIC* (μg/cm^3) 200	[243]
		Microsporum lanosum	200	
		Microsporum gypseum	400	
		Trichophyton gypseum	50	
		T. rubrum	200	
	3-Hydroxy-4′-methoxybibenzyl (**902**)	*Candida albicans*	200	
		Epidermophyton floccosum	12.5	
		Microsporum lanosum	25	
		Microsporum gypseum	25	
		Trichophyton gypseum	6.35	
		T. rubrum	6.35	
[L] *Heteroscyphus coalitus*	Heteroscyphic acid A (harziane diterpenoid) (**985**) Heteroscyphic acid B–I (a clerodane) (**936–993**) Heteroscyphins A–D (labdanes) (**994–997**) Heteroscyphin E (**998**) *ent*-Juncenic acid (**999**)	*Candida albicans* DSY 654 Suppression of biofilm formation of compound Modulation of transcription of related genes in *C. albicans*	*MIC* (μg/cm^3) 4–32 4 4	[341]
[M] *Homalia trichomanoides*	Atranorin (**358**)	*Candida albicans*	*MIC* (μg/cm^3) 2.0	[318]
	3α-Methoxyserrat-14-en-21β-ol (**975**)		2.0	
	Methyl 2,4-dihydroxy-3,6-dimethylbenzoate (**976**)		0.6	

(continued)

Table 13 (continued)

Species name[a]	Compound or solvent extract	Fungus name	Activity value	Refs.
[M] *Homalothecium lutescens* [M] *Hypnum cupressiforme* [M] *Pohlia nutans* [M] *Tortula muralis*	Essential oils	*Candida albicans* *Saccharomyces cerevisiae*	*MIC* (μg/cm^3) 165–1800	[348]
[M] *Hypnum andoi*	EtOH		Disk diffusion (mm/mm^3)	[313]
		Candida albicans	7/40, 8/60, 10/100	
[M] *Hypnum cupressiforme* [M] *Sematophyllum demissum*	*n*-Hexane		Inhibition zone (mm/500 μg/disk)	[304]
		Candida albicans	7.8	
		C. albicans	7.4	
[L] *Lepidozia fauriana*	Lepidozenolide (**367**)	*Candida albicans*	100 μg/cm^3	[185]
[L] *Lunularia cruciata* and many other liverworts	Lunularic acid (**196**)	*Alternaria brassicicola* *Septoria nodorum* *Uromyces fabae*	50 μg/cm^3 50 25	[38]

(continued)

Table 13 (continued)

Species name[a]	Compound or solvent extract	Fungus name	Activity value	Refs.
[L] *Marchantia polymorpha*	Marchantin A (**1**)	*Alternaria kikuchiana* *Aspergillus fumigatus* *A. niger* *Candida albicans* *Microsprorum gypseum* *Penicillium chrysogenum* *Pyricularia oryzae* *Rhizoctonia solani* *Saccharomyces cerevisiae* *Sporothrix schenckii* *Trichophyton* *mentagrophytes* *Trichophyton rubrum*	*MIC* (μg/cm^3) 100 100 25–100 100 100 100 12.5 50 100 100 3.13 100	[38]
[L] *Marchantia polymorpha*	MeOH	*Candia albicans/*		[349]
		Trichophyton mentagrophytes	*MIC* (mg/cm^3) 0.312/0.312 *MBC/MFC* (mg/cm^3) 0.625/0.62	
	Flavonoid-free extract	*C. albicans/T. mentagrophytes*	*MIC* (mg/cm^3) 0.156/0.312	
[L] *Marchantia polymorpha*	MeOH $CHCl_3$		*MIC/MFC* (mg/cm^3)	[298]
		Rhizoctonia solani	0.60/0.65	
		Sclerotium rolfsii	2.50/4.50	
[L] *Marchantia polymorpha*	Plagiochin D (**1120**)	*Candida albicans* strains QL-15 OL-28	16 μg/cm^3 32	[324, 325]
		SDEY-24R	16	
		DZFEY-09R	16	

(continued)

Table 13 (continued)

Species name[a]	Compound or solvent extract	Fungus name	Activity value	Refs.
	Plagiochin D (**1120**) + fluconazole	*C. albicans* strains	0.313–0.375	
		QL-15 OL-28 SDEY-24R DZFEY-09R		
[M] *Marchantia polymorpha*		*C. albicans*	*MID* (μg/cm^3)	[249] [324, 327] [328]
	Marchantin A (**1**)		2.5	
	Marchantin B (**195**)		4.0	
	Marchantin E (**199**)		2.5	
	Neomarchantin A (**212**)		0.3	
	Plagiochin E (**219**)		16–32	
	13,13′-*O*-Iso-propylidene-riccardin D (**934**)		0.4	
	Riccardin H (**935**)		4.0	
[L] *Marchantia polymorpha*	MeOH flavonoid extract	*Rhizoctonia solani/Fusarium oxysporum* (complete inhibition of mycelial/ germination of spore)	*MBC*/*MFC* (mg/cm^3) 0.156/0.312 5/10 (mg/cm^3)	[298]
[L] *Marchantia polymorpha*	DMSO (axenic culture)		*MIC*/*MFC* (mg/cm^3)	[350]
		Aspergillus fumigatus	0.25–0.5/0.5–1.0	
		A. versicolor	0.25–0.5/0.5–1.0	
		Penicillium funiculosum	0.5/1.0	
		P. ochrochloron	0.5/1.0	
		Trichoderma viride	0.1/0.25	
	DMSO (grown naturally)	*A. fumigatus*	0.5/1.0	

(continued)

Table 13 (continued)

Species name[a]	Compound or solvent extract	Fungus name	Activity value	Refs.
		A. versicolor	0.5/1.0	
		P. funiculosum	0.5/1.0	
		P. ochrochloron	1.0/2.0	
		T. viride	1.0/1.0	
[L] *Marchantia polymorpha*	MeOH		Inhibition zone (mm) (200 mg/cm^3)	[287]
		Candida albicans	18	
		Rhizopus oryzae	17	
[L] *Odontoschisma denudatum*			(%) at 100 ppm (growth inhibitory activity)	[345]
	(+)-8-Acetoxy-odontoschismenol (**1005**)	*Botrytis cinerea* *Rhizoctonia solani* *Pythium debaryanum*	22–39	
	6,12-Dihydroxy-dolabella-(3*E*,7*E*)-diene (**1006**)	*B. cinerea* *R. solani* *P. debaryanum*	24–44	
	6-Acetoxy-12,16-dihydroxydolabella-(3*E*,7*E*)-diene (**1007**)	*B. cinerea* *R. solani* *P. debaryanum*	5–13	
	6,16-Diacetpoxy-12-hydroxydolabella-(3*E*,7*E*)-diene (**1008**)	*B. cinerea* *R. solani* *P. debaryanum*	10–30	
	6-Hydroxy-3,4-epoxy-12-hydroxydolabell-(7*E*)-en-16-al (**1009**)	*B. cinerea* *R. solani* *P. debaryanum*	14–38	

(continued)

Table 13 (continued)

Species name[a]	Compound or solvent extract	Fungus name	Activity value	Refs.
[L] *Odontoschisma grosseverrucosum*	Odongrossin A (**1010**) Odongrossin G (**1016**)	*Candida albicans*	IC_{50} (μg/cm^3) Moderate antifungal activity	[346]
[L] *Paraschistochila pinnatifolia*	*ent*-1*β*-Hydroxykauran-12-one (**632**)	*Candia albicans*	IC_{50} (μg/cm^3) 15	[231]
[L] *Pallavicinia ambigua*	Pallamolides A–E (**980–984**), Tautomers I: **981** and **982** Tautomer II: **983** and **984**	*Candia albicans* DSY654	(1) Inhibition of virulence of efflux pump-deficient variety (2) Against *C. albicans*, inhibition of hyphal morphogenesis, adhesion and biofilm formation	[240]
[L] *Pellia endiviifolia*	Sacculatal (**128**)		*MIC* (μg/cm^3)	[288]
		Alternaria brassicicola	100	
		Aspergillus fumigatus	100	
		Candida albicans	25	
		Saccharomyces cerevisiae	25	
			MIC_{90} (μg/cm^3)	
		Botrytis cinerea	100	
[M] *Physcomitrella patens*	Lunularic acid (**196**) DMDO (axenic culture)		*MIC*/*MFC* (mg/cm^3)	[350]
		Aspergillus fumigatus	0.5/1.0	
		A. versicolor	0.5/1.0	
		Penicillium funiculosum	0.5/1.0	
		P. ochrochloron	0.5/1.0	
		Trichoderma viride	0.5/1.0	

(continued)

Table 13 (continued)

Species name[a]	Compound or solvent extract	Fungus name	Activity value	Refs.
	DMDO (grown in nature)	*A. fumigatus*	0.5/1.0	
		A. versicolor	1.0/2.0	
		P. funiculosum	0.5/1.0	
		P. ochrochloron	1.0/2.0	
		T. viride	1.0/1.0	
[L] *Plagiochasma appendiculatum*	80% EtOH		Inhibition zone (mm)	[299]
		Aspergillus niger	14/12	
[M] *Sphagnum fimbriatum*		*Fusarium solani*	12/11	
[L] *Plagiochasma appendiculatum*	$CHCl_3/(Me)_2CO/EtOH/H_2O$	*Candida albicans*	Inhibition zone (mm) 17/12/19/17	[316]
		Cryptococcus albidus	14/13/17/15	
		Aspergillus niger	15/14/18/11	
		A. flavus	13/10/16/12.7	
		A. spinulosus	12/10/15/10	
		A. terreus	15/14/14/16	
		A. nidulans	16/14/16/14	
		Trichophyton rubrum	14.3/25.7/25.7/18	
	Ketoconazole/fluconazole/metronidazole (10 μg/disk) controls	*Candida albicans*	24/23/17	
		Cryptococcus albidus	30/18/14	
		Aspergillus niger	21/20/20	
		A. flavus	18/19/21	
		A. spinulosus	22/21/22	

(continued)

Table 13 (continued)

Species name[a]	Compound or solvent extract	Fungus name	Activity value	Refs.
		A. terreus	28/30/24	
		A. nidulans	23/23/21	
		Trichophyton rubrum	32/24/23	
[L] *Plagiochasma appendiculatum*	$CHCl_3/(Me)_2CO//EtOH/H_2O$	*Candida albicans*	*MIC* (μg/disk) 100/200/10/100	[316]
		Cryptococcus albidus	200/200/100/200	
		Aspergillus niger	200/100/10/200	
		A. flavus	200/200/100/200	
		A. spinulosus	200/200/100/200	
		A. terreus	200/200/100/100	
		A. nidulans	100/200/100.200	
		Trichophyton rubrum	100/100/2.5/10	
[L] *Plagiochasma intermedium*	Pakyonol (**218**) Riccardin C (**182**) Neomarchantin A (**212**) Riccardin F (**243**) Isoriccardin C (**942**) Marchantin H (**940**)	*Candida albicans*	*MIC* (μg/cm^3) 8.0 *MIC* (μg/cm^3) 32–512	[330] [342]
	Riccardin C (**182**) + fluconazole	*Candida albicans*	*C. albicans* reduction (*MIC*: 256-fold more potent than **182** itself) *MIC* (μg/cm^3) 50–200 μg/cm^3	

(continued)

Table 13 (continued)

Species name[a]	Compound or solvent extract	Fungus name	Activity value	Refs.
	Pakyonol (**218**)	*Epidermophyton floccosum* *Microsporum gypseum* *Trichophyton gypseum*		
[L] *Plagiochila banksiana*	EtOH (0.05 mg bibenzyl)		Inhibition zone (mm)	[177]
		Candida albicans	5	
		Trichophyton mentagrophytes	1	
[L] *Plagiochila deltoidea*	EtOH (0.54)	*Candida albicans* *Trichophyton mentagrophytes*	1 Not determined (ND)	[177]
[L] *Plagiochila fasciculata*	EtOH (trace, invisible to resolve)	*Candida albicans* *Trichophyton mentagrophytes*	ND 7	[177]
[L] *Plagiochila stephensoniana*	EtOH (10.5)	*Candida albicans* *Trichophyton mentagrophytes*	6 12	[177]
[L] *Plagiochila suborbiculata*	EtOH (not detected)	*Candida albicans* *Trichophyton mentagrophytes*	1 ND	[177]
[L] *Plagiochila stephensoniana*	3-Methoxy-4′-hydroxybibenzyl (**286**)	*Candida albicans* *Trichophyton mentagrophytes* *Cladosporium resinae*	2 (12 $\mu g/cm^3$ dose) 10 (60) 10 (300) *MIC*/*MFC* ($\mu g/cm^3$) 125.5/125.5 7 (12) 7 (60) 9 (300) 62.50/62.50	[177]
	3-Methoxy-4′-hydroxy-(*E*)-stilbene (**287**) (synthetic)	*Candida albicans*	3 (60)	

(continued)

Table 13 (continued)

Species name[a]	Compound or solvent extract	Fungus name	Activity value	Refs.
		Trichophyton mentagrophytes	Inactive (12) 2 (60) 7 (300) 31.25/125.00	
		Cladosporium resinae	8 (12) 6 (60) 10 (300) 31.25/31.25	
	3-Methoxy-4′-hydroxy-(*Z*)-stilbene (**288**) (synthetic)	*Candida albicans*	2 (60)	
		T. mentagrophytes	2 (12) 7 (60) 7 (300) 31.25/31.25	
	(**286**/g)/dry liverwort peak		7 (12) 10 (60) 10(300) 15.62/15.62	
[L] *Plagiochila fasciculata*	2-Hydroxy-4,6-dimethoxyacetophenone (**767**)	*Candida albicans* *Cladosporium resinae* *Trichophyton mentagrophytes*	150 μg/disk	[258]
[L] *Plagiochila fruticosa*	Plagicosin F (**1025**)	*Candida albicans* DSY654	*MIC* (μg/cm^3) 16 (inhibition of hyphal morphogenesis, adhesion and biofilm formation)	[348]
[L] *Plagiochila nitens*	Plagiochilarin H (**459**)	*Candida albicans* DSY 654	*MIC* (μg/cm^3) 16	[236]

(continued)

Table 13 (continued)

Species name[a]	Compound or solvent extract	Fungus name	Activity value	Refs.
[L] *Pleurozia subinflata*	5β-Acetoxy-13-*epi*-neoverrucosanic acid (**1000**) 13-*epi*-Neo-verrucosan-5β-ol (**1001**)	*Lagenidium thermophilum*	*MIC* (μg/cm^3) 12.5	[343]
		Haliphthoros subahensis	50.0	
		L. thermophilum	100	
		H. subahensis	> 100	
[L] *Ptilidium pulcherrimum*	MeOH	*Aspergillus flavus* *A. versicolor* *A. ochraceus* *A. niger* *Penicillium finiculosum* *Trichoderma viride*	*MIC*/*MFC* (μg/cm^3) 0.5–2.5/2.5–5.0	[301]
[L] *Porella vernicosa* complex	Polygodial (**90**) Cinnamolide (**659**) Norpinguisone (**927**)	*Aspergillus fumigatus*	*MIC* (μg/cm^3) 100	[38]
		A. niger	25	
		Botrytis cinerea	100	
		Candida albicans	100	
		Trichophyton mentagrophytes	50	
		Microsporum gypseum	20	
		T. mentagrophytes	10	
		T. rubrum	20	
		Aspergillus niger	100	
[L] *Riccardia marginata*	2,4,6-Trichloro-3-hydroxybibenzyl (**977**) 2,4-Dichloro-3-hydroxybibenzyl (**978**) 2-Chloro-3-hydroxy-bibenzyl (**979**)	*Candida albicans* *Cladosporium resinae* *Trichophyton mentagrophytes*	μg/per disk 12–30	[339]

(continued)

Table 13 (continued)

Species name[a]	Compound or solvent extract	Fungus name	Activity value	Refs.
[L] *Riccardia polyclada*		*Cladosporium herbarum*	Inhibition zone (mm): dose applied per spot in parentheses	[336]
	2,6-Dichloro-3-hydroxy-4′-methoxy-bibenzyl (**967**)		2.2 (0.05 mg)	
	2,6,3′-Trichloro-4′-methoxybibenzyl (**968**)		1.2 (0.07 mg)	
	2,4,6,3′-Tetrachloro-3-hydroxy-4′-methoxy-bibenzyl (**970**)		2.0 (0.07 mg)	
	2,4,6,3′-Tetrachloro-3-hydroxybibenzyl (**969**)		1.5 (0.25 mg)	
[M] *Rhodobryum ontariense*	DMSO	*Aspergillus fumigatus* *A. versicolor* *Penicillium finiculosum* *P. ochrochloron* *Trichoderma viride*	*MIC* (μg/cm^3) 0.10–0.50	[302]
[L] *Scapania verrucosa*	Et_2O	*Aspergillus fumigatus*	*MIC* (μg/cm^3) 8	[206]
		Candia albicans	32	
		Cryptococcus neoformans	64	
		Pyricularia oryzae	> 128	
		Trichophyton rubrum	64	
[L] *Schistochila glaucescens*	Marchantin C (**197**)	*Trichophyton mentagrophytes*	*MIC* (μg/cm^3) 0.5	[141]
	Neomarchantin A (**212**)		1.0	
	Neomarchantin B (**213**)		0.5	
[L] *Targionia lorbeeriana*			*MIC*(μg/cm^3)	[254]
	Dehydrocostuslactone (**953**)	*Candida albicans*	5*	

(continued)

Table 13 (continued)

Species name[a]	Compound or solvent extract	Fungus name	Activity value	Refs.
		Cladosporium cucumerinum	40** 0.5* 20**	
	Acetyltrifloculoside lactone (**952**)	*C. cucumerinum*	10*	
	11α*H*-Dihydro-dehydrocostuslactone (**953**)	*C. cucumerinum*	3*	
	8,15-Acetyl-salonitenolide (**954**)	*C. cucumerinum*	5* *bioautographic TLC assay (minimum amount: (μg) **agar dilution assay	
[L] *Trichocolea tomentella* *T. mollissima* *T. lanata*	Tomentellin (**768**) Demethyl tomentellin (**769**) Trichocolein (**933**)	*Trichophyton mentagrophytes* *Candida albicans*	Inhibition zone (mm)/60 μg disk 1–2 1	[259]
[L] *Tritomaria quinquendentata*	(5*R*,7*S*,8*S*,9*S*)-*ent*-8-Hydroxy-4(15),11-eudesmadiene (**959**)	*Candida albicans* (with efflux pumps deficient strain DSY654) Inhibition of yeast-to-hyphal switch of DSY654 cells	*MIC* ($\mu g/cm^3$) 64	[335]
	11α-Hydroxydihydro-*ent*-isoalantolactone (**962**)		32	
	ent-Isoalantolactone (**963**)		16	
	Eudesm-4(15),11(13)-dien-6,12-diol (**964**)		64	
	ent-Codonolactone (**966**)		64	
	ent-Isoalantolactone (**963**)		4	

[a] [L] liverwort, [M] moss

MIC values of 31.3 and 31.3, and 31.3 and 15.6 μg/cm^3, respectively. The inhibition zones of **286** were 2, 10, and 10 mm (12, 60, and 300 μg/disk) against *C. albicans* and against *T. mentagrophytes* (7, 7, and 7 mm). The inhibition zones of the synthetic **287** and **288** were greater (7–10 mm/ 60–300 μg/disk) than that of the natural product **286**.

Two simple bibenzyls, 3,4′-dimethoxy-4-hydroxybibenzyl (**901**) and 3-hydroxy-4′-methoxybibenzyl (**902**), obtained from *Frullania muscicola*, showed antifungal properties against *Candida albicans* (*MIC* values of 0–200 μg/cm^3), *Epidermophyton flococcum* (200, 12.5), *Microsporum gypseum* (400, 25), *Microsporum lanosum* (200, 25), *Trichophyton rubrum* (200, 6.3), and *Trichophyton gypseum* (50, 6.3), respectively [243].

A disk diffusion assay performed on 2-hydroxy-3,4,6-trimethoxyacetophenone (**766**) and 2-hydroxy-4,6-dimethoxyacetophenone (**767**), from *Plagiochila fasciculata* at 150 μg/disk, showed both to have antifungal activity against *Cladosporium resinae* and *T. mentagrophytes.* Compound **767** also showed antifungal activity against *C. albicans* [258].

Tomentellin (**768**) and demethyl tomentellin (**769**) from *Trichocolea tomentella* and *T. mollissima*, showed mild antifungal activity against *C. albicans* and *T. mentagrophytes.* Trichocolein (**933**), isolated from *T. lanata*, showed only weak antifungal activity against *C. albicans* and *T. mentagrophytes*, as manifested by inhibition zones of 1 and 2 mm at 60 μg/disk [259].

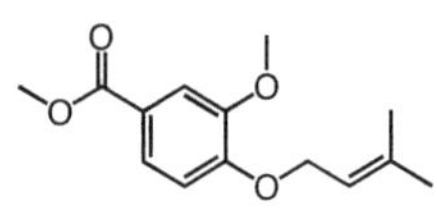

933 (trichocolein)

Marchantin A (**1**) also had antifungal activity against *Alternaria kikuchiana* (*MIC*/μg/cm^3), *Aspergillus fumigatus* (100), *A. niger* (25–100), *C. albicans*, *Microsprorum gypseum*, *Penicillium chrysogenum* (100), *Piricularia oryzae* (12.5), *Rhizoctonia solani* (50), *Saccharomyces cerevisiae, Sporothrix schenckii* (100), and the dermatophytes *Trichophyton mentagrophytes* (3.1) and *T. rubrum* (100) [38].

Marchantin C (**197**) and neomarchantins A (**212**) and B (**213**), isolated from *Schistochila glaucescens*, showed antifungal activity against *T. mentagrophytes*, with *MIC* values of 0.5, 1, and 0.5 μg/cm^3, respectively [141].

The antifungal activity of the ethanol extracts of five mosses, *Ceratodon purpureus*, *Dryptodon pulvinatus*, *Hypnum cupressiforme*, *Rhytidiadelphus squarrosus*, and *Tortula muralis*, were evaluated against *Candida glabrata.* Of these, only *D. pulvinatus* inhibited the growth of *Candida glabrata* at a concentration of 0.5 mg/cm^3 [275].

Plagiochin E (**219**), from *Marchantia polymorpha*, showed antifungal activity against four fluconazole-resistant *C. albicans* strains, QL-14, QL-28, SDEY-24R, and SDEY-09R, with *MIC* values of 16, 32, 16, and 16 μg/cm^3, respectively. When plagiochin E (**219**) was mixed with fluconazole, the antifungal activities were substantially more potent (*MIC* values in the range 0.31–0.38 μg/cm^3) [324]. The antifungal activity of plagiochin E (**219**) against *C. albicans* might be attributed to its inhibitory effect on cell wall chitin synthesis [325]. Plagiochin E (**219**) exerted its antifungal activity through the accumulation of mitochondrial dysfunction-induced reactive oxygen species in *C. albicans* [326]. Compound **219** also induced apoptosis in *C. albicans* through activating a metacaspase [327].

Reinvestigation of the antifungal activity of bis-bibenzyls from *Marchantia polymorpha* using a bioautographic method showed neomarchantin A (**212**), plagiochin E (**219**), and 13,13′-*O*-isopropylidene riccardin D (**934**) to possess antifungal activity against *C. albicans* with respective *MID* (minimum inhibitory dose) values of 0.25, 0.2, and 0.4 μg/cm^3, in comparison to the *MID* of 0.01 μg/cm^3 for the positive control, miconazole. In turn, marchantin A (**1**), marchantin B (**195**), marchantin E (**199**), and riccardin H (**935**), showed moderate growth inhibitory activities against this fungus, with *MID* values of 2.5, 4.0, 2.5, and 4.0 μg/cm^3, respectively [328]. Further bioassay-guided fractionation of the ether extract of the same *Asterella angusta* led to the isolation of two new bis-bibenzyls, riccardin I (**936**) and angustatin A (**245**), which showed the same antifungal activity [329].

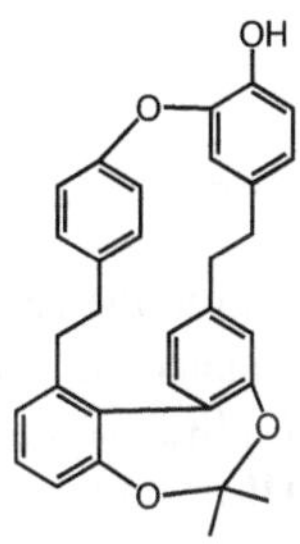

934 (13,13'-*O*-isopropylidene-riccardin D)

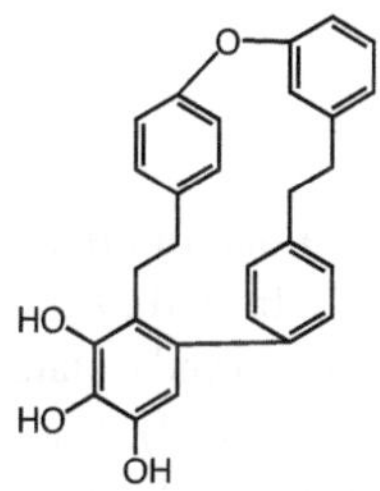

935 (riccardin H)

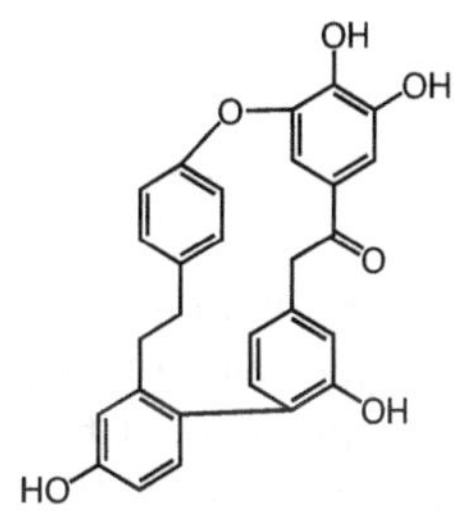

936 (riccardin I)

Direct TLC bioautographic detection of the antifungal activity of an ether extract of the liverwort *Asterella angusta* showed positive effects against *C. albicans*. Ten bis-bibenzyls, perrottetin E (**183**), riccardin B (**216**), marchantin M (**217**), plagiochin E (**219**), dihydroptychantol A (**228**), asterellin A (**937**), asterellin B (**938**), 11-demethylmarchantin I (**939**), marchantin H (**940**), and marchantin P (**941**) were tested individually against *C. albicans*. All of the compounds assessed showed antifungal activity, exhibiting *MIQ* (minimum inhibitory quantity) values between 0.3–15.0 μg/cm^3, and *MIC* (minimum inhibitory concentration) values in the range of 16 to 512 μg/cm^3. A free phenolic hydroxy group seems to play an important role in mediating antifungal activity, since bis-bibenzyls possessing a methoxy group displayed decreased potencies in this regard [329].

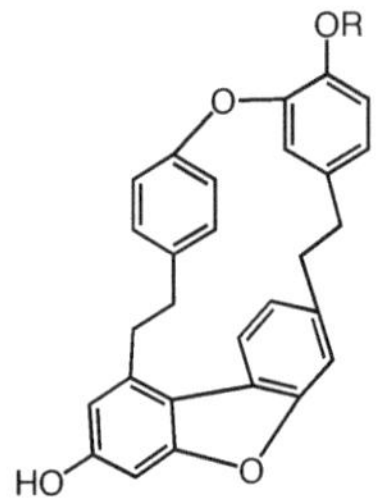

937 (asterellin A, R = H)
938 (asterellin B, R = Me)

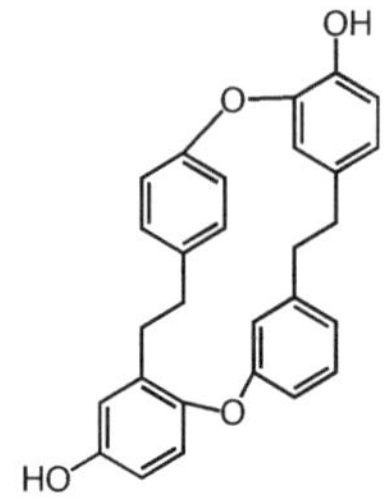

939 (11-demethyl-marchantin I)

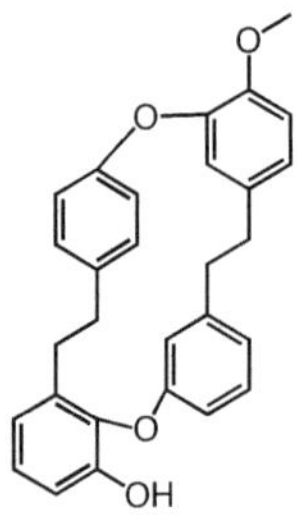

940 (marchantin H)

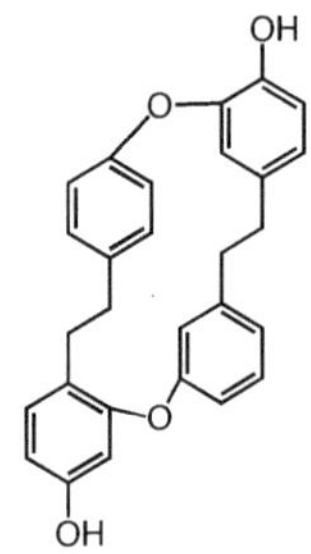

941 (marchantin P)

Plagiochasma intermedium is a thalloid liverwort belonging to the Aytoniaceae and a major source of bis-bibenzyls. In 2010, Xie et al., reinvestigated the phytochemical constituents of a Chinese specimen and isolated riccardin C (**182**), neomarchantin A (**212**), pakyonol (**218**), riccardin F (**243**), marchantin H (**940**), and isoriccardin C (**942**), which presented weak to no significant in vitro antifungal activity against fluconazole-sensitive and -resistant strains of *Candida albicans*, with MIC values ranging from 8.0 to > 512 μg/cm³. When riccardin C (**182**) was combined with fluconazole, its synergistic and/or additive activity resulted in the *MIC* values of fluconazole being reduced from 256 to < 8 μg/cm³ when evaluated against three resistant strains of *C. albicans* [330].

6′,8′-Dichloroisoplagiochin C (= bazzanin B (**943**)), isoplagiochin D (**944**), and 6′-chloro-*iso*-plagiochin D (= bazzanin S) (**945**) from *Bazzania trilobata*, each showed a discernible antifungal activity against *Pyricularia oryzae* using a microtiter plate test, with IC_{50} values of 3.9, 4.0, and 2.6 μg/cm³, and also against *Septoria tritici*, with IC_{50} values of 23.5, 15.9, and 4.5 μg/cm³. Compounds **943** and **944** also demonstrated inhibitory activities against *Botrytis cinerea* with IC_{50} values of 18.9 and 7.6 μg/cm³, respectively. Free phenolic hydroxy groups on the aromatic rings of the bis-bibenzyls were shown to play an important role in mediating inhibitory activity against fungi such as *Cladosporium cucumerinum* [331].

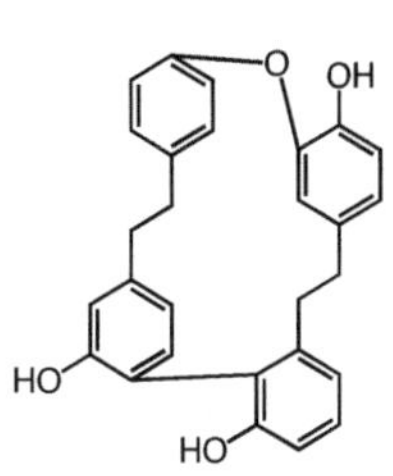

942 (isoriccardin C)

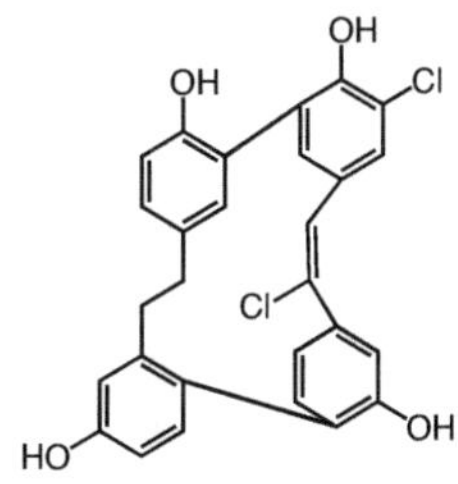

943 (bazzanin B)

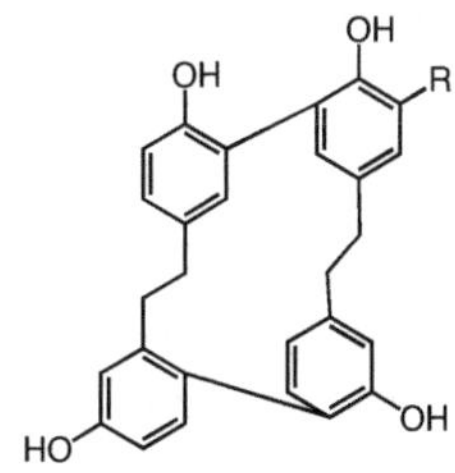

944 (isoplagiochin D, R = H
945 (bazzanin S, R = Cl)

Bazzania trilobata was found to produce six antifungal active sesquiterpenoids, viridiflorol (**946**), gymnomitrol (**947**), 5-hydoxycalamenene (**948**), 7-hydroxycalamenene (**949**), drimenol (**744**), and drimenal (**950**) [331]. Viridiflorol (**946**) was shown previously to have antifungal activity against *Cladosporium cucumerinum* [332]. It also showed weak antifungal activity against *Pyricularia oryzae*, with an IC_{50} value of 105.2 μg/cm^3.

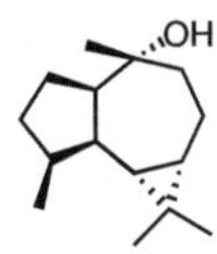

946 (viridiflorol)

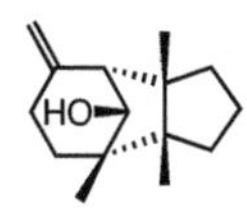

947 (gymnonitrol)

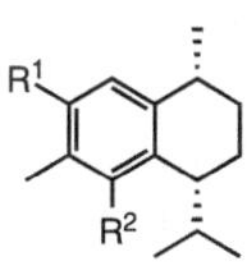

948 (5-hydroxycalamenene, R^1 = H, R^2 = OH)
949 (7-hydroxycalamenene, R^1 = OH, R^2 = H)

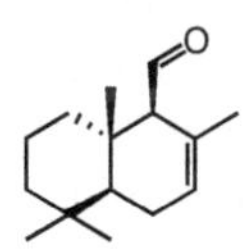

950 (drimenal)

Gymnomitrol (**947**) showed inhibitory effects against *Phytophthora infestans*, *Phytophthora oryzae*, and *Septoria tritici*, with IC_{50} values of 97.0, 1.7, and 53.0 μg/cm^3, respectively [331]. 5-Hydroxycalamenene (**948**) displayed inhibitory activity against *P. oryzae* with an IC_{50} of 1.7 μg/cm^3, while 7-hydroxycalamenene (**949**) had antifungal activities against *C. cucumerinum*, *P. oryzae*, and *S. tritici* with IC_{50} values of 97.0, 1.7, and 53.0 μg/cm^3. Compound **949** was tested for in vivo activity against *Plasmopara viticola* on grape vine leaves and showed an inhibitory effect at a concentration of 250 ppm. The infection was reduced from 100% in the control to 30% in the treated plants in a greenhouse [331]. 7-Hydroxycalamenene (**949**) from *Tilia europaea* is a phytoalexin [333].

Drimenol (**744**) was found to be less active than the calamenenes, and inhibited the growth of *C. cucumerinum* and *S. tritici* at concentrations of 6.6 and 80.1 μg/cm^3, respectively [331]. Drimenal (**950**) exhibited weak growth inhibitory activities against *B. cinerea* and *P. oryzae* with IC_{50} values of 81.8 and 61.6 μg/cm^3, and much more potent activities against *P. infestans* and *S. tritici* with IC_{50} values of < 0.03 and 17.6 μg/cm^3, respectively [331]. Cinnamolide (**659**) from the liverwort, *Porella* *vernicosa* complex, also showed moderate antifungal activity against *Microsporum gypseum, Trichophyton mentagrophytes*, and *T. rubrum* at *MIC* values of 20, 10, and 20 μg/cm^3, respectively [38].

Chiloscyphenol A (**403**), isolated from the Chinese liverworts *Bazzania albifolia* and *Chiloscyphus polyanthus*, demonstrated antifungal activity against ten *Candida albicans* strains, and also *C. krusei*, *C. glabrata*, and two *C. tropicalis* strains, with *MIC* values ranging from 8 to 32 μg/cm^3. It elicited fungicidal activity on *C. albicans* strains in both the planktonic state and in mature biofilms. Chiloscyphenol A (**403**) showed antifungal action leading to mitochondrial dysfunction and plasma membrane destruction [334].

As mentioned in the pungency (4.2) and cytotoxic activity (4.4) Sections of this volume, the genera *Chiloscyphus* and *Diplophyllum* are known sources of *ent*-eudesmane sesquiterpenoids. In 2023, Zhu et al., focused on the antifungal activity of the newly isolated eudesmanes **386–392** and seven known analogues (**376**, **378–380**, **384**, **393**, and **394**), from *C. polyanthus*. They were evaluated against *Candida albicans* wild type strain SC5314, with fluconazole as a positive control, and all these eudesmane sesquiterpenoids showed antifungal activities [188]. From *D. apiculatum*, three diterpenoids, diplapiculalins A–C (**741–743**) were obtained. The antifungal activities of these three diterpenoids were evaluated against *C. albicans* wild type strains SC5314 and DSY654. However, **741–743** were deemed inactive against the target fungi, as assessed using fluconazole as the positive control [251].

Dehydrocostus lactone (**951**), acetyltrifloculoside lactone (**952**), 11*αH*-dihydrodehydrocostus lactone (**953**) and 8,15-diacetylsalonitenolide (**954**) from *Targionia lorbeeriana* showed antifungal activity against *Cladosporium cucumerinum* with MIC values of 20, 10, 3, and 5 μg/cm^3, respectively, using a bioautographic TLC method. Dehydrocostus lactone (**951**) also exhibited the growth inhibitory activity against *Candida albicans*, having a *MIC* value of 5 μg/cm^3 [254].

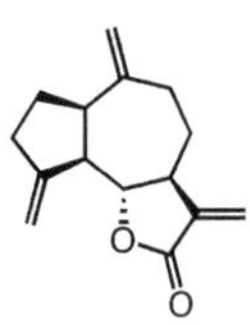

951 (dehydrocostus lactone)

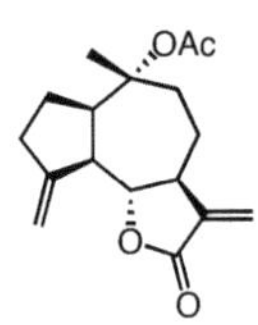

952 (acetyltriflocusolide lactone)

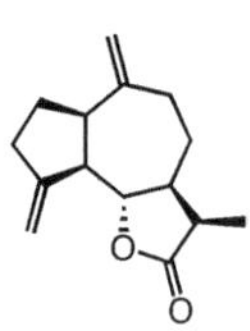

953 (11*αH*-dihydrodehydrocostus lactone)

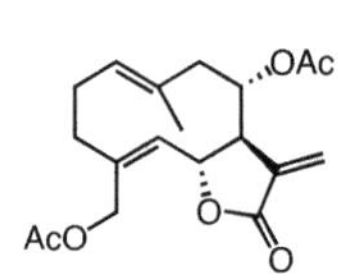

954 (8,15-diacetylsalonitenolide)

The European liverwort *Tritomaria quinquendentata* produces the two eudesmanes, (+)-7-*epi*-junenol (**955**) and (–)-7-*epi*-isojunenol (**956**) [3]. Reinvestigation of the 95% ethanol extract of the Chinese liverwort *Tritomaria quinquendentata* resulted in the isolation of several new eudesmanes, namely, *ent*-11*β*-ethoxy-13-hydroxy-*ent*-11,13-dihydroisoalantolactone (**957**), *ent*-(11*R*)-11,13-epoxyisoalantolactone (**958**), *ent*-8*β*-hydroxy-4(15),11-eudesmadiene (**959**), *ent*-2*α*,8*β*-dihydroxy-11(13)-cyperanen-4-one (**960**. Also obtained were the known compounds, *ent*-dihydroisoalantolactone (**961**), dihydro-isoalantolactone (**962**), *ent*-isoalantolactone (**963**), *ent*-6*β*,8*β*-dihydroeudesma-4(15),11-diene (**964**), *ent*-7*α*-hydroxy-2-oxoeudesm-(3*Z*)-ene (**760**), (–)-*ent*-dihydrodiplophyllolide (**965**), *ent*-codonolactone (**966**), and (–)-loliolide (**600**). Compounds **955** and **956** isolated from the European *T. quinquendentata* were not reported from this Chinese specimen. Compounds **959**, **962–964**, and **966** showed antifungal activity against *Candida albicans* double mutant strain DSY654 with MIC_{80} values of 64, 32, 16, 64, and 64 μg/cm^3, but were ineffective (IC_{50} > 128 μg/cm^3) against *C. albicans* strains SC5314, DSY448, DSAY465, and DSY653. *ent*-Isoalantolactone (**963**) inhibited the yeast-to-hyphal switch of DSY654 cells at 4 μg/cm^3 [335].

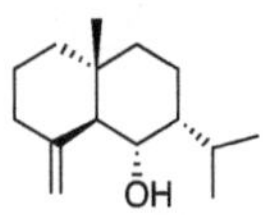

955 ((+)-7-*epi*-junenol)

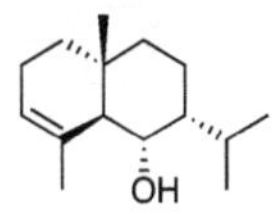

956 ((–)-7-*epi*-isojunenol)

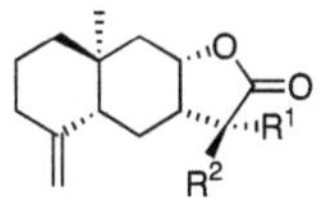

957 (*ent*-11*β*-ethoxy-13-hydroxy-dihydroisoalantolactone, R^1 = OEt, R^2 = CH_2OH)
961 (*ent*-dihydroisoalantolactone, R^1 = H, R^2 = Me)
962 (dihydroisoalantolactone, R^1 = Me, R^2 = CH_2OH)

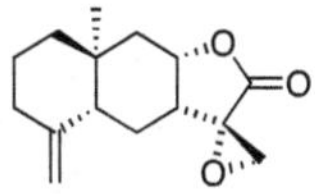

958 (*ent*-(11*R*),13-epoxy-isoalantolactone)

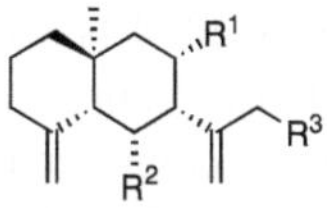

959 (*ent*-8*β*-hydroxy-4(15),11-eudesmadiene, R^1 = OH, R^2 = R^3 = H)
964 (*ent*-6*β*,8*β*-dihydroxyeudesma-4(15),11-diene, R^1 = R^3 = OH, R^2 = H)

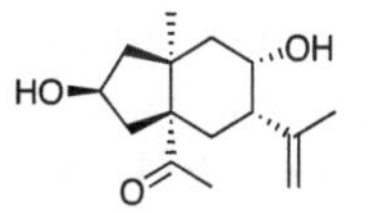

960 (*ent*-2*α*,8*β*-dihydroxy-11(13)-cyperanen-4-one)

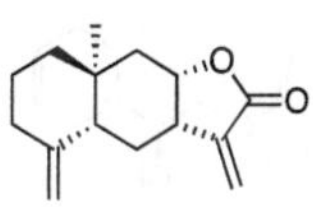

963 (*ent*-isoalantolactone)

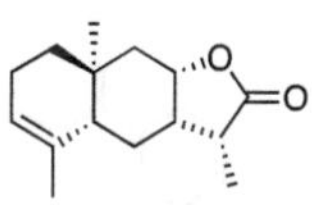

965 (–)-(*ent*-dihydrodiplophyllolide)

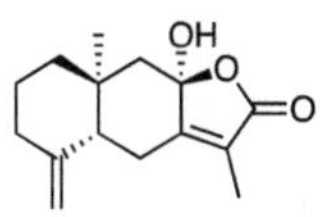

966 (*ent*-codonolactone)

Some species of the genus *Riccardia* elaborate high concentrations of the bioactive polychlorinated bibenzyls, 2,6-dichloro-3-hydroxy-4′-methoxybibenzyl (**967**), 2,6,3′-trichloro-3-hydroxy-4′-methoxy-bibenzyl (**968**), 2,4,6,3′-tetrachloro-3-hydroxybibenzyl (**969**), and 2,4,6,3′-tetrachloro-3-hydroxy-4′-methoxybibenzyl (**970**), which protect them from pathogens and herbivores. On TLC-bio-autography with cultures of the rotting fungus *Cladosporium herbarum*, compounds **967**, **968**, and **970** from *R. polyclada* showed fungicidal activities, as manifested by inhibition zones of 1.2–2.2 cm, which were larger than those obtained with the fungicide, ketoconazole. On the other hand, compound **969** was inactive against *C. herbarum* [336].

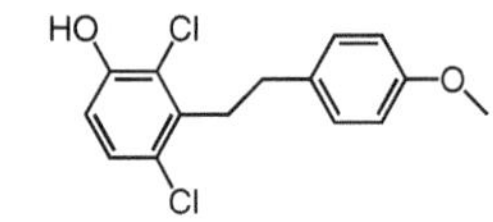

967 (2,6-dichloro-3-hydroxy-4'-methoxybibenzyl)

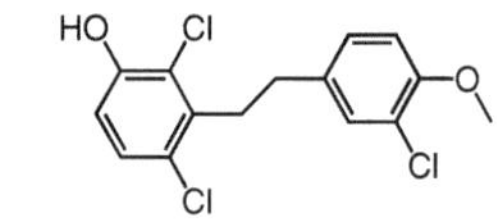

968 (2,6,3'-trichloro-3-hydroxy-4'-methoxybibenzyl)

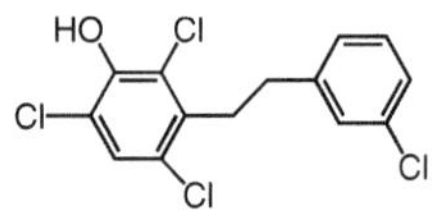

969 (2,4,6,3'-tetrachloro-3-hydroxybibenzyl)

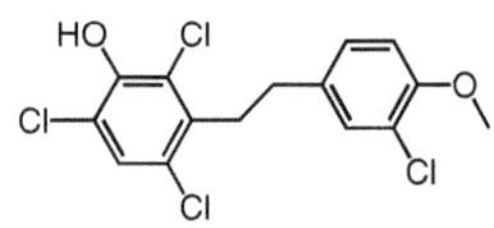

970 (2,4,6,3'-tetrachloro-3-hydroxy-4'-methoxybibenzyl)

The dichloromethane extract of the liverwort *Balantiopsis cancellata* demonstrated potent antifungal activity against *Cladosporium herbarum* at 0.01 mg/zone. Five aromatic esters, isotachin B (**14**), β-phenylethyl benzoate (**971**), (2*R*)-hydroxy-β-phenylethyl benzoate (**972**), β-phenylethyl (*E*)-cinnamate (**973**) and its (*Z*)-diastereomer (**974**) were purified, among which compound **14** showed the most potent antifungal activity against *C. herbarum* at 6 μg/zone, using TLC-bioautography. However, its activity was lower than those determined for either pure ketoconazole or the commercially available fungicide, captan, as observed in dilution experiments. The benzoate **972** also showed the same type of antifungal activity as mentioned above, at 50 μg/spot [337].

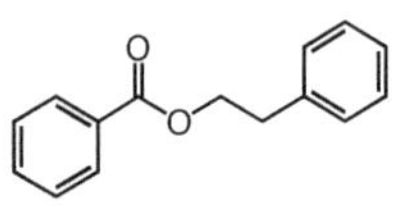

971 (β-phenylethyl benzoate)

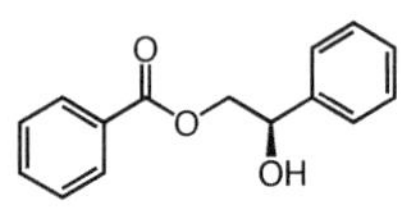

972 ((2*R*)-hydroxy-β-phenylethyl benzoate)

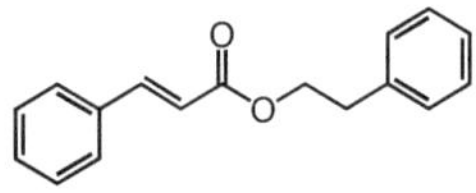

973 (β-phenylethyl (*E*)-cinnamate)

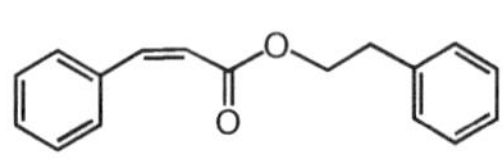

974 (β-phenylethyl (*Z*)-cinnamate)

Atranorin (**358**), 3*α*-methoxyserrat-14-en-21*β*–ol (**975**) and methyl 2,4-dihydroxy-3,6-dimethylbenzoate (**976**), isolated from the moss *Homalia trichomanoides*, showed antifungal activities against *C. albicans* with *MIC* values of 2.0, 2.0, and 0.6 μg/cm³ [338].

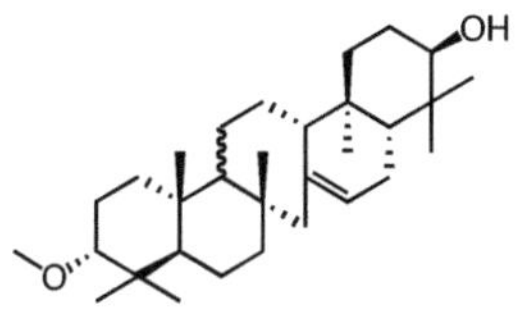

975 (3*α*-methoxyserrat-14-en-21*β*-ol)

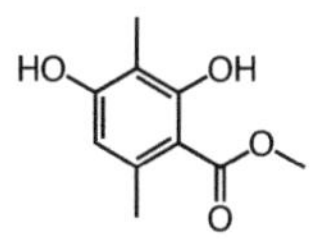

976 (methyl 2,4-dihydroxy-3,6-dimethylbenzoate)

A crude extract of *Riccardia marginata* displayed antifungal activity against the dermatophytic fungus *Trichophyton mentagrophytes*. The active compounds found were the chlorinated bibenzyls, 2,4,6-trichloro-3-hydroxybibenzyl (**977**), 2,4-dichloro-3-hydroxybibenzyl (**978**), and 2-chloro-3-hydroxybibenzyl (**979**), which inhibited the growth of both *C. albicans* and the plant pathogenic fungus, *Cladosporium resinae*, at a concentration of 30 μg/disk [334]. Compound **978** showed weak antifungal activity against *T. mentagrophytes* (12 μ*M* and 5 μ*M*/disk), while all of these chlorinated bibenzyls possessed antimicrobial properties against *Bacillus subtilis* [339].

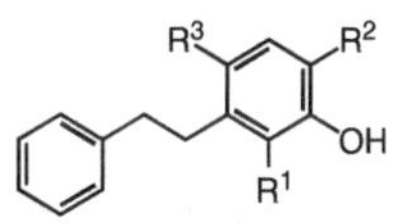

977 (2,4,6-trichloro-3-hydroxybibenzyl, $R^1 = R^2 = R^3 = Cl$)
978 (2,4-dichloro-3-hydroxybibenzyl, $R^1 = R^2 = Cl$, $R^3 = H$)
979 (2-chloro-3-hydroxybibenzyl, $R^1 = Cl$, $R^2 = R^3 = H$)

The liverwort *Pallavicinia ambigua* (Plate 20) produces pallamolides A–E (**980–984**), which are rearranged labdanes consisting of tautomers (tautomer I (**981–982**) and tautomer II (**983–984**)) that inhibit the hyphal morphogenesis, adhesion, and biofilm formation of *Candia albicans* strain DSY654 [340].

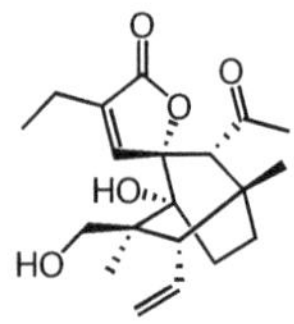

980 (pallamolide A)

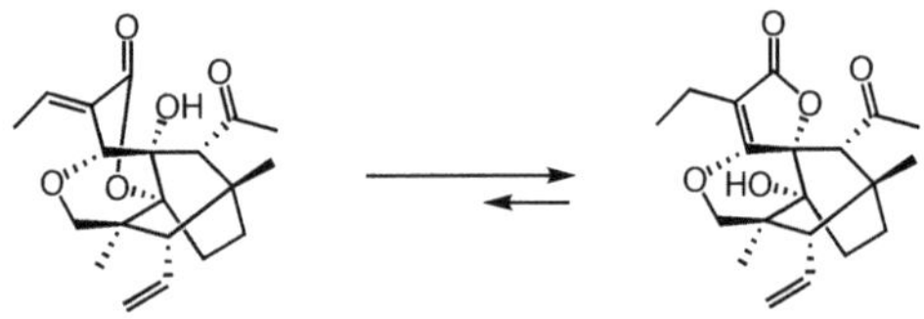

981 (pallamolide B) **982** (pallamolide C)

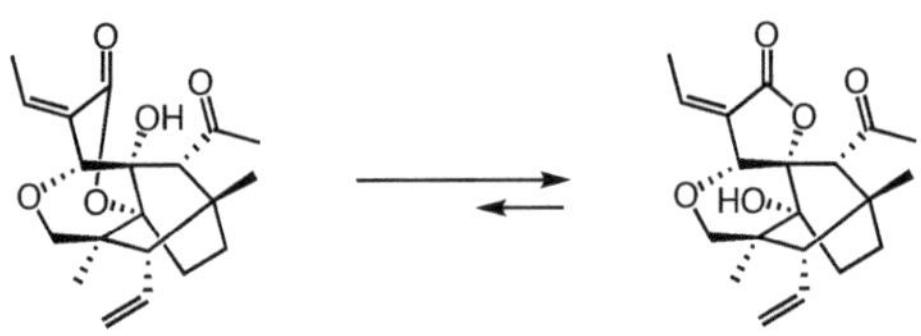

983 (pallamolide D) **984** (pallamolide E)

A 95% ethanol extract of the liverwort *Heteroscyphus coalitus* (Plate 29) was fractionated to give a harziane-type diterpenic acid, named heteroscyphic acid A (**985**), as well as eight clerodanes, heteroscyphic acids B–I (**986–993**) and four labdanes, heteroscyphins A–D (**994–997**), along with the guaiane sesquiterpene, heteroscyphin E (**998**) and the known clerodane, *ent*-juncenic acid (**999**). The absolute configurations of the new terpenoids were elucidated by ECD data interpretation and X-ray crystallographic analysis. All of the isolated terpenoids were evaluated for their antivirulent activities against *Candida albicans* strains DSY654 and SC5314, and nine compounds showed anti-filamentous activity against the former strain, but were ineffective against the latter strain. Most compounds blocked hyphal growth in the concentration range 4–32 μg/cm^3. Heteroscyphin D (**997**) possessed superior ability to inhibit the hyphal and biofilm formation of *C. candida* strain DSY654 [341].

Plate 29 *Heteroscyphus coalitus* (liverwort)

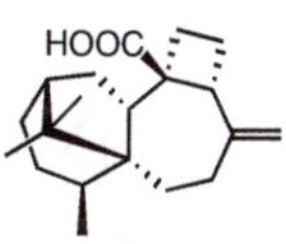

985 (heteroscyphic acid A)

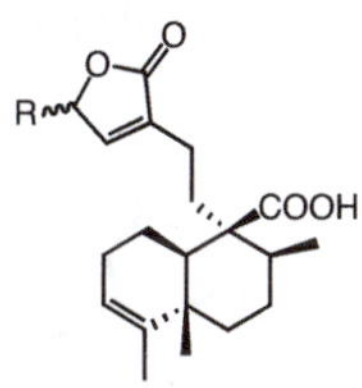

986 (heteroscyphic acid B, R = H)
987 (heteroscyphic acid C, R = OH)
988 (heteroscyphic acid D, R = OMe)
989 (heteroscyphic acid E, R = OEt)

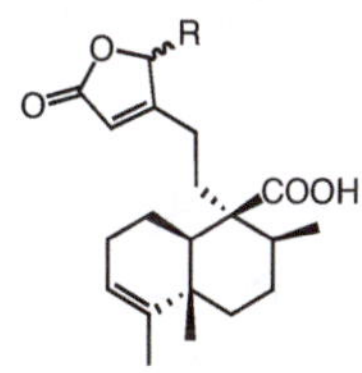

990 (heteroscyphic acid F, R = OH)
991 (heteroscyphic acid G, R = H)

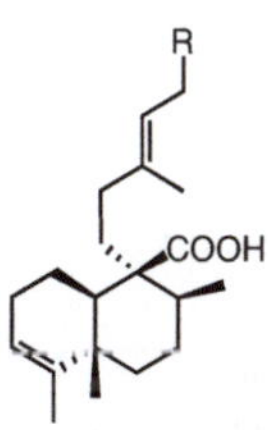

992 (heteroscyphic acid H, R = OH)
993 (heteroscyphic acid I, R = OAc)

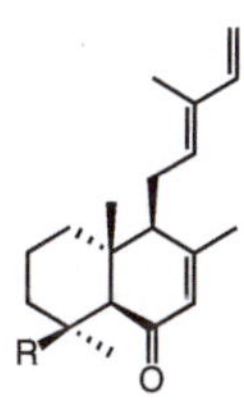

994 (heteroscyphin A, R = COOH)
995 (heteroscyphin B, R = CH_2OH)

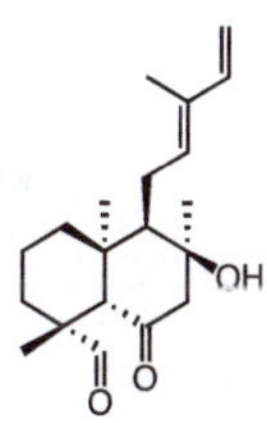

996 (heteroscyphin C)

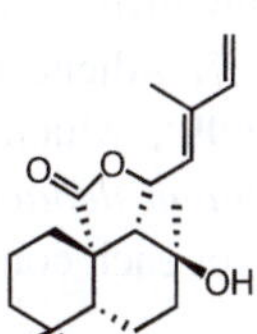

997 (heteroscyphin D)

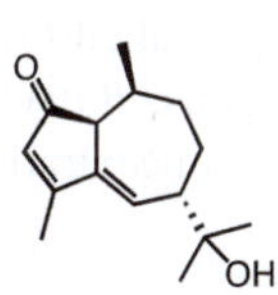

998 (heteroscyphin E)

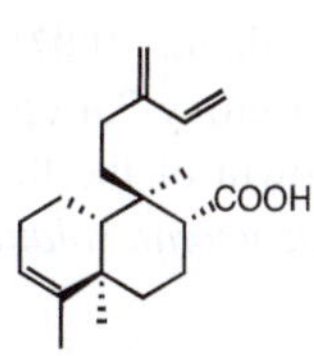

999 (*ent*-junceic acid)

Wang et al., reported in 2000 that the liverwort *Plagiochasma intermedium* elaborates the bis-bibenzyl, pakyonol (**218**), which possesses antifungal activity against *Epidermophyton floccosum*, *Microsporum gypseum*, and *Trichophyton gypseum* at a concentration range of 50–200 μg/cm^3 [342].

From the Bornean liverwort, *Pleurozia subinflata,* two cyathane diterpenoids were isolated, the new compound 5*β*-acetoxy-13-*epi*-neoverrucosanic acid (**1000**) and the previously known 13-*epi*-neoverrucosan-5*β*-ol (**1001**), along with chelodane (**1002**). Compounds **1000–1002** were evaluated against several marine fungal strains, *Lagenidium thermophylum* IPMB 1401, *Haliphthoros sabahensis* IPMB 1402, *Fusarium moniliforme* NJM 8995, *Fusariu oxysporum* NJM 0179, *Fusarium solani* NJM 8996, and *Ochroconis humicola* NJM 1503. Compound **1000** showed *MIC* values against *H. sabahensis* and *L. thermophylum* of 50 and 12.5 μg/cm^3, respectively. The other two diterpenes, **1001** and **1002** exhibited *MIC* values of 100 μg/cm^3 against *L. thermophylum* [343].

Cneorubin X (**1003**) and gymnomitr-3(15)-en-4*β*-ol (**1004**), isolated from a further Bornean liverwort, *Bazzania harpago*, showed antifungal activity against three fungi, *L. thermophilum, H. subahensis*, and *Haliphthoros milfordensis*, with *MIC* values of 50, 25, and 25 μg/cm^3, and 100, 100, and 100 μg/cm^3, respectively [344].

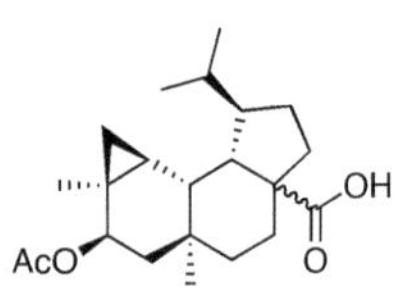

1000 (5*β*-acetoxy-13-*epi*-neovercosanic acid)

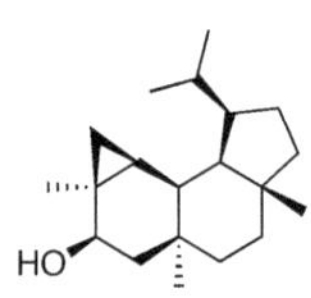

1001 (13-*epi*-neovercosan-5*β*-ol)

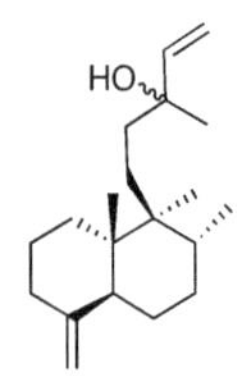

1002 (chelodane)

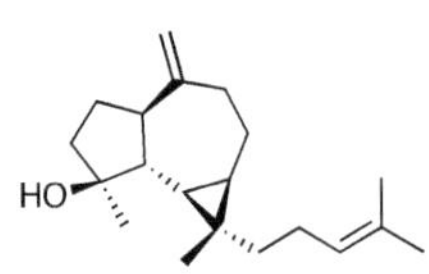

1003 (cneorubin X)

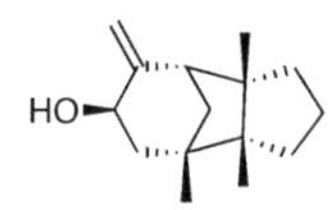

1004 ((–)-gymnomitr-3(15)-en-4*β*-ol)

The major secondary metabolites of the liverwort *Odontoschisma denudata* are the dolabellane diterpenoids, (+)-8-acetoxyodontoschismenol (**1005**), 6,12-dihydroxydolabella-(3*E*,7*E*)-diene (**1006**), 6-acetoxy-12,16-dihydroxydolabella-(3*E*,7*E*)-diene (**1007**), 6,16-diacetoxy-12-hydroxydolabella-(3*E*,7*E*)-diene (**1008**), and 6-acetoxy-3,4-epoxy-12-hydroxydolabell-(7*E*)-en-16-al (**1009**), which inhibited growth of the three plant pathogens, *Botrytis cinerea, Pythium debaryanum*, and *Rhizoctonia solani* by 5–44% at a concentration of 100 ppm for each compound [345].

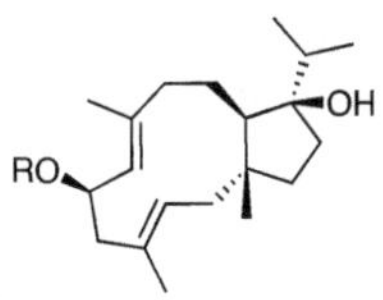

1005 ((+)-8-acetoxy-odontoschismenol, R = Ac)
1006 (6,12-dihydroxy-dolabella-(3*E*,7*E*)-diene, R = H)

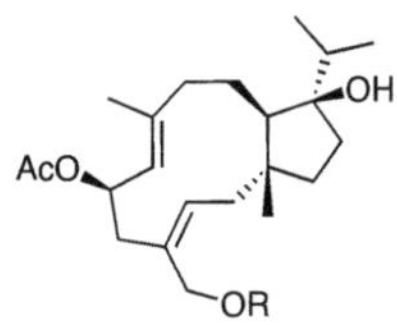

1007 (6-acetoxy-12,16-dihydroxydolabella-(3*E*,7*E*)-diene, R = H)
1008 (6,16-diacetoxy-12-hydroxydolabella-(3*E*,7*E*)-diene, R = Ac)

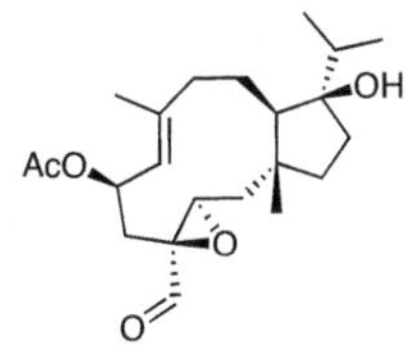

1009 (6-acetoxy-3,4-epoxy-12-hydroxydolabell-(7*E*)-en-16-al)

From the Chinese liverwort *Odontoschisma grosseverrucosum*, eight new terpenoids, odongrossins A–H (**1010**–**1017**), and the two known compounds, notolutesins A (**706**) and B (**707**) were isolated. Odongrossins A (**1010**) and G (**1016**) exhibited antifungal activity by inhibiting the hyphal formation of *Candida albicans* strain DSY654. Compound **1010** also inhibited adhesion and biofilm formation by this fungal strain [246].

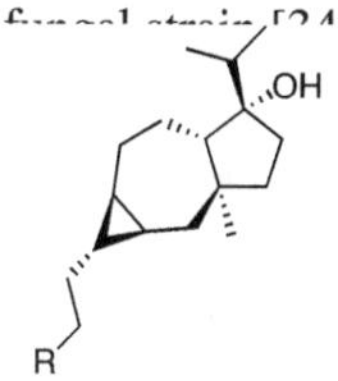

1010 (odongrossin A, R = COMe)
1011 (odongrossin B, R = OH)

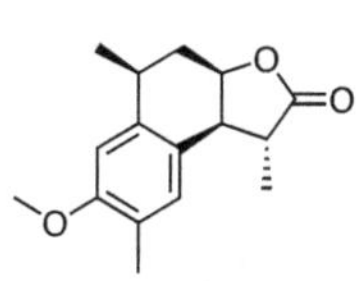

1012 (odongrossin C)

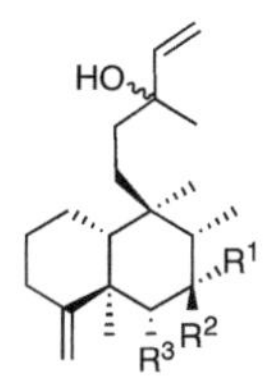

1013 (odongrossin D, $R^1 = R^2 = O$, $R^3 = H$)
1014 (odongrossin E, $R^1 = R^3 = H$, $R^2 = OH$)
1015 (odongrossin F, $R^1 = R^2 = H$, $R^3 = OH$)

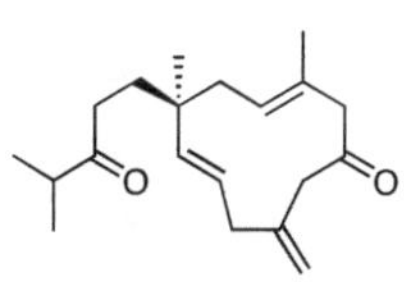

1016 (odongrossin G)

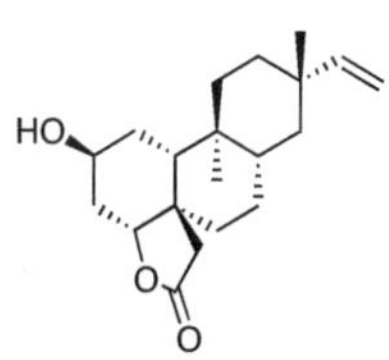

1017 (odongrossin H)

From the Japanese mosses, *Dicranum scoparium* and *D. japonicum*, the cyclopentanoyl fatty acids, 4-cyclopentenone-(2*S*),(2′*Z*)-pentenyl-(3*S*)-2″-octynoic acid (**1018**) and 2-cyclopentenone-2-[(2′*Z*)-pentenyl]-3-(1″,2″-octadienoic acid) (**1019**) were isolated. These two unsaturated fatty acids showed antifungal activity against *Botrytis cinerea* each with an *MIC* value of 60 μg/cm^3 [38].

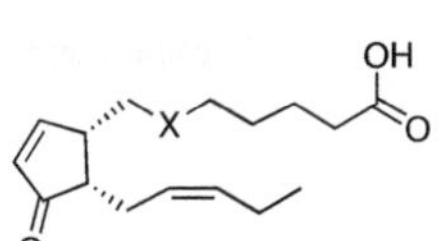

1018 (4-cyclopentenone-(2*S*),(2′*Z*)-pentenyl-(3*S*)-2″-octynoic acid, X = —C≡C—)

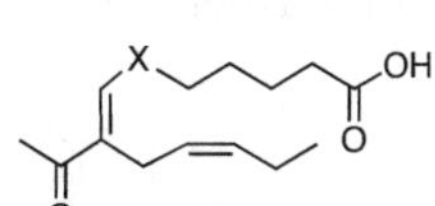

1019 (2-cyclopentenone-2-[(2′*Z*)-pentenyl]-3-(1″,2″-octadienoic acid, X = —CH=C=CH—)

More than 2000 *Plagiochila* species are known, especially in the southern hemisphere and a major chemical characteristic is the presence of highly oxidized 2,3-*seco*-aromadendrane sesquiterpenoids, as shown in 1995 by the present author and his associates. The Japanese *Plagiochila fruticosa* produces several 2,3-*seco*-aromadendranes, including the pungent plagiochiline A (**135**) and *ent*-bicyclogermacrene (**101**) [3]. Qiao et al., reported in 2020 that a Chinese specimen of *P. fruticosa* produces anadensin (**499**), loliolide (**600**), and cinnamolide (**659**), and six new bicyclogermacrane sesquiterpenoids, named plagicosins A–F (**1020–1025**). Also obtained were the 2,3-*seco*-aromadendrane plagicosin (**471**), five fusicoccanes, plagicosins H–L (**1026–1030**), and an eunicellene and two naturally rare gersemiane diterpenoids, plagicosins (**1031**) and N (**1032**), which were isolated from liverworts for the first time. Among these isolated compounds, plagicosin F (**1025**) showed potent antivirulence activity through the affecting the hyphal morphogenesis, adhesion, and biofilm formation of *Candida albicans* strain DSY654. Compound **1025** also regulated the genes related to hyphal formation [347].

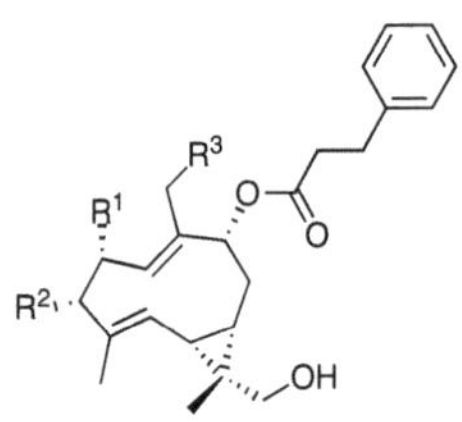

1020 (plagicosin A, $R^1 = R^3 = OH$, $R^2 = OAc$)
1021 (plagicosin B, $R^1 = OAc$, $R^2 = R^3 = OH$)
1022 (plagicosin C, $R^1 = OH$, $R^2 = R^3 = OAc$)

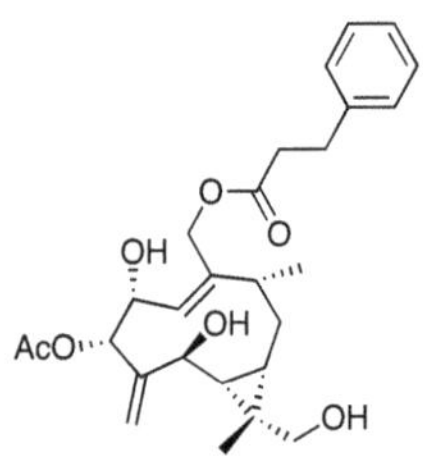

1023 (plagicosin D)

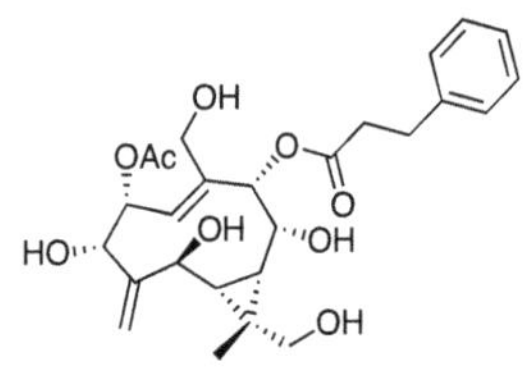

1024 (plagicosin E)

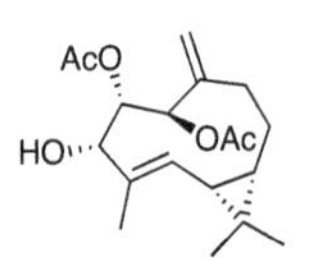

1025 (plagicosin F)

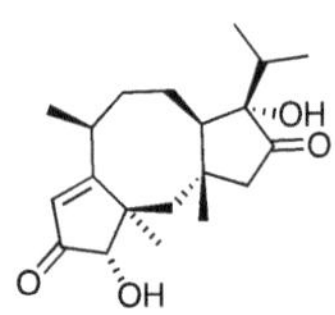

1026 (plagicosin H)

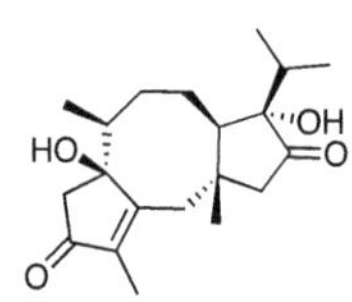

1027 (plagicosin I)

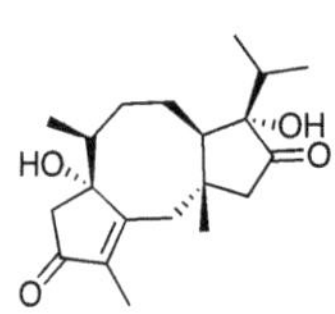

1028 (plagicosin J)

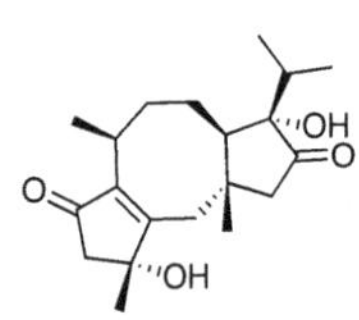

1029 (plagicosin K)

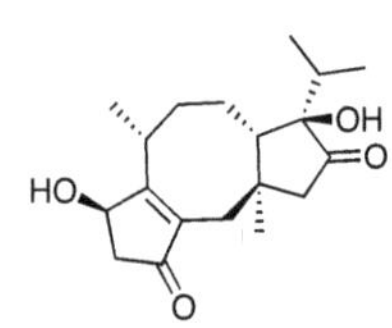

1030 (plagicosin L)

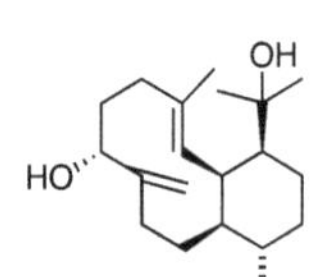

1031 (plagicosin M)

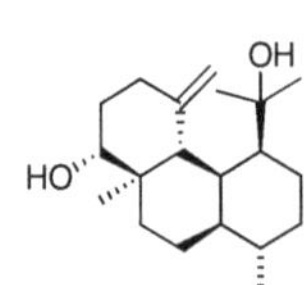

1032 (plagicosin N)

The antifungal activities of moss methanol and ethanol extracts have been studied by many researchers. Generally, the aqueous extracts show lesser antifungal activities than the alcoholic extracts. *Scapania* species are sources of lipophilic sesquiterpenoids and diterpenoids. Guo et al., evaluated in 2008 the antifungal effects of an ether extract of *Scapania verrucosa*, containing sesquiterpene hydrocarbons as well as fatty acids, against five fungal strains (*Aspergillus fumigatus*, *Candida albicans*, *Cryptococcus neoformans*, *Pyricularia oryzae*, and *Trichophyton rubrum*). Among these, *A. fumigatus* was sensitive to the ether extract, with an IC_{80} value of 8 μg/cm^3. The extract showed weaker antifungal effects against the other fungi evaluated (IC_{80} 32–64 μg/cm^3) [206].

An ethanol extract of the moss *Hypnum andoi* showed antifungal activity against *C. albicans* with inhibition zones of 7, 8, and 10 mm (40, 60, and 100 mm^3) [313]. The ethanol extracts of the two Turkish mosses (*Cinclidotus fontinaloides* and *Palustriella commutata*) had weak antifungal activity against *C. albicans* (*MIC* and *MIB*: > 100 and > 200 μg/10 cm^3) [314].

The antifungal activities of the essential oils of four mosses, *Homalothecium lutescens*, *Hypnum cupressiforme*, *Pohlia nutans*, and *Tortula muralis* were evaluated against the two fungi, *Candia albicans* and *Saccharomyces cerevisiae*. However, all of the essential oils exhibited only weak antifungal activities against *C. albicans* and *S. cerevisiae* at 165–1800 μg/cm^3 [348].

Mewari and Kumar examined the antifungal activity of a methanol extract of *Marchantia polymorpha* against four fungal strains, *Aspergillus flavus*, *A. niger*, *Candia albicans*, and *Trichophyton mentagrophytes*. This methanol extract showed the same level of antifungal activity against *C. albicans* and *T. mentagrophytes* with a *MIC* value of 0.3 mg/cm^3 for both, and *MBC*/*MFC* values of 0.6/0.6 mg/cm^3, respectively. However, *A. flavus* and *A. niger* were not inhibited by this extract [349].

This same group tested the antifungal activity of the methanol extract of *Marchantia polymorpha* against the fungal plant pathogens, *Alternaria solani*, *Fusarium oxysporum*, and *Rhizoctonia solani*. This extract completely inhibited the mycelial growth of *R. solani* when tested at the highest concentration used (5 mg/cm^3). The spore germination of *A. solani* and *F. oxysporum* was observed to be 100% by the extract at 10 mg/cm^3 [349].

The antifungal activities of the *n*-hexane, chloroform, methanol, and water extracts of the Indian *Marchantia polymorpha* were evaluated against *Candida albicans* and *Rhizopus oryzae* by Lakshmi and Rao. The highest level of antifungal activity was observed for the methanol extract of this liverwort against *C. albicans* (18 mm zone of inhibition) and *R. oryzae* (17 mm). Less effective antifungal activity was recorded for the *n*-hexane, chloroform, and water extracts against the two fungal strains [287].

In 2011, Gahtori and Chaturvedi evaluated the methanol and chloroform extracts of *Marchantia polymorpha* against four fungal strains, *Fusarium oxysporum*, *Rhizoctonia solani*, *Sclerotium rolfsii*, and *Tilletia indica* using the disk diffusion and microbroth dilution methods. Both extracts showed antifungal effects against *F. oxysporum* and *S. rolfsii*, with *MIC* values of 0.7 and 2.5 mg/cm^3, and *MFC* values of 0.7 and 4.5 mg/cm^3, respectively [298].

Joshi et al., evaluated in 2022 the antifungal activity of the 80% methanol extracts of the liverwort *Plagiochasma appendiculata* and the moss *Sphagnum fimbriatum* against the two fungal strains, *Fusarium solani* and *Aspergillus niger*. These extracts showed inhibition zones of 12 and 14 mm, and of 11 and 12 mm against *A. niger* and *F. solani*, respectively [299].

An ethanol extract of the Indian liverwort *Plagiochasma appendiculatum* showed potent antifungal activities against *Aspergillus niger* and *Trichophyton rubrum*, with *MIC* values of 10 and 2.5 μg/cm^3, respectively. The chloroform, acetone, and water extracts of this liverwort showed weak antifungal activity against *Candida albicans*, *Cryptococcus albidus*, *A. niger*, *A. flavus*, *A. nidulans*, *A. spinulosus*, *A. terreus*, and *Trichophyton rubrum*, with a *MIC* range of between 100 to 200 μg/ cm^3 [316].

Veljic and colleagues tested in 2010 the methanol extract of the Serbian liverwort *Ptilidium pulcherrimum* against six fungal strains, *Aspergillus flavus*, *A. versicolor*, *A. ochraceus*, *A. niger*, *Penicillium funiculosum*, and *Trichoderma viride*. The extract showed antifungal activities with *MIC* values between 0.5–2.5 mg/cm^3 and *MFC* values between 2.5–5.0 mg/cm^3, respectively [301].

This same group reported in 2012 that the DMSO extract of the moss, *Rhodobryum ontariense* showed antifungal activities against *Aspergillus fumigatus*, *A. versicolor*, *Penicillium finiculosum*, *P. ochrochloron*, and *Trichoderma viride* with *MIC* values in the range 0.1–0.5 mg/cm^3. These activities were more potent than that of the positive control, ketoconazole [302].

The antimicrobial activities of *n*-hexane, ethyl acetate, and methanol extracts of two Sri Lankan liverworts, a *Marchantia* and a *Plagiochila* species, and four mosses, *Calymperes motoleyi*, a *Fissidens* species, *Hypnum cupressiforme*, and *Sematophyllum demissum* were evaluated by Kirisanth et al., against *Candida albicans*. Only *H. cupressiforme* and *S. demissum* exhibited moderate zones of inhibition (7.8 and 7.4 mm, respectively), both at a 500 μg/disk concentration [304].

In turn, the antifungal activities of the DMSO extracts of *Atrichum undulatum*, *Physcomitrella patens*, and *Marchantia polymorpha* subsp. *ruderalis* from both natural sources and axenic cultures were evaluated against *Aspergillus versicolor*, *A. fumigatus*, *Penicillium funiculosum*, *P. ochrochloron*, and *Trichoderma viride*. All of the extracts possessed activity against all fungi evaluated with *MIC*/*MFC* levels of 0.1–2.0 mg/cm^3. Also, the extracts of the cultured species exhibited better antifungal activity than those from the species grown naturally. The extracts of the three species showed better inhibitory activities than the antifungal drugs, bifonazole (*MIC*/*MFC*: 0.2/0.3 mg/cm^3) and ketoconazole (2.5/3.0 mg/cm^3) against *T. viride* [350].

Six mosses, *Bryum capillare*, *Grimmia anodon*, *Orthotrichum rupestre*, *Pleurochaete squarrosa*, *Syntrichia ruralis*, and *Tortella tortuosa* were extracted with ethanol, methanol, acetone, and chloroform, and their antifungal effects were evaluated against the two fungi, *Candida albicans* and *Saccharomyces cerevisiae*. The methanol, acetone, and chloroform extracts of *B. capillare* and *G. anodon* showed the highest growth inhibitory activities against these two fungi, while the ethanol, methanol, and acetone extracts of *P. squarrosa* inhibited only the growth of *S. cerevisiae*. The growth of this same fungus was inhibited by the ethanol and methanol extracts of *O. rupestre*. However, the ethanol and methanol extracts of *S. ruralis* and the acetone extact of *T. tortuosa* had antifungal activity only against *C. albicans* [316]. The antifungal activity of the ethanol extract of *B. capillare* against *C. albicans* was weak ($MIC > 125$ mg/cm^3) [278].

As mentioned earlier, the European *Frullania dilatata* is a hazardous liverwort, since it causes potent allergic contact dermatitis due to the presence of several allergens, such as (+)-frullanolide (**3**). An ethanol extract of *F. dilatata*, containing **3**, 2,3-dimethyl anisole, and palmitic acid showed antifungal activities against *Candida albicans* and C. *tropicalis* with *MIC* values, in turn, of 0.257 and 0.514 mg/cm^3 [317].

The fractionation of the Chinese *Plagiochila nitens* led to the isolation of seven sacculatane diterpenoids, plagiochilarins A–G (**652–658**), along with the *seco*-aromadendrane, plagiochilarin H (**459**). All compounds were evaluated for their antifungal activity against *Candida albicans* strains SC5314 and CDR1, and the CDR2 double mutant strain DSY654 using the Alamar Blue assay. Only plagiochilarin H (**459**) showed antifungal activity against *C. albicans* DSY654 at an *MIC* value of 16 μg/cm^3 [236].

The antifungal activity of the ethanol, methanol, acetone, and chloroform extracts of five Turkish pleurocarpic mosses, namely, *Anomodon viculosus*, *Homalothecium sericeum*, *Hypnum cupressiforme*, *Leucodon sciuroides*, and *Platyhypnidium riparioides* were evaluated against two fungi, *Saccharomyces cerevisiae* and *Candida albicans.* The growth of *S. cerevisiae* was inhibited by all of the solvent extracts of *A. viculosus*, *H. sericeum*, and *L. sciuroides* (inhibition zone: 7–11 mm). The methanol and chloroform extracts of *H. cupressiforme* and the ethanol and methanol extracts of *P. riparioides* also inhibited these fungi (6.3–7.8 mm). In contrast, the growth of *C. albicans* was inhibited by the methanol and chloroform extracts of *A. viculosus* and the acetone extract of *H. cupressiforme*, and ethanol and acetone extracts of *L. sciuroides* (6.5–7.9 mm) [322].

4.10 Antiviral Activity

Sacculatal (**128**), isolated from *Pellia endiviifolia* showed antiviral activity against HIV with an ID_{50} value of 68.8 μg/cm^3 [351, 352]. In 2005, Xiao et al., reported the anti-hepatitis B virus activity of an 80% ethanol extract of *Marchantia convoluta.* The extract inhibited the proliferation of 2.2.15 cells derived from HepG2 murine hepatoma cells with an IC_{50} value of 30 μg/cm^3. This effect was caused by the induction of apoptosis. All concentrations of the extract also inhibited the secretion of HBsAg and HBeAg in the cultured medium of 2.215 cells [353] (see also Table 14).

In 2009, the H1N1 and H5N1 influenza A viruses caused pandemics throughout the world. Influenza A possesses an endonuclease within its RNA polymerase comprised of PA, PB1, and PB2 subunits. In order to obtain potential new anti-influenza compounds, 33 phytochemicals were evaluated using a PA endonuclease inhibition assay in vitro. Among them, the liverwort bis-bibenzyls marchantins A (**1**), B (**195**), D (**198**), and E (**199**), perrottetin F (**183**), and paleatin B (**202**) inhibited influenza PA endonuclease activity with IC_{50} values in the range from 5.3 to 23.7

Table 14 Antiviral compounds from liverworts

Species name	Compound or solvent extract	Virus	Activity value	Refs.
Blasia pusilla	Pusilatin B (**250**) Pusilatin C (**251**)	HIV-RT	no value given	[157]
Marchantia convoluta	80% EtOH (40 μg/cm^3) (containing apigenin (**482**), luteolin (**483**) and their glucuronides, quercetin (**851**))	Hepatitis B virus	IC_{50} (μg/cm^3) 30	[296, 353]
Marchantia paleacea subsp. *diptera* *Marchantia polymorpha*	Paleatin B (**202**)	HIV-1	IC_{50} (μg/cm^3) 22.1	[75, 76]
	Marchantin A (**1**)		11.5	
	Marchantin B (**195**)		9.3	
	Marchantin D (**198**)		23.7	
	Marchantin E (**199**)		21.2	
	Perrottetin F (**184**)		5.3	
Marchantia polymorpha *Plagiochila sciophila*	Marchantin A (**1**)	Influenza (PA endonuclease)	IC_{50} (μ*M*) 10	[354]
	Marchantin B (**195**)		10	
	Marchantin E (**199**)		10	
	Plagiochin A (**1056**)		10	
Pellia endiviifolia	Sacculatal (**128**)	HIV	ID_{50} (μ*M*) 68.8	[351, 352]

μg/cm^3 [354]. This was the first evidence that phytochemicals derived from liverworts could inhibit influenza A endonuclease. Marchantins A (**1**), B (**195**), and D (**198**), perrottetin F (**183**), and paleatin B (**202**) also showed anti-HIV-1 activities in the IC_{50} range from 5.3 to 23.7 μg/cm^3 [75, 76]. *Blasia pusilla* produces the bis-bibenzyl dimers, pusilatins A–D (**249–252**), among which pusilatin B (**250**) and C (**251**) showed weak HIV-RT inhibitory activities [157].

In 2005, Xiao et al., studied the antiviral activity of an 80% ethanol extract of *Marchantia convoluta*. Analysis by means of HPLC confirmed the presence of several flavonoids, among which apigenin (**482**), luteolin (**483**), apigenin 7,4′-di-*O*-glcronide (**1033**), apigenin 7-*O*-*β*-D-glucuronide (**1034**), luteolin 7,4′-di-*O*-glucuronide (**1035**) and quercetin (**851**), were identified. Luteolin (**483**) and its 7,4′-di-*O*-diglucuronide (**1035**) were the major flavonoids of the total extract. The same group further investigated the antibacterial, anti-inflammatory, and diuretic effects of the ethanol extract of *M. convoluta*. The investigators concluded that the above-mentioned biological activities were due to the presence of such a flavonoid mixture [296].

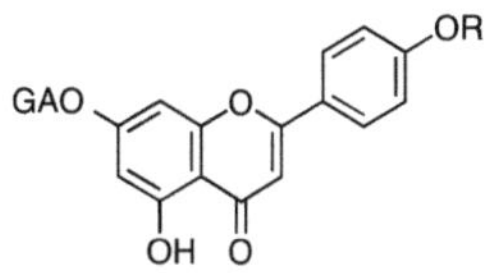

1033 (apigenin-di-*O*-glucuronide, R = GA)
1034 (apigenin-7-*O*-β-D-glucuronide, R = H)

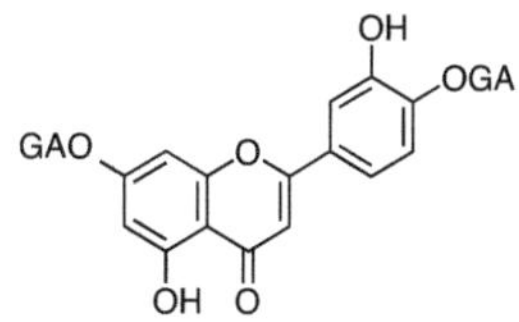

1035 (luteolin-7,4'-di-*O*-glucuronide)

4.11 Insect Antifeedant, Mortality, and Nematocidal Activities

As mentioned earlier, plagiochiline A (**135**), as found in several *Plagiochila* species, is a strong antifeedant effect against the African armyworm *Spodoptera exempta* (1–10 ng/cm^2 in a choice test for 2 h) [1, 2] (see overview in Table 15). Compound **135** showed nematocidal activity against *Caenorhabditis elegans* (111 μg/cm^3). Also, the pungent component sacculatal (**128**) was found to kill the tick species *Panonychus citri.* The pungent eudesmanolides **132** and **133** from *Chiloscyphus polyanthos*, tulipinolide (**134**) from *Wiesnerella denudata*, and a bitter-tasting diterpene furanolactone, gymnocolin A (**148**), from *Gymnocolea inflata*, also have antifeedant activity against the larvae of a Japanese *Pieris* species. However, their activities are less potent than that of plagiochiline A (**135**) [2].

The Portuguese liverwort, *Targionia lorbeeriana*, was found to contain the three guaianolides, dehydrocostuslactone (**951**), acetyltrifloculoside lactone (**952**) and 11-αH-dihydrodehydrocostuslactone (**953**), together with 8,15-diacetylsalonitenolide (**954**). Of these, compounds **951–953** showed larvicidal activity against *Aedes aegypti* at 12.5, 50, and 50 ppm, respectively [254]. Atisane 2 (**740**), obtained from *Lepidolaena clavigera*, showed insecticidal activity against blowfly larvae [250].

A series of natural drimanes and related synthetic compounds was tested for antifeedant activity against various aphids. Polygodial (**90**) from the *Porella vernicosa* complex and warburganal (**1036**) from the African tree *Warburgia ugandensis* (Canellaceae) were the most potent antifeedant metabolites against the aphids assessed [2, 355], and the natural (–)-polygodial (**90**) and its synthetic (+)-enantiomer (**1037**) showed similar levels of activity as aphid antifeedants. (–)-Polygodial (**90**) caused mosquito larvae lethality at a concentration of 40 ppm, and had mosquito repellent activity, which was more potent than the commercially available *N,N*-diethyl-*m*-toluamide (DEET). Plagiochilide (**442**) isolated from *Plagiochila* species killed brown plant hoppers, *Nilaparvata lugens* (Delphacidae), at a concentration level of 100 μg/cm^3 [2].

Table 15 Insecticidal and insect antifeedant compounds from bryophytes

Species name[a]	Compound	Insect name	Activity value	Refs
[L] *Balantiopsis cancellata*		*Spodoptera littoralis*	FR_{50}[b]	[337]
	CH_2Cl_2		0.42	
	β-Methylthioacrylate (= isotachin B) (**14**)		0.44	
	β-Phenylethyl benzoate (**971**)		0.56	
	(2*R*)-Hydroxy-β-phenylethyl benzoate (**972**)		0.50	
	β-Phenylethyl (*E*)-cinnamate (**973**)		0.96	
	β-Phenethyl (*Z*)-cinnamate (**974**)		0.50	
[L] *Bazzania japonica*	Fusumaoid A (**1046**) Fusumaoid B (**1047**)	*Sitophilus zeamais* (rice weevil)	ED_{50} (μg/cm^3) 25.0	[367]
[L] *Chiloscyphus polyanthos* [L] *Frullania dilatata* [L] *Frullania tamarisci* subsp. *tamarisci* [L] *Gymnocolea inflata* L] *Porella vernicosa* complex [L] *Wiesnerella denudata*	*ent*-Diplophyllolide (**132**) *ent*-7α-Hydroxy-diplophyllolide (**133**) *ent*-Frullanolide (**4**) Frullanolide (**3**) Gymnocolin A (**148**) Cinnamolide (**659**) Tulipinolide (**134**)	Larvae of *Pieris* species	(antifeedant)	[2]
[L] *Hymenophyton flabellatum*	1-(2,4,6-Trimethoxy-phenyl)-but-(2*E*)-en-1-one (**141**)	Larvae of the yellow butterfly (*Eurema hecabe mandarina*)	(antifeedant)	[357, 358]
[L] *Lepidolaena clavigera*	Atisane 1 (**739**) Atisane 2 (**740**)	Blowfly larvae	(insecticidal)	[250]

(continued)

Table 15 (continued)

Species name[a]	Compound	Insect name	Activity value	Refs
[L] *Lepidolaena clavigera*	Clavigerin A (**475**) Clavigerin D (**478**) Clavigerin B (**476**) Clavigerin C (**477**) Clavigerin A (**475**) Clavigerin B (**476**)	Larvae of the webbing clothes moth (*Tineola bisselliella*) Larvae of the Australian carpet beetle (*Anthrenocerus australis*) On wool of 0.05% *w/w* on wool of 0.026% *w/w* against carpet beetle larvae	(antifeedant) (antifeedant) (antifeedant)	[356]
[L] *Lepidolaena hodgsoniae*	Hodgsonox (**1039**)	Australian green blowfly larvae, *Lucilia cuprina*	LC_{50} (mg/cm^3) 0.27	[360]
[L] *Marchantia linearis*	60% MeOH (20 μ*M* sodium diethyldithio-carbamate)	*Spodoptera litura*	Antifeedant: 24.9–68% Larvicidal: 20.9–59% Pupicidal: 22–40.8% Mortality rate: 34–71%	[366]
[L] *Marchantia polymorpha*	Oil bodies Single ortholog gene, *Mp1HDZ*	Arthropod herbivores	Defended *M. polymorpha* against arthropod herbivores	[368]
[L] *Metacalypogeia alternifolia*	Metacalypogin (**1040**)	*Oncopeltus fasciatus*	(inhibition of metamorphosis)	[361]
[M] *Octoblepharum albidum*	2-Methoxy-4-methylphenol (**1049**) Methyl *p-tert*-butylphenyl acetate (**1050**)	*Epilachna sparsa* *Nilaparvata lugens*	Mortality	[369]
[L] *Plagiochila bursata*	Plagiochiline A (**135**)	*Spodoptera frugiperda*	Reduction of larval: growth 66% 25%	[362]

(continued)

Table 15 (continued)

Species name[a]	Compound	Insect name	Activity value	Refs
	Fusicogigantone A (**506**)		Larval mortality: 55% 75%	
	Plagiochiline A (**135**)		Pupal mortality: 20% 25%	
	Fusicogigantone A (**506**)			
	Plagiochiline A (**135**)		Abdomen and wing malformation in adults leading	
	Fusicogigantone A (**506**)			
	Plagiochiline A (**135**)		Inability to mate	
[L] *Plagiochila diversifolia*	Fusicogigantone B (**507**) (13*S*)-13-Hydroxy-labda-8,14-diene (**1042**) 13α,14-Diacetoxy-2-hydroxybicyclo-germacrene (**1045**)	*Spodoptera frugiperda*	Larval growth inhibition, 60% larval mortality and antifeedant effect (choice test) with a feeding ration Larval growth inhibition and 70% larval mortality Larval growth inhibition	[364]
[M] *Barbula lambarenensis* [M] *Bryum coronatum* [M] *Calymperes afzelii* [M] *Thuidium gratum*	H_2O	Maize stem borer	Toxic deterrent mortality: 80%	[365]

(continued)

Table 15 (continued)

Species name[a]	Compound	Insect name	Activity value	Refs
[L] *Plagiochila fruticosa* [L] *P. ovalifolia* [L] *P. yokogurensis*	Plagiochiline A (**135**) Plagiochilide (**442**)	*Spodoptera exempta* (=African army worm) Brown plant hopper (*Nilaparvata lugens*)	Antifeedant: 1–10 ng/cm^2 leaf disk choice test for 2 h Lethal: 100 μg/cm^3	[1, 2]
[L] *Porella vernicosa* complex	(–)-Polygodial (**90**) (+)-Polygodial (**1937**) (synthetic)	Aphid Mosquito Mosquito larvae	(antifeedant) (repellent) (lethal):100% (40 ppm)	[2, 355]
[L] *Riccardia polyclada*	2,4,6′,3′-Tetrachloro-3-hydroxy-4′-methoxy-bibenzyl (**970**) 2,4,6,3′-Tetrachloro-3-hydroxy-bibenzyl (**969**)	*Spodoptera littoralis*	(antifeedant)[c] FR_{50} 0.63 0.43	[336]
[L] *Targionia lorbeeriana*		*Aedes aegypti*	LC_{100} (concentration in ppm) lethal:	[254]
	Dehydrocostus lactone (**951**)		12.5	
	Acetyltrifloculoside lactone (**952**)		50	
	11*αH*-Dihydro-dehydrocostuslactone (**953**)		50	

[a] [L] liverwort, [M] moss
[b] FR_{50} feeding ratio in disk-choice assay: 10 μg/cm^2 in the leaf disk choice test (antifeedant)
[c] FR_{50} treated area eaten (when 50% of the untreated area had been eaten)/50% untreated area, at a dose of 10 μg/disk

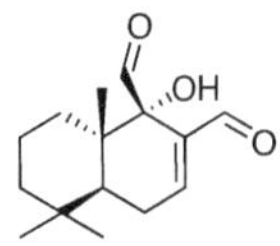

1036 (warburganal)

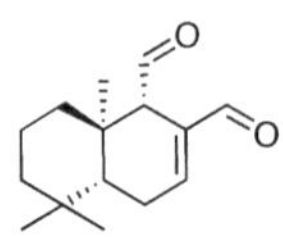

1037 ((+)-polygodial)

Lepidolaena clavigera produces the clavigerin group of bergamotane sesquiterpenoids. Of these, clavigerin A (**475**) and methoxyclavigerin D (**478**), showed less potent antifeedant activity against larvae of the webbing clothes moth, *Tineola bisselliella* (Lepidoptera), than either clavigerins B (**476**) or C (**477**). When the latter compounds were tested against larvae of the Australian carpet beetle, *Anthrenocerus australis* (Coleoptera), they again showed antifeedant activity. The insect antifeedant activity of the acetoxy acetals **260** and **261** could be due to their proposed facile hydrolysis to the corresponding aldehydes. The activity of plagiochiline A (**135**), possessing an acetoxy acetal, is also due to the fact that it can hydrolyze to a dialdehyde, which showed some biological activity [356]. Clavigerins A (**475**) and B (**476**) showed potent antifeedant activity against the larvae of the carpet beetle, with values on wool of 0.05% *w/w* for **260**, and of 0.03% *w/w* for **261**. Both compounds displayed significant insecticidal and antifeedant activities against the clothes moth larvae at 0.1% *w*/w. The positive control compound, azadirachtin, showed activity at 0.01% *w/w* [356].

Isotachin B (= β-phenethyl (*E*)-β-methylthioacrylate) (**14**), from *Balantiopsis cancellata*, showed moderate antifeedant activity against the larvae of *Spodoptera littoralis*, the fall armyworm, at an FR_{50} value of 0.44 μ*M* in a leaf disk assay [337]. 1-(2,4,6-Trimethoxyphenyl)-but-2(*E*)-en-1-one (**141**) from *Hymenophyllum flabellatum* [357] demonstrated antifeedant activity against larvae of the yellow butterfly, *Eurema hecabe mandarin* [358].

An in vitro nematode larval motility inhibition assay was developed to screen liverwort extracts for anthelmintic activity against the third-stage larvae of the sheep parasite, *Trichostrongylus colubriformis.* A crude ethanol (95%) extract from the New Zealand liverwort, *Plagiochila stephensoniana*, had larval motility inhibition activity against these larvae with an IC_{50} value of 1.9 mg/cm^3 (see Table 16). The major constituent, 3-methoxy-4′-hydroxybibenzyl (**286**) showed anthelmintic activity against the larvae with an IC_{50} value of 0.1 mg/cm^3 [359]. The compound 3,4′-dimethoxybibenzyl (**1038**) displayed anthelmintic activity (30% inhibition at 0.7 mg/cm^3), while the synthetic 3-methoxy-4′-hydroxy-(*E*)- (**287**) and 3-methoxy-4′-hydroxy-(*Z*)-stilbene (**288**) were more active (IC_{50} values of 0.07 and 0.06 mg/cm^3, respectively), than **286**. The same bioassay was carried out for natural (–)-polygodial (**90**), obtained from the higher plant *Pseudowintera colorata.* Compound **90** showed more potent anthelmintic activity (IC_{50} 0.07 mg/cm^3) when compared with **286** [359]. (–)-Polygodial (**90**) has also been found in liverworts, ferns, and the higher plants, *Polygonum hydropiper* [1–3], and even in *Dendrodoris* species, which are marine porostome nudibranchs. It was shown to inhibit the feeding of fish [2]. *Lepidolaena hodgsoniae* biosynthesizes hodgsonox (**1039**), which showed weak insecticidal activity against the larvae of the Australian green blowfly, *Lucilia cuprina*, with a LC_{50} value of 0.27 mg/cm^3, but was much less active than the standard insecticide, diazinon (LC_{50} 1.6 μg/cm^3 [360]. Metacalypogin (**1040**), from *Metacalypogeia alternifolia*, weakly inhibited the metamorphosis of *Oncopeltus fasciatus*, with four out of 30 insects in total not reaching adulthood [361].

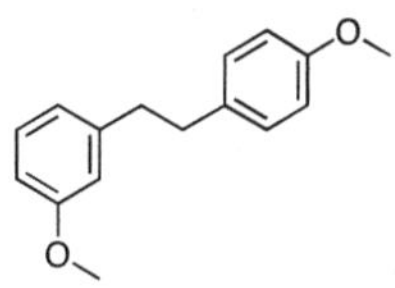

1038 (3,4'-dimethoxy-bibenzyl)

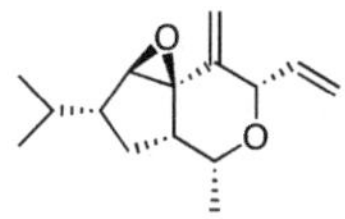

1039 (hodgsonox)

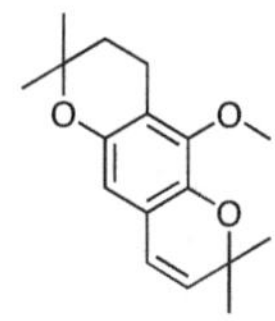

1040 (metacalypogin)

In 2010, Ramirez et al., studied the insecticidal constituents of the Argentine *Plagiochila bursata* and found the presence of a new *seco*-aromadendrane (**4**), together with the known plagiochilines A (**135**) and M (**425**), fusicogigantone A (**506**), and 1,4-dimethylazulene (**1041**), which is a significant chemical marker of the liverwort [362]. A few species belonging to the Plagiochilaceae contain **1041** and its analogues [1–3, 363]. Plagiochiline A (**135**) and fusicogigantone A (**506**) showed

Table 16 Nematocidal compounds from liverworts

Species name	Compound or solvent extract	Nematode	Activity value IC_{50}	Refs.
Plagiochila sp.	Plagiochiline A (**135**)	*Caenorhabditis elegans*	111 μg/cm^3	[2]
Plagiochila stephensoniana	95% EtOH 3-Methoxy-4′-hydroxy-bibenzyl (**286**)	Sheep parasite *Trichostrongylus colubriformis* (at the third larval tage)	1.9 mg/cm^3 (motility inhibition) (30% inhibition) 0.13 mg/cm^3	[359]
	3,4′-Dimethoxybibenzyl (**1038**) (synthetic)		0.6 mg/cm^3	
	3-Methoxy-4′-hydroxy-(*E*)-stilbene (**287**) (synthetic)		0.07	
	3-Methoxy-4′-hydroxy-(*Z*)-stilbene (**288**) (synthetic)		0.06	
Pseudowintera colorata (higher plant)	(–)-Polygodial (**90**)	Sheep parasite *Trichostrongylus colubriformis* (at the third larval stage)	(30% motility inhibition) 0.07	[359]

reduction of larval growth of *Spodoptera* species by 66 and 25%, larval mortality by 55 and 75%, and pupal mortality by 20 and 25%, respectively. Compound **135** also led to abdomen and wing malformations in the adult insects, leading to the impossibility of mating [362].

In 2017, Ramirez et al., studied the insecticidal activities of an ether extract of the Argentine *Plagiochila diversifolia* and of plagiochiline-15-yl-(4*Z*)-decenoate (**473**), together with the bicyclogermacrene, 3*α*,14-diacetoxy-2-hydroxybicyclogermacrene (**1043**), and three 2,3-*seco*-aromadendranes, plagiochiline B (**416**), furanoplagiochilal (**138**), and neofuranoplagiochilal (**441**), two fusicoccane diterpenoids, fusicogigantone A (**506**) and B (**507**), a labdane diterpene, (13*S*),13-hydroxylabda-8,14-diene (**1042**), as well as *trans*-nerolidol (**40**), spathulenol (**514**), globulol (**825**), 3α,14-diacetoxy-2*α*-hydroxybicyclogermacrene (**1045**), and α-tocopherol (**15**).

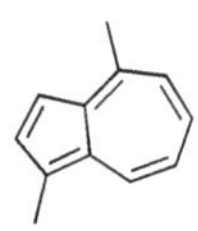

1041 (1,4-dimethyl-azulene)

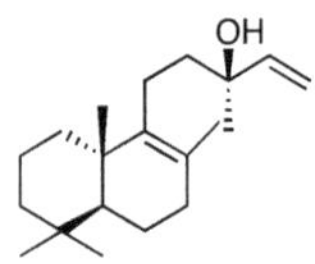

1042 ((13*S*)-13-hydroxy-labda-8,14-diene)

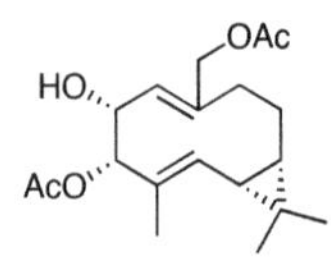

1043 (3*α*,14-diacetoxy-2*α*-hydroxybicyclo-germacrene)

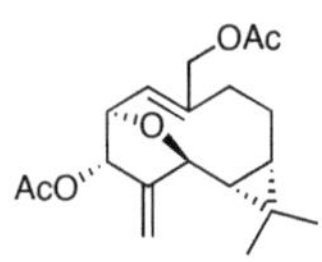

1044 (3*α*,14-diacetoxy-2*α*,5*β*-peroxybicyclo-germacra-1(10*E*),4(15)-diene)

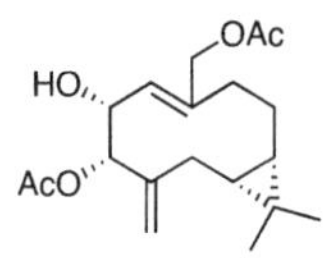

1045 (3*α*,14-diacetoxy-2α-hydroxybicyclo-germacra-1(10*E*),4(15)-diene)

(13*S*)-13-Hydroxylabda-8,14-diene (**1042**), fusicogigantone B (**507**) and 3*α*,14-diacetoxy-2*α*-hydroxbicyclogermacrene (= plagicosin F) (**1043**) incorporated into the diet (100 mg/g) reduced the larval growth of *Spodoptera frugiperda* by 70 ± 25, 57 ± 23, and 33 ± 16%, respectively. Treatment with **1042** and **507** produced 60 and 70% larval mortality in their early stages. Compound **1043** also exhibited antifeedant activity in a choice test with a feeding ratio of 0.54 ± 0.16. Treatment of compound **1042** produced abdomen and wing malformations in adult leading to an inability to mate [364].

Ande et al., evaluated in 2010 water extracts of four moss powders, *Bryum coronatum*, *Barbula lambarenensis*, *Calymperes afzelii*, and *Thuidium gratum* against maize stem borer moths. The extracts of *B. coronatum* and *C. afzelii* exhibited toxic and deterrent activities, that were comparable to the organophosphorus control insecticide, Tricel. The activity order of the tested moss solutions was *C. afzelii* > *B. coronatum* > *T. gratum* = *B. lambarenensis*. A distinctly higher stem borer mortality of 80% was oberved for *C. afzelii* while *B. coronatum* and the control Tricel gave a comparable mortality value of about 77%, and *T. gratum* and *B. lambarenensis* had somehat lower values of ca. 65% [365].

A cell suspension of cultured *Marchantia linearis* was extracted with 80% methanol to obtain a crude extract, which was analyzed by reversed-phase HPLC to identify apigenin (**482**), luteolin (**483**), rutin (**847**), quercetin (**851**), and kaempferol (**807**). Krishnan et al., tested the insecticidal activity of this methanol extract containing flavonoids. An in vivo examination showed significant antifeedant (24.9–68%), larvicidal (21–59%), pupicidal (22–41%), and mortality (34–71%) rates at all concentrations (1–5% of flavonoids) against the 5th instar larvae of *Spodoptera litura.* This extract also exhibited feeding deterrent activity, and the body weight of the insect 5th instar larvae when treated with 1–5% of this flavonoid-containing extract showed a significant reduction from 840 to 396 mg. However, the consumption of a basal diet with the incorporation of flavonoids by *S. litura* larvae was not significantly different compared to the consumption of the control diet by the larvae [366].

The genus *Bazzania,* belonging to the Lepidoziaceae, is an abundant source of sesquiterpenoids. *Bazzania* species produce various bis-bibenzyl derivatives [3]. Further investigation of the Japanese *Bazzania japonica* led to the isolation of two new eremophilane sesquiterpenoids, named fusumaols A (**1046**) and B (**1047**). This was the first isolation of eremophilanes from the genus *Bazzania.* Both compounds showed moderate insect repellent activities against the rice weevil *Sitophilus zeamis,* with an ED_{50} value of 25 μg/cm^3 each, using a filter paper disk method [367].

Romani et al., indicated in 2020 that the oil body formation in *Marchantia polymorpha* is controlled by only the orthologue gene, MpC1HDZ and that it serves as a defense against arthropod herbivores, but not for abiotic stress tolerance [368].

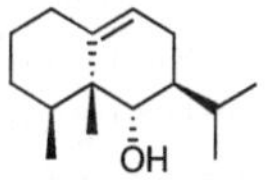

1046 (fusumaol A)

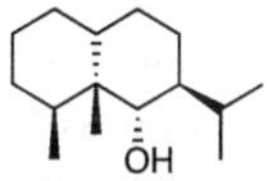

1047 (fusumaol B)

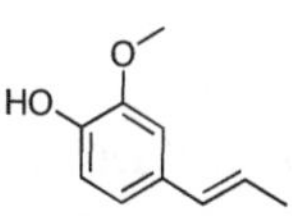

1048 ((*E*)-isoeugenol)

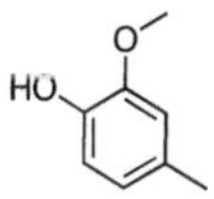

1049 (2-methoxy-4-methylphenol)

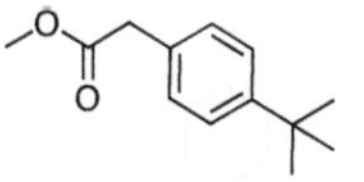

1050 (methyl-*t*-butyl-phenylacetate)

A *n*-pentane extract of the Brazilian moss, *Octoblepharum albidum,* obtained using a microscale simultaneous distillation apparatus, was analyzed by GC/MS to identify 12 volatile components, among which hexadecanoic, oleic and stearic acids, 1-octen-3-ol (**16**) and (*E*)-isoeugenol (**1048**) were detected as major components. (*E*)-Isoeugenol (**1048**) showed contact toxicity against *Sitophilus zeamis* and *Tribolium castaneum* and antifeedant activity against the larvae and/or adults of both species. 2-Methoxy-4-methylphenol (**1049**) and methyl-*p-tert*-butylphenyl acetate (**1050**), which are previously undescribed among bryophytes, were present in *O.*

albidum as minor constituents. These aromatic compounds resulted the mortality of *Epilachna sparsa* and *Nilaparvata lugens.* On the basis of these results, Alves et al., suggested in 2022 that the volatiles from *O. albidum* might participate in plant defense mechanisms against insects, causing mortality or developmental inhibition [369].

4.12 *Antitrypanosomal, Antileishmanial, and Antitrichomonal Activities*

Leishmaniasis is a group of infectious diseases caused by protozoal parasites representing more than 20 *Leishmania* species (Table 17). These parasites are transmitted to humans by the bite of an infected female phlebotomine sand-fly insect vector. There are more than 1 billion people in tropical areas where leishmaniasis is endemic. In turn, Chagas' disease is caused by the protozoan parasite *Trypanosoma cruzi,* which is transmitted to humans by the bite of a triatormine bug insect vector. In addition, African sleeping sickness is caused by *Trypanosoma brucei gambiense* or *Trypanosoma brucei rhodesiense* and is transmitted to humans by the bite of tsetse flies. The discovery and development of essential new drugs for such neglected diseases are urgent challenges.

The antitrypanosomal activities of 24 terpenoids from various plant sources was evaluated in 2011 by Otoguro et al., against *Trypanosoma brucei brucei* strain GUTat 3.1. Among them, 22 terpenes exhibited some level of antitrypanosomal activity in vitro. Of these, α-eudesmol (**1051**), isolated from the liverwort *Porella stephaniana*, and the other sesquiterpenoids, hinesol (**1052**), flornithine (**1053**), and 4α-hydroperoxy-1,2,4,5-tetrahydro-α-santonin (**1054**), showed potent and selective antitrypanosomal activities, with EC_{50} values of 0.1, 0.3, 0.6, and 0.5 μg/cm^3, respectively. Their activities were more potent than those of the commercially used antitrypanosomal drugs, eflornithine and suramin, which gave EC_{50} values of 2.3 and 1.5 μg/cm^3, respectively [370].

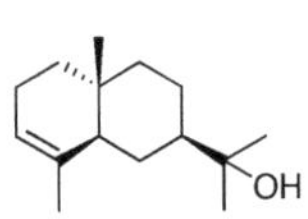

1051 (α-eudesmol)

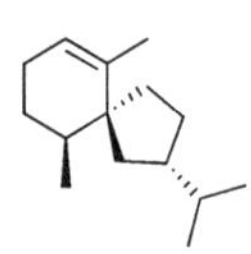

1052 (hinesol)

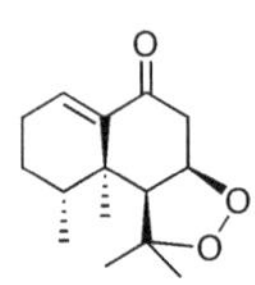

1053 (nardosinone)

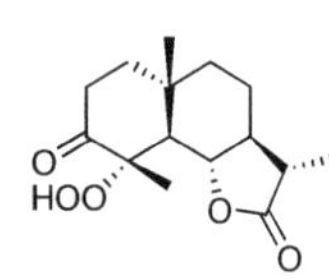

1054 (4α-hydroperoxy-1,2,4,5-tetrahydro-α-santonin)

Marchantin A (**1**), perrottetin D (**1055**), and plagiochin A (**1056**) also demonstrated selective and potent inhibitory activities against *Trypanosoma brucei brucei* strain GUTat 3.1 in vitro with IC_{50} values of 0.27, 0.44, and 0.93 μg/cm^3, respectively) [371].

Table 17 Antitrypanosomal, antileishmanial, and antitrichomonal compounds from liverworts

Species name	Compound	Microorganisms	Activity value	Refs
Marchantia polymorpha (Japanese specimen)	Marchantin A (**1**)		IC_{50} (μg/cm^3)	[371]
		Trypanosoma brucei	0.3 (antitrypanosomal) 3.4 (antiproliferation)	
		Plasmodium falciparum NF54	2.0	
		Plasmodium falciparum KI	2.1 (antitryposomal)	
		Trypanosoma brucei	14.9 (antitryposomal)	
		T. cruzi *Leishmania donovani*	1.6 (antileishmanial)	
Marchantia polymorpha (Iceland sp.)	Marchantin A (**1**)		IC_{50} (μ*M*)	[372]
		Plasmodium falciparum NF54	3.4	
		P. falciparum KI	2.0	
		Trypanosoma brucei rhodesiense	2.1	
		T. cruzi	14.9	
		Leishmania donovani	1.6	
		Recombinant PfFabZ enzyme (*P. falciparum*)	18.2	
Plagiochila disticha	Plagiochiline A (**135**)	*Leishmania amazonensis*	7.1 μ*M*(antileishmanial)	[200]
		Trypanosoma cruzi	14.5 (antitrypanosomal)	
Plagiochila fruticosa *Plagiochila sciophila*	Perrottetin D (**1055**) Plagiochin A (**1056**)	*Trypanosoma brucei*	0.4 (antitrypanosomal) 0.93 (antitrypanosomal)	[371]

(continued)

Table 17 (continued)

Species name	Compound	Microorganisms	Activity value	Refs
Plagiochila porelloides	Et_2O/Essential oil Et_2O: β-barbatene (**33**), bicyclogermacrene (**101**), globulol (**825**), 1,2-dihydro-4,5-dehydro-nerolidol Essential oil: β-barbatene (**33**), bicyclogermacrene (**101**), maalian-5-ol, *p*-menth-1-en-3-(2-methyl-(1*E*)-butenyl)-8-ol, *p*-menth-1-en-3-(2-methyl-(1*Z*)-butenyl)-8-ol	*T. brucei brucei* *L. mexicana mexicana* *T. brucei brucei* *L. mexicana mexicana*	IC_{50} μg/cm^3 2.0/16.0 5.2/2.0 Selectivity indices 11.7/1.5 0.6/0.3	[202]
Porella stephaniana		*Trypanosoma brucei brucei* GU Tat 3.1	EC_{50} (μg/cm^3)	[370]
	α-Eudesmol (**1051**)		0.1 (0.5 μ*M*) (antitrypanosomal)	
	Hinesol (**1052**) (from higher plants)		0.3	
	Nardosinone (**1053**) (from higher plants)		0.6	
	4*α*-Hydroperoxy-1,2,3,4-tetrahydro-α-santonin (**1054**)		0.5	
	Eflornithine, suramin (control drugs)		2.3, 1.5	
Ricciocarpos natans	2,5,4′-Trihydroxy-bibenzyl (= 14-hydroxy-lunularin, 2-hydroxy-lunularin) (**1057**) (synthetic)		(antileishmanial on promastigotes) IC_{50} (μ*M*)	[375]
		Leishmania donovani *L. brasiliensis* *L. amazonensis*	0.5 1.1 1.1 (antiepimastigote activity of *T. cruzi*)	

(continued)

Table 17 (continued)

Species name	Compound	Microorganisms	Activity value	Refs
		Trypanosoma cruzi Clone CL-B5	5.8 (antiamastigote activity leishmanial sp.)	
		L. amazonensis	1.5 (antiamastigote activity of *T. cruzi*)	
		Trypanosoma cruzi Clone CL-B5	17.3 (Leishmanicidal activity on promastigotes) 9.8 9.8	
	Pentamidine (control drug) *(IC_{50} μM)*	*Leishmania donovani*	9.8 (antiamastigote activity of *Leishmania* sp.)	
		L. brasiliensis	0.1 (antiepimastigote activity of *T. cruzi*)	
		L. amazonensis	54.7 (antiamastigote of *T. cruzi*)	
	Benznidazole (control drug) *(IC_{50} μM)*	*L. amazonensis*	192.1	
Lepidozia fauriana	Lepidozenolide (**367**)	*Trichomonas foetus*	100 μg/cm^3	[185]

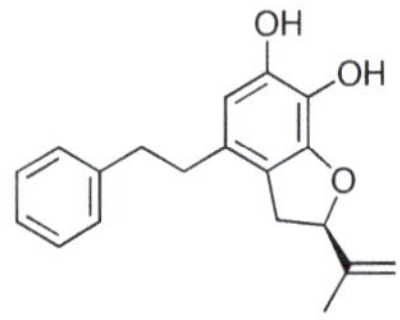

1055 (perrottetin D)

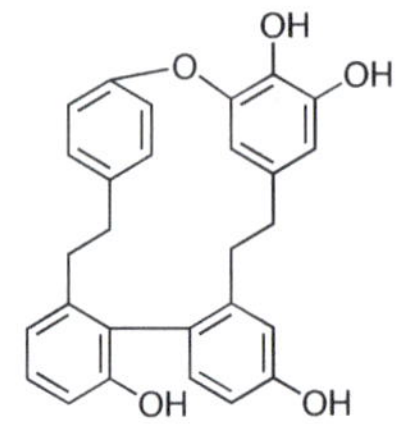

1056 (plagiochin A)

The in vitro antiplasmodial activity-guided fractionation of an ether extract of *M. polymorpha* (Plate 30) resulted in the isolation of marchantin A (**1**), and this inhibited the proliferation of the *Plasmodium falciparum* strains NF54 and KI, which cause malaria, with IC_{50} values of 3.4 and 2.0 μ*M*, respectively. This compound also showed inhibitory effects against *Trypanosoma brucei rhodesiense*, *T. cruzi* and *Leishmania donovani* (IC_{50} 2.1, 14.9, and 1.6 μ*M*, respectively). Marchantin A (**1**) was tested against three recombinant enzymes (*Pf*Fab1, *Pf*FabG, and *Pf*FabZ) of the *Pf*FS-II pathway of *P. falciparum* for its malaria prophylactic potential, and showed weak inhibition against PfFabZ with an IC_{50} value of 18.2 μ*M* [372].

The most characteristic compounds in *Plagiochila* species belonging to the Plagiochilaceae are the 2,3-*seco*-aromadedrane-type sesquiterpenoids, such as plagiochiline A (**135**). Plagiochiline A (**135**) isolated from *Plagiochila disticha* was found to possess antileishmanial activity against *Leishmania amazonensis* axenic amastigotes with an IC_{50} value of 7.1 μ*M* and trypanocidal activity against *Trypanosoma cruzi* trypomastigotes at an *MIC* 14.5 μ*M* [200].

Asakawa's group reported in 1994 that *Plagiochila porelloides* elaborated plagiochilines C (**420**) and D (**416**), along with plagiochiline A 15-yl octanoate (**460**), plagiochiline A decanoate (**461**), and plagiochiline A 15-yl-(4Z)-decenoate (**462**) [3]. Plagiochiline A 15-yl-octanoate (**460**) showed good cytotoxicity against

Plate 30 *Marchantia polymorpha* (♀; liverwort)

KB nasopharyngeal cancer cells [199]. Further work on the Corsican *P. poelloides* demonstrated that its essential oil contains β-barbatene (**33**), bicyclogermacrene (**101**), maalian-5-ol, *p*-menth-1-en-3-(2-methyl-(1*E*)-butenyl)-8-ol, and *p*-menth-1-en-3-(2-methyl-(1*Z*)-butenyl)-8-ol. Its ether extract contained the two sesquiterpene hydrocarbons **33** and **101**, globulol (**825**), and 1,2-dihydro-4,5-dehydronerolidol [202]. The in vitro antitrypanosomal and antileishmanial activities of both the liverwort essential oil and ether extract, were evaluated against *Trypanosoma brucei brucei* and *Leishmania mexicana mexicana*, respectively. The essential oil and ether extract exhibited moderate activities against *T. brucei brucei* and *L. mexicana mexicana* with IC_{50} values of 2.0 and 16.0 μg/cm^3 and 5.2 and 2.0 μg/cm^3, respectively. Only the essential oil showed a high selectivity against *T. brucei brucei* (SI = 11.7). However, 2,3-*seco*-aromadendranes were not reported in the Corsican sample investigated [202].

Roldos et al., reported that 2,5,4′-trihyroxybibenzyl (**1057**), which is a previously known natural product from the liverwort, *Ricciocarpos natans* [374], displayed antileishmanial and anti-Chagas' disease activity against three *Leishmania* species and against *Trypanosoma cruzi*, having IC_{50} values of 1.1–5.8 and 17.3 μ*M*, respectively, with an absence of cytotoxicity against various mammalian cells. These activities were more potent than those of the drugs pentamidine and benznidazole at the same concentration [375].

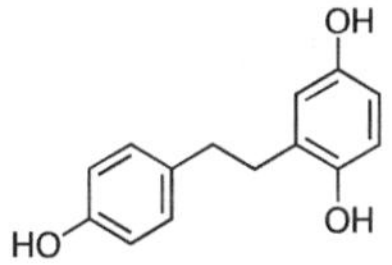

1057 (2,5,4'-trihydroxybibenzyl)

The liverwort *Lepidozia fauriana* collected in Taiwan produces lepidozane-type sesquiterpenoids, of which lepidozenolide (**367**) possessed not only cytotoxic, antimicrobial and antifungal effects, but also weak antitrichomonal activity against *Trichomonas foetus* (Br. M. stain) at 100 μg/cm^3 [185].

4.13 Brine Shrimp Lethality Activity

The chlorinated bibenzyls, **967–970** from *Riccardia polyclada* showed brine shrimp (*Artemia salina*) lethality activities, with LC_{50} values from 0.42 to 2.39 ppm, and were more active than the positive control compounds used, ketoconazole (LC_{50} 14.9 ppm) and Asuntol® (LC_{50} 10.8 ppm) [336].

A dichloromethane extract of *Balantiopsis cancellata* displayed lethality against the larvae of brine shrimp at a LC_{50} value of less than 2.5 ppm. All of the

Table 18 Brine shrimp (*Artemia salina*) larvae lethal compounds from liverworts

Species name	Compound	Activity value (lethality): LC_{50} (ppm)	Refs.
Balantiopsis cancellata	CH_2Cl_2	< 2.5	[337]
	β-Phenylethyl β-methylthioacrylate (**14**)	< 2.5	
	β-Phenylethyl benzoate (**971**)	< 2.5	
	(2*R*)-Hydroxy-β-phenylethyl benzoate (**972**)	7.4	
	β-Phenylethyl (*E*)-cinnamate (**973**)	< 2.5	
	β-Phenethyl (*Z*)-cinnamate (**974**)	< 2.5	
	Benzyl (*E*)-cinnamate (**1058**)	< 2.5	
Riccardia polyclada	2,6-Dichloro-3-hydroxy-4′-methoxybibenzyl (**967**)	0.4	[336]
	2,6,3′-Trichloro-3-hydroxy-4′-methoxybibenzyl (**968**)	0.4	
	2,4,6,3′-Tetrachloro-3-hydroxybibenzyl (**969**)	1.7	
	2,4,6,3′-Tetrachloro-3-hydroxy-4′-methoxybibenzyl (**970**)	2.4	

phenyl ethanol esters isolated from this species, β-phenylethyl benzoate (**971**), (2*R*)-hydroxy-β-phenylethyl benzoate (**972**), β-phenylethyl (*E*)-cinnamate (**973**) and β-phenylethyl (*Z*)-cinnamate (**974**), and benzyl (*E*)-cinnamate (**1058**) also showed inhibitory activities against brine shrimp larvae, at a concentration range of < 2.5 to 7.4 2.5 ppm (see Table 18) [337].

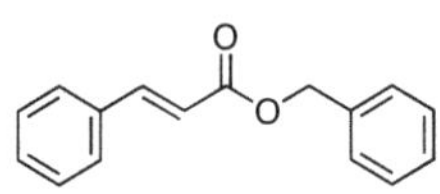

1058 (benzyl (*E*)-cinnamate)

4.14 Piscicidal Activity

The most active piscicidal compound discovered to date from bryophytes is the pungent (–)-polygodial (**90**) from the *Porella vernicosa* complex. Japanese killifish (*Oryzia latipes*) (see Table 19) were killed within 2 h by a 0.4 ppm solution of **90** [38]. Killifish were also killed within 2 h by a 0.4 ppm solution of synthetic (+)-polygodial (**1037**), which is also pungent. Hence, piscicidal activity is not affected by the chirality of polygodial. Natural (–)-polygodial (**90**) and its enantiomer **1037** are also very toxic to fresh-water bitterling fish, which were killed within 3 min by 0.4 ppm solutions [38].

Lobatiriccardia yakushimensis (= *Riccardia robata* var. *yakushimensis*), *Pallavicinia levieri*, *Pellia endiviifolia*, and *Trichocoleopsis sacculata* are known to produce large amounts of sacculatane diterpenoids, and their crude extracts were

Table 19 Piscicidal compounds from liverworts

Species name	Compound	Fish name	Activity value (ppm/min: concentration at which all fish were killed)	Refs.
Diplophyllum albicans	Diplophyllolide (**132**)	*Oryzia latipes*	6.7/240	[38]
Frullania dilatata	(+)-Frullanolide (**3**)	*Oryzia latipes*	0.4/240	[38]
Pellia endiviifolia *Trichocoleopsis sacculata*	Sacculatal (**128**) Isosacculatal (**1059**)	*Oryzia latipes*	0.4/120 Inactive	[2, 38]
Pellia endiviifolia	1β-Hydroxysacculatal (**129**)	*Oryzia latipes*	1.0/20	[38]
Plagiochila sp.	Plagiochiline A (**135**)	*Oryzia latipes*	0.4/240	[38]
Porella vernicosa complex	(–)-Polygodial (**90**) Isopolygodial (**1060**)	*Oryzia latipes*	0.4/120 0.4/3 Inactive	[38]
–	(+)-Polygodial (**1037**) (synthetic)	*Oryzia latipes*	0.4/120 0.4/3	[38]

found to be very toxic against killifish [2]. When the piscicidal activity of 14 sacculatals isolated from *Pellia endiviifolia* was tested in this manner, only sacculatal (**128**) and 1β-hydroxysacculatal (**129**) were lethal within 20 min at a concentration of 1 ppm, indicating that both the 8- and 9β-dialdehyde groups present in these molecules are responsible for this activity [109]. On the other hand, isosacculatal (**1059**) from *Pallavicinia, Pellia, Lobatiriccardia,* and *Trichocoleopsis* species and isopolygodial (**1060**), from the cultured cells of *Porella vernicosa* and the higher plant *Polygonum hydropiper*, exhibited neither piscicidal nor molluscicidal activity, even at a 1% level [38]. (+)-Frullanolide (**3**), diplophyllolide (**132**) and plagiochiline A (**135**) showed piscicidal activity against killifish at concentration levels of 0.4, 0.4, and 6.7 ppm/4 h, respectively [38].

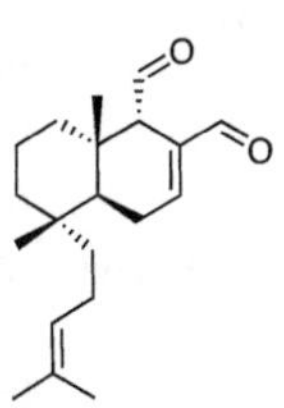

1059 (*iso*-sacculatal)

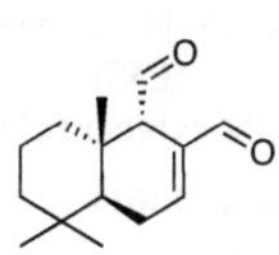

1060 (*iso*-polygodial)

4.15 5-Lipoxygenase, Calmodulin, Cyclooxygenase, Hyaluronidase, DNA Polymerase β, α-Amylase, and α-Glucosidase Inhibitory, Quinone Reductase-Inducing, and Anthocyanin Activities

Marchantin A (**1**), from several *Marchantia* species, showed 5-lipoxygenase inhibitory activities (89% at 10^{-5} *M*, 94% at 10^{-6} *M*, 45% at 10^{-7} *M*, 16% at 10^{-8} *M*), with (5*S*,12*R*)-dihydroxy-6,8,10,14-eicosatetraenoic acid and 5-hydroxy-6,8,11,14-eicosatetraenoic acid used for comparison. This bis-benzyl **1** displayed also calmodulin-inhibitory activity, with an ID_{50} of 1.85 μg/cm^3 [38] (Table 20).

The prenyl bibenzyls, perrottetins A (**1061**) and D (**1055**) from *Radula perrottetii* and 2-(3-methyl-2-butenyl)-3,5-dihydroxy-bibenzyl (**1062**), 3,5-dihydroxy-4(3-methyl-2-butenyl)-bibenzyl (**1063**), radulanin C (**1064**), H (**1065**), and 4′-hydroxyradulanin H (**1066**), also from a *Radula* species, riccardin A (**2**) from *Riccardia multifida,* and marchantins D (**198**) and E (**199**) from a *Marchantia* species, showed calmodulin-inhibitory activity (ID_{50} range 2.0–95.0 μg/cm^3) [38]. The simple bibenzyl, 3,4:3′,4′-dimethylenedioxybibenzyl (**1067**), from a *Frullania* species, also showed weak calmodulin-inhibitory activity (ID_{50} 100 μg/cm^3), as did

Table 20 5-Lipoxygenase and calmodulin inhibitory compounds from liverworts

Species name	Compound	Activity value	Refs
		5-Lipoxygenase Calmodulin (10^{-6} *M*) ID_{50} (μg/cm^3)	
Marchantia sp.	Marchantin A (**1**) Marchantin D (**198**) Marchantin E (**199**)	94% 1.9 40 6 0 36 7.0	[38]
Porella perrottetiana	Labda-12,14-diene-7,8-diol (**1068**)	82	[38]
Radula perrottetii	Radulanin A (**335**) 2-Geranyl-3,5-dihydroxybibenzyl (**899**) Perrottetin D (**1055**) 3,4′-Dimethoxybibenzyl (**1038**)	95.0 11 4.0 40 2.0 100	[38, 376]
	Perrottetin A (**1061**)	76 2.0	
	2-(3-Methyl-2-butenyl)-3,5-dihydroxy-bibenzyl (**1062**)	50 4.9	
	Radulanin H (**1065**)	15 17.0	
	4′-Hydroxyradulanin H (**1066**)	18.5	
	3,4:3′,4′-Dimethylenedioxybibenzyl (**1067**)	100	
Riccardia multifida	Riccardin A (**2**)	4 20.0	[38]

Table 21 Cyclooxygenase, DNA polymerase β, collagenase, hyaluronidase, *α*-amylase, *α*-glucosidase, tyrosinase, tyrosine phosphate, and lactate dehydrogenase inhibitory, anthocyanin activity, and quinone reductase-inducing compounds from bryophytes

Species name[a]	Compound or solvent extract	Activity value (cyclooxygenase)	Refs.
[L] *Marchantia* species	Marchantin A (**1**) Isoriccardin C (**942**) Marchantin B (**195**) Marchantin E (**199**)	IC_{50} 46.4 μ*M* 50.8 55.9 58.0	[376]
[L] *Marchantia polymorpha*	MeOH	100 μg/cm^3	[417]
[L] *Marchantia paleacea* subsp. *diptera*	Paleatin B (**202**)	45.2	[376]
[L] *Radula lindenbergiana*	Radulanin H (**1065**)	39.7	[376]
[L] *Radula perrottetii*	Perrottetin D (**1055**)	26.2	[376]
[L] *Reboulia hemisphaerica*	Riccardin C (**182**)	53.5	[376]
Species name[a]	Compound or solvent extract	Activity value (collagenase)	Refs.
[M] *Sphagnumgirgensohnii* *S. magellanicum* *S. palustre* *S. squarrosum*	EtOH/H_2O (1:1)	No activity value given	[281]
Species name[a]	Compound or solvent extract	Activity value given (hyaluronidase)	Refs.
[L] *Denotarisia linguifolia*		IC_{50} (mg//cm^3)	[262]
	n-Hexane	> 2	
	$CHCl_3$	> 2	
	EtOAc	0.45	
	EtOH	0.84	

(continued)

Table 21 (continued)

Species name[a]	Compound or solvent extract	Activity value given (hyaluronidase)	Refs.
[L] *Lunularia cruciata*	Lunularic acid (**196**)	IC_{50} 0.13 n*M*	[377]
[M] *Sphagnum girgensohnii* *S. magellanicum* *S. palustre* *S. squarrosum*	EtOH/H_2O (1:1)	(No activity value given) Hyaluronidase, and hyaluronic acid synthase inhibition (for NHDF cell lines)	[281]
Species name[a]	Compound or solvent extract	Activity value (DNA polymerase β) (μ*M*)	Refs.
[L] *Blasia pusilla*	Pusilatin B (**250**) Pusilatin C (**251**)	13.0 5.16	[157]
[L] *Frullania convoluta* [L] *Marchantia polymorpha* [L] *Marchantia paleacea* subsp. *diptera*	Perrottetin F (**184**) Marchantin A (**1**) Marchantin B (**195**) Marchantin D (**198**) Paleatin B (**202**)	14.4–97.5	[75]
Species name[a]	Compound or solvent extract	Activity value (α-amylase) (%)	Refs.
[M] *Fissidens* sp. [M] *Hypnum cupressiforme* [L] *Marchantia* sp. [L] *Plagiochila* sp.	EtOAc	39 8 23 12	[304]
Species name[a]	Compound or solvent extract	Activity value (α-glucosidase)	Refs.
[L] *Denotarisia linguifolia*		IC_{50} (mg/cm^3)	[262]
	n-Hexane	> 2	
	$CHCl_3$	0.03	
	EtOAc	0.10	
	EtOH	0.04	

(continued)

Table 21 (continued)

Species name[a]	Compound or solvent extract	Activity value (α-glucosidase)	Refs.
[M] *Hedwigia ciliata*	EtOAc (E-3 extract)	90 (% inhibition) 1000 μg/cm^3	[279]
	$EtOH/H_2O$ (E-2 extract)	40	
	96% EtOH (E-1 extract)	40	
	Acarbose (control)	95	
[M] *Hypnum cupressiforme*	EtOAc	100 (μg/ m^3)	[272]
	H_2O		
[L] *Marchantia polymorpha*	Marchantin C (**197**)	2.2% at 1 μ*M*	[380]
[L] *Marchantia polymorpha*	*n*-Hexane	50.2 μg/cm^3	[261]
	$CHCl_3$	21.5	
	EtOAc	10.3	
	EtOH	92.3	
	acarbose (control)	530.9	
[L] *Marchantia polymorpha*	In vitro cultured mass	IC_{50} μg/cm^3	[379]
	n-Hexane	11.9	
	$CHCl_3$	20.4	
	EtOAc	84.3	
	EtOH	361.4	
	Acarbose (control)	530.9	
Species name[a]	**Compound or solvent extract**	**Activity value (tyrosinase) (%)**	**Refs.**
[M] *Bryum argenteum*		% inhibition	[402]
	Field collected mass	25	
	n-Hexane	30	
	EtOAc	25	

(continued)

Table 21 (continued)

Species name[a]	Compound or solvent extract	Activity value (tyrosinase) (%)	Refs.
	EtOH	55	
	In vitro cultured mass	50	
	n-Hexane	35	
	EtOAc EtOH		
	Kojic acid	100 (200 μg)	
	Field collected mass *n*-Hexane EtOAc EtOH in vitro cultured mass *n*-hexane EtOAc EtOH	(mg/cm^3) 2 (moderate inhibition)	
	Kojic acid	200 $\mu g/cm^3$	
[L] *Denotarisia linguifolia*	*n*-Hexane	% inhibition 5.6	[262]
	$CHCl_3$	9.8	
	EtOAc	15.9	
	EtOH	18.3	
[M] *Hedwigia ciliata*	96% EtOH	(%) inhibition (1000 $\mu g/m^3$) 70%	[279]
	EtOAc	62	
	$EtOH/H_2O$ (1:1)	60	
	Kojic acid (control)	58	

(continued)

Table 21 (continued)

Species name[a]	Compound or solvent extract	Activity value (tyrosinase) (%)	Refs.
[M] *Hypnum cupressiforme*	96% EtOH EtOH/H_2O (1:1) EtOAc H_2O	10 μg/cm^3 2.5 times more potent than kojic acid	[272]
[L] *Marchantia polymorpha*	*n*-Hexane	(%) inhibition 4.9	[379]
	$CHCl_3$	22.1	
	EtOAc	65.0	
	EtOH	36.1	
	kojic acid	100.4	
[M] *Sphagnum palustre*	EtOH/H_2O (1:1)	Tyrosinase gene expression (no active value given)	[281]
Species name[a]	Compound or solvent extract	Protein tyrosine phosphatase	Refs.
[M] *Polytrichum alpinum*	Ohioensin F (**1069**)	IC_{50} (μ*M*) 3.5	[268, 378]
	Ohioensin G (**1070**)	5.6	
	Ohioensin A (**777**)	4.3	
	Ohioensin B (**778**)	7.6	

(continued)

Table 21 (continued)

Species name[a]	Compound or solvent extract	Activity value (lactate dehydrogenase)	Refs.
[L] *Conocephalum conicum*	(1*Z*,4*E*)-Lepidoza-1(10)-4-dien-14-ol (**484**), *rel*-(1(10)*Z*,4*S*,5*E*,7*R*)-germacra-1(10),5-dien-11,14-diol (**485**), and *rel*-(1(10)*Z*,4*S*,5*E*,7*R*)-humula-1(11),5-dien-8,12-diol (**486**)	10^{-8}–10^{-6} *M*	[386]
Species name[a]	Compound or solvent extract	Activity value (anthocyanin)	Refs.
[M] *Cinclidotus fontinaloides* [M] *Palustriella commutata*	EtOH	50 μg/cm^3 140 μg/cm^3	[314]
Species name[a]	Compound or solvent extract	Activity value (quinone reductase (Q R)-inducing)	Refs.
[L] *Frullania hamatiloba*	Fusicoauritone (**490**) Anadensin (**499**) Hamatilobene D (**1092**) Frullanian D (**1096**) 4*α*-Hydroxyfusicocca-3(7)-en-6-one (**1099**)	1.3–3.5 μ*M*	[384]
[L] *Scapania koponenii*	Scaparins A–C (**1100–1102**)	50–100 μ*M* (Hepa-1c1c7 cells)	[385]
[L] *Scapania verrucosa*	Scaparins D–G (**1103–1106**)	50–100 μ*M* (Hepa-1c1c7 cells)	[385]
[L] *Diplophyllum taxifolium*	Diplotaxifol A (**1070a**)	1.5 fold at 25 μ*M*	[381]
	Diplotaxifol B (**1070b**)	2.1 fold at 25 μ*M* sulforaphane (1.7 fold at 2.0 μ*M*) as positive control Dose dependent NQO1 induction (Hepa 1c1c7 cells)	

(continued)

Table 21 (continued)

Species name[a]	Compound or solvent extract	Activity value (quinone reductase (Q R)-inducing)	Refs.
	ent-Atractylenolide III	Dose dependent NQO1 induction (Hepa 1c1c7 cells)	
	Diplophyllolide (**132**) *ent*-11β-Hydroxydihydro-isoalantolactone (**1071**) *ent*-Isoalantolactone (**963**) *ent*-3-*epi*-Isotelekin (**1072**) *ent*-3α-Hydroxyeudesma-4,11-dien-12,8α-olide (**1073**) 3-Oxodiplophyllin (**1074**) 3αH-3,4-Epoxydiplophyllolide (**573**)	Quinone reductase induction	
[L] *Notoscyphus lutescens*	Notoscarins C–F (**1083–1086**)	50 μ*M* (Hepa 1c1c7 cells)	[382]
[L] *Pellia epiphylla*	Epiphyllins A–H (**1122–1129**) 1β-Hydroxysacculatanolide (**1130**) Pellianolactone B (**1131**)	50–100 μ*M*	[400]

[a] [L] liverwort, [M] moss

the labdane-type diterpene diol, labda-12,14-diene-7,8-diol (**1068**) (ID_{50} 82 μg/cm^3), isolated from *Porella perrottetiana* [38].

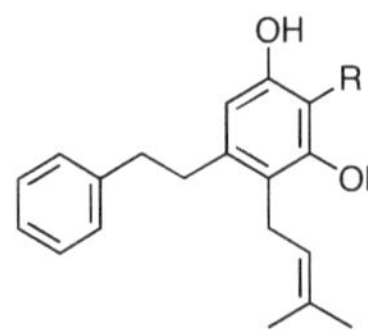

1061 (perrottetin A, R = OH)
1062 (2-(3-methyl-2-butenyl)-3,5-dihydroxybibenzyl, R = H)

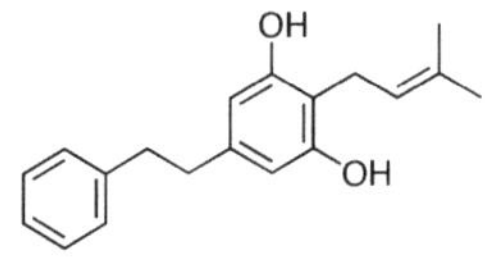

1063 (3,5-dihydroxy-4-(3-methyl-2-butenyl)bibenzyl)

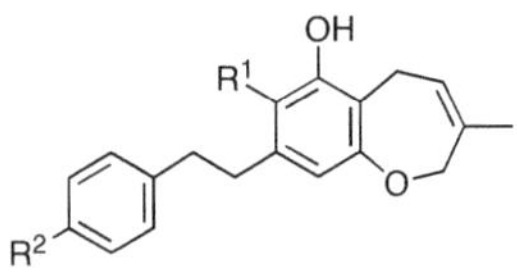

1064 (radulanin C, R^1 = R^2 = H)
1065 (radulanin H, R^1 = CO_2H, R^2 = H)
1066 (4'-hydroxyradulanin H, R^1 = COOH, R^2 = OH)

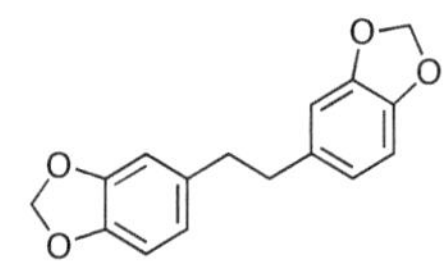

1067 (3,4:3',4'-dimethylenedioxybibenzyl)

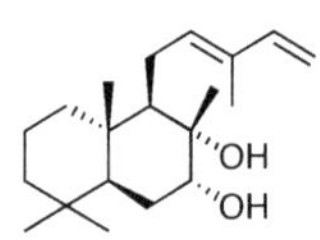

1068 (labda-12,14-dien-7α,8α-diol)

Riccardin A (**2**), marchantins D (**198**) and E (**199**), three prenyl bibenzyls (**1063**–**1065**), and perrottetin A (**1061**) inhibited 5-lipoxygenase (4–76% at 10^{-6} *M*) [38]. Moreover, a number of phenolic liverwort compounds showed the following IC_{50} values in terms of their cyclooxygenase-inhibitory activities: marchantin A (**1**) 46.4, riccardin C (**182**) 53.5, isoriccardin C (**942**) 50.8, marchantin B (**195**) 55.9, marchantin E (**199**) 58.0, paleatin B (**202**) 45.2, perrottetin D (**1055**) 26.2, and radulanin H (**1065**) 39.7 μ*M* [376] (Table 21).

Hyaluronic acid is a glucose-based polymer that can stimulate skin regeneration, retain moisture, increase viscosity, and lower the permeability of extracellular fluids. Hyaluronic acid synthesis naturally decreases while hyaluronidase increases, due to the aging process. Hyaluronidase is an endoglycosidase that breaks down hyaluronic acid into monosaccharides leading to the loss of skin strength, flexibility, and moisture, which causes wrinkles. Thus, anti-hyaluronidase agents may be used to preserve skin moisture.

Phan-Duy et al., in 2023 evaluated the inhibition of hyaluronidase by *n*-hexane, chloroform, ethyl acetate, and ethanol extracts of the Vietnamese liverwort, *Denotarisia linguifolia.* At a concentration of 2 mg/cm^3, all extracts apart from the *n*-hexane extract showed an inhibition percentage higher than 50%. Thus, the extracts were further examined for their IC_{50} values related to the inhibition of this enzyme, with the lowest IC_{50} value observed of 0.45 mg/cm^3, while the chloroform extract gave an IC_{50} value of 0.84 mg/cm^3 [262]. The chloroform extract of *D. linguifolia* was the most active against α-glucosidase with an IC_{50} value of 0.03 mg/cm^3. Moreover, all extracts of this species showed better activities against α-glucosidase, and thus had lower IC_{50} values than that of the control, acarbose [262].

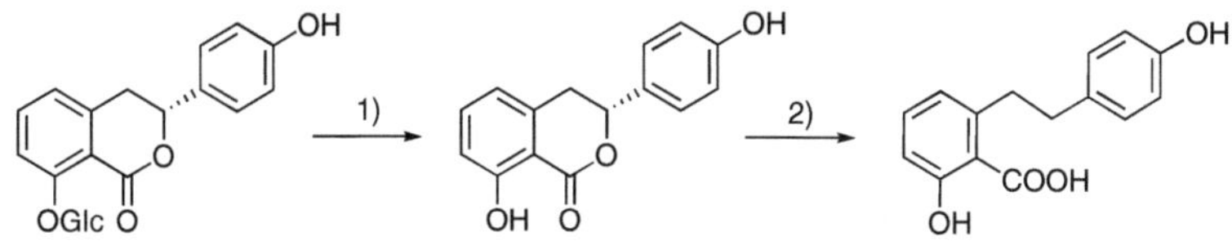

196a (hydrangenol glucoside) **196b** (hydrangenol) **196** (lunularic acid)

Scheme 12 Preparation of lunularic acid (**196**) from hydrangenol glucoside (**196a**) via hydrangenol (**196b**). (1) 1*N* H_2SO_4, reflux, (2) $NaBH_4$, $PdCl_2$/MeOH, reflux

Lunularic acid (**196**), which is found in almost all liverworts as a minor component, has potent anti-hyaluronidase activity, with an IC_{50} value of 0.13 n*M*. This activity is more potent than that of tranilast (*N*-3′,4′-dimethoxy-cinnamoylanthranilic acid), which is an antiallergenic agent developed in Japan for oral administration. While the content of **196** is very low in liverworts, a large amount was obtained from hydrangenol (**196b**), which may be prepared readily from hydrangenol glucoside (**196a**), isolated from the higher plant *Hydrangea macrophylla*. Hydrangenol (**196b**) was treated with sodium borohydride, followed by palladium chloride to afford lunularic acid in good yield (84%), as shown in Scheme 12 [377].

Marchantins A (**1**), B (**195**), D (**198**), perrottetin F (**184**) and paleatin B (**202**), isolated from *Marchantia polymorpha* and *M. paleacea* var. *diptera*, showed DNA polymerase β–inhibitory activity, with an IC_{50} range of between 14.4 and 97.5 μ*M* [75].

Blasia pusilla produces the bis-bibenzyl dimers, pusilatins A–D (**249–252**), of which pusilatin B (**250**) and C (**251**) were found to possess DNA polymerase *β*-inhibitory activities (IC_{50} values of 13.0 and 5.2 μ*M*, respectively) [157].

The inhibitory effects for *α*-amylase and *α*-glucosidase of the ethanol, ethanol/water (1/1), and ethyl acetate extracts of the moss, *Hedwigia ciliata*, were investigated to assess their antidiabetic potential. While none of these solvent extracts gave evidence of any significant *α*-amylase inhibition, they showed inhibitory activity against *α*-glucosidase. The most potent inhibition was determined for the ethyl acetate extract of *H. ciliata*. The ethanol and ethanol/water (1/1) extracts each showed a lower *α*-glucosidase-inhibitory activity, at a concentration of 1 mg/cm^3 [279].

Chemical studies on the methanol extract of the Antarctic moss *Polytrichum alpinum* led to the isolation of two new benzonaphthoxanthenones, named ohioensins F (**1069**) and G (**1070**), along with the known analogues, ohioensins A (**777**) and C (**779**) isolated previously from *Polytrichum ohioense* [268]. These four benzonaphthoxanthenones showed potent inhibitory effects against the therapeutically useful protein tyrosine phosphatase 1B (PTP1B) in a dose-dependent manner at an IC_{50} concentrations range from 3.5 to 7.6 μ*M*. Kinetic analysis of the inhibition by the four benzonaphthoxanthenones showed inhibition of PTP1B activity in a non-competitive manner. This was the first report of PTP1B inhibitory activity of the benzonaphthoxanthenones from bryophytes [378].

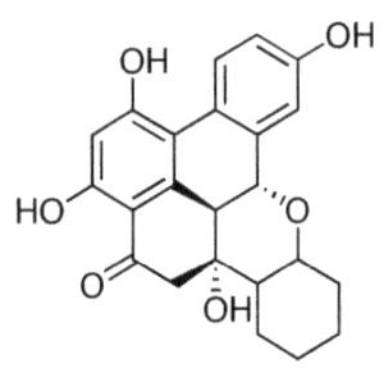

1069 (ohioensin F)

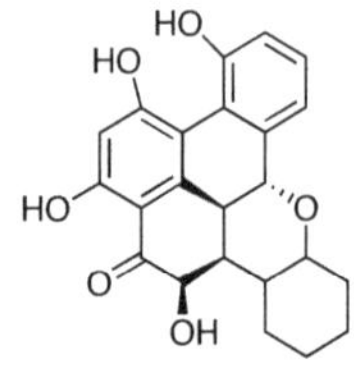

1070 (ohioensin G)

Amylase inhibitory activity of the ethyl acetate extracts of two Sri Lankan liverworts, a *Marchantia* species and a *Plagiochila* species and four mosses, *Calymperes motoleyi*, a *Fissidens* species, *Hypnum cupressiforme*, and *Sematophyllum demissum,* were evaluated by Kirisanth et al., in 2020. Among the those tested, the ethyl acetate extract of *Fissidens* species showed a 39% inhibition of α-amylase at 5 mg/cm^3. The extracts of *H. cupressiforme, Marchantia* sp., and *Plagiochila* sp. elicited weaker activities of 8, 23 and 12%, respectively, at this same concentration [304].

Inhibitory activity against α-glucosidase has been associated with antiobesity and antidiabetic propensitiess. The ethanol (96%), ethanol/water (1/1), ethyl acetate and water extracts of the moss *Hypnum cupressiforme* were prepared, and shown by LC-MS to contain various flavonoids and their glycosides, aromatic acids such as gallic (**796**), caffeic (**800**), and *p*-hydroxybenzoic (**799**) acids, along with quinic acid (**798**). The ethyl acetate and aqueous extracts of this moss both displayed α-glucosidase inhibitory activity at a concentration level of 10 μg/cm^3 [272].

The *n*-hexane, chloroform, ethyl acetate, and ethanol extracts of the Vietnamese cultured *Marchantia polymorpha* exhibited moderate α-glucosidase inhibitory effects Of these, the inhibitory effect of an *n*-hexane extract, with an IC_{50} value of 11.9 μg/cm^3, was over 40 times times more potent than that of the control, acarbose [379]

Among the bis-bibenzyls found in liverworts, only marchantin C (**197**) thus far has shown inhibitory activity against α-glucosidase (52.2% at 1 μ*M*). However, the activity found was lower than that of 1-deoxynojirimycin (100% at 0.4 μ*M*). This was the first report of the α-glucosidase inhibitory activity of a macrocyclic bis-bibenzyl derivative [380].

Wang et al., investigated in 2016 the phytochemical constituents of *Diplophyllum taxifolium,* and isolated two new prenyl aromadendranes, named diplotaxifols A (**1070a**) and B (**1070b**). The stereostructure of compound **1070a** was determined by X-ray crystallographic analysis. These were obtained together with 13 eudesmane sesquiterpenoids, *ent*-diplophyllolide (**132**), e*nt*-*iso*-alantolactone (**963**), dihydrodiplophyllolide (**965**), *ent*-codonolactone (**966**), *ent*-11β-hydroxydihydroisoalantolactone (**1071**), *ent*-3-*epi*-*iso*-telekin (**1072**), *ent*-3α-hydroxyeudesma-4,11-dien-12,8α-olide (**1073**), *ent*-3-oxodiplophyllin (**1074**), *ent*-*iso*-telekin (**1075**), *ent*-11,13-dihydroisoalantolactone (**1076**), *ent*-6β,12-hydroxyeudesm-4(15),11,(13)-diene (**1077**), *ent*-3-oxo-7,8,11αH-eudesm-4-en-12,8α-olide (**1078**), *ent*-5α-hydroxydiplophyllolide (**1079**), and 3αH-3,4-epoxydiplophyllolide (**1080**). Among the isolated compounds, diplotaxifols A (**1070a**) and B (**1070b**) as aromadendanes are rare and were found for the first time in liverworts. However, *ent*-eudesmane sesquiterpene lactones are very characteristic metabolites of *Diplophyllum* species, as documented by Asakawa and his associates [2]. Among the isolated products, compounds **1070a** and **1070b** and the eudesmanolides **963**, **966**, **1071–1074**, and **1080** showed quinone reductase-induction activity using murine Hepa1c1c7 hepatoma cells. Diplotaxifol B (**1070b**) was the most potent quinone reductase-inducer among these isolated compounds, and gave a 2.1-fold enhancement in enzymatic activity at 25 μ*M*. The α-methylene γ-lactone moiety of the eudesmanolides plays an important role in modulating this type of biological activity [381].

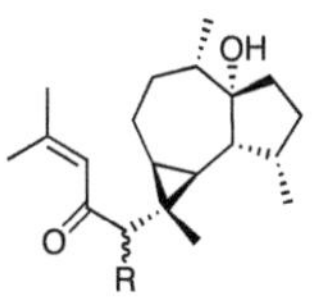

1070a (diplotaxifol A, R = β-OH)
1070b (diplotaxifol B, R = H)

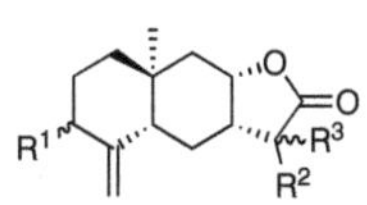

1071 (*ent*-11β-hydroxy-dihydroisoalantolactone, R^1 = H, R^2 = α-Me, R^3 = OH)
1072 (*ent*-3-*epi*-isotelekin, R^1 = α-OH, R^2 = R^3 = CH_2)
1075 (*ent*-isotelekin, R^1 = β-OH, R^2 = R^3 = CH_2)
1076 (*ent-11β-hydroxydihydro-isoalantolactone* R^1 = H, R^2 = α-Me R^3 = OH)

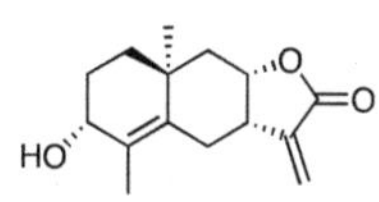

1073 (*ent*-3β-hydroxy-eudesma-4,11-dien-12,8α-olide)

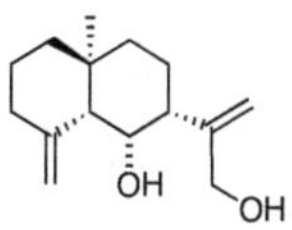

1077 (*ent*-6β,12-dihydroxy-eudesma-4(15),13-diene)

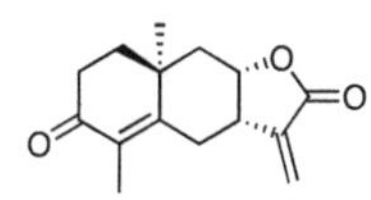

1074 (*ent*-3-oxodiplophyllin)

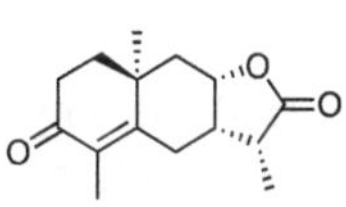

1078 (*ent*-3-oxo-11αH-eudesm-4-ene-12,8α-olide)

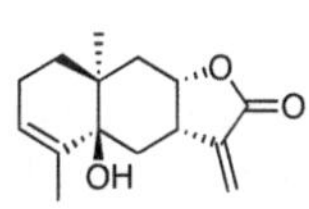

1079 (*ent*-5α-hydroxy-diplophyllolide)

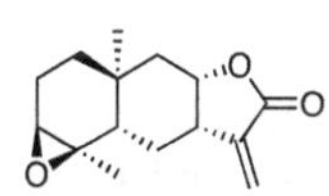

1080 (3α*H*-3,4-epoxy-*ent*-diplophyllolide)

Further fractionation of *Notoscyphus lutescens* led to the isolation of nine new dolabranes named notoscarins A–I (**1081–1089**) and the new butyrolactone derivative **1090**. Notoscarin A (**1081**), in having a 6,18-cyclodolabrane structure formed by intramolecular cyclization, and the 19-*nor*-dolabranes **1082** and **1083** were the first compounds of these types to have been isolated from the liverworts [382]. The absolute stereostructures of notolutesin A (**707**), and notoscarins A (**1081**) and B (**108**) and their analogues were established using a combination of NMR spectroscopy, X-ray crystallographic analysis, and ECD calculations.

Notolutesin A (**707**) exhibited cytotoxicity against the PC-3 cell line, as mentioned earlier, and notoscarins C–F (**1083–1086**) possessed moderate to weak quinone reductase-inducing activity in Hepa1c1c7 cells at a concentration of 50 μ*M* [382].

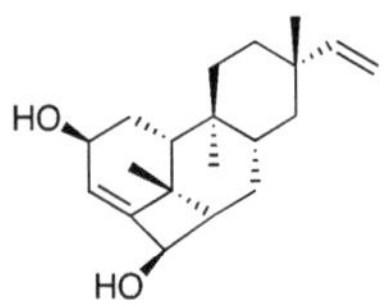

1081 (notoscarin A)

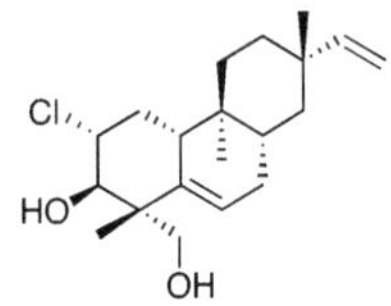

1082 (notoscarin B)

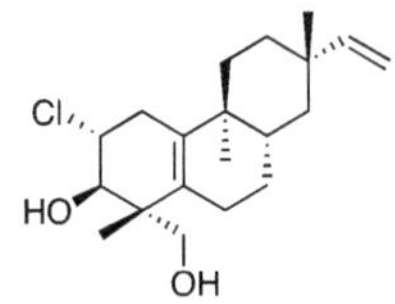

1083 (notoscarin C)

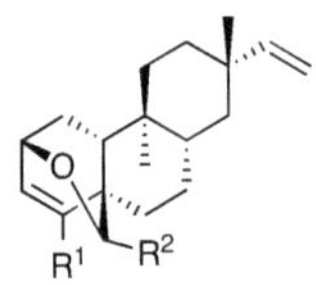

1084 (notoscarin D, R^1 = CH_2OH, R^2 = OMe)
1085 (notoscarin E, R^1 = CHO, R^2 = OH)

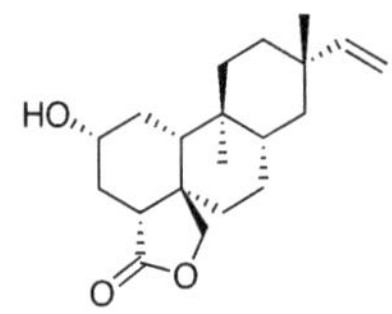

1086 (notoscarin F)

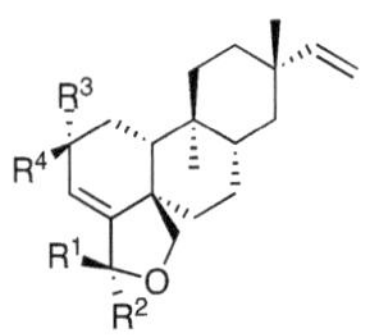

1087 (notoscarin G, R^1 = OMe, R^2 = R^3 = H, R^4 = OH)
1088 (notoscarin H, R^1 = R^3 = H, R^2 = OMe, R^4 = OH)

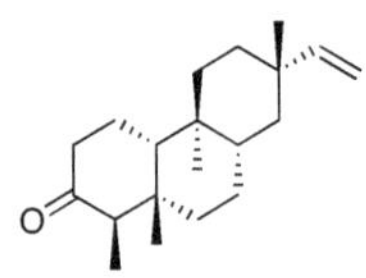

1089 (notoscarin I)

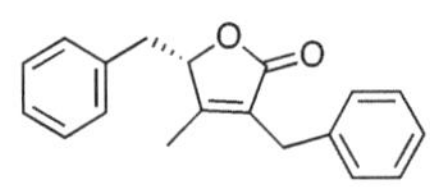

1090 (notoscarin J)

There are about 500 *Frullania* species that are divided into several chemotypes. Types I and II are very characteristic chemically since they produce eudesmane- and/or eremophilane-type sesquiterpene lactones and sesquiterpene lactone/bibenzyl derivatives, respectively. The *Frullania* species belonging to the above two types cause potent allergenic contact dermatitis [120]. *Frullania hamatiloba* contains neither sesquiterpene lactones nor bibenzyls, but it produces mainly labdane diterpenoids [383]. Qiao et al., reported in 2019 the isolation of six new labdanes named frullanians A–F (**1093–1098**) from this Chinese species together with two known labdanes, hamatilobene A (**1091**) and hamatilobene D (**1092**), and three known fusicoccane diterpenoids, 4α-hydroxyfusicocca-3(7)-en-6-one (**1099**), fusicoauritone (**490**), and anadensin (**499**). The stereostructure of frullanian A (**1093**) was established by X-ray crystallographic analysis [384]. All isolated diterpenes were tested in the quinone reductase-induction assay using Hepa-1c1c7 cells, and frullanian

D (**1096**) and compounds **490**, **499**, **1092**, and **1099** demonstrated dose-dependent quinone reductase-inducing activity with activities ranging from 1.34- to 3.47-fold. Frullanian D (**1096**) afforded the protection of MOVAS cells against hydroperoxide-induced cytotoxicity, and activated the Nrf2 (nuclear factor-erythroid 2-related factor 2) signaling pathway in MOVAS cells. Also, compound **1096** upregulated the expression of the antioxidant protein NQ1 (NAD(P)H quinone oxidoreductase 19 and γ-GCS (γ-glutamylcysteine synthase). Docking analysis further confirmed the activation of the Nrf2 pathway by compound **1096** [384].

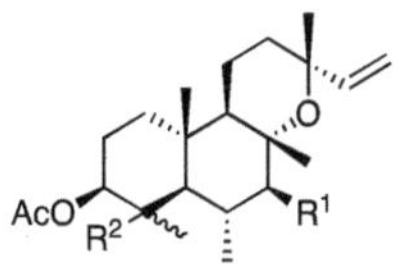

1091 (hamatilobene A, R^1 = OH, R^2 = OAc)
1092 (hamatilobene D, R^1 = OAc, R^2 = OH)

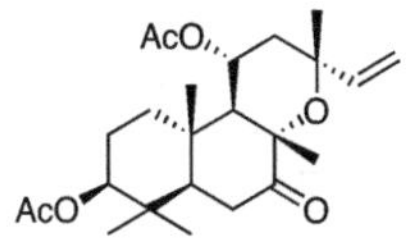

1093 (frullanian A)

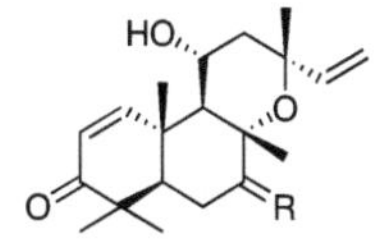

1094 (frullanian B, R = α-OH, β-H)
1095 (frullanian C, R = α-H, β-OH)

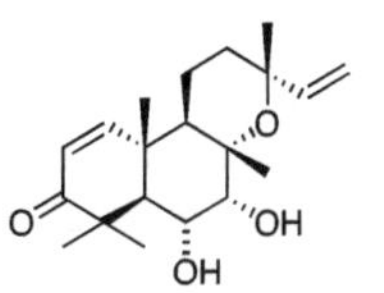

1096 (frullanian D)

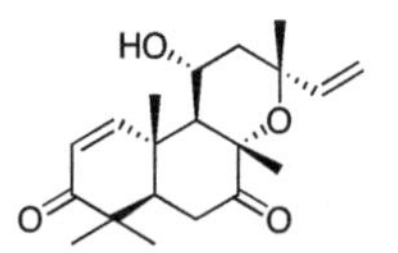

1097 (frullanian E)

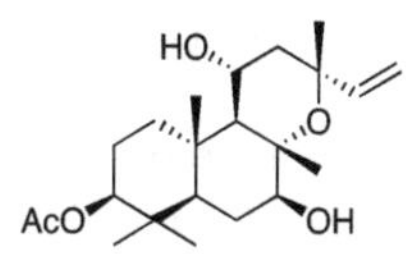

1098 (frullanian F)

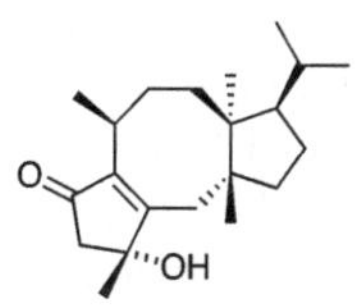

1099 (4α-hydroxyfusicocca-3(7)-en-6-one)

Han et al., isolated the rearranged clerodane diterpenoids, scaparins A–C (**1100–1102**) from a 90% ethanol extract of *Scapania koponenii* and scaparins D–F (**1103–1105**) and a trachylobane diterpenoid, scaparin G (**1106**) from *Scapania undulata*, along with acetoxymarsupellone (**415**), chandonanthone (**519**), and chandonanone F (**529**). Scapanins A–C (**1100–1102**) exhibited moderate to weak quinone reductase-inducing activity in Hepa-1c1cF cells (MQI concentrations: 1.30, 1.26, 0.94, and 1.08 (100 μ*M*) for compounds **1100–1102** and **1106**) and 2.17, 1.65, and 2.06 (50 μ*M*) for compounds **1103–1106**), respectively [385].

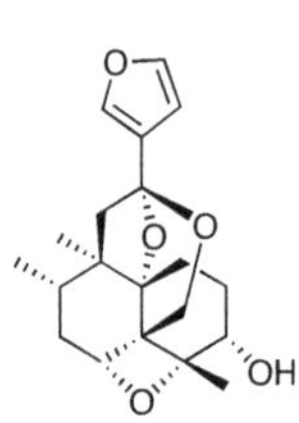

1100 (scaparin A)

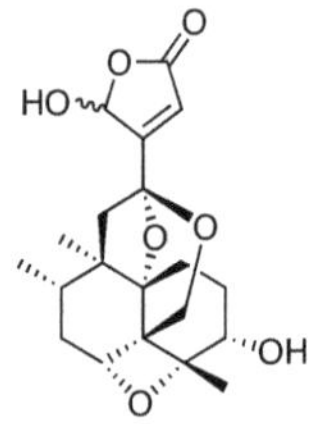

1101 (scaparin B)

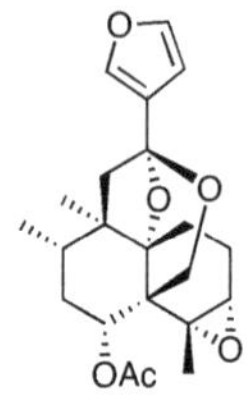

1102 (scaparin C)

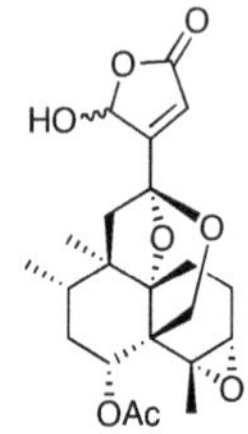

1103 (scaparin D)

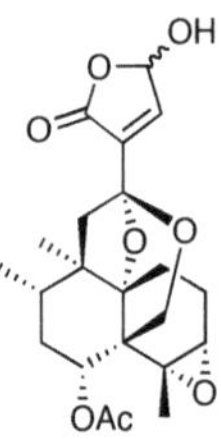

1104 (scaaparin E)

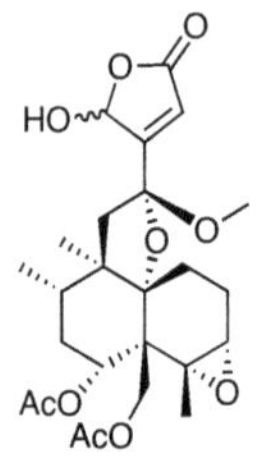

1105 (scaparin F)

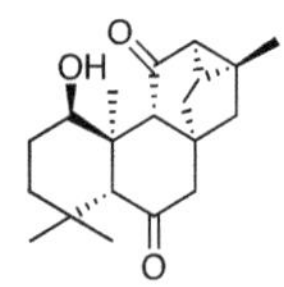

1106 (scaparin G)

Yayintas et al., reported total anthocyanin activities of the ethanol extracts of the two mosses, *Cinclidotus fontinaloides* and *Palustriella commutata*, as 50 and 144 μg/cm^3, respectively [314].

The EtOH/H_2O extracts of four *Sphagnum* species, *S. girgensohnii*, *S. magellanicum*, *S. palustre*, and *S. squarrosum*, indicated various levels of activity on the expression of genes encoding collagen matrix metalloproteinases (MMPs). Thus, they increased the expression of MMP1 and inhibited MMP8 and MM13, when compared with cells not treated with the extracts. However, statistically significant differences between the *Sphagnum* extracts in regard to their effects on the expression of these genes were not observed. On the basis of the data obtained, Zych et al., concluded that the use of extracts obtained from the four *Sphagnum* species in anti-aging cosmetics may not be beneficial and that further studies would be necessary to clarify their impact on human skin [281].

Filipović et al., reported in 2022 that the three sesquiterpene alcohols, (1*Z*,4*E*)-lepidoza-1(10)-4-dien-14-ol (**484**), *rel*-(1(10)*Z*,4*S*,5*E*,7*R*)-germacra-1(10),5-dien-11,14-diol (**485**), and *rel*-(1(10)*Z*,4*S*,5*E*,8*R*)-humula-1(10),5-diene-7,14-diol (**486**), isolated from the Serbian liverwort, *Conocephalum conicum*, are potential anticancer agents since they possess membrane-damaging properties as well as lactate dehydrogenase (LDH) inhibitory activity, with both being desirable features for this type of drug [386].

Table 22 Melanin production inhibitory and sedative effective compounds from liverworts

Species name	Compound	Activity	Activity value	Refs.
Cyathodium foetidissimum	4-Methoxystyrene (**31**)	Antimelanoma production	10 times more potent than kojic acid	[103]
Cyathodium foetidissimum	A mixture of 4-methoxystyrene (**31**) (24%) 3,4-Dimethoxystyrene (**96**) (29%) Skatole (**11**) (16%)	Sedative	No value given	[103]

4.16 Melanin Production Inhibitory and Sedative Effects

4-Methoxystyrene (**31**), which is one of the major components of the Tahitian *Cyathodium foetidissimum,* is known to be ten times more potent in terms of anti-melanin production than kojic acid [103]. A triethyl citrate solution (0.005%) of a mixture that consisted of three major compounds (skatole (**11**), 4-methoxystyrene (**31**), and 3,4-dimethoxystyrene (**96**)), as identified in *C. foetidissimum* and detected in the ratio shown in Table 22, also exhibited a sedative effect for human subjects [103].

4.17 Antidiabetic Nephropathy Activity

Chiloscyphus polyanthos is an abundant source of eudesmane sesquiterpenoids [2, 3] (Table 23). Fractionation of a 95% ethanol extract of the Chinese species led to the isolation of two new cyperanes, (3*S*,5*S*,7*R*,10*S*)-3,7-dihydroxycypyran-4-one (**1107**) and (3*R*,5*R*,7*R*,*10S*)-3,7-dihydroxycyperan-4-one (**1108**), along with five known eudesmanes, (3*S*,6*R*,7*S*,10*S*)-3,6,7-trihydroxyeudesmene (**1109**), (3*S*,7*R*,10*S*)-3,7-dihydroxyeudesmene (**1110**), (3*S*,6*R*,7*S*,10*S*)-3,7-epoxy-6-acetoxyeudesmene (**1111**), (9*S*)-4,5-*seco*-4-*nor*-3-carboxy-9-oxoeudesm-4-ene (**1112**), and (6*R*,7*S*,10*R*)-6,7-dihydroxy-3-oxoeudesm-(4*E*)-ene (**385**). The absolute configurations of compounds **1107** and **1108** were determined by X-ray crystallographic and ECD analysis. Cyperane sesquiterpenoids are very rare in Nature, and this is the first example of the isolation of such compounds among the liverworts. The two *ent*-cyperanes **1107** and **1108** inhibited cell proliferation and extracellular matrix accumulation in high-glucose cultured mesangial cells, in a dose-dependent manner [387].

Table 23 Antidiabetic nephropathy active compounds from liverworts

Species name	Compound	Activity	Ref.
Chiloscyphus polyanthos	(3*S*,5*S*,7*R*,10*S*)-3,7-Dihydroxycypyran-4-one (**1107**) (3*R*,5*R*,7*R*,10*S*)-3,7-Dihydroxycypyeran-4-one (**1108**)	Inhibited cell proliferation and extracellular matrix accumulation in high glucose cultured mesangial cells	[387]

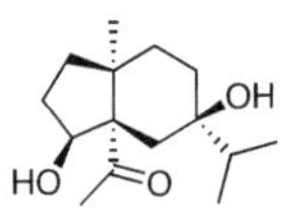

1107 ((3*S*,5*S*,7*R*,10*S*)-3,7-dihydroxycypyran-4-one)

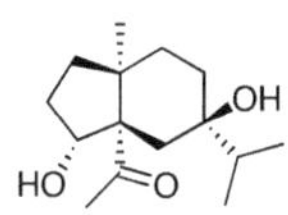

1108 ((3*R*,5*R*,7*R*,10*S*)-3,7-dihydroxycyperan-4-one)

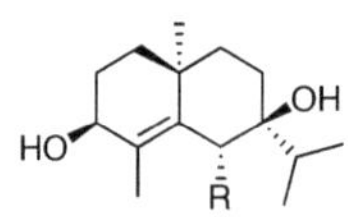

1109 ((3*S*,6*R*,7*S*,10*S*)-3,6,7-trihydroxy-eudesmene, R = OH)
1110 ((3*S*,7*R*,10*S*)-3,7-dihydroxy-eudesmene, R = H)

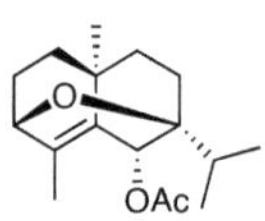

1111 ((3*S*,6*R*,7*S*,10*S*)-3,7-epoxy-6-acetoxyeudesmene)

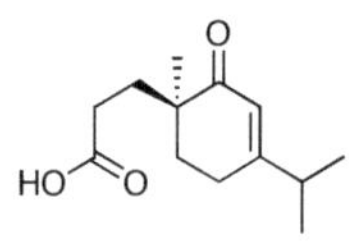

1112 ((9*S*)-4,5-*seco*-4-*nor*-3-carboxy-9-oxo-eudesm-4-ene)

4.18 Superoxide Release Inhibitory Activity

Excess superoxide anion radical (O_2^-) in organisms causes various types of angiopathy, such as cardiac infarction and arterial sclerosis. Polygodial (**90**) and sacculatal (**128**) each showed superoxide anion radical release inhibition at 4.0 μg/cm^3 from guinea pig peritoneal macrophages (GPPMs). Plagiochilal B (**137**) from *Plagiochila* species and infuscaic acid (= cleroda-3,13(16),14-trien-17-oic acid) (**1113**) from *Jungermannia infusca* inhibited superoxide release from rabbit peritoneal macrophages and guinea pig peritoneal macrophages at IC_{50} values of 6.0 μ*M* and 25 μg/cm^3, respectively, and from guinea pig peritoneal macrophages induced by formyl methionyl leucyl phenylalanine, respectively [38] (see Table 24).

Table 24 Superoxide release inhibitory compounds from bryophytes

Species name[a]	Compound or solvent extract	Activity value (μg/cm^3)	Refs
[L] *Bazzania japonica*	Cyclomyltaylyl 10-caffeate (**1115**)	GPPM[b] PMN[c] 7.5	[1]
[L] *Conocephalum conicum*	Bicyclogermacrenal (**1117**)	12.5 50	[38]
[L] *Jungermannia infusca*	Infuscaside A (**148**) Infuscaside B (**150**) Infuscaic acid (**1113**)	50 50 50 50 0.07 25	[38]
[M] *Hypnum cupressiforme*	EtOH/H_2O (1:1) EtOAc and H_2O extracts (flavonoids: eriodictyol (**805**), apigenin (**836**), naringenin (**806**), kaempferol (**807**), acacetin (**808**), quercetin 3-*O*-glucoside (**803**) and rutinoside (**802**), isorhamnetin 3-*O*-glucoside (**804**), and gallate (**796**), 5-*O*-caffeoylquinic (**798**), *p*-hydroxybenzoic (**799**), caffeic (**800**), *p*-coumaric (**801**) acids)	10 μg/cm^3 (superoxide anion scavenging potential)	[272]
[L] *Marchantia polymorpha* (archegoniophore)	60% EtOH (flavonoids)	Superoxide anion scavenging potential	[388]
[L] *Mastigophora diclados*	Herbertene-1,2-diol (**406**) Herbertene 2,3-diol (**1118**)	25 25 25 25	[38]
[L] *Pellia endiviifolia*	Sacculatal (**128**)	4.0	[38]
[L] *Plagiochila* sp.	Plagiochilal B (**137**)	40 6.0	[38]
[L] *Plagiochila* sp.	Plagiochilide (**442**)	25 25	[38]
[L] *Porella elegantula*	Norpinguisone methyl ester (**929**)	35	[38]
[L] *Porella perrottetiana*	Perrottetianal A (**158**)	25	[38]
[L] *Porella vernicosa* complex	Polygodial (**90**) Norpinguisone (**927**)	4.0 25 50	[38]
[L] *Radula javanica*	Radulanin K (**1119**)	6	[40]

[a] [L] liverwort, [M] moss
[b] Guinea pig peritoneal macrophages induced by formylmethionine leucyl phenylalanine
[c] Rabbit polymorphonuclear leucocyte

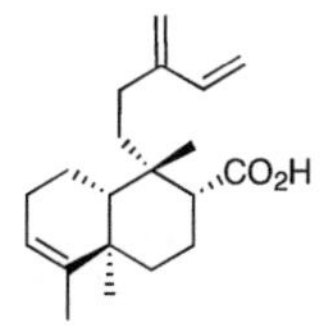

1113 (infuscaic acid)

Norpinguisone methyl ester (**929**) isolated along with norpinguisone methyl ester (**1114**) and norpinguisone (**1116**) from *Porella elegantula*, exhibited 50% inhibition of the release of superoxide from guinea pig peritoneal macrophages at 35 μg/cm^3. The same type of activity (IC_{50} = 7.5 μg/cm^3) was found for cyclomyltaylyl 10-caffeate (**1115**) from *Bazzania japonica*. Other sesquiterpenoids, including plagiochilide (**442**) isolated from *Plagiochila fruticosa*, *P. ovalifolia*, and *P. yokogurensis*, norpinguisone (**927**) from *Porella vernicosa*, bicyclogermacrenal (**1117**) from *Conocephalum conicum*, herbertene-1,2-diol (**406**), and herbertene-2,3-diol (**1118**) from *Mastigophora diclados*, the diterpenoids, infuscasides A (**148**) and B (**150**) from *Jungermannia infusca*, and perrottetianal A (**158**) from *Porella perrottetiana*, also inhibited superoxide release from GPPMs, with IC_{50} values ranging between 12.5 and 50 μg/cm^3 [1, 38]. Radulanin K (**1119**), from *Radula javanica*, also inhibited the release of superoxide anion radical from guinea pig peritoneal macrophages, with an IC_{50} value of 6 μg/cm^3 [40].

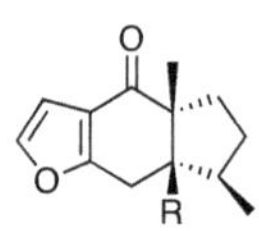

1114 (norpinguisone methyl ester, R = CO_2Me)
1116 (norpinguisone, R = Me)

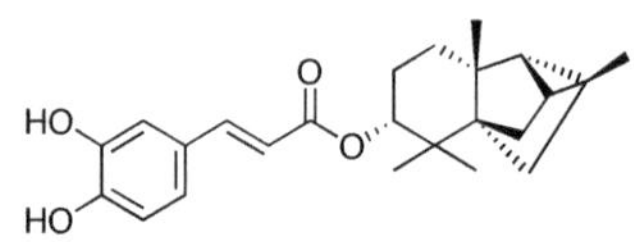

1115 (cyclomyltaylyl 10-caffeate)

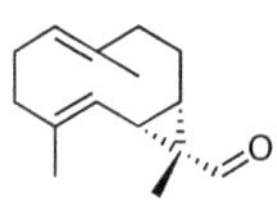

1117 (bicyclogermacrenal)

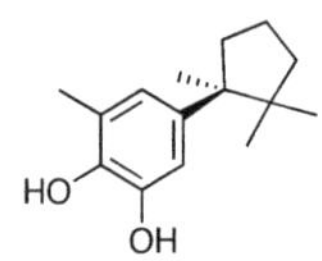

1118 (herbertene-2,3-diol)

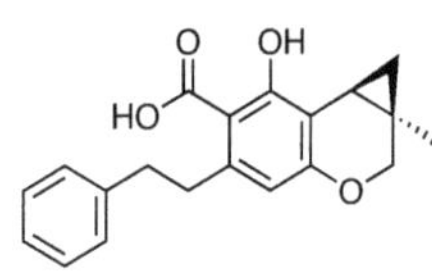

1119 (radulanin K)

An extract of the moss, *Hypnum cupressiforme*, which was found to contain simple aromatic acids, including gallic (**796**), caffeic (**800**), benzoic (**799**), and quinic (**798**) acids, and a mixture of flavonoids and their glycosides, inhibited superoxide anion radical release at a concentration of 10 μg/cm^3 [272].

Wang et al., analyzed *Marchantia polymorpha* and confirmed that the total flavonoid content in the archegoniophore was about ten times higher than that of the gametophytes [388]. While a flavonoid-containing 60% ethanol extract from the archegonoiophore of *M. polymorpha* displayed a superoxide anion scavenging effect, the extract of its gametophytes did not show any superoxide anion release inhibitory activity.

4.19 Antioxidant Activity

Antioxidants are compounds that inhibit oxidation, a series of chemical reactions that can produce free radicals and chain reactions and may damage the cells of organisms.

The ethanol extracts, containing sterols, terpenoids, and phenolic substances as well as fatty and amino acids of thirteen Latvian mosses, including *Aulacomnium*, *Climacium*, *Hylocomium*, *Polytrichum*, and four *Sphagnum* species, were tested for 2,2-diphenyl-1-picrylhydrazyl (DPPH) radical-scavenging activity. All of the extracts evaluated showed moderate antioxidant activity, as documented in Table 25 [265].

Mohandas and Kumaraswamy reported in 2018 that terpenoids from the Indian moss, *Thuidium tamariscellum*, displayed antioxidant activity, although the active compounds were not isolated [389].

An acetone extract of the moss *Leptodictyum riparium* stressed with lead, cadmium, heat shock, and salinity, was tested for antioxidant activity on human whole blood leukocytes using a chemiluminescence assay. Chemiluminescence represents a simple sensitive method to study the oxidative metabolism on phagocytes and this test is useful to identify compounds with antioxidant and antiinflammatory activities. The types of stress on the moss extract induced significant increases of chemiluminescence inhibition, with heavy metals being the most potent enhancers of antioxidant activity. The different stresses induced antioxidant activity in the following order: Cd > Pb > salinity > heat shock [390].

An aqueous extract of *Cryphaea heteromalla*, which contained benzoic (**799**), caffeic (**800**), and coumaric (**801**) acids, showed a protective effect against the generation of reactive oxygen species (ROS) generation induced by *tert*-butyl hydroperoxide using the murine NIH-3T3 fibroblast cell line isolated from a mouse NIH/Swiss embryo [391].

Ielpo et al., reported in 2000 that an acetone extract of *Lunularia cruciata* inhibited the emission of chemiluminescence, stimulated by phorbol myristate acetate, in human leucocytes. It was found that the stimulated cells were were inhibited to a greater degree than the non-stimulated ones. The antioxidant activity of *L. cruciata* may depend on the presence of various compounds, such as the flavonoid quercetin (**851**) and terpenes [392].

In 2022, Joshi et al., evaluated the antioxidant activity of extracts of the liverwort *Plagiochasma appendiculata* and the moss *Sphagnum fimbriatum* and found that their IC_{50} values against DPPH and nitric oxide-scavening activity (NOSA) were 56.1 and 54.0, and 54.0 and 53.0 μg/cm^3, respectively [299].

The Chilean moss *Sphagnum magellanicum* containing several phenolic acids, including caffeic (**800**), chlorogenic (**843**), *p*-coumaric (**801**), gallic (**796**), salicylic (**850**), and vanillic (**905**) acids, showed antioxidant capacity at 0.18 m*M* of DPPH reduced/cm^3 of the extract, equivalent in oxygen radical absorbance capacity assay units to 841 μ*M* Trolox/100 g units [305].

Wang et al., evaluated in 2016 the antioxidant effect of a 60% ethanol extract of the archegoniophora and gametophytes of *Marchantia polymorpha* in a DPPH assay. With increasing doses from 5 to 40 cm^3, the DPPH free radical-scavenging potential observably increased, and 40 cm^3 of the extract from the archegoniophore

Table 25 Antioxidant compounds from bryophytes

Species name[a]	Compound or solvent extract	Activity value	Refs.
[M] *Bryum argenteum*	Field collected mass	IC_{50} (mg/cm^3) (DPPH)	[402]
	n-Hexane	> 5	
	EtOAc	3.0	
	EtOH	1.8	
	In vitro cultured mass *n*-Hexane	> 5	
	EtOAc	2.0	
	EtOH	> 5	
	Ascorbic acid (control)	0.01	
	H_2O/50% EtOH/96% EtOH	(μ*M*/TE/L) (DPPH)	[403]
Brachythecium rutabulum		87.7/46.76/8.06	
Callicladium haldinianum		79.7/130.55/32.21	
Hypnum cupressiforme		15.0/60.10/8.56	
Orthodicranum montanum		29.0/101.59/5.41	
Polytrichastrum formosum		117.8/128.47/23.00	
		(μ*M*/TE/L) (ABTS)	
Brachythecium rutabulum		171.5/202.53/28.46	
Callicladium haldinianum		88.5/176.46/16.48	
Hypnum cupressiforme		79.0/101.21/4.60	
Orthodicranum montanum		44.6/109.66/ 8.13	
Polytrichastrum formosum		215.6/190.19/10.27	

(continued)

Table 25 (continued)

Species name[a]	Compound or solvent extract	Activity value	Refs.
	H_2O/50% EtOH/96% EtOH	Total phenolic acids (μg/CAE/cm^3)	[403]
Brachythecium rutabulum		7.6/11.04/12.38	
Callicladium haldinianum		6.2/12.99/11.74	
Hypnum cupressiforme		4.3/8.68/6.52	
Orthodicranum montanum		5.6/13.24/10.83	
Polytrichastrum formosum		8.2/17.65/7.21	
	H_2O/50% EtOH/96% EtOH	Total flavonoids (μg/QE/cm^3)	[403]
Brachythecium rutabulum		1.3/6.92/6.52	
Callicladium haldinianum		1.0/7.09/4.65	
Hypnum cupressiforme		3.9/5.63/0.68	
Orthodicranum montanum		2.8/8.75/1.12	
Polytrichastrum formosum		3.0/9.38/2.12/	
	H_2O/50% EtOH/96% EtOH	Total flavonoids (μg QE/cm^3)	[403]
Brachythecium rutabulum		0.92/2.59/2.37	
Callicladium haldinianum		0.37/4.06/1.74	
Hypnum cupressiforme		0.27/2,67/1.68	
Orthodicranum montanum		0.53/3.66/1.12	
Polytrichastrum formosum		0.97/1.10/3.50 TE: Trolox equivalent CAE: caffeic acid equivalent QE: quercetin equivalent	

(continued)

Table 25 (continued)

Species name[a]	Compound or solvent extract	Activity value	Refs.
Brachythecium rutabulum *Callicladium haldinianum* *Hypnum cupressiforme* *Orthodicranum montanum* *Polytrichastrum formosum*	50% EtOH	Inhibition of formation of AOPP	[403]
[M] *Bryum capillare*	EtOH	IC_{50} (mg/cm^3) (DPPH) 8.41 0.3	[278]
	BHA (control)	IC_{50} (mg/ cm^3) (metal chelating) 49.7 % plasma lipid peroxidation 33.56	
[M] *Bryum moravicum*	H_2O Ascorbic acid (control)	356.4 (μg/cm^3) (TPC) 84.6 (eq/mg) (ABTS)	[60]
[L] *Denotarisia linguifolia*		IC_{50} (mg/cm^3) (DPPH)	[262]
	n-Hexane	> 2	
	$CHCl_3$	> 2	
	EtOAc	0.66	
	EtOH	0.05	
	Ascorbic acid	0.88	
		IC_{50} (mg/cm^3) (lipid peroxidation)	
	n-Hexane	2.9	
	$CHCl_3$	1.9	
	EtOAc	2.4	
	EtOH	2.4	

(continued)

Table 25 (continued)

Species name[a]	Compound or solvent extract	Activity value	Refs.
	Quercetin (**851**)	0	
	EtOH (phenolic acid and flavonoid rich)	ABTS/FRAP (μ*M*)	[275]
[M] *Ceratodon purpureus*		20.98/771.33	
[M] *Dryptodon pulvinatus*		12.40/549.4	
[M] *Hypnum cupressiforme*		7.59/476	
[M] *Rhytidiadelphus squarrosus*		6.4/412	
[M] *Tortula muralis*		7.99/469	
[M] *Cinclidotus fontinaloides*	EtOH (flavonoid containing)	26 ABTS (μ*M*)	[314]
[M] *Palustriella commutata*		10	
[M] *Cryphaea heteromalla*	MeOH/H_2O (1:4) EtOH/H_2O (1:4)	NIH-3T3 murine fibroblast ROS production	[391]
[L] *Frullania dilatata*	EtOH (frullanolide (**3**) 2,3-dimethylanisole (**1133**) linoleic, palmitic (**1132**), valerenic (**1134**) acid-containing)/ascorbic acid (control)	89.3% (200 μg/cm^3) (DPPH) 85.654/94.515 (200) 85.645/93.653 (100) 65.456/ 92.878(50) 43.009/90.501 (25) 35.813/68.095 (12.5) 16.224/47.973 (6.25) 10. 436/28.418 (3.125) 10.002/ 20.788 (1.075)	[317]
[M] *Hedwigia ciliata*	EtOH/H_2O (1:1)	(%) inhibition (1000 μg/cm^3) 96	[279]

(continued)

Table 25 (continued)

Species name[a]	Compound or solvent extract	Activity value	Refs.
	96% EtOH	80	
	EtOAc	70	
	BHT (= 3,5-di-*tert*-4-hydroxytoluene)	100	
	BHA (= 2-*tert*-butyl-4-hydroxyanisole)	110	
	AA (= ascorbic acid)	100	
[M] *Hedwigia ciliata*	96% EtOH EtOH/H_2O (1:1) EtOAc	μg/cm^3 (DPPH) 1000	[279]
	LPS (LPS-treated BV2 cells	% β-carotene/linoleic acid	
	96% EtOH EtOH/H_2O (1:1) EtOAc	80	
[M] *Hypnum cupressiforme*	EtOAc H_2O EtOAc H_2O	23.8%/1 mg/cm^3 19.2%/1 mg/cm^3 84.1%/BHT 85.2%/BHA 88.1%/vitamin C β-Carotene bleaching assay: four-fold higher than vitamin C	[401]
[M] *Hypnum plumaeforme* [M] *Leucobryum juniperoideum* [M] *Thuidium kanedae*	EtOH/H_2O (7:3)	47.2–119.9 mg/g AE mg) (DPPH, ABTS, FRAP)	[60]
[M] *Leptodictyum riparium*	Acetone	Whole blood chemiluminescence	[390]

(continued)

Table 25 (continued)

Species name[a]	Compound or solvent extract	Activity value	Refs.
[L] *Lunularia cruciata*	MeOH	48% (650 μg/cm^3) (DPPH) 35% (350) 22% (250)	[60]
[L] *Lunularia cruciata*	$(Me)_2CO$ (flavonoids and terpenoids containing)	98% (2 mg/cm^3) (ABST) Phagocytes Chemiluminescence	[392]
Species name[a]	Compound	Activity value (DPPH) IC_{50} (μ*M*), (μg/cm^3)	Refs.
Almost all liverworts	α-Tocopherol (**15**)	Antioxidant activity	[75]
[L] *Jungermannia subulata*	Subulatin (**1121**)	The same activity as α-tocopherol (**15**)	[399]
[L] *Marchantia diptera*	Marchantin H (**940**)	0.5 μ*M* (non-enzymatic iron-induced lipid peroxidation in rat brain homogenates)	[395]
[L] *Marchantia paleacea*	MeOH	IC_{50} 20 (μg/cm^3) (DPPH)	[60]
[L] *Marchantia paleacea* subsp. *diptera*	Paleatin B (**202**)	11.7 μg/cm^3 (arachidonic acid oxidation)	[60, 396]
L] *Marchantia polymorpha*	MeOH	4.62 μg/cm^3 (DPPH) IC_{50} 1.3 μg/cm^3 (DPPH)	[388]
	60% EtOH	IC_{50} 3.0 (ABTS)	
[L] *M. polymorpha*	EtOH	75.4% (DPPH)	[393]
[L] *M. polymorpha*	MeOH	IC_{50} 449.5 μg/cm^3 (DPPH) IC_{50} 244.1 (ABTS)	[394]
	EtOAc	IC_{50} 275.6 (DPPH) IC_{50} 212.6 (ABT	

(continued)

Table 25 (continued)

Species name[a]	Compound	Activity value (DPPH) IC_{50} (μ*M*), (μg/cm^3)	Refs.
[L] *M. polymorpha*		EC_{50} μg/cm^3 (DPPH)	[261]
	n-Hexane	> 5000	
	$CHCl_3$	356.35	
	EtOAc	578.35	
	EtOH	693.82	
[L] *Marchantia polymorpha*	Marchantin A (**1**)	20 μg/cm^3 (DPPH)	[134]
[L] *Marchantia polymorpha*	Marchantin A (**1**)	20 μg/cm^3 (DPPH)	[396]
	Marchantin B (**195**)	0.4 (arachidonic acid oxidation)	
	Marchantin D (**198**)	0.4	
	Marchantin E (**199**)	5.6	
	Marchantin H (**940**)	2.7	
	Isomarchantin C (**264**)	10.0 (DPPH) 5.3 (arachidonic acid oxidation)	
[L] *Radula perrottetii*	Perrottetin D (**1055**)	0.72 (arachidonic acid oxidation)	
[L] *Mastigophora diclados*	Et_2O/MeOH (–)-Herbertene-1,2-diol (**406**) (–)-Mastigophorene C (**407**) (–)-Mastigophorene D (**408**)	21.4 /20.8 μ*M* 1.9 μg/cm^3 2.7 2.0	[182]
[L] *Pellia epiphylla*	Epiphyllin B (**1123**) Epiphyllin C (**1124**) Epiphyllin D (**1125**) 1*β*-Hydroxysacculatanolide (**1130**) Pellianolactone B (**1131**)	Protective effects on H_2O_2-induced oxidative stress and apoptosis in PC12 cells	[400]

(continued)

Table 25 (continued)

Species name[a]	Compound	Activity value (DPPH) IC_{50} (μ*M*), (μg/cm[3])	Refs.
[L] *Plagiochasma appendiculatum*	Crude extract ($CHCl_3$, acetone, EtOH, H_2O): 250 mg.kg/control	0/93/1.49 (n*M*/protein):TBARS 4,5/6.36 (μg/mg protein):GHS level) 252.11/116.3 (catalase activity)	[360]
[L] *Plagiochasma appendiculata*	80% MeOH	(FRAP)	[299]
[M] *Sphagnum fimbriatum*			
[L] *Plagiochila asplenioides*	MeOH (palmitic, stearic, oleic, linoleic acid-rich)	59.0% (DPPH) 24.1 (ABTS) 4729 mg/$FeSO_4$/kg (FRAP)	[320]
[L] *Plagiochila ovalifolia*	2-(3-Methyl-2-butenyl)-3,5-dihydroxybibenzyl (**306**)	50–200 mg/cm^3, 28–48%	[297]
	Plagiochin D (**462**)	50–200 μg/cm^3, 62–78% inhibition	
[M] *Polytrichastrum alpinum*	MeOH	96.8 μg/cm^3 (DPPH) 103.93 (ABTS) 145/6 (NO) 10.0 μg (FRAP)	[398]
	Ohioensin F (**1069**)	10 μg/cm^3 (DPPH) 14.3 (ABTS) 63 (NO) 9.8 (FRAP)	
	Ohioensin G (**1070**)	10 (DPPH) 14.3 (ABTS) 63 (NO) 9.6 (FRAP)	

(continued)

Table 25 (continued)

Species name[a]	Compound	Activity value (DPPH) IC_{50} (μ*M*), (μg/cm^3)	Refs.
[L] *Radula lindenbergiana*	Radulanin H (**1065**)	15.7 (arachidonic acid oxidation)	[396]
[L] *Radula perrottetii*	Perrottetin D (**1055**)	(arachidonic acid autoxidation assay)	[396]
[L] *Reboulia hemisphaerica*	Riccardin C (**182**)	12.7 (arachidonic acid oxidation)	[396]
[M] *Sphagnum magellanicum* [M] *S. squarrosum* [M] *Sphagnum girgensohnii* [M] *S. palustre*	EtOH/H_2O (1:1)	91.8 μ*M* TE/g dry extract (ABTS) 63.8 μ*M* TE/g dry extract (DPPH) 98.0 m*M* Fe2 + /dry extract (FRAP) 6.1% (lipid peroxidation inhibition) 41.9 μ*M* (ABTS) 20.6 μ*M* (DPPH) 45.9 μ*M* (FRAP) 9.0% (lipid peroxidation inhibition) 36.1 μ*M* (ABTS) 10.1% (lipid peroxidation inhibition) 3.3 μ*M* ChTE/cm^3 (protein oxidation inhibition) 52.5 μ*M* (ABTS) 11.5% (lipid peroxidation inhibition) 2.6 μ*M* ChTE/cm^3 (protein oxidation inhibition) on PL creation AOPP creation	[281]
[M] *Sphagnum magellanicum*	EtOH (containing caffeic (**800**), chlorogenic (**843**), *p*-coumaric (**801**), 3,4-dihydroxybenzoic (**903**), gallic (**796**), salicylic (**850**), and vanillic (**905**) acids)	0.18 m*M* (DPPH) equivalent in ORAC (Oxygen radical absorbance capacity assay) units to 841 μ*M* Trolox/100 g units	[305]

(continued)

Table 25 (continued)

Species name[a]	Compound	Activity value (DPPH) IC_{50} (μ*M*), (μg/cm^3)	Refs.
[M] *Thuidium tamariscellum*	EtOH (terpenoid containing)	IC_{50} 16 μg/cm^3 (DPPH) 34.5 (H_2O_2) 18.5 (ABTS) 40 (FRAP)	[389]
[M] *Leptodictyum riparium*	Metal (cadmium, lead) salinity heat shock	Antioxidant inducing scale (on human leucocyte): Cd > Pb > salinity > heat shock	[390]
[L] *Denotarisia linguifolia*		IC_{50} (μg/cm^3) (lipid peroxidation)	[262]
	n-Hexane	2.9	
	$CHCl_3$	1.9	
	EtOAc	2.4	
	EtOH	2.4	

[a] [L] liverwort, [M] moss

scavenged about 65% of the free radicals. The scavenging effect of the gametophyte extract was lower than from the archegoniophore. The DPPH and ABTS (2,2′-azino-bis(3-ethylbenzthaiazoline-6-sulfonic acid) radical-scavening activities of the archegoniophore were determined as IC_{50} 1.3 and 3.0 μg/cm^3 [388].

Rana et al., reported in 2018 the presence of tannins and phenolic compounds, carbohydrates, steroids, hexose sugars, and organic acids along with the absence of proteins, alkaloids, and amino acids in an ethanol extract of *Marchantia polymorpha*. This crude extract showed moderate antioxidant activity (75.4%) using a DPPH assay. With respect to the anti-inflammatory activity also observed for this extract, the active compounds still remained to be clarified [393].

In 2012, Gokbulut et al., tested the antioxidant effects of the methanol and ethyl acetate extracts of *M. polymorpha* using the DPPH and ABTS methods. The methanol extract showed antioxidant activity with IC_{50} values of 0.5 and 0.4 mg/cm^3, while the ethyl acetate extract exhibited slightly more potent activities, with IC_{50} values of 0.3 and 0.2 mg/cm^3, respectively, than the former extract. The authors suggested that the activity found was due to the presence of the flavone, luteolin (**483**) [394].

The ethanol extracts containing flavonoids of the two Turkish mosses, *Cinclidotus fontinaloides* and *Palustriella commutata*, were found to possess weak antioxidant activity. The total antioxidant activities of the moss extracts were determined by the Trolox (6-hydroxy-2,5,7,8-tetramethylchroman-2-carboxylic acid) method. The Trolox equivalent values of the moss extracts were calculated, in turn, as 26 and 10 mg/g. In addition, the anthocyanin activities of the mosses were determined as 50 and 144 mg/dm^3 [314].

The antioxidant activities of the phenolic acid- and flavonoid-rich ethanol extracts of five mosses, *Ceratodon purpureus*, *Dryptodon pulvinatus*, *Hypnum cupressiforme*, *Rytideadelphus squarrosus*, and *Tortula muralis*, were evaluated using the ABTS+ and the ferric reducing antioxidant power (FRAP) assays. Both the radical-scavening activity and the reducing properties in the extracts decreased in the following order: *C. purpureus* > *D. pulvinatus* > *T. muralis* > *H. cupressiforme* > *R. squarrosus*. The Trolox equivalent value for *C. purpureus* was 40% higher than that for *D. pulvinatus* and 60–70% higher in comparison to *H. cupressiforme*, *R. squarrosus*, and *T. muralis*. A similar tendency was observed using the FRAP method [275].

The ether and methanol extracts of the Tahitian *Mastigophora diclados* and an unidentified Indonesian *Marchantia* species showed radical-scavening activities with IC_{50} values of 21.4 and 20.8, and 22.8 and 20.8 μM, respectively. Several terpenoid and phenanthrene derivatives of previously known structure were detected by GC-MS, but these were not tested for their antioxidant activities individually [182]. Marchantin A (**1**) also showed free-radical scavenging activity with an IC_{50} values of 20 μg/cm^3 [134].

Marchantin H (**940**) inhibited non-enzymatic, iron-induced lipid peroxidation in rat brain homogenates, and gave an IC_{50} value of 0.51 μM, with this effect being more potent than that of desferrioxamine. Marchantin H (**940**) also suppressed NADPH-dependent, microsomal lipid peroxidation with an IC_{50} value of 0.32 μM, while not affecting the microsomal electron transport of NADPH-cytochrome P450 reductase. It also inhibited the copper-catalyzed oxidation of human low-density lipoprotein.

Hsiao et al., concluded in 1996 that marchantin H (**940**) is a potentially effective and versatile antioxidant, and can be used as a chaperone to protect biologically active macrocyclic molecules against peroxidative damage [395].

The formation of *o*-quinone radicals from compounds in the marchantin series was established by Schwartner et al., in 1996. Using pulse radiolysis and EPR techniques, in addition to kinetic data, the bis-bibenzyls, marchantins A (**1**), B (**195**), D (**198**), E (**199**), and H (**940**) as well as isomarchantin C (**264**) from *M. polymorpha* were confirmed as effective antioxidants at 0.4–20 μg/cm^3. Riccardin C (**182**) from *Reboulia hemisphaerica*, perrottetin D (**1055**) and radulanin H (**1065**) from *Radula* species also showed potent activity in an arachidonic acid autoxidation assay [396].

Bioactivity-guided fractionation of an ether extract of *Plagiochila ovalifolia* using the DPPH-radical scavenging assay resulted in the isolation of 2-(3-methyl-2-butenyl)-3,5-dihydroxybibenzyl (**1062**) and plagiochin D (**1120**), and both displayed antioxidative activity (28–48% inhibition at 50–200 μg/cm^3 and 62–78% inhibition at 50–200 μg/cm^3, respectively) [397].

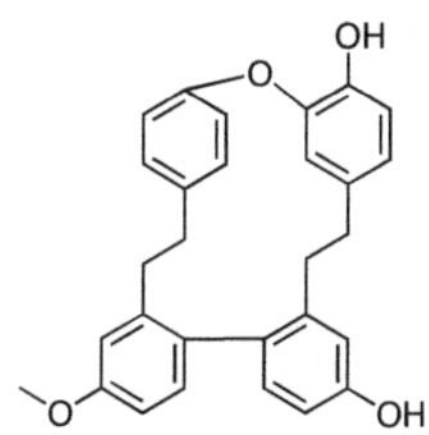

1120 (plagiochin D)

The Antarctic moss, *Polytrichastrum alpinum*, is similar chemically to the well-known *Polytrichum commune*, since both species produce benzonaphthoxanthenones. Ohioensins F (**1069**) and G (**1070**) isolated from the former species showed antioxidant activities, having IC_{50} values of 10 μg/cm^3 (DPPH), 14 (ABTS), 62–63 (NO), and 9.6–9.8 μg (FRAP), respectively [398].

An antioxidant assay on subulatin (**1121**), obtained from *Jungermannia sublata*, was carried out using an erythrocyte membrane ghost system. Compound **1121** showed the same level of activity as that of the positive control, α-tocopherol (**15**). Compound **1121** might play a role in the detoxification of oxygen generated by photo-oxidation in liverworts [399]. α-Tocopherol (**15**), which has been found in almost all liverworts, is postulated as a significant antioxidant for the oil bodies of these plants [75].

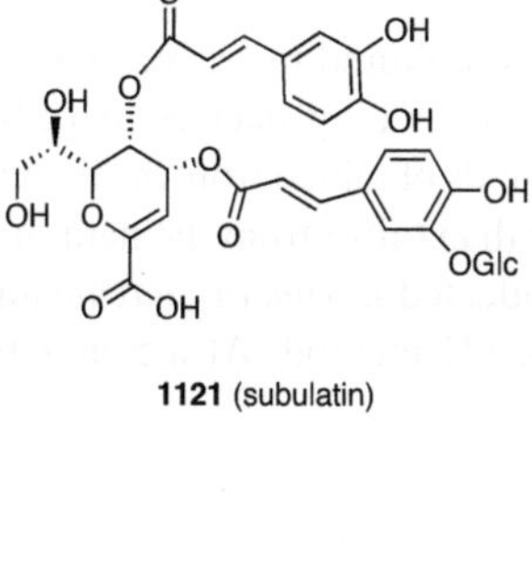

1121 (subulatin)

The liverwort *Pellia* species are abundant sources of sacculatane diterpenoids that have a pungent taste. Li et al., reinvestigated *Pellia epiphylla* in 2019 to identify eight sacculatanes, named epiphyllins A–H (**1122–1129**), together with 1*β*-hydroxysacculatanolide (**1130**) and pellianolactone B (**1131**). Epiphyllins B–D (**1123–1125**), and compounds **1130** and **1131** showed moderate NAD(P)H quinone reductase-inducing effects using Hepa 1c1c7 murine hepatoma cells. Compound **1131** also demonstrated a protective effect against hydrogen peroxide-induced PC12 cellular apoptosis [400].

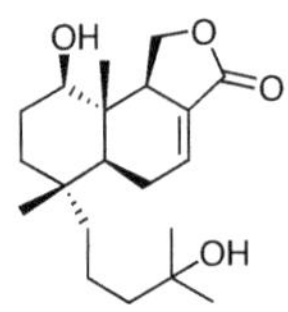

1122 (epiphyllin A)

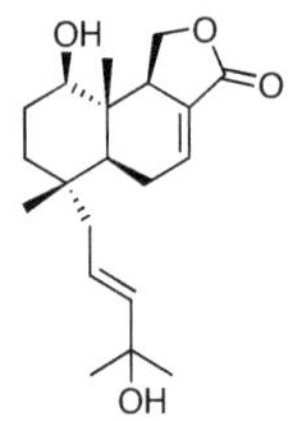

1123 (epiphyllin B)

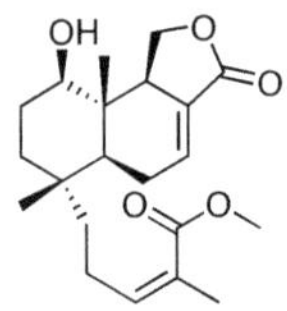

1124 (epiphyllin C)

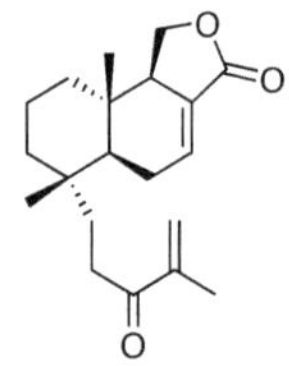

1125 (epiphyllin D)

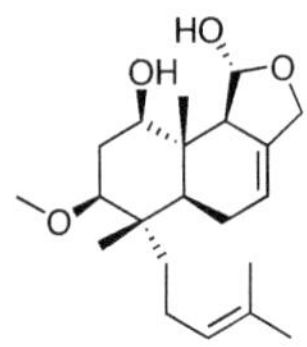

1126 (epiphyllin E)

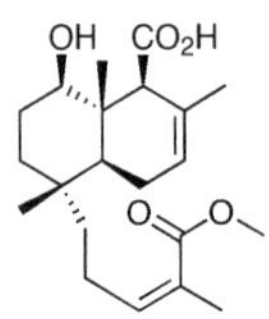

1127 (epiphyllin F)

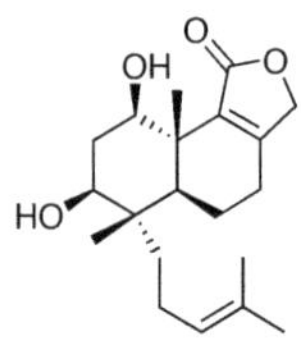

1128 (epiphyllin G)

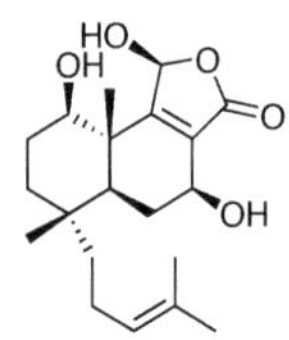

1129 (epiphyllin H)

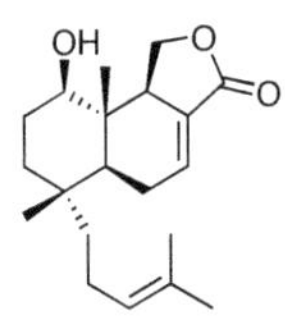

1130 (1*β*-hydroxy-sacculatanolide)

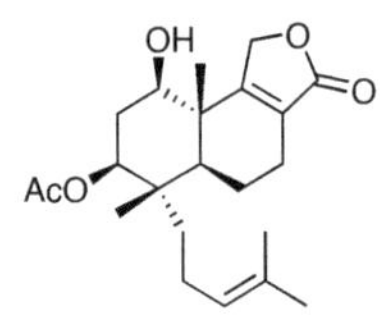

1131 (pellianolactone B)

Lunić et al., reported in 2020 and 2022 that in BHT, BHA, and ascorbic acid assays of the ethyl acetate and water extracts of *Hypnum cupressiforme*, including flavones and simple aromatic acids, such as caffeic (**800**) and gallic (**796**) acids, showed moderate scavenging activity (23.8 and 19.2%) at a concentration of 1 mg/cm^3 [401].

Field-collected and in vitro-cultured moss specimens of the moss *Bryum argenteum* were extracted with *n*-hexane, ethyl acetate, and ethanol, and each extract yield was compared. The extract yields obtained from the in vitro-cultured moss were three to one hundred times higher than those from the natural *B. argenteum.* The antioxidant activities of the field-collected *Bryum argenteum* and its in vitro-cultured mass were evaluated using the DPPH method. At a 5 mg/cm^3 concentration, the ethyl acetate and ethanol extracts from the natural sample and the ethyl acetate extract from the cultured moss showed more than 50% DPPH inhibition [402].

In 2022, Smolinska-Kodla et al., studied the DPPH and ABTS radical scavenging activities and the inhibition of advanced oxidation protein product formation of three solvent extracts (aqueous, 50% ethanol, and 96% ethanol) of four Polish mosses, *Brachythecium rutabulum, Callicladium haldinianum, Hypnum cupressiforme, Orthodicranum montanum,* and *Polytrichastrum formosum.* Among the aqueous extracts, the most potent advanced oxidation protein product inhibitory effects were shown by *O. monyanum* and *P. formosum* and also their activities were comparable to that exhibited by 1 m*M* ascorbic acid solution.

For these three extracts, the 50% ethanol extract showed the most potent radical-scavenging activities. Each 50% ethanol extract also inhibited the formation of advanced oxidation protein product, comparable to a 1 m*M* ascorbic acid solution. The 50% and the 96% ethanol extracts of *H. cupressiforme* exhibited the lowest free-radical scavenging activities, although *B. rurabulum* had inhibitory activity on the formation of advanced oxidation protein product [403].

Antioxidant activity testing of an ethanol extract of the Turkish moss *Bryum capillare* was carried out using the DPPH, metal chelating, and plasma lipid peroxidation methods. The extract showed weak antioxidant activities with IC_{50} values of 8.4 and 49.7 mg/cm^3, and 33.6%, respectively [278].

Among four extracts prepared from *Denotarisia linguifolia,* the ethyl acetate and ethanol extracts at a 2 mg/cm^3 concentration showed moderate DPPH-scavenging effects, with IC_{50} values of 0.7 and 0.9, respectively, with the *n*-hexane and chloroform extracts being less active. At a 4 mg/cm^3 concentration, all extracts inhibited antioxidant activity against α-glucosidase, with the chloroform extract having the highest observed effect ($IC_{50} = 1.9$ mg/cm^3) [262].

An ethanol extract of *Frulania dilatata,* containing frullanolide (**3**), 2,3-dimethylanisole (**1133**) and palmitic (**1132**), linoleic, and valerenic (**1134**) acids, showed weak antioxidant activity with 10–85% (1–200 μg/cm^3, DPPH assay), when compared to ascorbic acid (20–95%) as control [317].

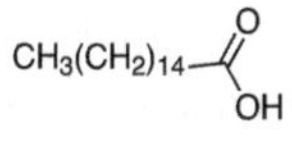

1132 (palmitic acid)

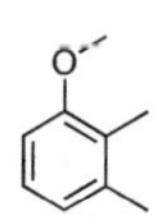

1133 (2,3-dimethylanisole)

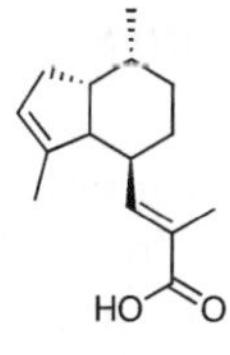

1134 (valerenic acid)

An antioxidant activity evaluation of the 96% ethanol, ethanol/water (1/1) and ethyl acetate extracts of the moss *Hedwigia ciliata* was carried out using the β-carotene-linoleic acid assay. The best activities were seen for the ethanol and ethanol/water (1/1) extracts, with 96% inhibition each, followed by the 96% ethanol and ethyl acetate extracts with 80 and 70% inhibition, respectively [279].

In 2018, Tan et al., extracted the Vietnamese *Marchantia polymorpha* with *n*-hexane, chloroform, ethyl acetate, and ethanol, and evaluated the antioxidant activity of each extract. Except for the *n*-hexane extract, the other three extracts showed moderate antioxidant effects in the DPPH assay [261].

The crude solvent extract of *Plagiochasma appendiculatum* showed strong antioxidant effects in inhibiting lipid peroxidation and increasing the superoxide dismutase and catalase activities [299]. On application of a *P. appendiculatum*-containing ointment, the epithelization period decreased, and a visibly decreased scar area occurred. A significant increase in the tensile strength and hydroxyproline content was observed when compared to the control group and was comparable to the standard drug, nitrofurazone. Singh et al., concluded in 2006 that a *P. appendiculatum* extract might be useful as a wound-healing agent [319].

The Turkish liverwort, *Plagiochila asplenioides*, exhibited moderate antioxidant activities, when evaluated in the DPPH, ATS, and FRAP assays [320].

The antioxidant activities of the water/ethanol (1:/1) solvent extracts of four *Sphagnum* species, *S. magellanicum*, *S. girgensohnii*, *S. palustre*, and *S. squarrosum*, which contain *p*-coumaric acid and rutin were examined using the ABTS, DPPH, and FRAP methods, with respect to the peroxidation of lipids and advanced oxidation protein product creation. Among these four species, *S. magellanicum* was found to display antioxidant effects using the three above-mentioned methods, with IC_{50} values of 91.8, 63.8, and 98 μM, respectively. In turn, *S. girgensohnii* and *S. palustre* extracts exhibited the most potent inhibitory activities on the oxidation of linoleic acid (inhibition of lipid peroxidation). Their inhibition percentages were 10.1 and 11.5, respectively. The activity in terms of protein oxidation inhibition, as demonstrated by the reduced production of advanced oxidation protein product, was found for the *S. palustre* extract at a 2.6 μM concentration, with the least potency in this regard found for the *S. girgensohnii* extract [281].

4.20 Anti-inflammatory and NO Production Inhibitory Activities

In the course of an investigation of the structure-activity relationships between perrottetinene (**279**) from a *Radula* liverwort species, which contains the compounds 2,2-dimethyl-7-hydroxy-5-(2-phenylethyl)chromene (**1135**) and 1-methoxy-4-(2-methyl-propenyl)benzene (**1136**) [173, 174], and (–)-*trans*-Δ^9-tetrahydrocannabinol (**281**) from *Cannabis sativa*, Chicca and his group found that **281** showed a higher anti-inflammatory activity than **279** [404] (see Table 26).

Asakawa et al., reported in 1995 that a primitive liverwort, *Haplomitrium mnioides* (Plate 31), produced three new labdane lactones, haplomitrenolides A–C (**1137**–**1139**) [2]. Zhou et al., [405] reinvestigated the same species to isolate three further

Table 26 Antiinflammatory and cytoprotective compounds from bryophytes

Species name[a]	Compound or solvent extract	Activity	Refs.
[L] *Haplomitrium mnioides*	Haplomitrenolide A (**1137**) Haplomitrenolide C (**1139**) Hapmnioide A (**1140**)	Inhibition of IL-6 production	[405]
[L] *Marchantia polymorpha*	EtOH	Inhibition of protein (bovine serum albumin) denaturation 75.5%/1 mg/cm^3	[393]
[L] *Radula campanigera* [L] *R. chinensis* [L] *R. laxiramea* [L] *R. marginata* [L] *R. perrottetii* [L] *Radula* species (Peruvian unidentified sp.)	*cis*-Perrottetinene (**279**)	Anti-inflammatory (rat)	[173, 174]
		Anti-inflammatory effect against carrageenan-induced paw edema (%)	[414]
[L] *Plagiochasma rupestre*	Et_2O Eudesma-3,7(11)-dien-8-one, β-elemene (**37**), 5-*epi*-aristolochene	27.2	
[L] *Porella cordaeana*	Pinguisanine, perrottetianal (**158**), spiropinguisanine,	25.4	
[L] *Reboulia hemisphaerica*	β-Microbiotene, grimaldone (**65**), β-caryophyllene (**35**), β-elemene (**37**)	29.4	
		Suppressing effect on *p*-benzoquinone-induced abdominal constriction (%)	
[L] *Plagiochasma rupestre*		44.6	
[L] *Porella cordaeana*		27.6	
[L] *P. platyphylla*		31.5	
[L] *Radula hemisphaerica*		41.1	
[L] *Plagiochasma appendiculatum*	Fresh paste with water	To cure skin eruptions caused by sun light, burns, boils and blisters	[415]
		Incision wound	

(continued)

Table 26 (continued)

Species name[a]	Compound or solvent extract	Activity	Refs.
[L] *Plagiochasma appendiculatum*	Crude extract ointment (100 mg/599 mm^2)/nitrofurazone (2%) ointment (standard drug)	Epithelization period (days) 18.0/13.3 Tensile strength (g) 386/411.2 Scar area (mm^2): 35.4/30.7	[319]
[M] *Tortula muralis*	EtOH	Wound healing	[275]
		Cytoprotective effect	
[M] *Mylia nuda*	Mylnudone D (**1186**)	Glutamic acid-induced neurological cells (SH-SY5Y): 10 μ*M*	[416]

[a] [L] liverwort, [M] moss

new rearranged labdanes, hapmnioides A–C (**1140–1142**). The absolute configuration of compound **1140** was established as (4*R*,5*S*,9*S*,10*R*,12*R*) by X-ray crystallographic analysis using CuK_α radiation. Accordingly, compound **1140** is a highly rearranged labdane-type diterpenoid with a 1,10:5,6-di-*seco*-12,20-olide-1,5:6,10-dicyclolabdane skeleton. The stereostructures of hapmnioides B (**1141**) and C (**1142**) were also determined by X-ray crystallographic analysis and comparison of their NMR data. Plausible biogenetic pathways of the hapmnioides (**1140–1142**) also were proposed. Initiated by oxidation at C-1, compounds **1140–1142** might be formed from haplomitrenolides (**1138**) or C (**1139**), which coexist in *H. mnioides*, through a cascade diradical rearrangement reaction. Compounds **1137** and **1140** suppressed the expression of IL-6 in LPS-induced cells, indicating a potential anti-inflammatory activity. Compound **1139** also showed an inhibitory activity on IL-6 expression. Immunofluorescence analysis revealed that compounds **1137** and **1140** inhibited phosphorylation and nuclear translocation of p65 in PC3 cancer cells following a two-hour treatment. These results showed that compounds **1137** and **1140** are potential anti-inflammatory compounds [405].

Plate 31 *Haplomitrium mnioides* (liverwort)

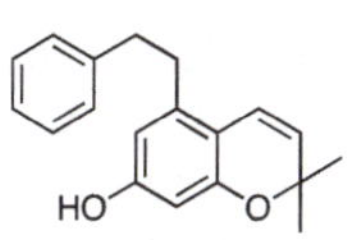

1135 (2,2-dimethyl-7-hydroxy-5-(2-phenylethyl)chromene)

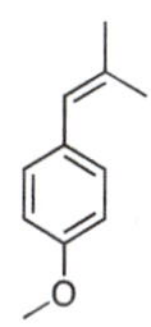

1136 (1-methoxy-4-(2-methylpropenyl)benzene)

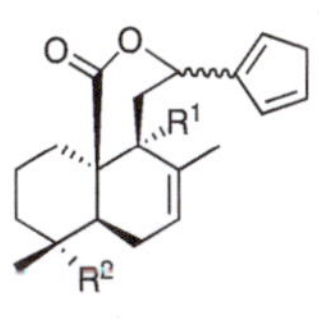

1137 (haplomitrenolide A, $R^1 = H$, $R^2 = Me$)
1138 (haplomitrenolide B, $R^1 = OH$, $R^2 = Me$)
1139 (haplomitrenolide C, $R^1 = H$, $R^2 = CO_2Me$)

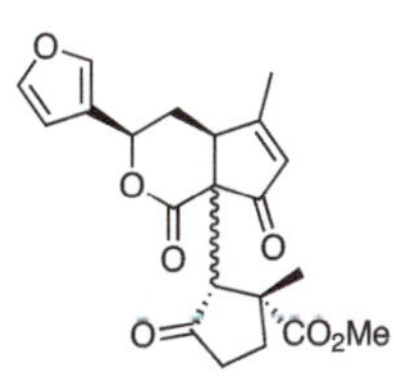

1140 (hapmnioide A)

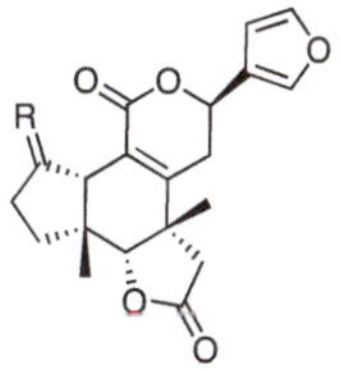

1141 (hapmnioide B, R = O)
1142 (hapmnioide C, R = α-OH, β-H)

In 2018, Rana et al., evaluated the potential anti-inflammatory activity of an ethanol extract of *Marchantia polymorpha* using bovine serum albumin. At a concentration range of 200–400 mg/cm^3, the extract significantly protected against protein denaturation, while at a concentration of 1 mg/cm^3, the extract showed a 75.5% inhibition of the protein denaturation [393].

Over-production of nitric oxide (NO) is involved in inflammatory response-induced tissue injury and the formation of carcinogenic *N*-nitrosamines [406]. Large amounts of NO are expressed and generated by induced inducible nitric oxide synthase (iNOS) on the stimulation endotoxins or cytokines involved in pathological responses. Thus, inflammatory disease can be induced by the over-production of NO by iNOS. Accordingly, finding new agents that inhibit NO production from natural sources has beome a quite a common drug discovery approach.

In 2022, Marques et al., analyzed the 70% ethanol extracts of 29 mosses and three liverworts to determine those with NO production inhibitory effects. They found that extracts of *Dicranum majus* and *Thuidium delicatulum* inhibited NO production in a concentration-dependent manner, with IC_{50} values of 1.0 and 1.5 mg/ cm^3, respectively [407].

The inhibition of lipopolysaccharide (LPS)-induced NO production in cultured RAW 264.7 cells by 17 natural bis-bibenzyls, marchantins A (**1**), B (**195**), C–E (**197–199**), and **940**, isomarchantin C (**264**), riccardin A (**3**), C (**182**), and F (**243**), isoplagiochins (**265**, **266**, and **944**), two perrottetins (**184**, **1145**), ptychantol (**1146**), compound **1056**, synthetic marchantin A trimethyl ether (**1143**), and marchantin B tetramethyl ether (**1144**), have been evaluated, and the IC_{50} values of these bis-bibenzyls are shown in Table 27 [408]. Among these bis-bibenzyls, marchantin A (**1**) showed the most potent NO production inhibitory activity, with an IC_{50} value of 1.4 μM. Concerning the structural requirements for the effective inhibition of NO production by bis-bibenzyls, the presence of C-1–C-2′, and C-14–C-11′ diaryl ether bonds seems to be important in mediating this type of inhibitory activity. Compounds with a C-7 and C-8 unsaturation exhibited a greatly decreased inhibition of NO, while the introduction of a hydroxy group at C-7′ resulted in a slightly decreased activity. Marchantin A trimethyl ether (**1143**) and marchantin B tetramethyl ether (**1144**) showed weaker activities than the parent marchantins A (**1**) and B (**195**) [408].

Table 27 NO production-inhibitory compounds from bryophytes

Species name	Compound or solvent extract	Activity value	Refs
[M] *Dicranum majus* [M] *Thuidium delicatulum*	70% EtOH	1.04 μg/cm^3 1.54	[407]
[M] *Hypnum cupressiforme*	96% EtOH/EtOAc gallic acid (**796**), procatechuic acid (**797**), caffeic acid (**800**), and *p*-coumaric acid (**801**), apigenin (**482**), acacetin (**808**), eriodictyol (**805**), isorhamnetin, kaempferol (**807**), and quercetin (**851**)-containing)	10 μg/cm^3 LPS-treated BV2 cells (normal murine microglia cell lines)	[272]
[M] *Hedwigia ciliata*	96% EtOH EtOH/H_2O (1:1) EtOAc	10 μg/cm^3 LPS-treated BV2 cells 20–30 μg/cm^3 MDA-MB-232 cell	[279]
[L] *Lepidozia reptans*	(1*R*,11*R*,12*R*)-6,12-Dihydroxydolabella-(3*E*,7*E*)-diene (**1006**)	IC_{50} (μ*M*) (estimated by the model of LPS-induced NO production with macrophage cells) 75.8	[413]
	(1*S*,3*E*,7*E*,11*S*,12*S*)-12-Hydroxy-dolabella-3,7-dien-6-one (**1170**)	64.8	
	(1*S*,3*E*,7*Z*, 11*S*,12*S*)-12-Hydroxy-dolabella-3,7-dien-6-one (**1171**)	37.9	
	13*α*-Hydroxy-*ent*-16-kaurene-3,5-dione	71.0	
	ent-3*β*-Hydroxy-(16*S*)-kauran-15-one	> 100	
	Vitamin E quinone (**1180**)	> 100	

(continued)

Table 27 (continued)

Species name	Compound or solvent extract	Activity value	Refs
[L] *Bazzania nitida* [335]	Myltayl-4(12)-enyl 2-caffeate (**1149**)	IC_{50} (μM) 6.3	[409]
[L] *Marchantia polymorpha* *M. palmata* [L] *Marchantia pappeana* [L] *Plagiochila barteri*	Marchantin A (**1**) Marchantin B (**195**) Marchantin C (**197**) Marchantin D (**198**) Marchantin E (**199**) Isomarchantin C (**264**) Marchantin H (**940**) Marchantin A trimethyl ether (**1143**) (synthetic) Marchantin B (tetramethyl ether (**1144**) (synthetic)	1.4 4.1 13.3 10.2 62.2 > 100 15.3 42.5 42.5	[408]
[L] *Marchantia polymorpha*	MeOH	6.99 μM/100 µg/cm^3 extract LPS (1 µg/cm^3-treated HaCaT cells	[417]
	EtOAc fraction of MeOH extract	20 µg/30 µg/cm^3 extract 3/50 2/100 LPS (1 µg/cm^3-treated HaCaT cells	
	n-BuOH fraction of MeOH	40/30 20/50 1/100 LPS (1 µg/cm^3-treated HaCaT cells	
[M] *Racomitrium* sp.	MeOH	11.6 μM/100 µg/cm^{3}extract LPS (1 µg/cm^3-treated HaCaT cells	

(continued)

Table 27 (continued)

Species name	Compound or solvent extract	Activity value	Refs
[L] *Marchantia convoluta*	80% EtOH	Inhibition rate of auricle tympanites: 21.8%	[296]
[L] *Mastigophora diclados* [L] *Bazzania decrescens*		IC_{50} (μ*M*)	[409]
	α-Herbertenol (**405**)	76.0	
	Herbertene-1,2-diol (**406**)	8.0	
	Mastigophorene C (**407**)	10.2	
	Mastigophorene D (**408**)	15.2	
	β-Herbertenol (**886**)	12.2	
	1,2-Dihydroxyherberten-12-al (**886**)	34.0	
	1,2-Cuparenediol (**1147**)	9.2	
	2-Hydroxy-4-methoxy-cuparene (**1148**)	4.1	
	L-(N_6-iminoethyl)lysine (positive control)	18.6	
[L] *Plagiochila fruticosa*	Isoplagicchin A (**265**) Isoplagiochin B (**266**) Isoplagiochin D (**944**)	> 100 > 100 14.3	[408]
[L] *Pellia endiviifolia*	Perrottetin E 11′-methyl ether	49.9	[408]
[L] *Plagiochila sciophila*	Plagiochin A (**265**)	10.0	[408]
[L] *Ptychanthus striatus*	Ptychantol A	> 100	[408]
[L] *Porella densifolia*	Norpinguisone methyl ester (**929**) Norpinguisone (**927**) *ent*-16-Kauren-15-one (**1150**)	45.5 1.7 69.4	[410]
[L] *Radula appressa*	Radulanin A (**335**)	20.0	[412]
	2-Geranyl-3,5-dihydroxybibenzyl (**899**)	4.5	
	Radulanin L (**336**)	15.3	

(continued)

Table 27 (continued)

Species name	Compound or solvent extract	Activity value	Refs
	6-Hydroxy-4-(2-phenylethyl)-benzofuran (**1154**)	12.7	
	(2*S*)-2-Methyl-2-(4-methyl-3-pentenyl)-7-hydroxy-5-(2-phenylethyl)chromene (**1255**)	16.1	
	o-Cannabicyclol (**1256**)	13.2	
	L-(N_6-iminoethyl)lysine (positive control)	18.6	
[L] *Radula perrottetii*	Perrottetin F (**184**)	7.4	[409]
[L] *Blasia pusilla*	Riccardin C (**182**)	> 100	[409]
[L] *Mastigophora diclados*	Riccardin F (**243**)	5	[409]
[L] *Plagiochila sciophila*	Plagiochin A (**1056**)	9.07	[409]
[L] *Riccardia multifida*	Riccardin A (**2**)	2.50	[409]
[L] *Thysananthus spathulistipus*	3β,4β:15,16-Diepoxy-13(16), 14-clerodadiene (**1151**)	20.1	[411]
	Thysaspathone (**1152**)	11.6	
[L] *Trocholejeunea sandvicensis*		Maximum inhibition rate (*MIR*)	[412]
	Tropinguisanolide A (**1157**)	83.2%/100 μ*M*	
	Ptychanolactone (**1167**)	83.5	
	Tropinguisanolide B (**1166**)	96.3	
		IC_{50} (μ*M*)	
	Tropinguisanolide A (**1157**)	100	
	Ptychanolactone (**1167**)	100	
	Tropinguisanolide B (**1166**)	100	

[a] [L] liverwort, [M] moss

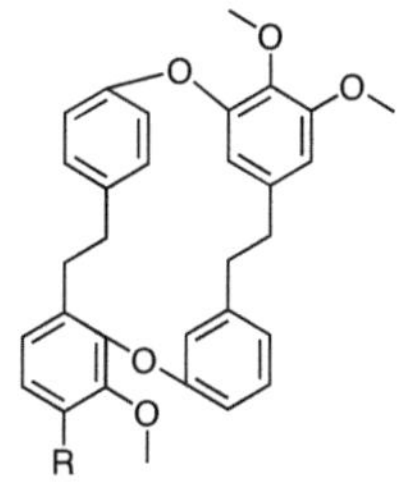

1143 (marchantin A trimethyl ether, R = H)
1144 (marchantin B tetramethyl ether, R = OMe)

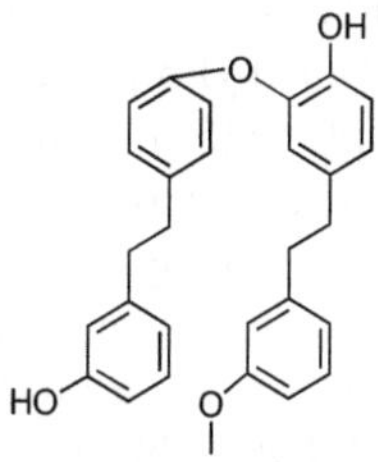

1145 (perrottetin E 11'-methyl ether)

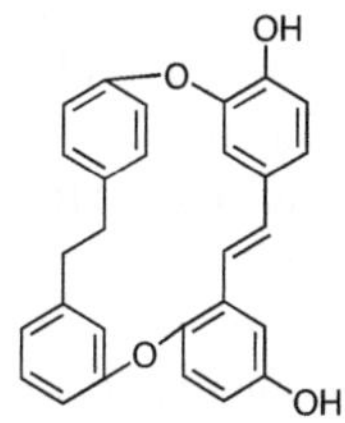

1146 (ptychantol A)

The herbertane monomers α-herbertenol (**405**), β-herbertenol (**886**), and herbertene-1,2-diol (**406**), 1,2-dihydroxyherberten-12-al (**889**), their two dimers from *Mastigophora diclados*, and 1,2-cuparenediol (**1147**) and 2-hydroxy-4-methoxycuparene (**1148**) from *Bazzania decrescens*, showed inhibition of lipopolysaccharide (LPS)-induced production of nitrite, and their individual IC_{50} values in the range 4.1–76.0 μ*M* are detailed in Table 27. Potent inhibition was observed for the herbertanes and cuparenes when the hydroxy group is located at the *meta* and/or *para* positions to C-12. An aromatic methyl group also seems to be important for this type of inhibition, while oxidation of the aromatic methyl to a formyl group decreases the inhibition of NO production [409].

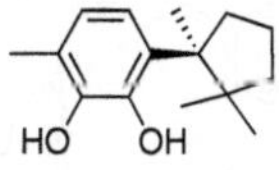

1147 (1,2-cuparenediol)

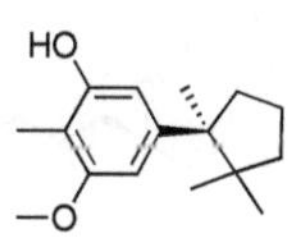

1148 (2-hydroxy-4-methoxy-cuparene)

Myltayl-4(12)-enyl 2-caffeate (**1149**), from *Bazzania nitida,* displayed the inhibition of NO production with an IC_{50} value of 6.3 μ*M* [409]. Norpinguisone methyl ester (**1114**), norpinguisone (**1115**), and *ent*-16-kauren-15-one (**1150**) isolated from *Porella densifolia* showed inhibitory activities against NO production in LPS-stimulated RAW 264.7 cells at IC_{50} values of 45.5, 1.7, and 69.4 μ*M*, respectively [410].

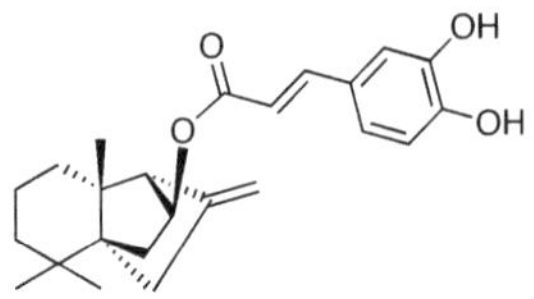

1149 (myltayl-4(12)-enyl 2-caffeate)

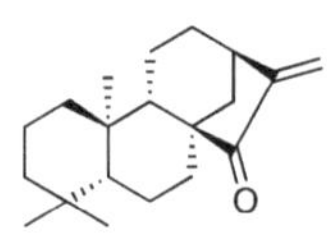

1150 (*ent*-16-kauren-15-one)

3β,4β:15,16-Diepoxy13(16),14-clerodadiene (**1151**) and thysapathone (**1152**) from *Thysananthus spathulistipus* and the bibenzyl derivatives, radulanin A (**335**), radulanin L (**336**), gymnomitr-3(15)-en-5α-ol (**1153**), 6-hydroxy-4-(2-phenylethyl)benzofuran (**1154**), (2*S*)-2-methyl-2(4-methyl-3-pentenyl)-7-hydroxy-5-(phenylethyl)chromene (**1155**), *o*-cannabicyclol (**1156**) and 2-geranyl-3,5-dihydroxybibenzyl (**899**) from *Radula appressa*, were tested for the inhibition of NO production in cultured RAW 264.7 cells in response to lipopolysaccharide (LPS). All of the bibenzyls evaluated inhibited NO production in LPS-stimulated RAW 264.7 cells with 50% inhibitory concentrations in the range of 4.5–20 µ*M*, with the best activity shown by 2-geranyl-3,5-dihydroxybibenzyl (**899**). This type of activity was suggested to be due to the antioxidant properties of these compounds [411].

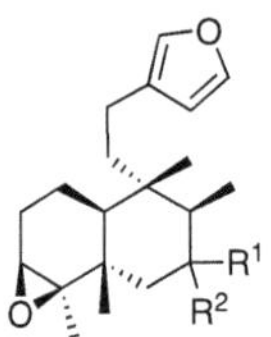

1151 (3β,4β:15,16-diepoxy-13(16),14-clerodadiene, $R^1 = R^2 = H$)
1152 (thysaspathone, $R^1 = R^2 = O$)

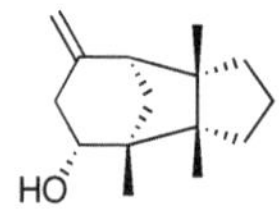

1153 (gymnomitr-3(15)-en-5α-ol)

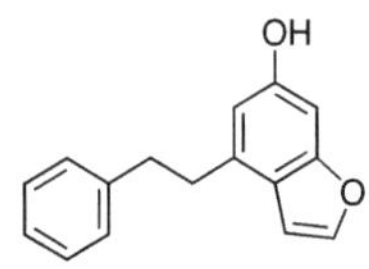

1154 (6-hydroxy-4-(2-phenylethyl) benzofuran)

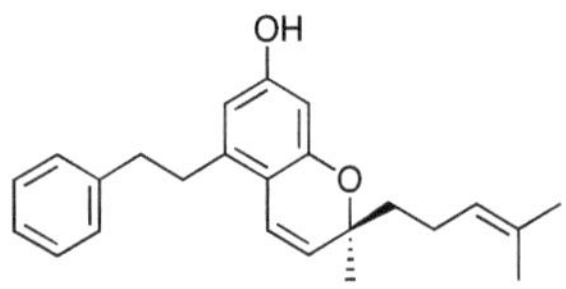

1155 ((2*S*)-2-methyl-2-(4-methyl-3-pentenyl)-7-hydroxy-5-(2-phenylethyl)chromene)

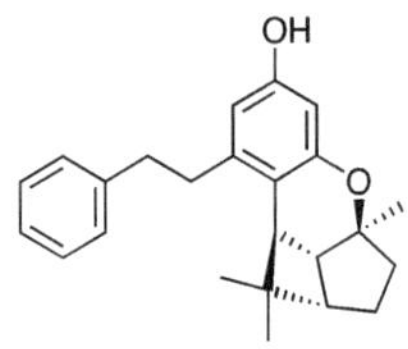

1156 (*o*-cannnabicyclol)

The liverwort *Trocholejeunea sandvicensis* is an abundant source of pinguisane sesquiterpenoids. Asakawa's group isolated 10 pinguisanes, including **1158** and **1169** from a Japanese collection of this species. Further fractionation of a Chinese sample of *T. sandvicensis* led to the isolation of nine new pinguisanes, tropinguisanolide A (**1157**), tropinguisenes A (**1159**) and B (**1160**), tropinguisanic acid (**1161**), tropinguisanols A (**1162**) and B (**1163**), tropinguisanins A (**1164**) and B (**1165**), and ptychanolactone (**1166**), along with three known pinguisane sesquiterpenoids, ptychanolactone methyl ether (**1167**), tropinguisabnolide B (**1168**) and ptychanolide (**1169**), which were isolated from different Lejeuneaceae species [2, 3]. The newly isolated compounds **1157** and **1166** and the previously known ptychonolactone methyl ether (**1167**) showed NO production inhibitory activity at an IC_{50} value of 100 μg/cm³ and with maximum inhibition rates of 83.2, 83.5, and 96.3%, respectively [412].

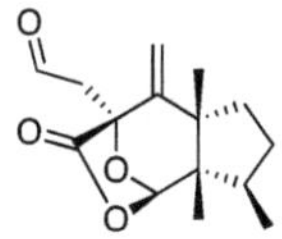

1157 (tropinguisanolide A)

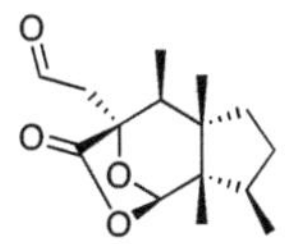

1158 (lejeuniapinguisanolide)

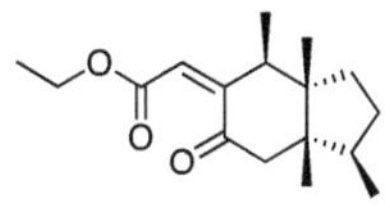

1159 (tropinguisanene A)

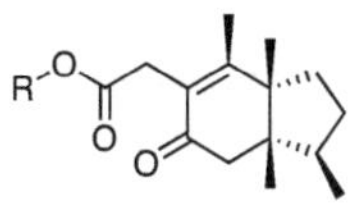

1160 (tropinguisanene B, R = H)
1161 (tropinguisanic acid, R = Et)

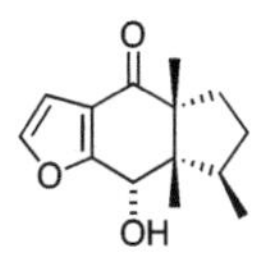

1162 (tropinguisanol B)
(= norpinguisone-7α-ol)

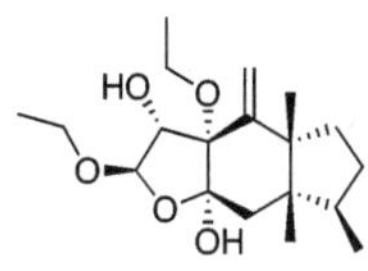

1163 (tropinguisanol A)

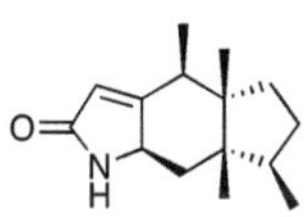

1164 (tropinguisanin A)

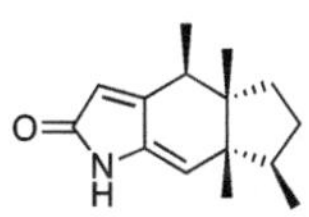

1165 (tropinguisanin B)

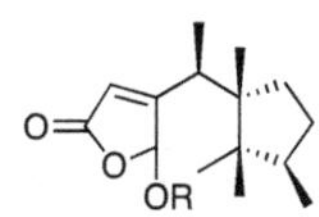

1166 (ptychanolactone, R = H)
1167 (ptychanolactone methyl ether, R = Me)
1168 (tropinguisabnolide B, R = Et)

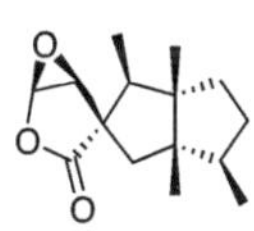

1169 (ptychanolide)

Lepidozia reptans belonging to the Lepidoziaceae elaborates various sesquiterpenoids [2, 3]. Reinvestigation of a Chinese specimen resulted in the isolation of five new diterpenoids (1*S*,3*E*,7*Z*,11*S*,12*S*)-12-hydroxydolabella-3,6-dien-6-one (**1170**), (3*E*,7*Z*,11*S*,12*S*)-12-hydroxydolabella-3,6-dien-6-one (**1171**), and (3*S*,5*S*,8*R*,10*R*,13*S*,16*R*)-3,13-dihydroxy-*ent*-kauran-15-one (**1174**) and two new sesquiterpenes, (4*R*,5*R*,7*S*)-7-hydroxy-7-isopropyl-1,4-dimethylspiro[4.4]non-1-ene-2-carbaldehyde (**1177**) and (6*E*,7*S*,10*R*)-dihydroxy-3-oxo-eudesm-(4*E*)-ene (**385**), together with several known compounds, (1*R*,6*R*,11*R*,12*R*)-6,12-dihydroxydolabella-(3*E*,7*E*)-diene (**1006**), 13α-hydroxy-*ent*-kaur-16-ene-3,15-dione (**1172**), *ent*-kaur-3,15-dione (**1175**), 15-oxo-*ent*-(16*S*)-kaurane (**1173**), *ent*-3α-hydroxykauren-15-one (**1176**), (5*S*,6*R*,7*S*,10*R*)-6,7-dihydroxy-2-oxoeudesm-(3*Z*)-ene (**376**), 1,10-epoxylepidozenal, 6,7-epoxy-3-methenyl-11,11,7-trimethyl-[8,1,0]-2-hendecene (**1178**), 6,7-dimethylmethylene-4-aldehyde-1β-hydroxy-10(15)-ene-(4*Z*)-bicyclodecylene (**1179**), and vitamin E quinone (**1180**). The absolute configurations of the two new dolabellanes, **1170** and **1171**, were determined by X-ray crystallographic analysis and ECD calculations. Compounds **1170**,

1171, **383**, **1172**, **1173**, and **1180** were tested for their potential anti-inflammatory activities in the LPS-induced model of NO production with macrophage cells using the flavonoid luteolin (**483**) as the positive control, and with the mechanism of the active compounds **1170** and **1171** then further explored. Each compound showed potential anti-inflammatory effects with an ED_{50} value between 38 and 100 μ*M*. Compound **1171** displayed a higher inhibitory activity of LPS-induced inflammatory responses than **1170** by evaluating the mRNA levels of IL-6 and IL-β [413].

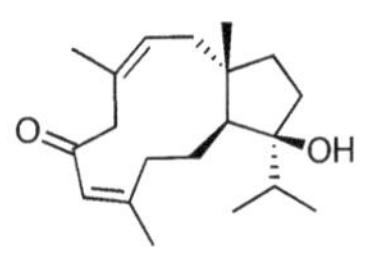

1170 (1*S*,3*E*,7*E*,11*S*,12*S*)-12-hydroxydolabella-3,7-dien-6-one)

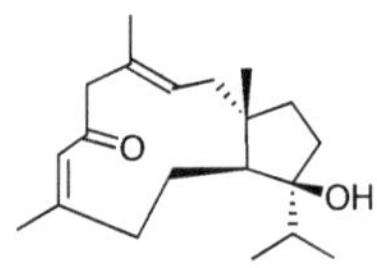

1171 (1*S*,3*E*,7*Z*,11*S*,12*S*)-12-hydroxydolabella-3,7-dien-6-one)

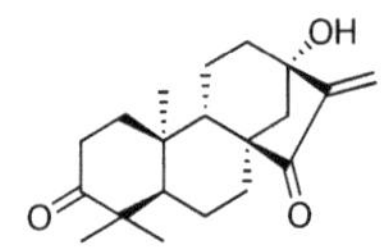

1172 (*ent*-13*β*-hydroxy-kaurene-3,15-dione)

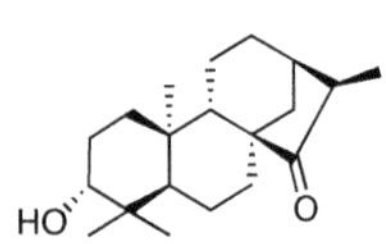

1173 (*ent*-3*β*-hydroxy-(16*S*)-kauran-15-one)

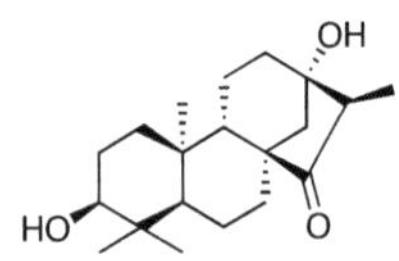

1174 (*ent*-3*α*,13*β*-dihydroxy-(16*S*)-kauran-15-one

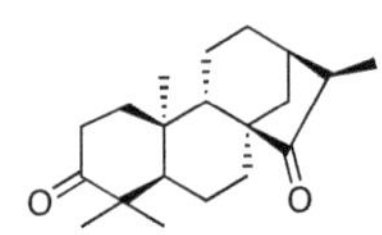

1175 (*ent*-16-kaurene-3,15-dione)

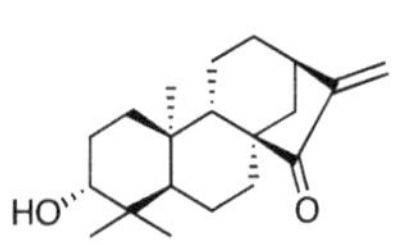

1176 (*ent*-3*β*-hydroxykaurene-15-one)

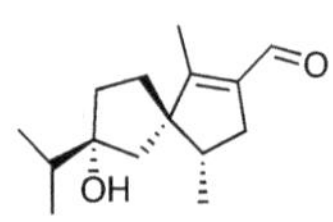

1177 (4*R*,5*R*,7*S*)-7-hydroxy-7-isopropyl-1,4-dimethyl-spiro[4.4]non-1-ene-2-carbaldehyde)

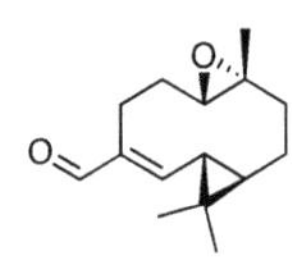

1178 (6,7-epoxy-3-methenyl [8.1.0]-2-hendecene) (= 9*α*,10*β*-epoxy-4-formylbicyclogermacrene)

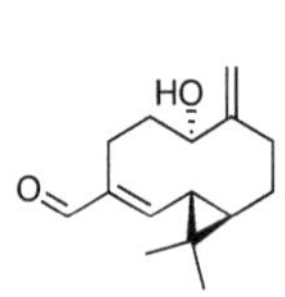

1179 (6,7-dimethyl-methylene-4-formyl-1*α*-hydroxy-10(15)-ene-(4*Z*)-bicyclo-decylene) (= 4-formyl-10*α*-hydroxy-bicyclogermacra-(3*E*),9(15)-diene)

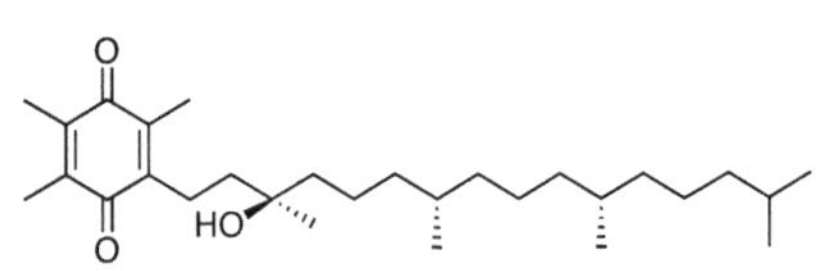

1180 (vitamin E quinone)

The ethanol (96%) and ethyl acetate extracts from the moss, *Hypnum cupressiforme*, were found to contain gallic (**796**), protocatechuic (**797**), 5-*O*-caffeoylquinic (**798**), caffeic (**800**), and *p*-coumaric (**801**) acids, and the flavonoids, apigenin (**482**), eriodictyol (**805**), naringenin (**806**), kaempferol (**807**), and acacetin (**808**) as well as quercetin (**851**) and isorhamnetin 3-*O*-glucoside (**804**). They inhibited NO production of LPS-stimulated BV2 cells, as measured by the nitrite levels in cell-free supernatants [272].

In 2021, the ethanol extracts of the mosses, *Ceratodon purpureus*, *Dryptodon pulvinatus*, *Hypnum cupressiforme*, *Rhytidiadelphus squarrosus*, and *Tortula muralis* were tested against human foreskin fibroblast HFF-1 proliferation and for their wound-healing propensities by Wolski et al., Among these five mosses, the ethanol extract from *T. muralis* showed the best wound- healing effects [275].

An 80% ethanol extract containing flavonoids from *Marchantia convoluta* showed protective effects against the acute hepatic injury of mice caused by CCl_4. High doses of the extract (0.2 g/kg) and of the positive control (aspirin 0.2 g/kg) inhibited the auricle tympanites of mice caused by dimethylbenzene. The inhibition rate of this crude extract was 21.8% [296].

Antiinflammatory and antinociceptive effects of eight Turkish liverworts, *Corsinia coriandrina*, *Mannia androgyna*, *Plagiochasma rupestre*, *Porella cordaeana*, *P. platyphylla*, *Reboulia hemisphaerica*, *Riccia fluitans*, and *Targionia hypophylla* were investigated in connection with the major volatile components present, especially the sesquiterpenoids. Carrageenan-induced paw edema and *p*-benzoquinone-induced abdominal constriction animal models were used for the activity assessment. The ether extracts of *P. cordaeana*, *P. rupestre*, and *R. hemisphaerica* exhibited significant inhibitory activities on carrageenan-induced paw edema in mice with inhibition values of 25.4, 27.2, and 29.4%, respectively. Also, the *P. cordaeana*, *P. platyphylla*, *P. rupestre*, and *R. hemisphaerica* ether extracts showed promising suppressive effects in a *p*-benzoquinone-induced abdominal constriction model, with values of 27.6, 31.5, 44.6, and 41.1%, respectively [414].

In Himachal Pradesh, Kangra, India, the fresh thallus of *Plagiochasma appendiculatum* has been used ethnomedically for the treatment of burns, boils, and blisters on the body and skin eruptions caused by the hot sunlight in summer. The active compounds have not yet been identified. This anti-inflammatory activity might result from the presence of bis-bibenzyls, like members of the marchantin series found in this species [415].

The liverwort *Mylia nuda* belonging to the Myliaceae, produces not only several mono-, sesqui-, and diterpenoids, but also the bis-bibenzyl, isomarchantin C (**264**) and two flavone di-*C*-glucosides, apigenin-6,8-di-*C*-glucoside (**1181**) and luteolin-6,8-di-*C*-glucoside (**1182**) [2, 3]. Fractionation of the ethanol extract of the Chinese *M. nuda* resulted in the isolation of seven 1,10-*seco*-aromadendrane-benzoquinone-type heterodimers, mylnudones A–G (**1183–1189**), and the rearranged aromadendrane sesquiterpenoid **1190**. Also, four known aromadendrane sesquiterpenoids were obtained, myltayloriones A (**1191**) and B (**1192**), α-taylarione (**1193**) and myliol (**1194**), which were previously isolated from the Japanese *Mylia taylorii* [2]. The structures of the new compounds **1183–1190** were established by NMR and ECD spectroscopy and single-crystal X-ray diffraction analysis. All of the isolated compounds (**1183–1194**) were evaluated for their cytoprotective effects in SH-SY5Y cells using a glutamic acid-induced injury model. Among the test compounds, **1183–1189** showed cytoprotective effects, while **1186** exhibited the most potent effect.

Pretreatment with DMSO (control), glutamic acid (5 m*M*), or compound **1186** (10 μ*M*), and their combinations showed compound **1186** to improve cell morphology and survival viability as well as decreased ROS levels in glutamic acid-induced SH-SY5Y nerve cells. These results suggested that compound **1186** evokes protection against glutamic acid-induced neurological deficits [416].

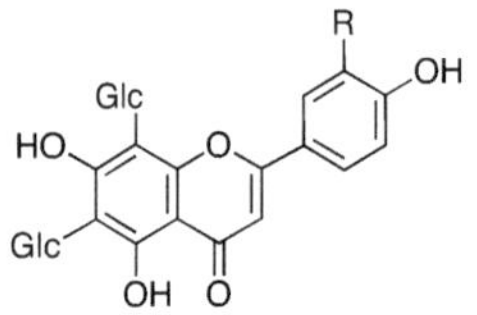

1181 (apiginin-6,8-di-*C*-glucoside, R = H)
1182 (luteolin-6,8-di-*C*-glucoside, R = OH)

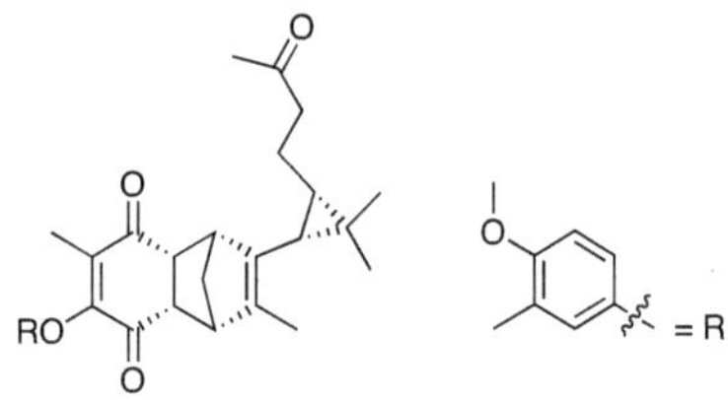

1183 (mylnudone A)

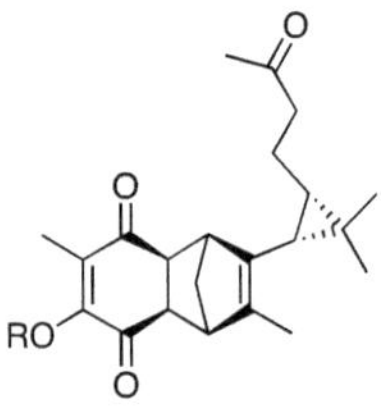

1184 (mylnudone B)

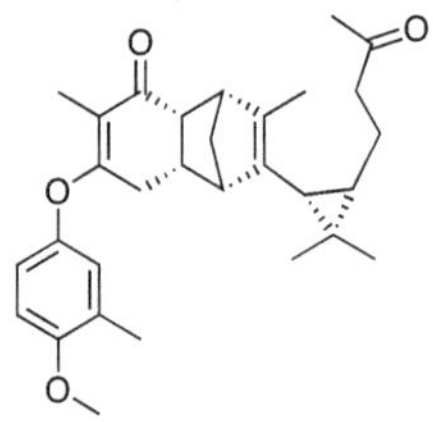

1185 (mylnudone C)

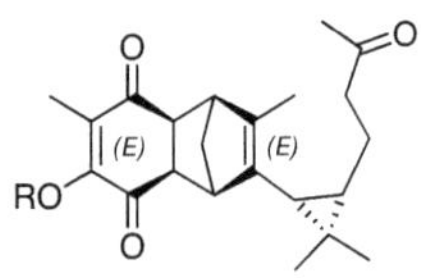

1186 (mylnudone D)

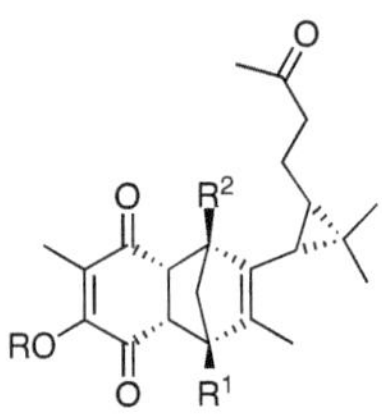

1187 (mylnudone E, R^1 = H, R^2 = Me)
1188 (mylnudone F, R^1 = Me, R^2 = H)

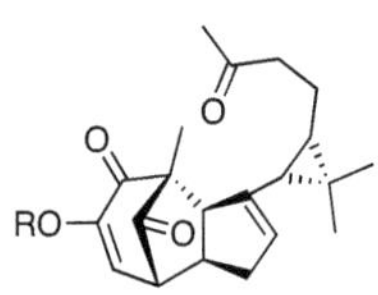

1189 (mylnudone G, R = 3-methoxy-4-methylbenzene)

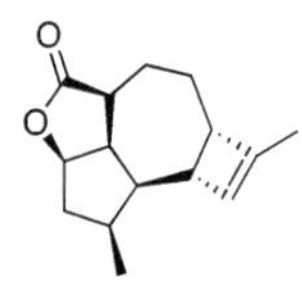

1190 (rearranged aromadendrane)

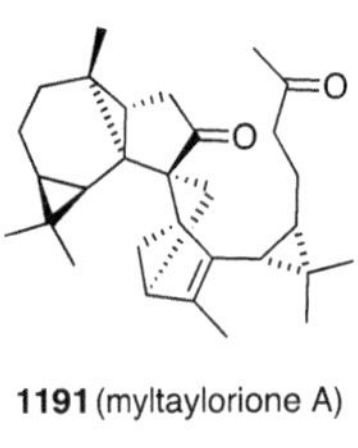

1191 (myltaylorione A)

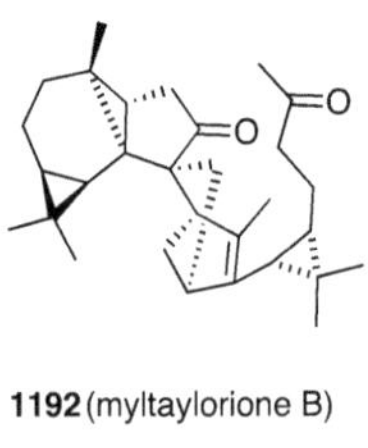

1192 (myltaylorione B)

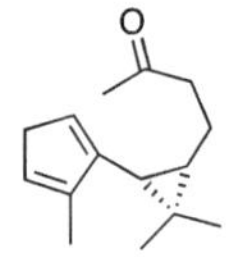

1193 (α-taylarione)

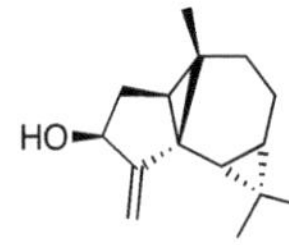

1194 (myliol)

Three solvent extracts (96% ethanol, ethanol/water 1/1, ethyl acetate) of the moss, *Hedwigia ciliata* each showed reduction of NO production by LPS-stimulated BV-2 cells and hence suggested their potential antineuroinflammatory activity. The ROS production also increased in LPS- stimulated BV-2 cells after treatment with the solvent extracts at a selected concentration (10 μg/cm^3) [279]. Inhibition was also observed in the MDA-MB-231 cell line at selected concentrations (20–30 μg/cm^3). These results suggested that selected solvent extracts of *H. ciliata* have the potential for increasing antioxidant mechanisms in MDA-MB-231 cells including mediating ROS production [401].

In 2021, Kim et al., studied the potential anti-inflammatory effects of the methanol extracts of the Korean liverwort, *Marchantia polymorpha* and the moss, *Racomitrium canescens*, in LPS-induced HaCaT human keratinocyte cells. To evaluate their effects, the levels of nitric oxide production (NO), and the mRNA expression of inducible nitric oxide synthase (iNOS), cyclooxygenase-2 (COX-2) and tumor necrosis factor-α (TNF-α), and interleukins (IL)-6 and -1β in LPS-induced HaCaT cells were measured. The methanol extracts of *M. polymorpha* and *R. canescens* decreased the production of NO by LPS-stimulated HaCaT cells in the medium to 7.0 and 11.6 μ*M*, respectively, at a concentration of 100 mg/cm^3. Production of NO was also inhibited by ethyl acetate and *n*-butanol fractions of the methanol extract of *M. polymorpha* at concentration ranges of 2–20 μg and 1–40 μg/30–100 μg/cm^3 of the extracts by LPS-stimulated HaCat cells, respectively [417]. The cellular factors iNOS and COX-2 and their reaction products are connected closely with inflammatory diseases. The LPS-induced production of all of TNF-α, IL-6 and IL-1β in HaCaT cells was suppressed effectively after treatment with a methanol extract of *M. polymorpha.* In constrast, the methanol extract of *R. canescens* did not show any significant effect on the production of TNF-α. When taken altogether, these data showed that the methanol extracts of both bryophytes can protect HaCaT cells against LPS-induced cell injury. Although the active compounds from the methanol extracts of *M. polymorpha* and *R. cancescens* were not isolated, marchantin G (**200**), and the di-*O*-*β*-D-glucopyranoside **1196** of pinoresinol (**1195**) were identified by LC-MS [417]. As mentioned earlier, compounds in the marchantin series, marchantins A–E (**1, 195, 197–199**), H (**940**), and isomarchantin C (**264**), as well as other marchantin A derivatives, have shown potent NO production inhibition [408].

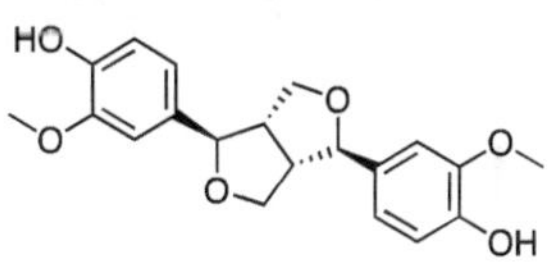

1195 (pinoresinol)

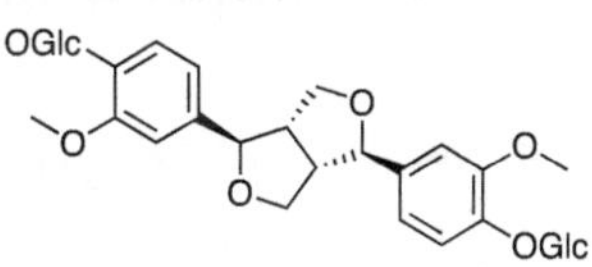

1196 (pinoresinol-di-*O*-*β*-D-glucopyranoside)

4.21 Acetylcholinesterase and Tyrosinase Inhibitory Activities

Acetylcholinesterase (AChE) is a major enzyme that hydrolyzes acetylcholine, a key neurotransmitter for synaptic transmission, into acetic acid (**45**) and choline. Mild inhibition of this enzyme has been shown to possess therapeutic relevance in Alzheimer's disease. Tyrosinase is a copper-containing enzyme that is distributed widely in various organisms and plays an important role in melanogenesis. Thus, the discovery of tyrosinase inhibitors is attractive in the cosmetic field for their role as depigmentation agents. In order to obtain AChE and tyrosinase inhibitors, certain solvent extracts and purified compounds from bryophytes have been bioassayed (see Table 28).

ent-Longipinane-type sesquiterpenoids have been found among liverworts in *Marsupella* species belonging to the Marsupellaceae [2, 3]. Fractionation of a 95% ethanol extract of the Chinese *Marsupella alpina* led to the isolation of six longipinanes, marsupellins A–F (**1197–1202**), along with the known acetoxymarsupellone (**415**), (–)-*ent*-12*β*-hydroxylongipinan-3-one (**1203**), and 9-acetoxymarsupellol (**1204**). The absolute structure of marsupellin A (**1197**) was determined by a combination of X-ray crystallographic analysis and ECD calculations. All of the longipinanes (**1197–1204**) were evaluated for their acetylcholinesterase inhibitory activity using a bioautographic TLC assay. The minimum quantities of **1197** and **1198** required for AChE inhibition were 0.6 and 0.8 μg, respectively, as compared to galanthamine (0.004 μg) used as a control. The other compounds (**1199–1202**) possessed weak AChE inhibition (4 to 10 μg). In a microplate test, marsupellins A (**1197**) and B (**1198**) showed moderate AChE inhibition (23.1% and 25.9% at 5 μ*M*) [418].

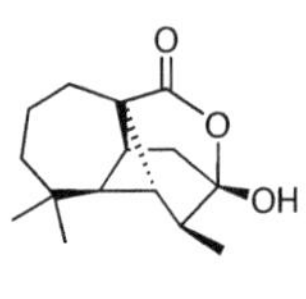

1197 (marsupellin A)

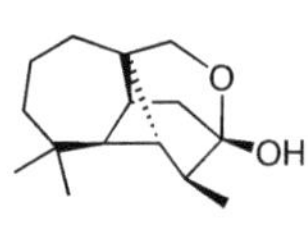

1198 (marsupellin B)

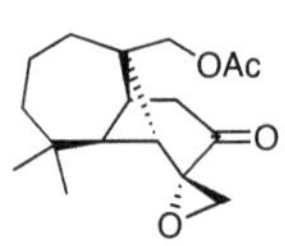

1199 (marsupellin C)

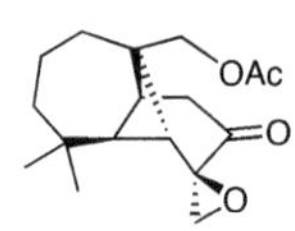

1200 (marsupellin D)

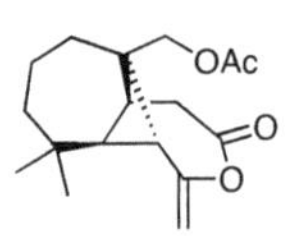

1201 (marsupellin E)

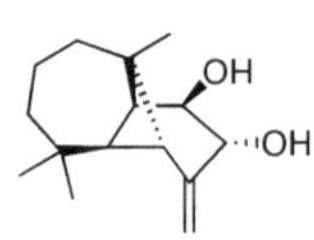

1202 (marsupellin F)

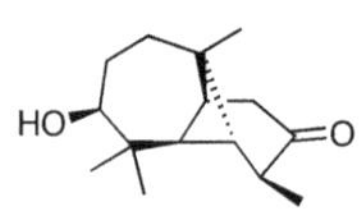

1203 ((–)-*ent*-12*β*-hydroxylongipinan-3-one)

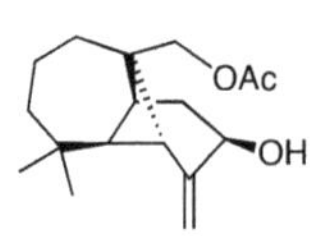

1204 (9-acetoxymarsupellpellol)

Marchantin A (**1**) isolated from *Marchantia polymorpha* did not affect dopachrome formation during an enzymatic reaction using tyrosinase. Thus, subject to additional testing, **1** might prove useful as an active compound in a chemopreventive formulation for melanoma, in having no adverse effects on skin hyperpigmentation [136]. In 2016, Wang et al., identified tentatively the flavonoid content of 80%

Table 28 Acetylcholinesterase, tyrosinase, ROS, TNF-α, and IL-1β and IL-6 secretion inhibitory compounds from bryophytes

Species name[a]	Solvent extract	Activity	Activity value	Refs.
[M] *Hedwigia ciliata*		AChE inhibitory	(% inhibition at 1000 μg/cm^3)	[279]
	EtOAc		60	
	EtOH/H_2O (1:1)		15	
	96% EtOH		10	
	Galanthamine (control)		80	
[M] *Hypnum cupressiforme*	EtOAc H_2O	AChE inhibitory	10 mg/cm^3	[272]
[M] *Hypnum cupressiforme*	EtOAc	ROS TNF-α IL-6 secretion inhibitory		[401]
[L] *Marsupella alpina*	Marsupellin A (**1197**)	AChE inhibitory	Bioautography (μg)/ microplate (5 μ*M*) 0.5/23.0%	[418]
	Marsupellin B (**1198**)		0.8/25.9%	
	Marsupellins C–F (**1199–1202**)		4–10	
[L] *Marchantia polymorpha* (Archegoniophore)	Flavonoids	AChE inhibitory	IC_{50} (μg/cm^3) 125.6	[388]
[L] *Marchantia polymorpha* [M] *Racomitrium canescens*	EtOH	iNOS IL-6 IL-1β TN F-α	100 mg/cm^3	[417]
[L] *Scapania undulata*	Scapaundulin A (**685**)	AChE inhibitory	*MIQ* (minimum inhibitory quantity) (ng) 240	[239]

(continued)

Table 28 (continued)

Species name[a]	Solvent extract	Activity	Activity value	Refs.
	Scapaundulin C (**687**)		250	
	5α,8α,9α-Trihydroxy-(13*E*)-labdan-12-one (**688**)		250	
	3α,8α-Dihydroxy-(13*E*)-labden-12-one (**689**)		240	
	(13*S*)-15-Hydroxy-labd-8(17)-en-19-oic acid (**690**)		500	

[a] [L] liverwort, [M] moss

ethanol extracts of the archegoniophore and the gametophyte of *Marchantia polymorpha* using LC-DAD-ESI/MS data. The former sample contained apigenin (**482**), kaempferol 3-*O*-rutinoside (**844**), apigenin-7-*O*-*β*-D-glucoside (**1205**), baicalein-6,7-di-*O*-*β*-D-glucoside (**1206**), chrysoeriol-7-*O*-neohesperioside (**1207**), luteolin-3′-glucuronide (**1208**), and the latter sample produced chrysoeriol (**800**), kaempferol (**807**), tricin-7-*O*-rutinoside (**1209**), and chrysoeriol (**1210**). Both extracts were evaluated against AChE and each gave an IC_{50} value of 125.6 μg/cm^3 [388].

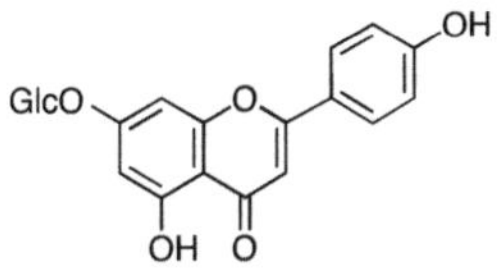

1205 (apigenin-7-glucoside)

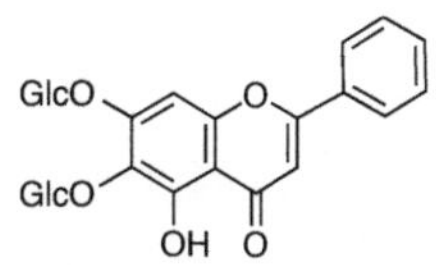

1206 (baicalein-6,7-di-*O*-glucoside)

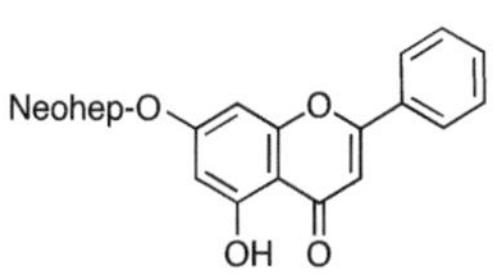

1207 (chrysoeriol-7-*O*-neohesperioside)

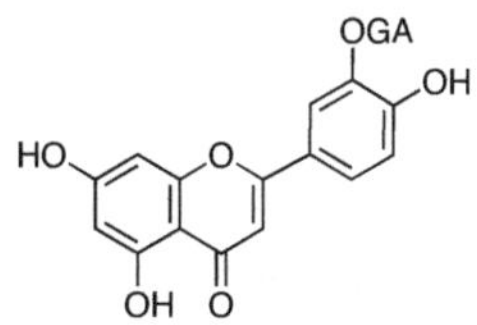

1208 (luteolin-3'-glucuronide)

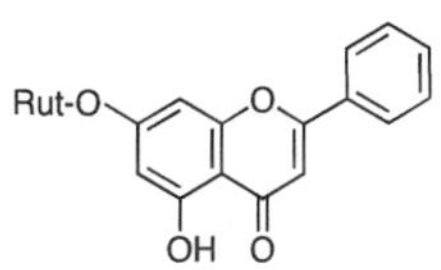

1209 (tricin-7-*O*-rutinoside

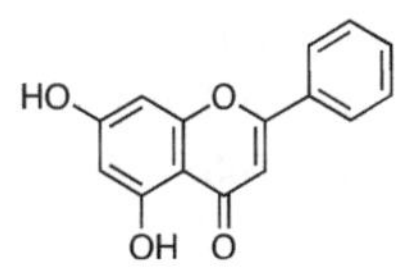

1210 (chrysoeriol)

The total phenolic, phenolic acid, flavonoid, and triterpenoid constituents of the moss, *Hypnum cupressiforme*, were measured using a Multiskan Sky Thermo Scientific microtiter plate reader, and AChE- and tyrosinase-inhibitory activities of a crude extracts containing phenolic compounds and triterpenoids were evaluated. The ethyl acetate and water extracts of *H. cupressiforme* (Plate 32) had AChE (10 μg/cm^3) inhibitory activity. The ethanol, ethanol/water (1/1), ethyl acetate and water extracts of this moss possessed tyrosinase inhibitory activity at this same concentration, which was was 2.5 times more potent than the inhibition by a control, kojic acid [272].

In 2022, Lunic et al., performed a study on the neuroprotective activity of *Hypnum cupressiforme*. The antioxidant activity was measured using a β-carotene bleaching assay, while the MTT, NBT, ELISA, and Griess assays were carried out to explore the anti-neuroinflammatory and neuroprotective potential of the ethyl acetate extract of this moss. Inhibitory activities against acetylcholinesterase (AChE) and tyrosinase were assessed experimentally and by molecular docking analysis. It was found that the secondary metabolites present in the ethyl acetate extract inhibited the secretion of ROS, NO, TNF-α, and IL-6, alleviating the inflammatory potential of H_2O_2, and LPS in microglial and neuronal cells. Potent inhibitory activities against AChE and

Plate 32 *Hypnum cupressiforme* (moss)

tyrosinase were observed in vitro. Molecular docking simulations suggested that the flavonoids present exhibited inhibitory potential by interacting with the active sites in AChE, and the hydroxy cinnamic acid derivatives that occurred showed the best affinities against tyrosinase [401].

Scapania undulata biosynthesizes labdane monomers and dimers, some of which possess a potently bitter taste [2, 3]. A Chinese collection of this species elaborated scapaundulin A (**685**) and scapaundulin C (**687**), together with the three known cytotoxic labdanes, 5α,8α,9α-trihydroxy-(13*E*)-labdan-12-one (**688**), 5α,8α-dihydroxy-(13*E*)-labden-12-one (**689**), and (13*S*)-15-hydroxylabd-8(17)-en-19-oic acid (**690**). These were evaluated for their AChE inhibitory activity using a bioautographic TLC assay. The minimum inhibitory quantities of **685** and **687**–**690** required for the inhibition of AChE inhibition were determined as 240, 250, 250, 240, and 500 ng, respectively [239].

The *n*-hexane-, ethyl acetate-, and ethanol-soluble solvent extracts of naturally grown and cultured samples of the moss, *Bryum argenteum*, were evaluated for their inhibitory activities of tyrosinase. At a concentration of 2 mg/cm^3, all the solvent extracts showed inhibitory activities that were compared with a positive control, kojic acid, at a 200 μg/cm^{3}concentration. The extracts from the cultured samples showed more potent activities than those from the field samples. The *n*-hexane extract of the cultured sample was the most potent inhibitor, followed by those obtained with ethyl acetate and ethanol [402].

Denotarisia linguifolia was extracted with *n*-hexane, chloroform, ethyl acetate, and ethanol. At a concentration of 2 mg/cm^3, inhibition of the solvent extracts against tyrosinase was below 50%, indicating that the tyrosinase inhibitory activity of the extracts was weak. The ethanol extract showed an inhibitory percentage of 18.2% [262].

The ethyl acetate, 96% ethanol, and ethanol/water (1/1) extracts of *Hedwigia ciliata* showed moderate to low anti-AChE inhibitory activities in comparison with the positive control galanthamine. Among these, the ethyl acetate extract possessed the highest inhibitory effect (60%) at 1 g/cm^3 [279]. The same three solvent extracts of *H ciliata* were also evaluated for their tyrosinase inhibitory activities. Among them, the 96% ethanol extract possessed a significant effect against tyrosinase with a higher inhibition potency at a concentration of 1 mg/mg^3 observed than for the positive control kojic acid [279].

Marchantia polymorpha was extracted with *n*-hexane, chloroform, ethyl acetate, and ethanol, respectively, and the extracts were evaluated their tyrosinase inhibitory activities. The ethyl acetate extract showed the highest inhibitory activity (65%), when compared to the positive control used (kojic acid 100%) [261].

The ethanol/H_2O extracts of four *Sphagnum* species, *Sphagnum girgensohnii*, *S. magellanicum*, *S. palustre* and *S. squarrosum* were evaluated for their tyrosinase inhibitory activities. Tyrosinase gene expression was inhibited in a statistically significant manner by all of the tested *Sphagnum* extracts, in comparison to the relative expression of this gene in cells that were not treated by the extracts. The highest inhibition of tyrosinase gene expression was seen for the *S. palustre* extract [281].

4.22 *Neurotrophic Activity*

Mastigophorenes A (**885**), B (**1211**), and D (**408**) that occurred in *Mastigophora diclados*, along with other compounds such as 4-*epi*-sandaracopimaric acid (**1212**), exhibited neurotrophic properties at 10^{-5}–10^{-7} *M*, greatly accelerating neuritic sprouting and network formation in a primary neuritic cell culture derived from a fetal rat cerebral hemisphere [419, 420]. Plagiochilal B (**137**) and plagiochilide (**442**)

Table 29 Neurotrophic compounds from liverworts

Species name	Compound	Activity value	Refs.
Jungermannia infusca		Neurite bundle formation	[75]
	Infuscaside A (**148**)	10^{-7} *M*	
	Infuscaside B (**150**)	10^{-7} *M*	
Mastigophora diclados	Mastigophorene A (**885**) Mastigophorene B (**1212**) Mastigophorene D (**408**)	10^{-5}–10^{-7} *M* (accelerating neurite sprouting and network formation in a primary neuritic cell culture derived from the fetal rat hemisphere)	[419, 420]
Plagiochila fruticosa	Plagiochilal B (**137**) Plagiochilide (**442**) Plagiochin A (**1056**)	10^{-5} *M* 10^{-6} *M* 10^{-5} *M*	[421]

from *Plagiochila fruticosa* showed acceleration of neurite sprouting, and enhancement of acetylcholine transferase activity in a neuronal cell culture of a fetal rat cerebral hemisphere at 10^{-5} *M*. Plagiochin A (**1056**) showed the same activity at 10^{-6} *M* [421] (Table 29).

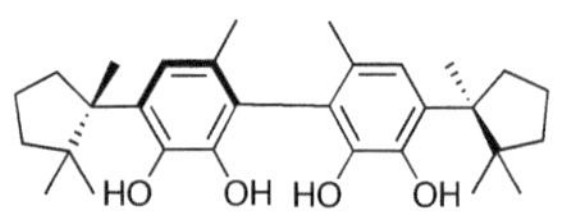

1211 (mastigphorene B)

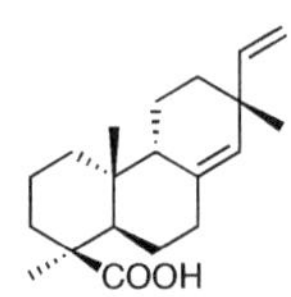

1212 (4-*epi*-sandarocopimaric acid)

Two bitter-tasting diterpene glucosides, infuscasides A (**149**) and B (**150**), obtained from *Jungermannia infusca,* exhibited neurite bundle formation activities at a concentration of 10^{-7} *M* [75].

4.23 Muscle Relaxant and Calcium Inhibitory Activities

Marchantin A (**1**) and related macrocyclic bis-bibenzyls are structurally similar to the bisbibenzylisoquinoline alkaloid, *d*-tubocurarine, which has clinically useful skeletal muscle relaxant activity. It was found that marchantin A (**1**) and its trimethyl ether (**1143**) also show muscle relaxant activities [422]. Although the mechanism of action of marchantin A (**1**) and its ether **1143** in effecting muscle relaxation is still unknown, it is interesting that these macrocyclic bis-bibenzyls, in possessing no nitrogen atoms in their molecules, cause a concentration-dependent decrease of contraction of the rectus abdominus in frogs.

Marchantin A and its trimethyl ether also had muscle relaxant activity using a frog model. Nicotine in Ringer's solution causes a maximum contraction of the rectus abdominus in frogs at a concentration of 10^{-6} *M*. After pre-incubation of marchantin A trimethyl ether (**1143**) (at a concentration of 2×10^{-7}–2×10^{-4} *M*) in Ringer's solution, nicotine (10^{-5}–10^{-4} *M*) was added. At a concentration of 10^{-6} *M*, the contraction of the rectus abdominus in frogs decreased by about 30% [422]. *d*-Tubocurarine exhibited similar effects to **1443** [423]. Using MM2 calculations, this indicated that the conformation of marchantin A (**1**) and its trimethyl ether and the presence of an *ortho* hydroxy group in **1** and an *ortho* methoxy group in the trimethyl ether contribute to the muscle relaxant activity [422]. In contrast, marchantin A triacetate and 7′,8′-dehydromarchantin A, and acyclic bis-bibenzyls such as perrottetin E (**183**) and F (**184**), did not show muscle-relaxant activity [422].

A structural similarity between marchantin A (**1**) and the therapeutically important bisbibenzylisoquinoline alkaloid, cepharanthine (**1213**), was also pointed out by Keseru and Nogradi in 1995 [423]. They predicted that the similar biological properties of **1** and cepharathine could be attributed to their binding to a common receptor. Included were antibacterial activity against Gram-negative and positive

Table 30 Muscle-relaxant and calcium inhibitory compounds from liverworts

Species name	Compound	Active value (muscle contraction decrease for rectus abdominus frogs (RAF) and mice)	Refs.
Marchantia polymorpha	Marchantin A (**1**) Marchantin A trimethyl ether (**1143**) (synthetic)	(for RAF mice) 30% at 10^{-6} *M*	[422]
		Calcium inhibitory activity (μg/cm^3)	
Marchantia polymorpha	Marchantin A (**1**) Riccardin A (**2**)	1.9 20.0	[423–425]

bacteria and against human H37Rv *Mycobacterium tuberculosis, M. avium,* and *M. nocardia*, cytotoxic activity against various human and murine cancer cells, and the inhibition of 5-lipoxygenase and calmodulin. The wide range of biological activities of marchantin A (**1**) could also be interpreted by a mechanism of action based on calcium-ion binding [424]. Conformational analysis of marchantin A (**1**) and riccardin A (**2**) was carried out by a systematic unbounded multiple minimum search. The mobility of the macrocyclic rings of both compounds was analyzed in a variable temperature (20–100°C) ^{1}H NMR study. The results indicated the restricted mobility of the macrocyclic ring of riccardin A (**2**) and gave further evidence for its more rigid nature in comparison to marchantin A (**1**) (Table 30). Comparison of the calcium inhibitory activity of **1** and **2** (ID_{50}: 1.9 and 20 μg/cm^3) implied a reduced affinity of **2** for calcium ions, which was consistent with the calculated differences in steric and electrostatic properties of **1** and **2**. Thus, introduction of a biphenyl linkage to the macrocyclic ring decreases its mobility and this might be responsible for the above-mentioned reduction of biological activity of **2** when compared to **1** [425].

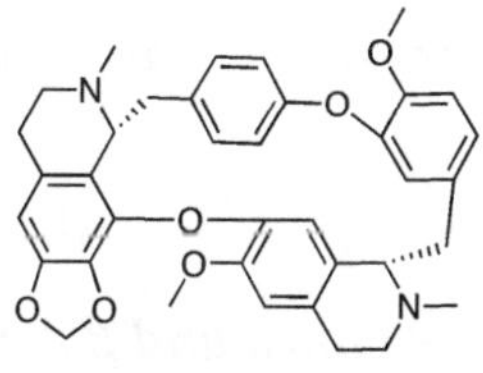

1213 (cepharanthine)

4.24 Cardiotonic and Vasopressin Antagonist Activities

Marchantin A (**1**) shows cardiotonic activity in frogs via an increased coronary blood flow of 2.5 cm^3/min at 0.1 mg. 3-Hydroxy-5-methoxy-4-(3-methyl-2-butenyl)-bibenzyl (**900**) from *Radula complanata* exhibited vasopressin antagonist activity

Table 31 Cardiotonic compound from liverworts

Species name	Compound	Activity value	Ref.
Marchantia polymorpha	Marchantin A (**1**)	2.5 cm^3/min at 0.1 mg (coronary blood flow)	[38]

Table 32 Vasopressin-antagonist compounds from liverworts

Species name	Compound	Activity value (ID_{50} $\mu g/cm^3$)	Refs.
Radula complanata	3-Hydroxy-5-methoxy-4-(3-methyl-2-butenyl)bibenzyl (**900**)	27	[38]
	3-Hydroxy-5-methoxy-4-geranylbibenzyl (**1214**) (synthetic)	17	[38]
	2-Geranyl-3-hydroxy-5-methoxy-bibenzyl (**1215**) (synthetic)	57	

(ID_{50} = 27 $\mu g/cm^3$) (Table 31). The two synthetic prenyl bibenzyls, 3-hydroxy-5-methoxy-4-geranylbibenzyl (**1214**) and 3-hydroxy-5-methoxy-2-geranylbibenzyl (**1215**) also showed vasopressin antagonist activities, with ID_{50} values of 17 and 57 $\mu g/cm^3$, respectively (Table 32) [38].

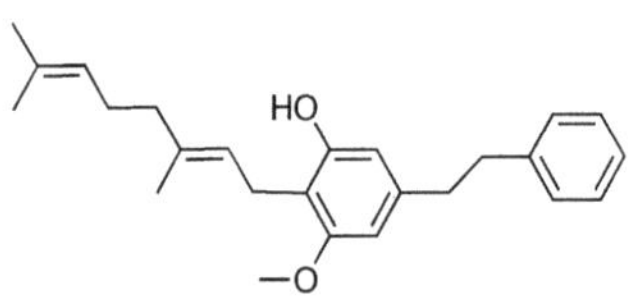

1214 (3-hydroxy-5-methoxy-4-geranylbibenzyl)

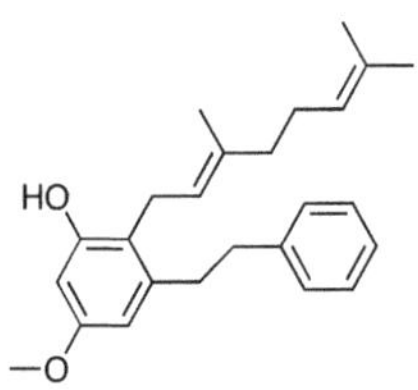

1215 (3-hydroxy-5-methoxy-2-geranylbibenzyl)

4.25 Liver X-Receptor α Agonist and β Antagonist Activities

The plasma high density lipoprotein level is inversely related to the risk of atherosclerotic cardiovascular disease. In a search for agents that increased high density lipoprotein production, riccardin C (**182**) and riccardin F (**243**), from *Reboulia hemisphaerica* (Plate 33) and *Blasia pusilla*, were shown to function as a liver X-receptor α agonist and a liver X-receptor β antagonist, respectively. Riccardin C (**182**) increases the plasma HDL level without elevating triglyceride levels in mice and may provide a novel tool for identifying subtype-function and for drug development against obesity [426]. From 1.25 kg of the dried liverwort *Blasia pusilla*, 1 g of pure riccardin C (**182**) was obtained [75].

Plate 33 *Reboulia hemisphaerica* (liverwort)

Table 33 Liver X-receptor (LXR)α-agonist and (LXR)β–antagonist active compound from liverworts

Species name	Compound	Effect	Ref.
Reboulia hemisphaerica	Riccardin C (**182**)	Plasma high density, lipoprotein increasing effect	[426]

Seven *O*-methyl derivatives of riccardin C (**182**) were synthesized, including riccardins A (**2**) and F (**243**). According to a preliminary structure-activity relationship study of these seven *O*-methylated riccardins, the three phenolic hydroxy groups of riccardin C (**182**) were determined as being responsible for binding to the LXRα receptor (Table 33) [427].

4.26 Cathepsins B and L Inhibitory Activities

Cathepsin B belongs to a family of lysosomal cysteine proteases known as the cysteine cathepsins, and it plays an important role in intracellular proteolysis, while cathepsin L is correlated with osteoporosis [428] and allergy [429]. The laboratory of the author of this volume currently is searching for enzyme inhibitors from natural sources to develop potential chemopreventive drugs for these diseases. The marchantin series showed inhibitory activities for both cathepsins B and L. Isomarchantin C (**264**), which is found in *Marchantia* and *Dumortiera* species, was the most potent inhibitor against both these enzymes (95% for cathepsin L and 93% for cathepsin B, at 10^{-5} *M*). In contrast, infuscaic acid (**1113**) exhibited lower activity levels (63% and 32% at 10^{-5} *M*, respectively) [430]. Bioassay-guided fractionation of the crude extract of *Porella japonica*, which inhibited both cathepsins B and L, afforded the five new guaianolides **1216–1220**. Of these compounds, only 11,13-dehydoporelladiolide (**1219**) showed weak inhibitory activities against cathepsin B (13.4% at 10^{-5} *M*) and cathepsin L (24.7% at 10^{-5} *M*) (Table 34) [430].

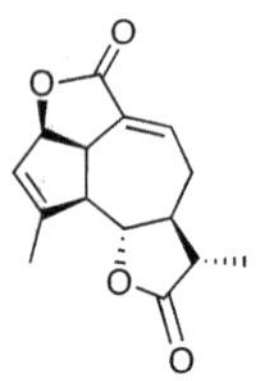

1216 (porelladiolide)

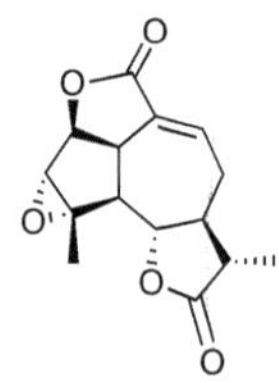

1217 (3α,4α-epoxyporelladiolide)

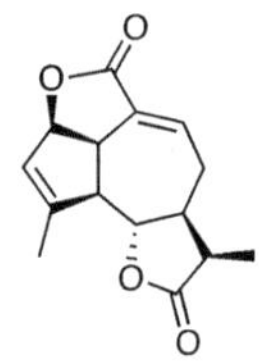

1218 (11-*epi*-porelladiolide)

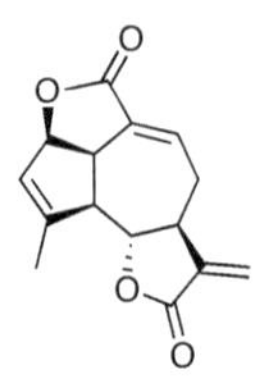

1219 (11,13-dehydroporelladiolide)

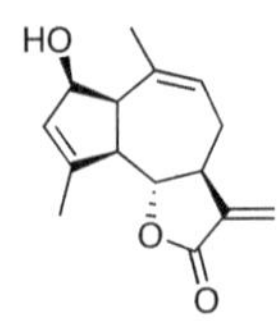

1220 (porellaolide)

Table 34 Cathepsin L and cathepsin B inhibitory compounds from liverworts

Species name	Compound	Activity value	Refs.
Bryopteris filicina *Marchantia paleacea* subsp. *diptera*	Isomarchantin C (**264**)	95 and 93% for cathepsins L and B (at 10^{-5} *M*)	[430]
Jungermannia infusca	Infuscaic acid (**1113**)	63 and 32% for cathepsins L and B (at 10^{-5} *M*)	[430]
Porella japonica	11,13-Dehydroporelladiolide (**1219**)	24.7 and 13.4% for cathepsins L and B (at 10^{-5} *M*)	[430]

4.27 Antiplatelet, Antithrombin, and Thromboxane Synthase Inhibitory Activities

Marchantiaquinone (**1221**), obtained from *Reboulia hemisphaerica*, showed 100% inhibition of antiplatelet activity at a concentration of 100 μg/cm^3 in washed rabbit platelets, when induced by each of thrombin, arachidonic acid, collagen, and platelet-activating factor (PAF) [431].

Plagiochiline C (**420**) exhibited significant antiplatelet effects on arachidonate- (100 μ*M*) (95% and 45% inhibition, respectively, at 100 and 50 μg/cm^3) and collagen-induced (10 μg/cm^3; 100% inhibition at 100 μg/cm^3 level) aggregations of washed rabbit platelets. In turn, isoplagiochilide (**443**) displayed less potent effects than plagiochiline C (**420**) [432].

Table 35 Antiplatelet, antithrombin, and thromboxane synthase inhibitory compounds from liverworts

Species name	Compound	Activity value	Refs.
Plagiochila elegans	Plagiochiline C (**420**)	Antiplatelet 100% arachidonate-induced (100 μ*M*), 95% and 45% inhibition at 100 and 50 μg/cm^3 collagen-induced (10 μg/cm^3) (100% inhibition at 100 μg/cm^3) aggregation of washed rabbit platelets	[432]
	Isoplagiochilide (**443**)	Antiplatelet inhibition: less active than plagiochiline C (**420**)	
Lepidozia fauriana	(–)-5*β*-Hydroperoxylepidozenolide (**1222**)	50% arachidonate-induced 100 μ*M*), 93% (collagen induced 10 μg/cm^3) (for rabbit platelets)	[185]
Lepidozia vitrea	Lepidozenolide (**367**)	100% arachidonate-induced 100 μ*M*) (100%, 87% (collagen induced 10 μg / cm^3) (for rabbit platelets)	[185]
Reboulia hemisphaerica	Marchantiaquinone (**1221**)	100 μg/cm^3	[431]
Jungermannia comata	Perrottetin E (**183**)	Antithrombin IC_{50} (μ*M*) 18	[433]
Lunularia cruciata and many other liverworts	Lunularic acid (**196**)	Thromboxane synthase inhibition IC_{50} 5.0 × 10^{-3} *M*	[38]

Again using washed rabbit platelets, lepidozenolide (**367**) and (–)-5*β*-hydroperoxylepidozenolide (**1222**) from *Lepidozia vitrea* and *L. fauriana* demonstrated antiplatelet effects on both arachidonate- (100 μ*M*) (100 and 50% inhibition, respectively) and collagen- (10 μg/cm^3) (87 and 93% inhibition, respectively), induced aggregations, at the 100 μg/cm^3 level. Compound **367** was the more potent of these two compounds when evaluated against the effects induced by PAF (2 ng/cm^3; 100% inhibition) [185].

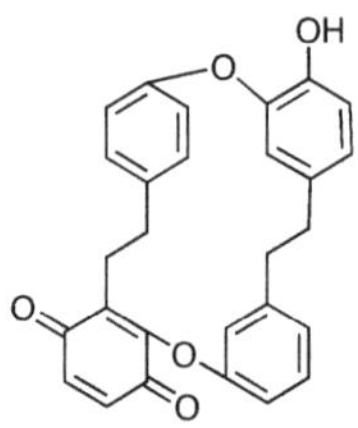

1221 (marchantiaquinone)

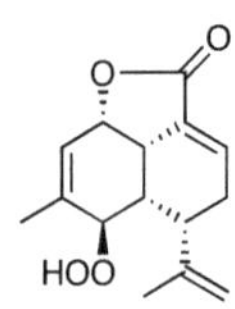

1222 ((–)-5*β*-hydroperoxylepidozenolide)

Perrottetin E (**183**), obtained from *Jungermannia comata,* showed inhibitory activity with an IC_{50} value of 18 μ*M* against thrombin, which is associated with blood coagulation [433]. Lunularic acid (**196**) exhibited thromboxane synthetase inhibitory activity (ID_{50} 5.0 × 10^{-3} *M*) (Table 35) [38].

4.28 Farnesoid X-Receptor Activation Effect

The farnesoid X-receptor, a member of the nuclear-receptor super-family, controls the expression of critical genes in bile acid and cholesterol homeostasis. Marchantins A (**1**) and E (**199**) activated farnesoid X-receptor in a receptor-binding assay at a high potency level, comparable to that of the most potent endogenous bile acid, chenodeoxycholic acid (Table 36) [434].

Table 36 Farnesoid X-receptor-active compounds from liverworts

Species name	Compound	Activity	Ref.
Marchantia polymorpha	Marchantin A (**1**) Marchantin E (**199**)	Higher activation potency in activation than chenodeoxycholic acid	[434]

4.29 Vasorelaxant Activity

Vasodilators are useful for the treatment of cerebral vasospasm and hypertension, and for improvement of peripheral circulation. Several endothelium-dependent vasodilators, such as acetylcholine, bradykinin, and histamine, elevate Ca^{2+} levels in endothelial cells and activate NO release leading to vasorelaxation. *Lepidozia fauriana* afforded as constituents lepidozenolide (**367**) and (–)-5*β*-hydroperoxylepidozenolide (**1222**), and *L. vitrea* produced the latter sesquiterpene hydroperoxide. Both compounds caused vasorelaxation of rat thoracic aorta, in terms of phasic and tonic contractions induced by norepinephrine (3 μ*M*), when evaluated at 100 μg/cm^3. Compound **367** also inhibited potassium- (80 μ*M*) and, in particular, calcium-induced (1.9 μ*M*) vasoconstriction [185].

In 2001, Morita et al., examined the vasorelaxant activity of 21 bis-bibenzyls isolated from liverworts, and found that plagiochin, riccardin, isoplagiochin, and compounds in the perrottetin series (30 μ*M*) showed potent vasorelaxation effects on rat aortic rings (85–95%). The mode of action for riccardins A (**2**), C (**182**), F (**243**), and plagiochin A (**1056**), was deduced to be mediated through the increased release of NO from endothelial cells. A vasorelaxant effect of isoplagiochin B (**266**) may be modulated by opening K^+ channels and Ca^{2+} influx through ROCs (receptor–operated Ca^{2+} channels). The activity of perrottetin F (**184**) and isoplagiochin D (**462**) may be due to internal inhibitory effects such as VDCs (voltage-dependent Ca^{+2} channels) and/or ROCs. This was the first report of vasorelaxant activity of a series of bis-bibenzyls on the rat aortic artery (Table 37) [435].

Table 37 Vasorelaxation active compounds from liverworts

Species name	Compound	Activity	Refs.
Lepidozia fauriana	Lepidozenolide (**367**)	(rat thoracic aorta) Potassium (80 μ*M*)- and calcium (1.9 μ*M*)-induced vasoconstriction	[185]
Lepidozia fauriana *Lepidozia vitrea*	5*β*-Hydroperoxylepidozenolide (**1222**)	100 μg/cm^3 (vasorelaxation of the rat thoracic aorta)	[185]
Blasia pusilla *Plagiochila sciophila* *Reboulia hemisphaerica* *Riccardia multifida* *Plagiochila fruticosa* *P. fruticosa* *Radula perrottetii*	Riccardin F (**243**) Plagiochin A (**1056**) Riccardin C (**182**) Riccardin A (**2**) Isoplagiochin B (**266**) Isoplagiochin D (**462**) Perrottetin F (**184**)	85–95% (rat aortic rings) Increased release of NO from endothelial cells Opening of K^+ channels and Ca^{2+} influx through ROCs Internal inhibitory effects of VDCs and/or ROCs	[435]

4.30 Psychoactivity

Radula species are very tiny liverworts and about 500 species are known globally. Almost all *Radula* species produce mainly bibenzyl and prenyl bibenzyls, while a few species also elaborate bis-bibenzyl derivatives [172, 173]. It is noteworthy that some *Radula* species, such as the Japanese *R. perrottetii* (Plate 34) and the New Zealand *R. marginata* biosynthesize *cis*-perrottetinene (**279**) and perrottetinenic acid (**280**), the structures of which are very similar to that of the well-known psychoactive compound, (–)-*trans*-Δ^9-tetrahydrocannabinol (**281**), obtained from the higher plant *Cannabis sativa* [436, 437]. Chicca and associates [404] found that **279** showed the same type of psychoactivity and anti-inflammatory activity as **281**. It is also interesting that the psychoactivity of **279** is weaker than **281**, although the anti-inflammatory activity of **279** is more potent than **281** (Table 38).

Plate 34 *Radula perrottetii* (liverwort)

Table 38 Psychoactive compounds from liverworts

Species name	Compound	Activity	Refs.
Radula campanigera *R. chinensis* *R. laxiramea* *R. marginata* *R. perrottetii* *Radula* species (Peruvian unidentified species)	*cis*-Perrottetinene (**279**)	Equivalent activity to that of tetrahydrocannabinol (rat)	[404, 436, 437]

Table 39 Diuretic liverwort extract

Species name	Solvent extract	Activity value	Ref.
Marchantia polymorpha	80% EtOH	Increases urine volume (100 and 200 mg/kg of *M. polymorpha*)	[296]

4.31 Diuretic Activity

The diuretic activity of the Chinese *Marchantia polymorpha* was suggested by Ding in 1982 [35]. Later, Xiao et al., in 2004 documented that an 80% ethanol extract of *Marchantia convoluta* containing the flavones, apigenin (**482**), luteolin (**483**), their glucuronides, and quercetin (**851**) showed a significant increase of the volume of urine of rats treated with 100 and 200 mg/kg, p.o. of the extract. The effect of a high dose of the extract was comparable to that of urea (750 mg/kg, p.o.) (Table 39) [296].

4.32 Irritancy and Tumor-Promoting Activities

12-*O*-Tetradecanoyl-phorbol-13-acetate (**1223**) and teleocidin (**1224**), which are potent tumor promoters, cause irritation on the mouse ear, induction of ornithine decarboxylase, and adhesion of cultured HL-60 human promyelocytic leukemia cells. Two pungent sesquiterpenoids, polygodial (**90**) and plagiochiline A (**135**), isolated from the liverworts, *Porella* and *Plagiochila* species, cause a similar type of irritancy to the mouse ear as those of the above-mentioned tumor promoters, although neither of them showed ornithine decarboxylase induction. On the other hand, the pungent sacculatal (**128**) from *Pellia* and the non-pungent plagiochiline C (**420**) from the *Plagiochila* species, which are a slightly irritant di- and sesquiterpenoid, respectively, caused ornithine decarboxylase induction. Thus, these terpenoids are potential tumor promoters. Ornithine decarboxylase induction is completely inhibited by 13-*cis*-retinoic acid (**1225**). (+)-Frullanolide (**3**) obtained from *Frullania dilatata* inhibited tumor promotion (Table 40) [38].

Table 40 Irritancy and ornithine decarboxylase-inducing compounds from liverworts

Species name	Compound	Irritancy (100 μg/ear of mouse)	ODC activity (nmol CO_2/mg protein)	Refs.
Pellia endiviifolia Trichocoleopsis sacculata	Sacculatal (**128**)	+	1.7	[38]
Plagiochila fruticosa Plagiochila ovalifolia Plagiochila semidecurrens	Plagiochiline A (**135**)	++++	0	[38]
	Plagiochiline C (**420**)	+	2.5	
Porella vernicosa complex	Polygodial (**90**)	+++	0	[38]
Sapium sebiferum[a]	TPA[b] (**1223**)	++++	7.2	[38]
Streptomyces mediocidicus[c]	Teleocidin (**1224**)	++++	3.6	[38]

[a] Higher plant (Euphorbiaceae)
[b] 12-*O*-tetradecanoyl-phorbol-13-acetate
[c] Bacterium

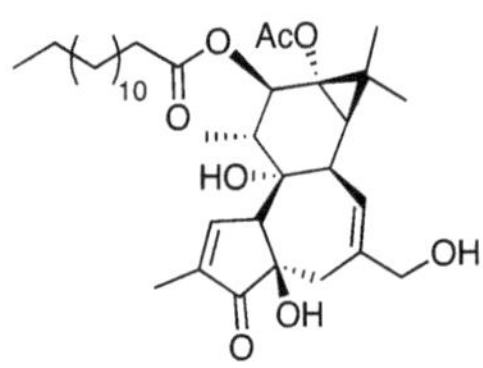

1223 (12-*O*-tetradecanoyl-phorbol-13-acetate)

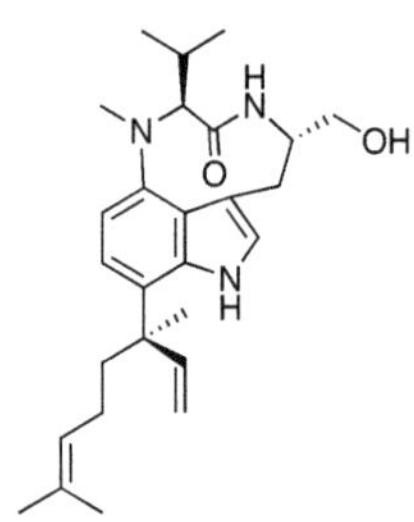

1224 (teleocidin)

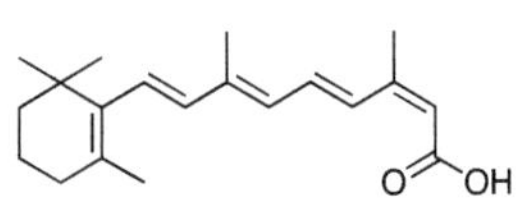

1225 ((13*Z*)-retinoic acid)

4.33 Sex Pheromones of Brown Algae from Liverworts

It is known that female gametophytes of marine brown algae, such as *Ectocarpus siliculosus, Dictyopteris membranacea,* and *Cutleria multifida,* release the sex pheromones, dictyotene (**92**), multifidene (**93**), dictyopterene (**94**), (*E*)-ectocarpene (**95**), and their related aliphatic hydrocarbon analogs, which attract their conspecific

Table 41 Sex pheromones of brown algae from liverworts

Species name	Compound	Effects	Refs.
Chandonanthus hirtellus (Tahitian sp.) (Japanese sp.)	Dictyotene (**92**) (*E*)-Ectocarpene (**95**)	Attraction of conspecific males of several brown algae, such as *Dictyopteris* sp.	[100]
Fossombronia angulosa	Dictyotene (**92**) Multifidene (**93**) Dictyopterene (**94**)	Attraction of conspecific males of several brown algae, such as *Dictyopteris* sp.	[100]

males [438]. Interestingly, dictyopterene (**94**) is also responsible for an ocean-like smell [439]. These pheromones are lipophilic and volatile acetogenins consisting of C_8 or C_{11} linear or monocyclic hydrocarbons or their epoxides [440]. Surprisingly, two liverworts produce exactly the same sex pheromones as those found in certain brown algae. *Fossombronia angulosa* from Greece produces dictyopterene (**94**) as the major component, as well as dictyotene (**92**) and multifidene (**93**) [100]. Dictyotene (**92**) and (*E*)-ectocarpene (**95**) were also found in Tahitian *Chandonanthus hirtellus* as the major volatile components (Table 41) [101]. However, the same species collected in Japan contained both pheromones only as very minor components [441]. These chemical identifications suggests that some families of liverworts and algae may have an evolutionary relationship [3].

4.34 Plant Growth-Regulatory Activity

It is known that higher plants often do not grow in places inhabited by bryophytes, especially liverworts. The present author has suggested that some liverworts produce allopathic compounds. In fact, most crude extracts from liverworts showed inhibitory activity against the germination, root elongation, and the second coleoptile growth of rice in husk, wheat, lettuce, and radish. Pungent or non-pungent sesquiterpene lactones, and the pungent sesqui- and diterpenoids, and 2,3-*seco*-aromadendrane sesquiterpene hemiacetals showed plant growth-inhibitory activities at a concentration of 25–500 ppm against rice in husk as shown in Table 42 [38].

(–)-Polygodial (**90**) inhibits the germination and root elongation of rice in the husk at 100 ppm. At a concentration of less than 25 ppm, it markedly promotes the root elongation of rice [1, 3] The synthetic (+)-isomer (**1037**) of **90** showed almost the same activity as those of the natural (–)-isomer. 3*α*,4*α*-Epoxy-5*α*-acetoxy-8*α*-hydroxy-sphenoloba-(13*E*,16*E*)-diene (**1226**), a naturally occurring but rare diterpene, as found in *Anastrophyllum minutum*, completely inhibited the shoot and root elongation of rice in husk at a concentrations between 20 and 500 ppm [2]. The four guaianolides, zaluzanins C (**1227**) and D (**368**), and 8*α*-acetoxyzaluzanins C (**369**) and D (**370**) isolated from *Wiesnerella denudata*, also inhibited the germination and root elongation of rice in husk at 100 and 50 ppm, respectively. In addition,

Table 42 Plant growth-regulatory compounds from liverworts I

Species name	Compound	Activity value	Refs.
		Germination and growth of rice in husk) Complete inhibition: inhibition: germination root elongation (ppm) (ppm)	
Anastrophyllum minutum	3α,4α-Epoxy-5α-acetoxy-8α-hydroxy-sphenoloba-(13*E*,16*E*)-diene (**1226**)	20–500	[2, 38]
Chiloscyphus polyanthos *Diplophyllum albicans*	Diplophyllolide (**132**)	200 100	[38]
Chiloscyphus polyanthos *Diplophyllum albicans*	3-Oxodiplophyllin (**1074**)	200 100	[38]
Frullania dilatata	(+)-Frullanolide (**3**) (+)-Epoxyfrullanolide (**1228**) (synthetic)	200 100	[38]
Plagiochila sp.	Plagiochiline A (**135**)	100 100	[38]
Porella perrottetiana	Perrottetianal (**158**)	500 not tested	[38]
Porella vernicosa complex	(–)-Polygodial (**90**) (+)-Polygodial (**1037**) (synthetic)	> 100 ppm < 25 ppm (root elongation of rice husk)	[38]
Wiesnerella denudata	Zaluzanin D (**368**) 8α-Acetoxyzaluzanin C (**369**) 8α-Acetoxyzaluzanin D (**370**) Zaluzanin C (**1227**)	100 50 200 50 200 50 200 50	[38]
Frullania tamarisci		Onion seeds germination (%): concentrations (mm^3/cm^3) control: 100%	[443]

(continued)

Table 42 (continued)

Species name	Compound	Activity value	Refs.
	Essential oil: alcohol-rich sample oil	100: 100 250: 100 500: 100 1000: 86	
	Essential oil: lactone-rich sample oil	100: 100% 250: 100 500: 93 1000: 82 Growth of onion roots (length of root cm) Concentrations (mm^3/cm^3)	
	Essential oil: alcohol-rich sample oil	Control: 5.7 100: 7,2 250: 7.8 500: 5.6 1000: 3.4	
	Essential oil: lactone-rich sample oil	Control: 4.0 100: 6.9 250: 6.9 500: 5.1 1000:3.2	
		Activity value (ppm) (complete inhibition of growth of leaf and root of rice seedlings)	
Lepidozia vitrea	(+)-Vitrenal (**1232**) Lepidozenal (**488**) Isobicyclogermacrenal (**1233**)	25 250 50	[38]

(continued)

Table 42 (continued)

Species name	Compound	Activity value	Refs.
Plagiochila ovalifolia	Plagiochiline A (**135**) Plagiochiline C (**420**) Ovalifolienalone Ovalifolienal (**447**)	50 25 500 50	[38]
Plagiochila semidecurrens	9*α*-Acetoxy-ovalifoliene (**449**)	25	[38]
Lunularia cruciata and many other liverworts	Lunularic acid (**196**)	1 μ*M* (germination and growth inhibition against *Lactuca sativa* (lettuce) and *Lepidium sativum* (cress) 120 μ*M* (induction of α-amylase in embryo-less barley seeds) 40 μ*M* (inhibition of the growth of the liverwort, *Lunularia cruciata* strain callus)	[38, 446]
Haplomitrium mnioides	Haplomitrenolide A (**1237**)	IC_{50} (μg/cm^3) 19.1	[442]
	Haplomitrin B (**1236**)	44.6 (inhibition of root elongation of *Arabidopsis thaliana*)	
Radula species	Radulanin A (**335**)	50 μ*M* (inhibition of seedling of *Arabidopsis thaliana*)	[444]
Plagiochila sciophila	Fusicosciophins A–E (**726–730**)	Activity value (μ*M*) (lettuce seed dormancy breaking activity) 1.5–37.5	[248]

3-oxodiplophyllin (**1074**) showed this same type of phytotoxic effect against the germination of rice in husk and its root elongation at 200 and 100 ppm. Epoxyfrullanolide (**1228**), derived from (+)-frullanolide (**3**), inhibited the germination of rice in husk, and was ten-fold more potent than the parent γ-lactone [38].

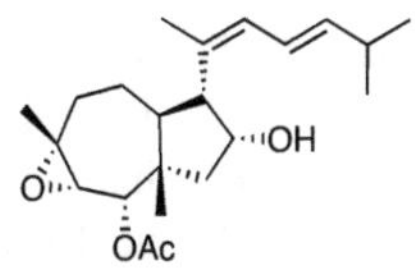

1226 (3α,4α-epoxy-5α-acetoxy-8α-hydroxy-sphenoloba-(13*E*,16*E*)-diene)

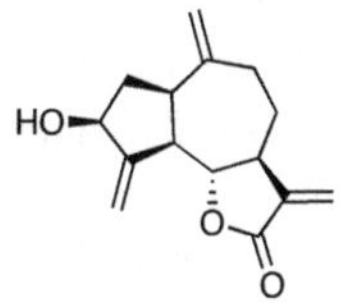

1227 (zaluzanin C)

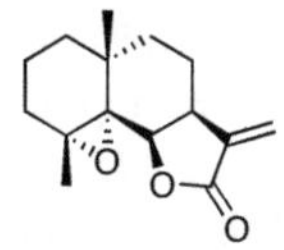

1228 (epoxyfrullanolide)

Gymnocolin A (**148**), lunularic acid (**196**), drimenol (**744**), longiborneol (**1229**), and 16α-kauranol (**1230**) inhibited the root elongation of cress. In contrast, gymnocolin A (**148**), scapanin A (**155**), drimenol (**744**), and longiborneol (**1229**) promoted the germination of wheat seeds at lower concentrations. Longifolene (**109**) promoted the coleoptile growth of wheat. In contrast, pinguisone (**1231**), an antifeedant sesquiterpene ketone, inhibited growth of wheat coleoptiles. Several members of the plagiochiline series, plagiochilines A (**135**) and C (**420**), ovalifolienal (**447**), and 9α-acetoxyovalifoliene (**449**) isolated from *Plagiochila* species, and the three sesquiterpene aldehydes, (+)-vitrenal (**1232**), lepidozenal (**488**), and isobicyclogermacrenal (**1233**) from *Lepidozia vitrea*, were found to inhibit the growth of rice seedlings [38].

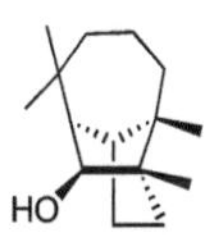

1229 (longiborneol)

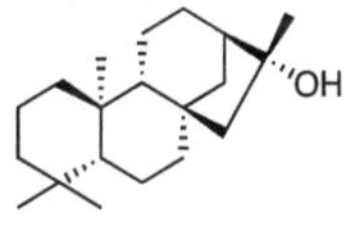

1230 (16α-kauranol)

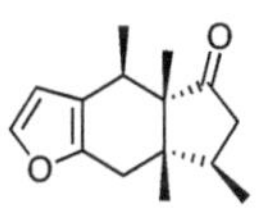

1231 (pinguisone)

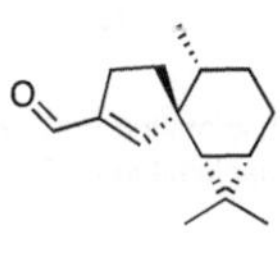

1232 ((+)-vitrenal)

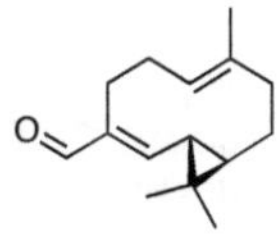

1233 (isobicyclogermacrenal)

The primitive liverwort, *Haplomitrium mnioides*, produces labdane lactones, such as haplomitrenolide A (**1237**) [2], which exhibit antiinflammatory activity, as described earlier [398]. In 2015, Zhou et al., investigated a Chinese collection of this species and isolated two new labdanes, haplomitrins A (**1235**) and B (**1236**), together with haplomitrenolide D (**1234**), along with the known haplomitrenolides A–C (**1237–1239**) [442]. A light-driven reaction of haplomitrenolides C (**1239**) and A (**1237**) gave haplomitrins A and B (**1235** and **1236**), respectively. Intramolecular cyclization of **1237** converted this to more complex congeners like haplomitrins F and G (**1240** and **1241**). The formation of **1235** and **1236** from **1239** and **1237**, raises the interesting postulate that a photochemical reaction is involved in the biosynthesis pathway. These structural features can be used as molecular markers of *H. mnioides*. The haplomitrins and haplomitorenolides are significant chemosystematic markers of *H. mnioides*, which although recognized as a very simple liverwort, [3], is chemically very complex. Haplomitrenolide A (**1237**) and haplomitrin B (**1236**) showed allelopathic activity against *Arabidopsis thaliana*, and inhibited its root elongation with IC_{50} values of 19.1 and 44.6 μg/cm^3, respectively [442].

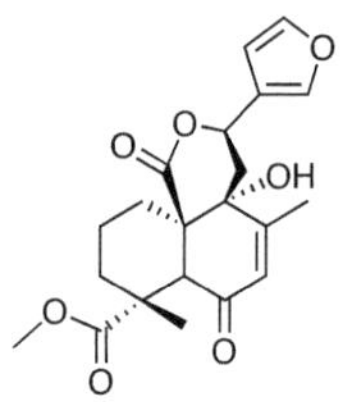

1234 (haplomitrenolide D)

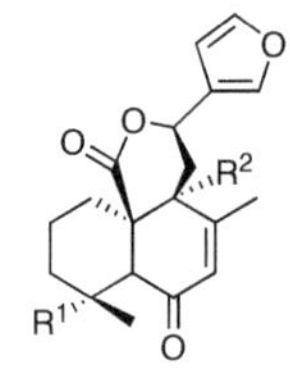

1237 (haplomitrenolide A, R^1 = Me, R^2 = H)
1238 (haplomitrenolide B, R^1 = Me, R^2 = OH)
1239 (haplomitrenolide C, R^1 = COOMe, R^2 = H)

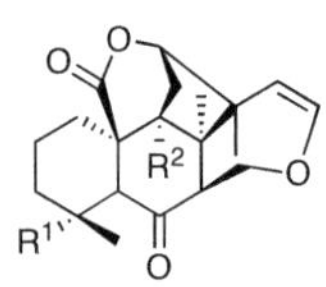

1235 (haplomitrin A, R^1 = CO_2Me, R^2 = H)
1236 (haplomitrin B, R^1 = R^2 = H)

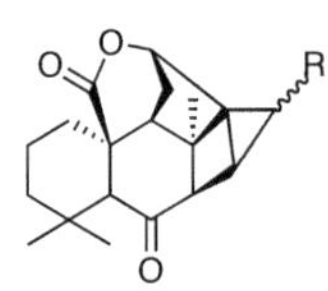

1240 (haplomitrin F, R = α-CHO)
1241 (haplomitrin G, R = β-CHO)

Further investigation of the essential oil of the Corsican *Frullania tamarisci*, collected in different localities and seasons, using GC/MS and NMR spectroscopy, led to identification of several sesquiterpenoids. These included (–)-frullanolide (**4**), tamariscol (**52**), pacifigorgiol (**1242**), α-cyclocostunolide (**1247**), and γ-cyclocostunolide (**1248**), γ-dihydrocyclocostunolide (**1249**), and germacra-1(10)(*E*,5*E*)-dien-11-ol (**1250**), as the major components, together with, as minor constituents, α-selinene (**41**), pacifigorgia-1,6(10)-diene (**1243**), pacifigorgia-1,10-diene (**1244**), pacifigorgia-1(9),10-diene (**1245**), pacifigorgia-2(10),11-diene (**1246**), and eremophilene (**1251**). Collection differences due to the season, locality, and altitude for the identified components were not seen in the Corsican sample of *F. tamarisci*. This species containing sesquiterpene alcohols and sesquiterpene lactones was tested for its effects on the germination and root elongation of onion seeds. The inhibition of onion seed germination was seen predominantly in a sesquiterpene alcohol-rich sample, while the growth stimulation of onion roots was observed for the sesquiterpene lactone-rich sample [443].

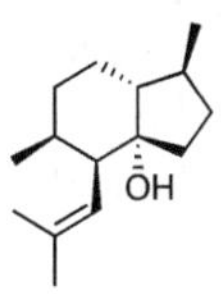

1242 (pacifigorgiol)

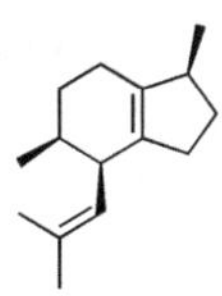

1243 (pacifigorgia-1(6),10-diene)

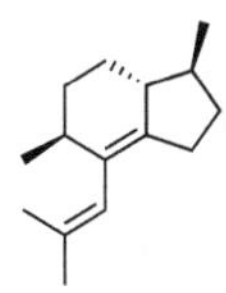

1244 (pacifigorgia-1,10-diene)

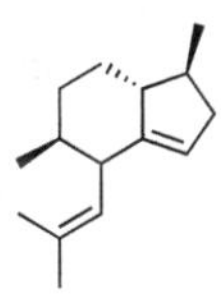

1245 (pacifigorgia-1(9),10-diene)

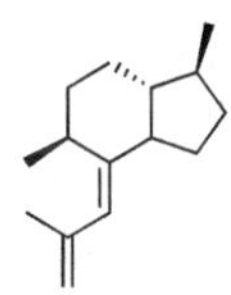

1246 (pacifigorgia-2(10),11-diene)

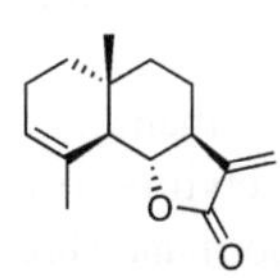

1247 (α-cyclocostunolide)

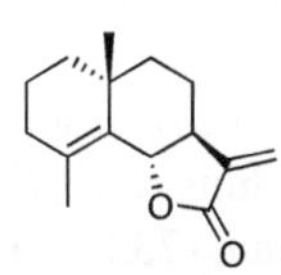

1248 (γ-cyclocostunolide)

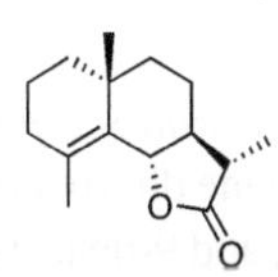

1249 (γ-dihydrocyclo-costunolide)

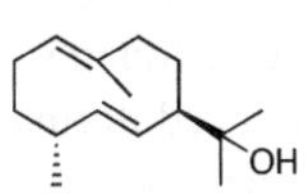

1250 (germacra-1(10)(*E*,5*E*)-dien-11-ol)

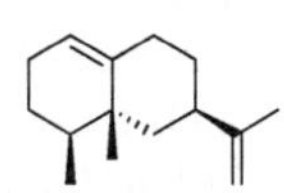

1251 (eremophilene)

Radulanin A (**335**) is a natural 2,5-dihydrobenzoxepin (accessible also by synthesis [444]) produced by several liverwort species belonging to the *Radula* genus (Radulaceae) [2, 3]. Thuillier et al., found in 2023 that radulanin A (**335**) shows phytotoxic activity on *Arabidopsis thaliana* seedlings and this activity was associated with cell death and depended partially on light exposure [445]. Radulanin A (**335**) from *Radula* species also showed potent phytotoxic effects against *A. thaliana*. It inhibited seedlings at 50 μ*M*. In order to obtain insights into the structural features required for the biological activity of **335**, 11 analogues (**1252–1262**) were synthesized and their phytotoxicity was compared using chlorosis, as a proxy for plantlet death, and this was observed with compound **335** after treatment for seven days, with concentration levels above 100 μ*M*. Prenylbibenzyls **1252** and **1253**, found in some *Radula* species, as well as the dihydro products **1255** from **1352** and **1256** from **335**, each exhibited a similar effect. On the other hand, dihydropinosylvin (**1261**) and a prenylated chromene (**1259**) were less active, with a minimum concentration of 150 μ*M* being required to induce chlorosis. Chromenes **1257** and **1260** induced bleaching only by using a higher test dose. The *O*-methylated dihydropinosylvin (**1261**), the *O*-substituted compounds **1254** and **1258**, and the chromene **1260** were not regarded as phytotoxic. On the basis of the above obervations, the bioactivity of radulanin A (**335**) requires the presence of a hydroxy group and is modulated by the its heterocyclic ring and aliphatic sidechain (Table 43).

Photosynthesis measurements based on chlorophyll a fluorescence indicated that radulanin A (**335**) and the *Radula* chromene, 2,2-dimethyl-7-hydroxy-5-(2-phenylethyl)chromene (**1135**), inhibited photosynthetic electron transport, with IC_{50} values of 95 and 100 μ*M*, respectively. Thermo-luminescence studies showed that compound **335** targeted the Q_8 site of the photosystem II, having a similar mode of action to 3-(3,4-dichorophenyl)-1,1-dimethylurea. The same authors suggested that the identification of an easy-to-synthesize analog of radulanin A (**335**), with an equivalent mode of action and efficacy, might be useful for future herbicide development [445].

Plagiochila sciophila is a rich source of sesquiterpenoids, bis-bibenzyls and the fusicoccane diterpenoids, fusicosciophins A–E (**726–730**), 8-deacetylfusicosciophin E (**731**), and 9-deacetylfusicosciophin E (**732**). The lettuce seed dormancy breaking-activities of compounds **726–732** were evaluated in the presence of the dormancy inducing phytohormone abscisic acid (**1263**). The drimenol compounds **726–732**, when evaluated in the presence of compound **1263**, were found to possess moderate activity (1.5–37.5 μ*M*). Of these, compound **731** induced dormancy breaking at a high concentration of 37.5 μ*M*, after 60 h of incubation [248].

Table 43 Plant growth regulatory compounds from liverworts II

Species name	Compound	Inhibition of cress root (*M*)	Promotion of wheat seeds (*M*)	Promotion of wheat coleoptiles (*M*)	Refs.
Bazzania trilobata	Drimenol (**744**)	10^{-3}–10^{-4}	10^{-6}–10^{-7}		[38]
Gymnocolea inflata	Gymnocolin (**148**)	10^{-3}–10^{-4}	10^{-6}–10^{-7}		[38]
Lunularia cruciata	Lunularic acid (**196**)	10^{-3}–10^{-4}	10^{-6}–10^{-7}		[38]
Lunularia cruciata	Lunularic acid (**196**)	1 μ*M* (cress and lettuce germination and growth)			[446]
Scapania undulata	Scapanin A (**155**) Longiborneol (**1229**) Longifolene (**109**)	10^{-3}–10^{-4}	10^{-6}–10^{-7} 10^{-6}–10^{-7}	10^{-4}–10^{-7}	[38]
Aneura pinguis	Pinguisone (**1231**)	Seedling	–	Inhibition of wheat coleoptiles	[38]
Radula sp.		*Arabidopsis thaliana*		IC_{50} (μ*M*) (inhibition of photosynthetic electron transport)	[445]
	Radulanin A (**335**)			95	
	2,2-Dimethyl-7-hydroxy-5-(2-phenylethyl)chromene (**1135**)			100	

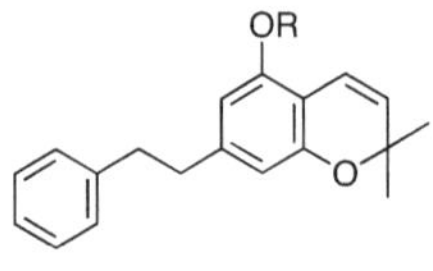

1252 (2,2-dimethyl-5-hydroxy-7-(2-phenylethyl) chromene, R = H)
1254 (2,2-dimethyl-5-methoxy-7-(2-phenylethyl) chromene, R = Me)

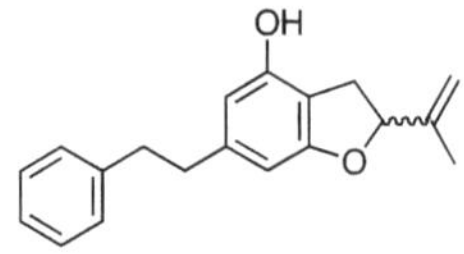

1253 (tylimantin B)

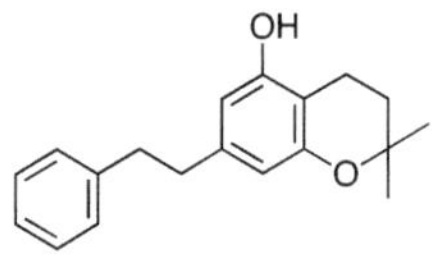

1255 (2,2-dimethyl-5-hydroxy-7-(2-phenylethyl)chromane)

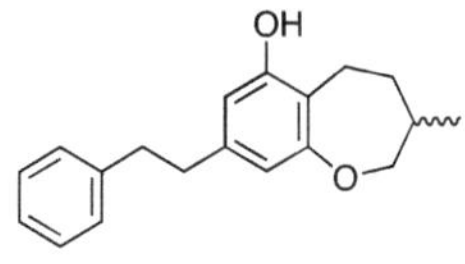

1256 (dihydroradulanin A)

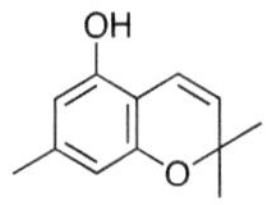

1257 (2,2-dimethyl-5-hydroxy-7-methyl)chromene)

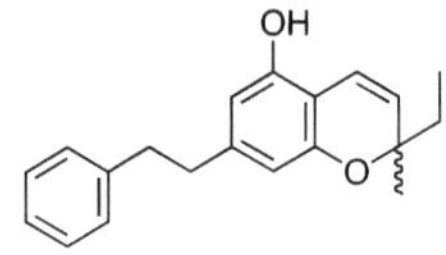

1258 (2-methyl-2-ethyl-5-hydroxy-7-(2-phenylethyl)chromene)

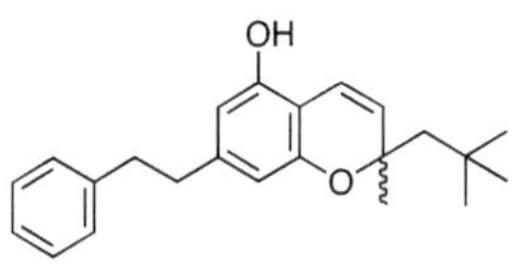

1259 (2-methyl-2-(2,2-dimethylpropyl)-5-hydroxy-7-(2-phenylethyl)chromene)

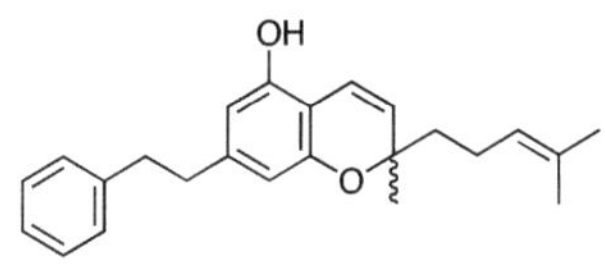

1260 (2-methyl-2-(4-methyl-3-pentenyl)-5-hydroxy-7-(2-phenylethyl)chromene)

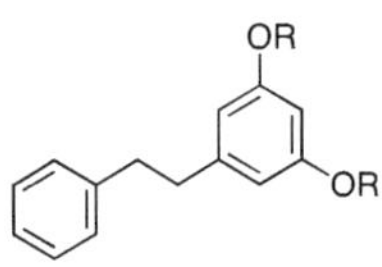

1261 (dihydropinosylvin, R = H)
1262 (dihydropinosylvin dimethyl ether, R = Me)

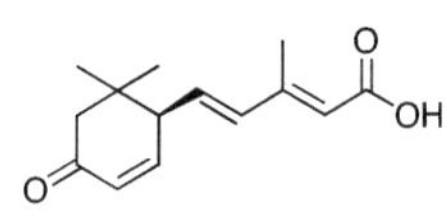

1263 (abscisic acid)

Lunularic acid (**196**) was found to inhibit the germination and growth of cress (*Lepidium sativum*) and lettuce (*Lactuca sativa*) at 1 μ*M*, and gibberellic acid-induced α-amylase induction in embryo-less barley seeds at 120 μ*M*, which is recognized as a specific activity of abscisic acid (**1263**) [446]. Compounds **196** and **1263** inhibited equally the growth of the liverwort *Lunularia cruciata* callus at 40 and 129 μ*M*. Superimposition between the stable conformers of **196** and those of **1263** obtained by computational analysis (molecular mechanics calculations), has been used to explain why **196** has abscisic acid-like activity in higher plants. This indicates that both **196** and **1263** probably bind to the same receptor. On the basis of the distribution pattern of **196** and **1263**, a hypothesis has been proposed that higher plants may have altered their endogenous growth regulator from lunularic acid (**196**) to abscisic acid (**1263**) in their evolutionary process [446].

5 Synthesis of Bioactive Compounds

As mentioned in each section provided on biological activity, there are many structurally different terpenoids and aromatic compounds in the bryophytes and those from other organisms, and several bioactive compounds have been synthesized, in particular several sesqui- and diterpenoids, bi-benzyls, and bis-bibenzyls.

5.1 Synthesis of Sesquiterpenoids

The cuparane-type sesquiterpenoid grimaldone (**65**), possessing a strong pleasant odor, isolated from the thalloid liverwort, *Mannia fragrans*, has an interesting carbon framework with trimethylcyclopropane and bicyclo[3.1.0]hexane sub-units, and three quaternary chiral carbon atoms. Thus, this compound has attracted the interest of organic chemists, and, in 2000, Srikrishna and Ramachary carried out the total synthesis of racemic grimaldone (**65**) starting from Hagemann's ester, as shown in Scheme 13 [447].

Racemic frullanolide (**3** + **4**), which causes potent allergenic contact dermatitis, was isolated from the liverworts, *Frullania tamarisci*, *F. asagrayana*, and *F. dilatata.* It was synthesized by Yoshikoshi et al., in 1990, as shown in Scheme 14 [448].

Since pinguisane-type sesquiterpenoids were found in the liverwort *Aneura pinguis*, more than 60 pinguisanes and their analogs were found not only in the Metzgeriales, but also in the Jungermanniales of the Marchantiophyta [1–3]. To date, these have not yet been found in any other organisms. The pinguisane sesquiterpenoids are of interest for the total synthesis of natural products, since they have a novel tricyclic furan ring with a *cis*-junction between the five- and six-membered rings and the four adjacent *cis*-located methyl groups. Pinguisone (**1231**) exhibts insect antifeedant and plant growth-inhibitory activities. Bernasconi et al., reported in 1981 the stereoselective total synthesis of pinguisone (**1231**) starting from the known (+)-(*S*)-2,3,7,7α-tetrahydro-7α-methyl-6*H*-indene-1,5-dione, as shown in Scheme 15 [449].

In 1985, Uyehara et al., also accomplished the total synthesis of racemic pinguisone (**1231**) and racemic deoxopinguisone (**1264**), by photochemical transformation of bicyclo[3.2.2]non-6-en-1-one to bicyclo[4.3.0]-non-4-en-7-one [450].

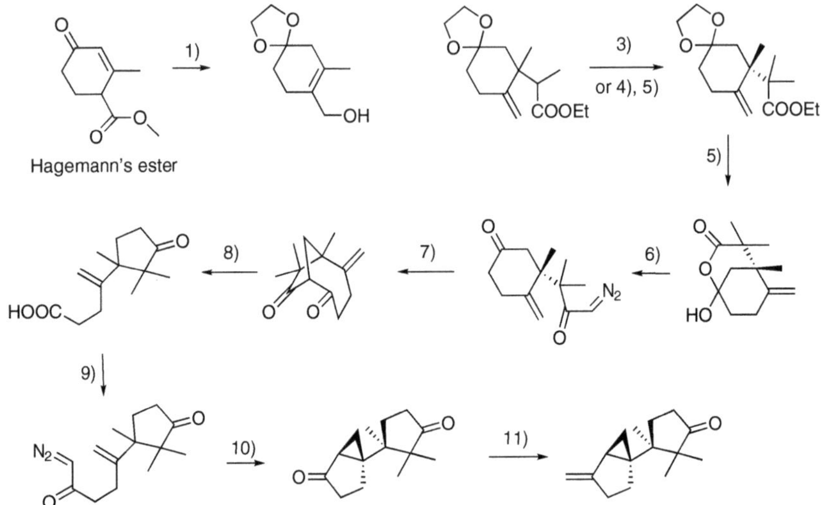

65 (racemic grimaldone)

Scheme 13 Total synthesis of racemic grimaldone (**65**). (1) $(CH_2OH)_2$, (2) $LiAlH_4$, (3) $EtC(OEt)_3$, (4) $MeC(OEt)_3$, (5) MeI, (6) KOH, (7) CH_2N_2, (8) BF_3, (9) NaOH, (10) Cu, $CuSO_4$, (11) Ph_3P^+MeI, *t*-AmO^-K^+

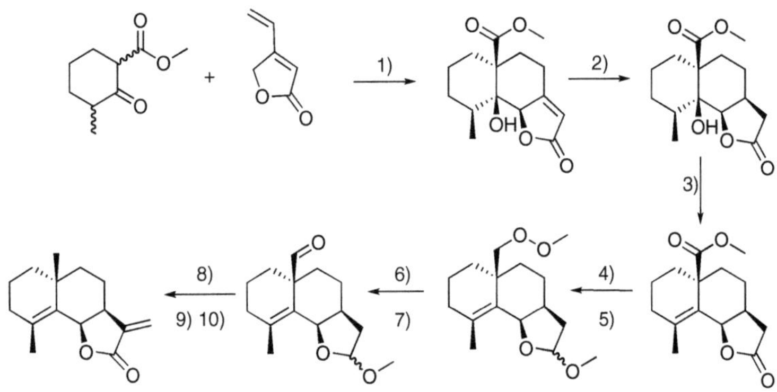

3 + **4** (racemic frullanolide)

Scheme 14 Synthesis of racemic frullanolide (**3+4**). (1) THF, (2) $NaBH_4$-$NiCl_2$, (3) $SOCl_2$/Py, (4) diisobutylaluminum hydride, (5) trimethyl *ortho*-formate, (6) $LiAlH_4$, (7) Collin's reagent, (8) Huang-Minlon reagent, (9) Jones′ reagent, (10) Grieco-Hiroi reagent (lithium diisopropylamide-formaldehyde, methanesulfonyl chloride, DBU)

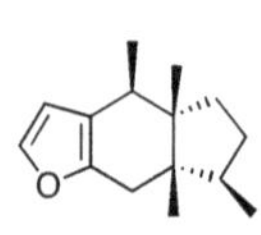

1264 (deoxopinguisone)

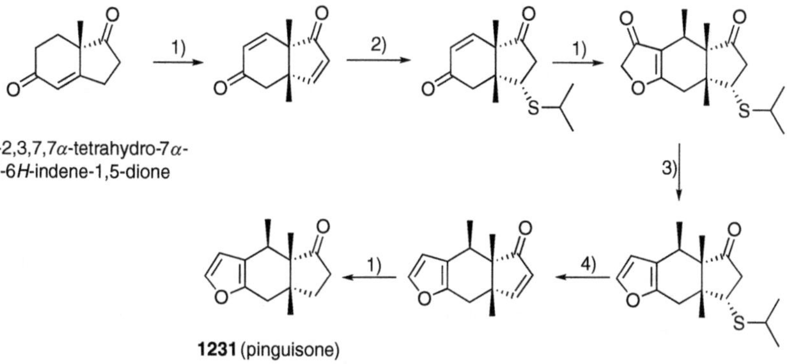

(+)-(S)-2,3,7,7α-tetrahydro-7α-methyl-6H-indene-1,5-dione

1231 (pinguisone)

Scheme 15 Synthesis of pinguisone (**1231**). (1) Me_2CuLi, (2) Me_2CHSH, (3) 9-borabicyclononane, (4) $NaIO_4$

Mastigophora diclados produces the herbertane dimers, mastigophorenes A (**885**) and B (**1212**), possessing neurotrophic activity. Both compounds were obtained through the biotransformation of herbertene-1,2-diol (**406**) using *Penicillium sclerotiorum* [451]. Fukuyama et al., accomplished in 2001 the total synthesis of mastigophorenes A (**885**) and B (**1212**) [452].

The *Plagiochila* genus of liverworts are abundant sources of 2,3-*seco*-aromadendrane sesquiterpenoids, of which several possess potent cytotoxicity against human cancer cell lines, and display insect-antifeedant, piscicidal, and allopathetic activities. Their structures are synthetically quite interesting since they contain hemiacetal and epoxy functionalities in their tricyclic and pentacyclic frameworks. Several organic chemists have noted the challenge represented by their total synthesis, but, to date, only one example of a total synthesis of this series of compounds has been reported in the literature. Thus, (+)-plagiochiline N (**431**) was synthesized from santonin (**1265**) as the starting material, as shown in Scheme 16 [453].

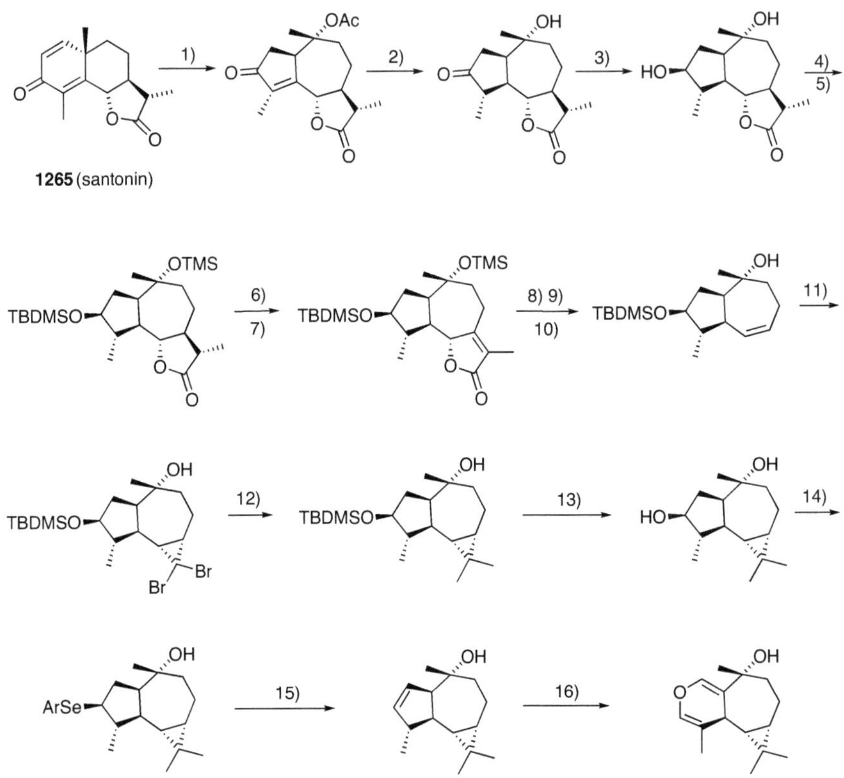

Scheme 16 Synthesis of (+)-plagiochilin N (**431**) from santonin (**1265**). (1) *hv*, AcOH, (2) a: NaTeH, b: KOH, c: HCl, (3) $NaBH_4$, (4) TBDMsCl-imidazole, (5) $TMSCl\text{-}Et_3N$, (6) $Na(Me_3Si)_2$-Davis'-reagent, (7) a: MsCl, b: DBU, (8) a: O_3, b: Me_2S, (9) Red Al reagent, (10) a: Ph_2PCl-I2, b: H_2O-AcOH-THF, (11) $HCBr_3$-Na, *t*-amylate, (12) MeLi-MeI, (13) HF/THF, (14) *o*-$NO_2C_6H_4SeCN\text{-}Bu_3P$, (15) H_2O_2, (16) a: O_2, b: MesS, c: TsOH

5.2 *Synthesis of Diterpenoids*

The diterpene sacculatal (**128**) is a potent pungent diterpene dialdehyde possessing cytotoxic, antimicrobial, and antiviral effects that occurs in liverworts such as *Pellia endiviifolia* and *Trichocoleopsis sacculata.* Its synthesis was completed by Hagiwara and Uda in 1989, starting from the optically active 5*β*,8*aβ*-dimethyl-5*α*-(4-methyl-3-pentenyl)-3,4,4*α*,5,6,7,8,8*α*-octahydro-1(2*H*)-naphthalenone, an advanced common intermediate for sacculatane synthesis [454].

The enantioselective total synthesis of perrottetianal A (**158**), a bitter-tasting diterpene dialdehyde obtained from several liverworts, was accomplished by Hagiwara and Uda, beginning with an optically active Wieland-Mischer ketone analogue [455].

The Indian medicinal plant, *Coleus forskolii,* which has been used to treat disorders of the digestive organs, produces the highly oxygenated labdane diterpenoid, forskolin (**1266**). It shows blood pressure-lowering and cardioprotective properties,

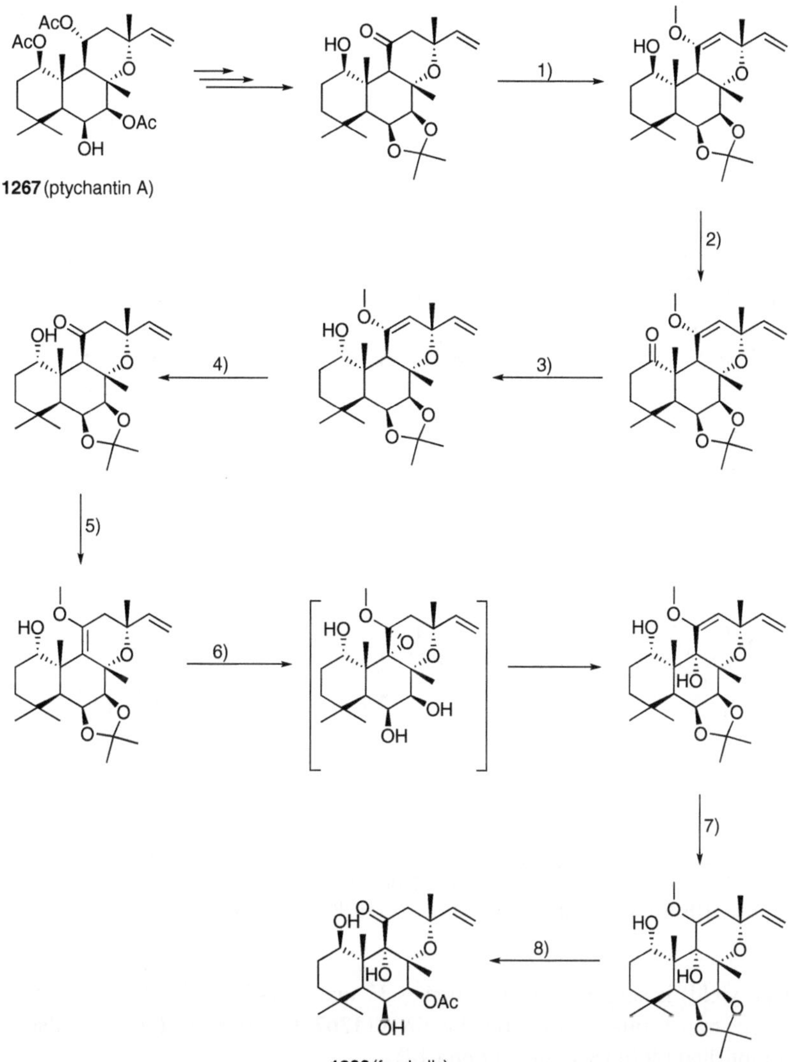

Scheme 17 Semisynthesis of forskolin (**1266**) from ptychantin A (**1267**) isolated from the liverwort *Ptychanthus striatus*. (1) KH, Me_2SO_4, THF, (2) CrO_3, Py, CH_2Cl_2, (3) Na, *t*-BuOH, (4) 1% HCl, THF, (5) KH, Me_2SO_4, THF, (6) MCPBA, CH_2Cl_2, K_2CO_3, (7) 10% $HClO_4$, THF, (8) Ac_2O, Py

and has therapeutic potential in glaucoma, congestive heart failure, and bronchial asthma [456]. A highly oxygenated labdane diterpene triacetate, ptychantin A (**1267**) was isolated from the Japanese liverwort *Ptychanthus striatus* (Lejeuneaceae) as the major component [457], and shows a high similarity to **1266** and its congener, 1,9-dideoxyforskolin (**1268**). The synthesis of ptychantin A (**1267**) to forskolin (**1266**) in 12 steps (12% overall yield) and to 1,9-dideoxyforskolin (**1268**) in 8 steps (37%

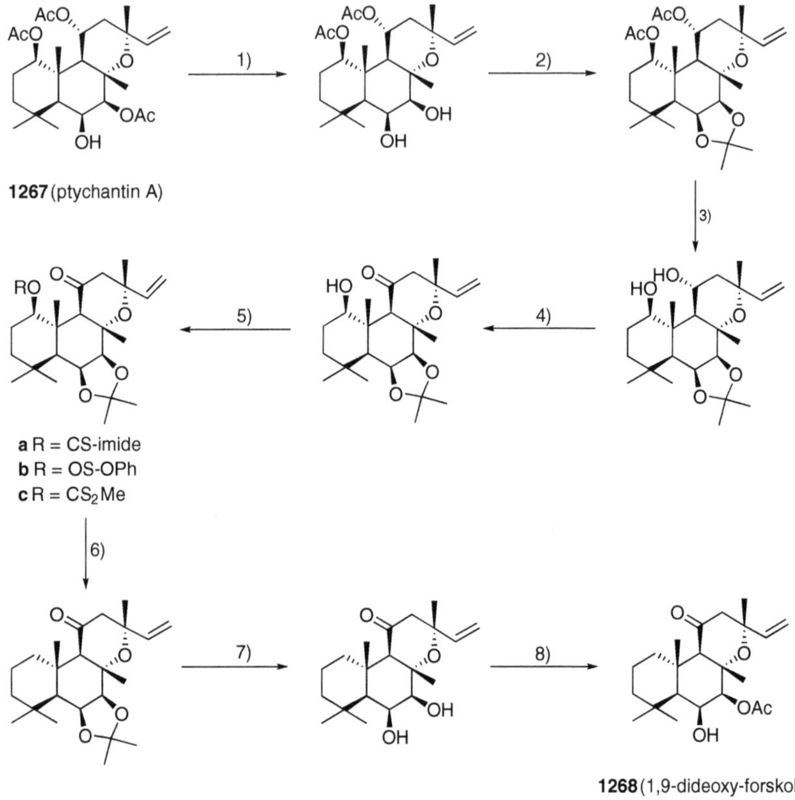

Scheme 18 Semisynthesis of 1,9-dideoxyforskolin (**1268**) from ptychantin A (**1267**). (1) KOH, MeOH, (2) 2,2-dimethoxypropane, PTSA, (3) $LiAlH_4$, Et_2O, (4) PCC, AcONa, CH_2Cl_2, (5) thiocarbonyldiimidazolide, DMAP, for **a**: phenylchlorothionoformate, DMAP, CH_2Cl_2 for **b**: *n*-BuLi, CS_2, MeI, THF, (6) AlBN, *n*-Bu_3SnH, toluene, (7) $HClO_4$, THF, (8) Ac_2Py, DMAP

overall yield) is summarized in Schemes 17 and 18 [458, 459]. A more expedient synthetic transformation from ptychantin A (**1267**) to forskolin (**1266**) has also been accomplished [460] (Schemes 17 and 18).

5.3 *Synthesis of Ambrox*

Ambrox (**1269**) is an expensive aroma compound originating from mammals. It was semi-synthesized from a large amount of labda-12,14-dien-7α,8α-diol isolated from *Porella perrottetiana* in seven steps in high yield as shown in Scheme 19 [461].

Note added during preparation of this volume: Benjamin List and his group have very recently published a diastereoselective and enantioselective synthesis of (–)-ambrox (**1269**) (see also Scheme 19). Luo N, Turberg M, Leutzsch M, Mitschke

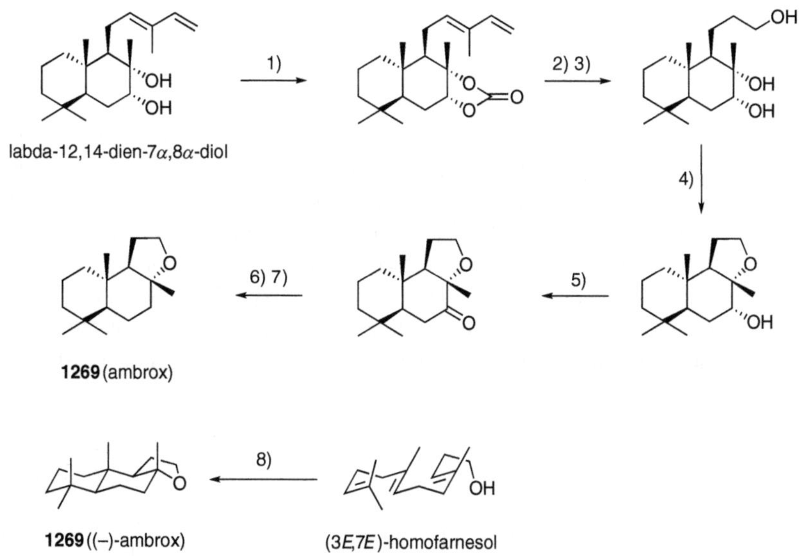

Scheme 19 Semisynthesis of ambrox (**1269**) from labda-12,14-dien-7α,8α-diol isolated from the liverwort, *Porella perrottetiana*, and dia-and enatioselective synthesis from *(3E*,7*E*)-homofarnesol. (1) $MeCOCOOCCl_3/Py/CH_2Cl_2$, (2) O_3/CH_2C12, (3) $LiAlH_4$, (4) $H^+/MeNO_2/p$-TsOH, (5) CrO_3-H_2SO_4, (6) $TsNHNH_2$, (7) $NaBH_3CN$, (8) 2 mol% of a chiral imidodiphosporimidate Brønsted acid catalyst

B, Brunen S, Wakchaure VN, Nöthling N, Schelwies M, Pelzer R, List B (2024) The catalytic asymmetric polyene cyclization of homofarnesol to ambrox. Nature 632:795.

5.4 *Synthesis of Psychoactive* cis- *and* trans-*Perrottetinene*

The total synthesis of the natural psychoactive *cis*-perrottetinene (**279**) and its *trans*-isomer (**1270**) was carried out by Chicca and associates [404]. An efficient total synthesis of perrottetinene (**279**) was accomplished independently by Son et al., in 2008 [462] and Park and Lee in 2010 [463].

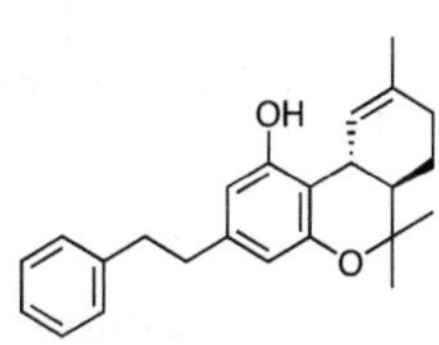

1270 (*trans*-perrottetinene)

5.5 Synthesis of Bibenzyls

Numerous bibenzyls have been isolated from many liverworts, in particular from *Radula* and *Frulania* species. Perrottetin A (**1061**) possesses a calmodulin-inhibitory effect, and perrottetin D (**1055**) from *Radula perrottetii*, which shows calmodulin-inhibitory and selective and potent antitrypanosomal activities, were synthesized in 1990 by Asakawa [38] (Scheme 20). 2-(3-Methyl-2-butenyl)-3,5-dihydroxybibenzyl (**306**) and 2-geranyl-3,5-dihydroxybibenzyl (**899**) from a *Radula* sp. were also prepared by prenylation of 3-hydroxy-5-methoxybibenzyl (**320**) (Scheme 21) [38].

A one-pot total synthesis of allelopathic radulanin A (**335**) isolated from a *Radula* species was carried out by Zhang et al., [444] from a cyclic 1.3-diketone

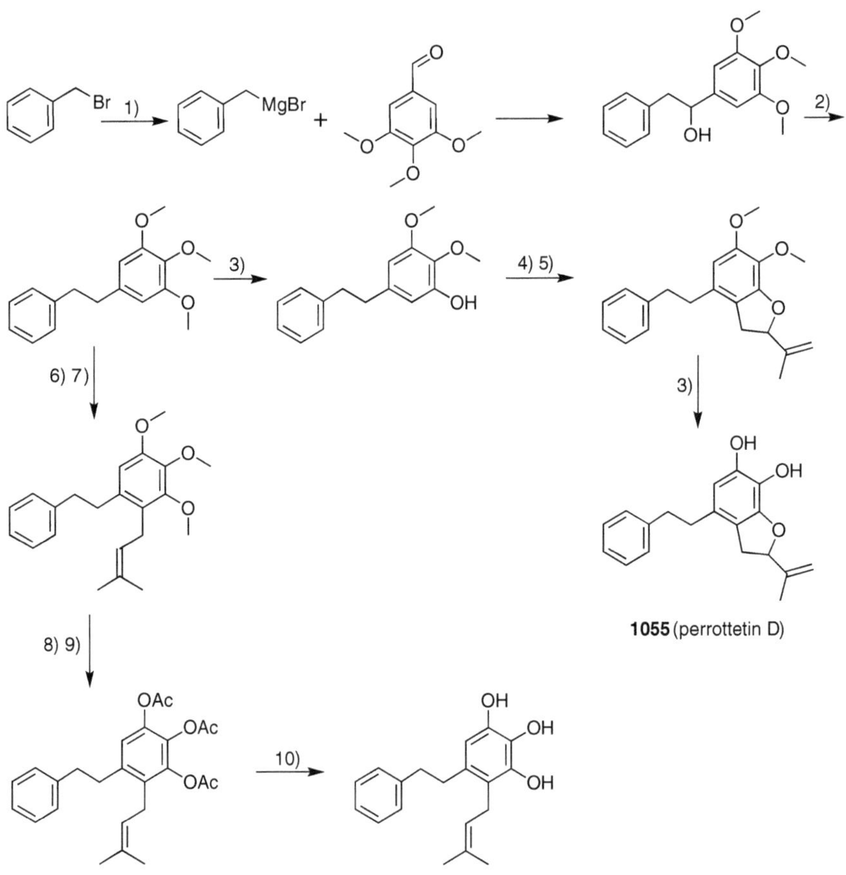

1055 (perrottetin D)

1061 (perrottetin A)

Scheme 20 Total synthesis of perrottetins A (**1061**) and D (**1055**) (1) Mg, (2) H_2/20% Pd-C, (3) EtSH/NaH, (4) MeONa/DMF, (5) (Z)-1,4-dibromo-2-methyl-2-butene, (6) *n*-BuLi, (7) (Z)-1,4-dibromo-2-methyl-2-butene, (8) BBr_3, (9) Ac_2O/Py, 10) $LiAlH_4$

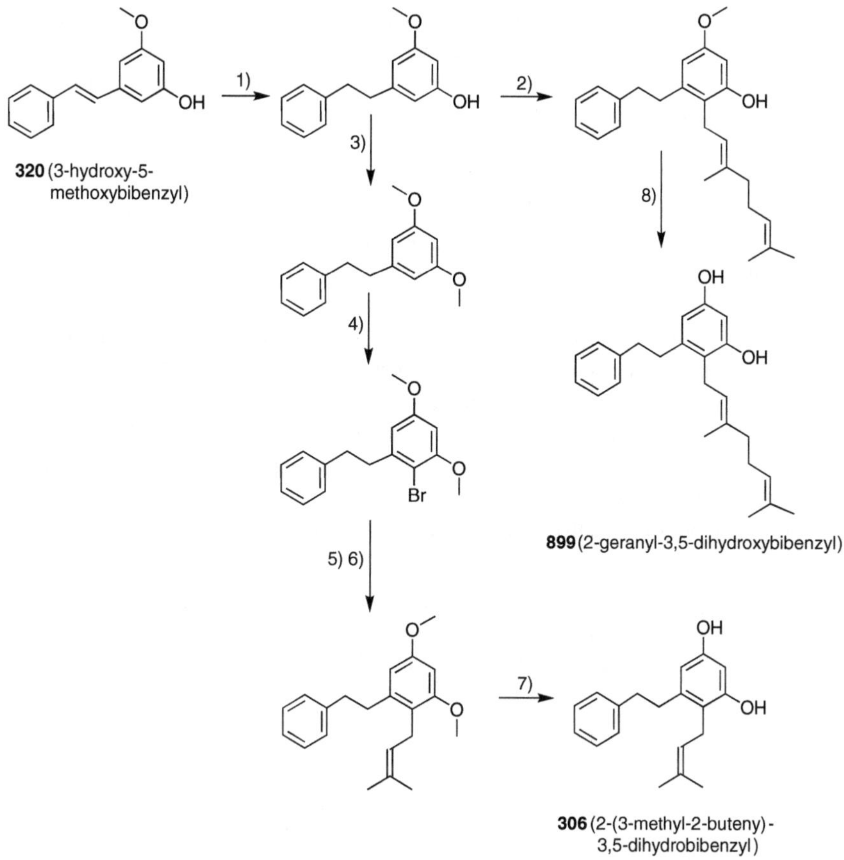

Scheme 21 Synthesis of 2-(3-methyl-2-buteny)-3,5-dihydroxybibenzyl (**306**) and 2-geranyl-3,5-dihydroxybibenzyl (**899**). (1) H_2/10% Pd-C, (2) MeONa/MeOH/1-bromo-3,7-dimethyl-2,6,-octadiene, (3) MeI/K_2CO_3, (4) NBS/DMF, (5) *n*-BuLi, (6) (Z)-1,4-dibromo-2-methyl-2-butene, (7) BBr_3, (8) EtSNa/DMF

and 1,4-bromo-2-butane, through a retro-Claisen [3,3]-sigmatropic rearrangement of the *syn*-2-vinylcyclopropyl diketone intermediate. An efficient large-scale production of radulanin A (**335**) by a photochemical ring expansion reaction of a 2,2-dimethylchromene derivative also was reported [464].

5.6 Synthesis of bis-Bibenzyls

The total synthesis of the acyclic bis-bibenzyl, perrottetin E (**183**), possessing cytotoxicity and genotoxicity, isolated from *Radula perrottetii*, was accomplished by

Asakawa in 1990 using 3-hydroxy-4-benzyloxybenzaldehyde, prepared from 3,4-dihydroxbenzaldehyde, as the starting material. This was followed by a Wittig reaction, to give monobenzyloxyperrottetin E, which was debenzylated to afford perrottetin E (**183**) (Scheme 22) [38].

Almost all of the liverwort bis-bibenzyls possess interesting biological activities and several structures exhibit optical activity, although asymmetric carbon atoms are absent in these molecules. Thus, these naturally rare compounds have attracted organic chemists to synthesize not only chiral but also nonchiral bis-bibenzyls. The first isolated bis-bibenzyl from liverwort was marchantin A (**1**) from the Japanese *Marchantia polymorpha* and then riccardin A (**2**) from *Riccardia multifida* subsp. *decrescens*, with the first total synthesis of **1** accomplished by using the intramolecular Wadsworth-Emmons olefination and Wittig reactions [465, 466]. Riccardins A (**2**), B (**216**), C (**182**), and marchantin C (**197**) were synthesized in 1990 by Gottsegen et al., [467] and Speicher and Holz [468], respectively. Furthermore, the total synthesis of all of the biologically active bis-bibenzyls, isoriccardin D (**881**), perrottetins E (**183**) and F (**184**), 10′-hydroxyperrottetin E (**247**), and 10,10′-dihydroxyperrottetin E (**248**), marchantins C (**197**), H (**940**), O (**1271**), and P (**941**),

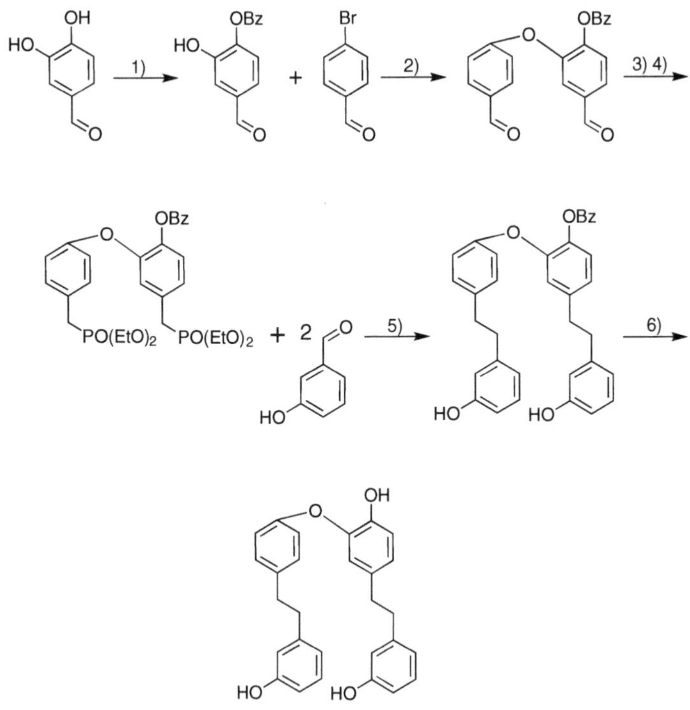

183 (perrottetin E)

Scheme 22 Total synthesis of perrottetin E (**183**). (1) BzBr/K_2CO_3, (2) CuO/K_2CO_3, (3) $NaBH_4$, (4) SBr_2, (5) $P(OEt)_3$, (6) H_2/10% Pd-C

dihydroptychantol A (**228**), asterellin A (**937**), plagiochins A (**1056**), B (**1272**), C (**1273**), and D (**1120**), and marchantiaquinone (**1221**), have been reviewed [76].

In addition, the total syntheses of the potentially bioactive bis-bibenzyls marchantins I (**1274**) and F (**1275**), isoplagiochin C (**1276**), bazzanins A (**1277**) and J (**1278**), 12-chloroisoplagiochin D (**1279**), polymorphatin A (**1280**), perrottetin G (**1281**), the optically active cavicularin (**1282**), and 12,12′-bis(10′-dihydroxyperrottetin E) (**1283**) were accomplished by different organic chemists [76].

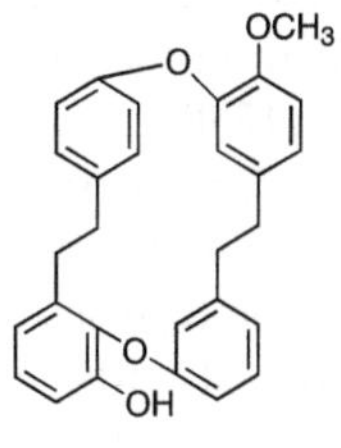

1271 (marchantin O)

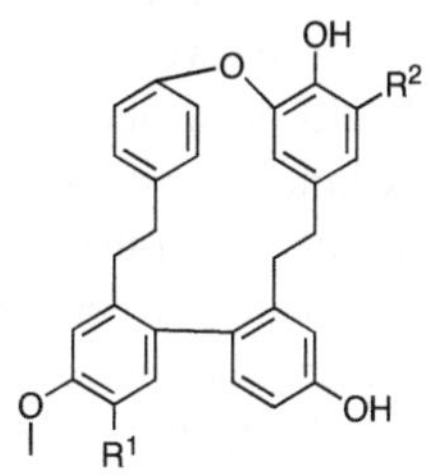

1272 (plagiochin B, R1 = OH, R2 = H)
1273 (plagiochin C, R1 = R2 = H)

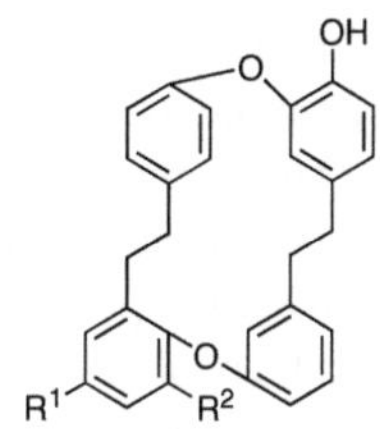

1274 (marchantin I, R1 = OMe, R2 = H)
1275 (marchantin F, R1 = H, R2 = OMe)

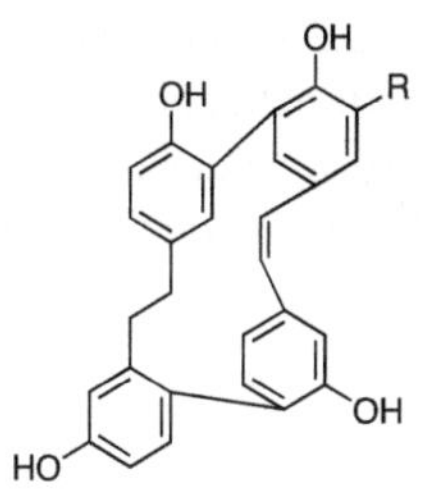

1276 (isoplagiochin C, R = H)
1277 (bazzanin A, R = Cl)

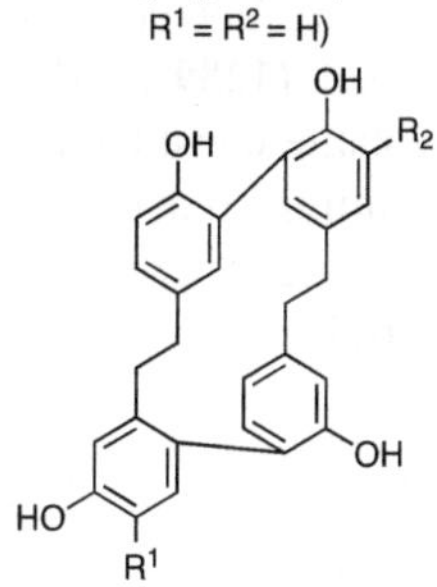

1278 (bazzanin J, R1 = H, R2 = Cl)
1279 (12-chloroisoplagiochin D, R1 = Cl, R2 = H)

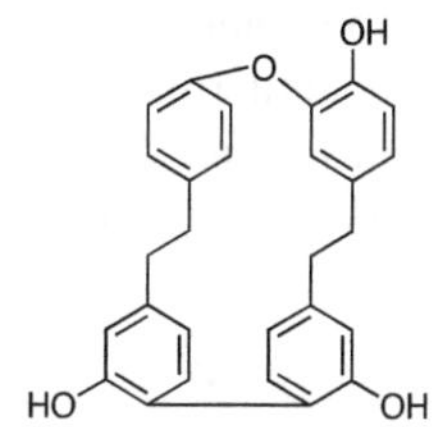

1280 (polymorphatin A)

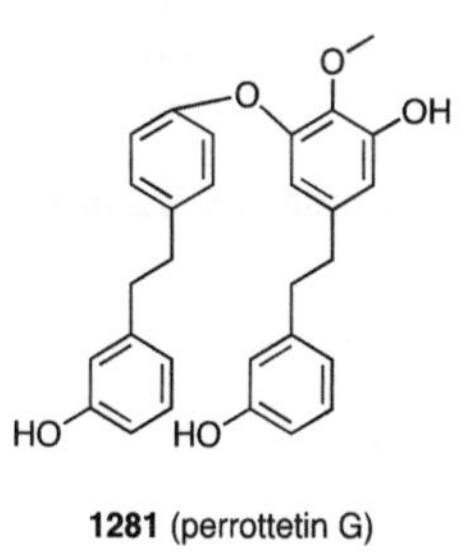

1281 (perrottetin G)

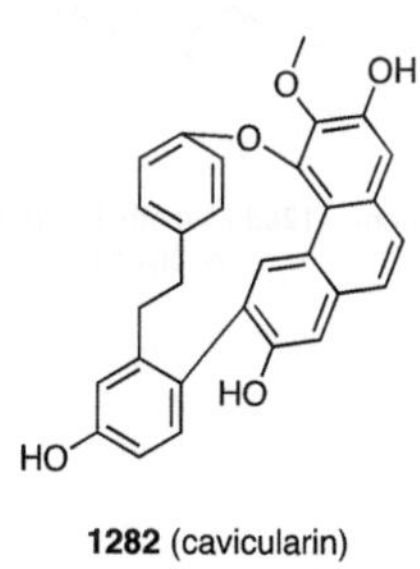

1282 (cavicularin)

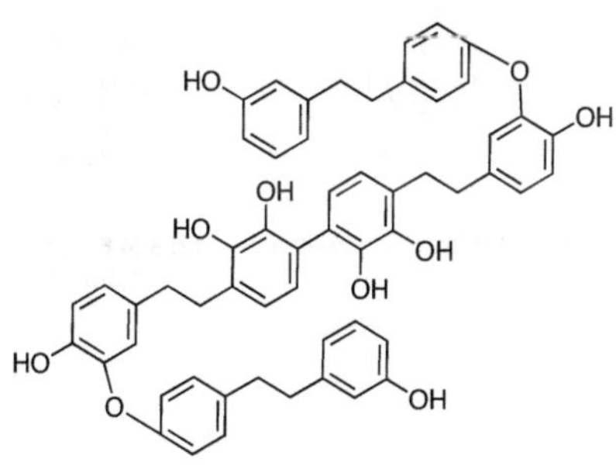

1283 (12′,12″-*bis*(10′-hydroxyperrottetin E)

6 Endophytic Constituents from Bryophytes

6.1 Cytotoxic Activity

The ether extracts of the Chinese liverwort *Scapania verrucosa* and an endophytic fungus *Chaetomium fusiforme*, obtained from *S. verrucosa*, were analyzed by GC/MS and 49 volatile components, including aromadendrene (**816**) β-bourbonene (**1284**), calarene (**1285**), hexadecanoic acid, eudesma-4,11(12)-dien-2-ol (**1286**), 1,1,2,2-tetrachloroethane (**1287**), and acetic acid (**45**) were identified from the liverwort as major components, together with 4,8,13-duvatrien-1,3-diol (**1288**), as a minor component. The chemical profile of the Chinese *S. verrucosa* was different from that of the same species when collected in Europe and from other *Scapania* species, since the former liverwort produces calarene-, maaliene-, and aristolane-type sesquiterpenoids, which are absent in the latter species. The endophytic fungal extract exhibited a wider range of antiproliferative activity than the host liverwort extract. A549 lung, LOVO colon, HL-60 leukemia, and QGY liver human cancer cells were evaluated for this fungal extract, and gave IC_{50} values of > 100, 9.1, 4.1, and 31.2 μg/cm^3, respectively. In turn, the host liverwort ether extract showed weak cytotoxic activities in the IC_{50} range of 42.9– > 100 μg/cm^3. The cultured fungus contained acetic acid (**45**), 3-methylvaleric acid methyl ester (**1289**), and butane-2,3-diol (**1290**), as the major compounds. Although the volatile constituents of both extracts demonstrated little similarity, they displayed antifungal activity with IC_{80} values of 8–64 μg/cm^3 against *Candida albicans*, *Cryptococcus neoformans*, *Trichophyton rubrum*, and *Aspergillus fumigatus* (Table 44) [469].

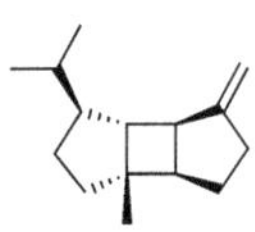

1284 (ß-bourbonene)

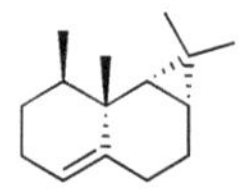

1285 (calarene)

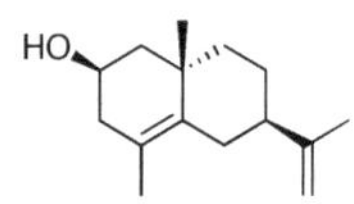

1286 (eudesm-4,11(12)-dien-2-ol)

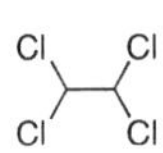

1287 (1,1,2,2-tetrachloroethane)

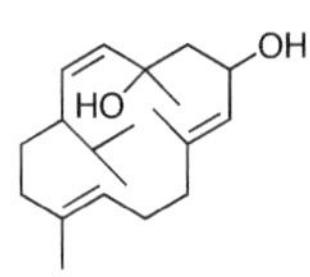

1288 (4,8,13-duvatrien-1,3-diol)

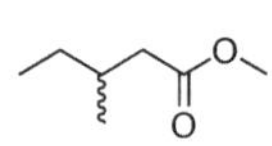

1289 (3-methyl valeric acid methyl ester)

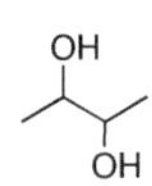

1290 (butan-2,3-diol)

Table 44 Cytotoxic active compounds and solvent extracts from endophytes of bryophytes I

Species names[a]	Compound or solvent extract	Activity values Cell lines[b]	Refs.
Aspergillus fumigatus/[L] *Heteroscyphus tener*		IC_{50} (μ*M*) PC3: PC3D: A549: NCIH1460	[472]
	Asperfumigatin (**1310**)	30.6 > 40 > 40 > 40	
	Brevianamide (**1319**)	> 40 32.0 > 40 > 40	
	Demethoxyfumitremorgin (**1311**)	32.0 > 40 > 40 > 40	
	Fumitremorgin C (**1312**)	28.9 39.4 > 40 > 40	
	Cyclotryprostatin C (**1313**)	33.9 > 40 > 40 > 40	
	12,13-Dihydroxyfumitremorgin C (**1314**)	36.2 39.6 > 40 > 40	
	Verruculogen TR-2 (**1315**)	38.9 > 40 > 40 > 40	
	20-Hydroxycyclotryprostatin B (**1316**)	32.5 > 40 > 40 > 40	
	Chaetominine (**1317**)	30.1 > 40 > 40 > 40	
	Isochaetominine (**1318**)	32.2 > 40 > 40 > 40	
	Fumiquinazoline J (**1319a**)	> 40 > 40 > 40 > 40	
	13-Dehydroxycyclotryptostatin C (**1320**)	27.8 > 40 > 40 > 40	
	Spirotryprostatin B (**1321**)	26.2 > 40 > 40 > 40	
	Fumigaclavine C (**1322**)	23.4 > 40 > 40 > 40	
	Pyripyropene A (**1323**)	23.4 27.7 38.3 > 40	
	Pseurotine (**1324**)	> 40 > 40 > 40 > 40	
	Trypacidin (**1325**)	19.9 > 40 > 40 > 40	
	12-*seco*-Trypacidin (**1326**)	> 40 > 40 > 40 > 40	
	Sulochrin (**1327**)	> 40 > 40 > 40 > 40	
	Questin (**1328**)	> 40 > 40 > 40 > 40	
	8′-*O*-Methylasterric acid (**1329**)	> 40 > 40 > 40 > 40	
	Fumiquinazolin C (**1330**)	> 35.9 > 40 > 40 > 40	
Aspergillus niger/ [L] *Heteroscyphus tener*		IC_{50} (μ*M*) A2780	[471]
	Asperazine (**1302**)	56.7	
	Asperazine A (**1301**)	56.3	
Aspergillus niger/ [L] *Heteroscyphus niger*	EtOAc (*R*)-Malformin (**1300**)	IC_{50} (μ*M*) HepG-2: 48.2 A2780: 0.14 H1699: 1.02 K562: 0.12 M231: 0.45 PC3: 0.24	[470]

(continued)

Table 44 (continued)

Species names[a]	Compound or solvent extract	Activity values Cell lines[b]	Refs.
Xylaria sp. NC1214/[M] *Hypnum* sp.		IC_{50} (μ*M*)	[476]
	Cytochalasin C (**1376**)	PC-3 M: 1.65 NCI-H460: 1.06 SF268: 0.96 MCF-7: > 5 NDA-MB-231: 1.72	
	Cytochalasin D (**1377**)	PC-3M: 1.03 NCI-H460: 1.06 SFT268: 0.22 MCF-7: 1.44 NDA-MB-231: 1.01	
	Cytochalasin Q (**1378**)	PC-3M: 1.53 NCI-H460: 0.51 SF268: 1.31 MCF-7: > 5 NDA-MB-231: 1.32	
Smardaea sp./[M] *Ceratodon purpureus*		IC_{50} (μ*M*)	[475]
	Sphaeropsidin A (**1354**)	NCI-H460: 1.9–2.8	
	Sphaeropsidin D (**1357**) 6-*O*-Acetoxy-sphaeropsidin A (**1362**)	SF264: 2.1–4.2 NCF-7: 2.0–3.0 MDA-MB-231: 1.4−3.7 PC-3: 2.5–9.0 PC-3 M: 2.4–2.7 MIAPPaCCa-2: 2.0–9.0 WI-38: 3.7–> 10	
Bacterial endophytes/[L] *Marchantia polymorpha*	EtOAc (containing anthranilic acid (**1331**), *N*-phenylethyl-acetamide (**1332**), gancidine W (**1333**), cyclo(phenyl-alanylprolyl) (**1336**), 2,2-dimethyl-*N*-phenylethyl-propionamide (**1335**), *N*-(phenylethyl) phenyl-acetamide (**1334**))	CC_{50} (μg/cm^3)/SI HeLa: 26.1/14.7 FaDu: 55.5/6.9 SCC-25: 0.7	[473]
Bacterial endophytes/[L] *Marchantia polymorpha*		CC_{50} (μg/cm^3) CC_{50} (μg/cm^3)/SI	[474]
	EtOAc fraction 1	VERO: 792.1 HeLa:106.2/7.5 RKO:71.4/11.1 FaDu: 158/5.0	
	EtOAc fraction 2	VERO: 714 HeLa: 116.4/6.0 RKO: 98.5/7.3 FaDu: 155.6/3.7	
	EtOAc fraction 3	VERO: 666.8 HeLa: 141.5/4.6 RKO: 139.7/4.6 FaDu: 147.5/4.4	

(continued)

Table 44 (continued)

Species names[a]	Compound or solvent extract	Activity values Cell lines[b]	Refs.
	EtOAc fraction 4	VERO: 226.8 HeLa: 54.5/4.2 RKO: 43.4/5.2 FaDu: 106.1/2.1	

[a] [L] liverwort, [M] moss
[b] FaDu: hypopharyngeal squamous cell carcinoma, HeLa: cervical adenocarcinoma, MCF-7: human breast adenocarcinoma, MIAPaCa-2: human pancreatic cancer, MDA-MB-231: metastatic breast adenocarcinoma, NCI-H460: human non-small cell lung cancer, PC-3: human prostate adenocarcinoma, PC- SF-268: human CNS glioma, VERO: normal kidney fibroblast, and WI-38: normal human primary fibroblast cell lines

As mentioned earlier, the liverwort, *Heteroscyphus tener* is an abundant source of sesquiterpenoids. In 2013, Li et al., studied the secondary metabolites of the endophytic fungus, *Aspergillus niger*, which was obtained from *H. tener*. The fungus mycelium was extracted with ethyl acetate to give a weakly cytotoxic crude extract against the HepG-2 cell line (IC_{50} 42.8 μg/cm^3). Further fractionation of this extract led to the isolation of five new naphtha-γ-pyrones, namely, rubrofusarin-6-*O*-*α*-D-ribofuranoside

Table 45 Cytotoxic active compounds and solvent extracts from endophytes of bryophytes II

Species name[a]	Compound or solvent extract	Cell lines[b]	Activity value	Refs.
Chaetomium fusiforme/[M] *Scapania verrucosa*	Et_2O		IC_{50} (μg/cm^3)	[469]
		A549	> 100	
		LOVO	9.11	
		HL-60	4.06	
		QGY	31.23	
Xylaria sp.NC1214/ [M] *Hypnum* sp.	Cytochalasin C (**1376**)	PC-3 M NCI-H460 SF268 MCF-7 MDA-MB-231	IC_{50} (μ*M*) 1.65 1.06 0.96 > 5	[476]
	Cytochalasin D (**1377**)	PC-3 M NCI-H460 SFT268 MCF-7 MDA-MB-231	1.72 1.03 1.06 0.22 1.44 1.01 1.53 0.51 1.31 > 5 1.32	

(continued)

Table 45 (continued)

Species name[a]	Compound or solvent extract	Cell lines[b]	Activity value	Refs.
	Cytochalasin Q (**1378**)	PC-3 M NCI-H46 SF268 MCF-7 MDA-MB-231		
Smardaea sp./[M] *Ceratodon purpureus* [95]	Sphaeropsidin A (**1354**) Sphaeropsidin D (**1357**) 6-*O*-Acetyl-sphaeropsidin A (**1362**)		IC_{50} (μ*M*)	[475]
		NCI-H460	1.9–2.8	
		SF264	2.1–4.2	
		NCF-7	2.0–3.0	
		MDA-MB-231	1.4–3.7	
		PC-3	2.5–9.0	
		PC-3 M	2.4–2.7	
		MIAPPaCCa-2	2.0–9.0	
		WI-38	3.7– > 10	
Endophytes bacteria/ [L] *Marchantia polymorpha*	EtOAc (containing anthranilic acid (**1331**), *N*-phenylethyl-acetamide (**1332**), gancidine W (**1333**), cyclo(phenyl-alanylprolyl) (**1336**), 2,2-dimethyl-*N*-phenylethyl-propionamide (**1335**), *N*-(phenylethyl) phenyl-acetamide (**1334**))		CC_{50} (μg/ cm^3)/SI	[473]
		HeLa	26.14/ 14.72	
		FaDu:	55.52/6.93	
		SCC-25	0.66	
Endophytes bacteria/ [L] *Marchantia polymorpha*			CC_{50} (μg/ cm^3) CC_{50} (μg/ cm^3)/SI	[474]
	EtOAc fraction 1	VERO HeLa RKO FaDu	792.1 106.2/7.5 71.4/11.1 158/5.0	
	EtOAc fraction 2	VERO HeLa RKO FaDu	714 116.4/6.0 98.5/7.3 155.6/3.7	
	EtOAc fraction 3	VERO HeLa RKO FaDu	666. 8 141.5/4.6 139.7/4.6 147.5/4.4	
	EtOAc fraction 4	VERO HeLa RKO 2 FaDu	226.8 54.5/4.2 43.4/5.2 106.1/2.1	

[a] [L] liverwort, [M] moss

[b] FaDu: hypopharyngeal squamous cell carcinoma, HeLa: cervical adenocarcinoma, MCF-7: human breast adenocarcinoma, MIAPaCa-2: human pancreatic cancer, MDA-MB-231: metastatic breast adenocarcinoma, NCI-H460: human non-small cell lung cancer, PC-3: human prostate adenocarcinoma, PC-3M: metastatic human prostate adenocarcinoma, SCC-25: tongue squamous cell carcinoma, SF-268: human CNS glioma, VERO: normal kidney fibroblast, and WI-38: normal human primary fibroblast cell lines

(**1291**), (*R*)-10-(succinimid-3-yl)-TMC-256A1 (**1292**), (*R*)-asperpyrone E (**1293**), (*R*)-isoaurasperone A (**1294**), and (*R*)-isoaurasperone F (**1295**), along with five known compounds, (*R*)-dianhydroaurasperone C (**1296**), (*R*)-aurasperone D (**1297**), (*R*)-asperpyrone D (**1298**), (*R*)-asperpyrone A (**1299**), and malformin A (**1300**). The absolute configurations of the new compounds were elucidated by a combination of NMR and CD spectroscopic analysis. The isolated products were tested for their cytotoxic activities using a MTT assay. Malformin A (**1300**) demonstrated significant cytotoxic effects against the A2780, H1688, K562, M231, and PC3 human cancer cell lines in vitro with IC_{50} values of 0.14,1.0, 0.12, 0.45, and 0.24 μ*M*, respectively. The other compounds tested did not show any discernible cytotoxic effects against these cell lines (Table 45) [470].

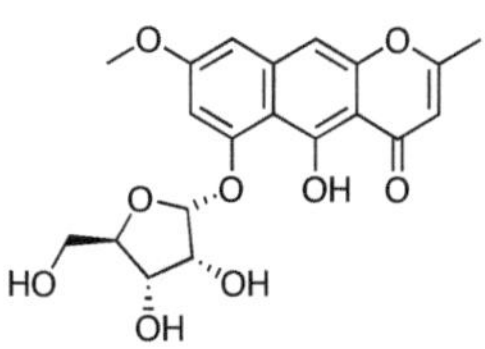

1291 (rubrofusarin-6-*O*-α-D-ribofuranoside)

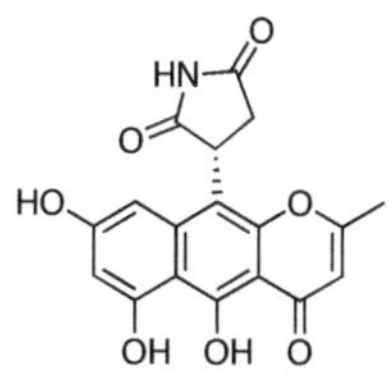

1292 ((*R*)-10-(3-succinimidyl)-TMC-256A1)

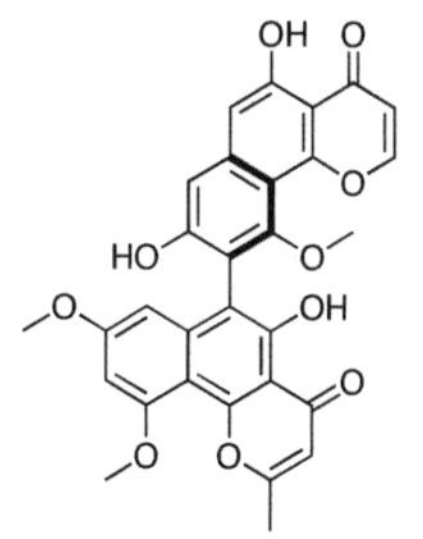

1293 ((*R*)-asperpyrone E)

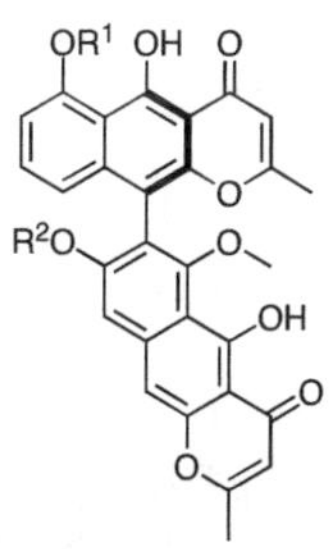

1294 ((*R*)-isoaurasperone A, $R^1 = R^2 = H$)
1296 ((*R*)-dianhydroaurasperone C, $R^1 = Me$, $R^2 = H$)
1297 ((*R*)-aurasperone D, $R^1 = H$, $R^2 = Me$)

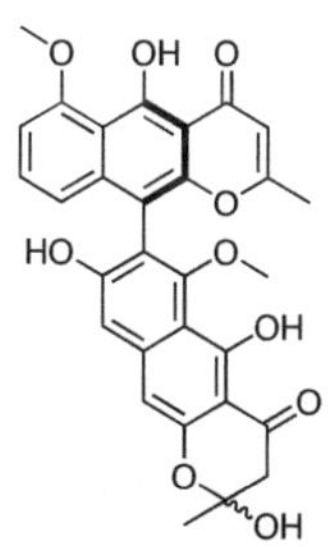

1295 ((*R*)-isoaurasperone F)

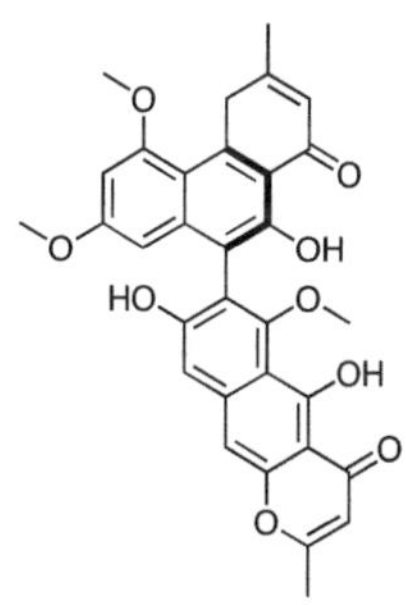

1298 ((*R*)-asperpyrone D)

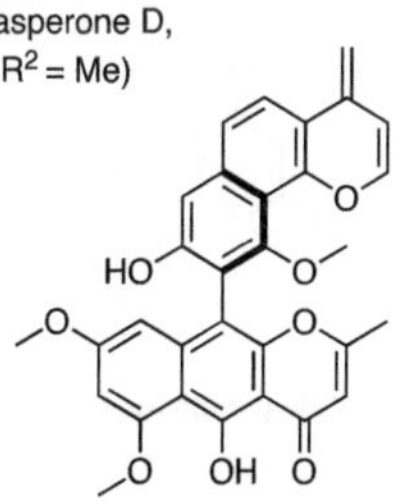

1299 ((*R*)-asperpyrone A)

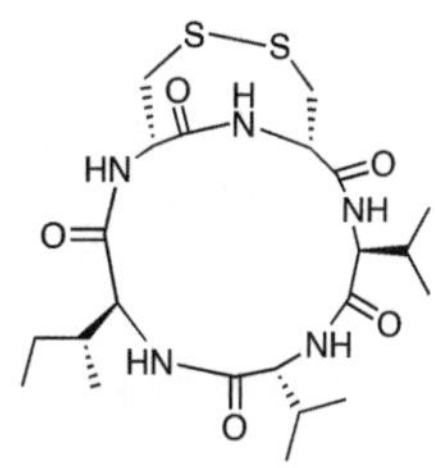

1300 (malformin A)

In 2015, the same group as mentioned above reinvestigated the chemical constituents of the endophytic fungus, *A. niger*, obtained from the liverwort, *Heteroscyphus tener*. The ethyl acetate extract of the fermented fungus was chromatographed on silica gel to isolate a new diketopiperazine, named asperazine A (**1301**), and eight known components, asperazine (**1302**), *cyclo*(D-Phe-L-Trp) (**1303**), *cyclo*(L-Trp-Trp) (**1304**), 4-hydroxymethyl-5,6-dihydro-pyran-2-one (**1305**), walterolactone A (**1306**), and campyrones A–C (**1307–1309**), and all were evaluated for cytotoxicity against the PC3, A2780, K562, MDA-MB-231, and NCI-H1688 cell lines. Compounds **1301** and **1302** showed weak cytotoxicity against A-2780 cells with IC_{50} values of 56.7 and 56.3 μ*M*, respectively [471].

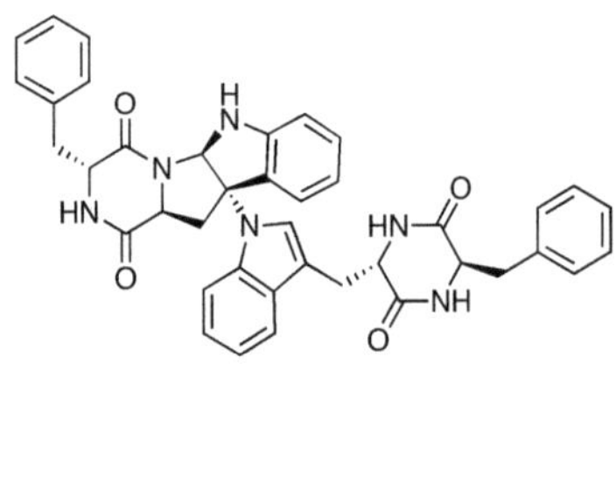

1301 (asperazine A)

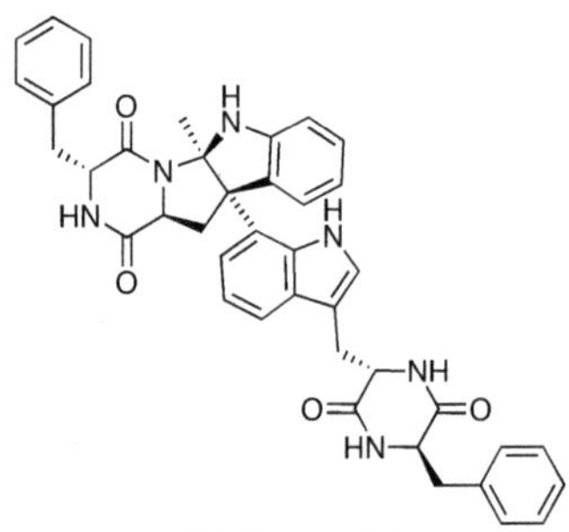

1302 (asperazine)

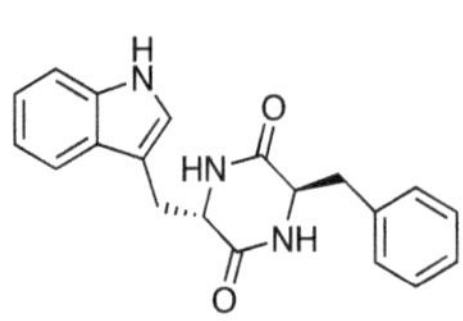

1303 (*cycb*(D-Phe-L-Trp))

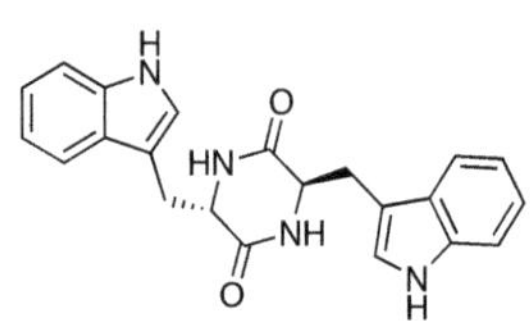

1304 (*cycb*(L-Trp-L-Trp))

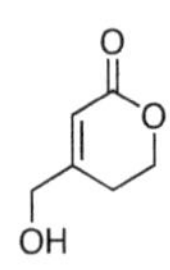

1305 (hydroxymethyl-5,6-dihydropyran-2-one)

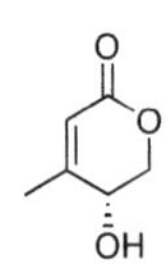

1306 (walterolactone)

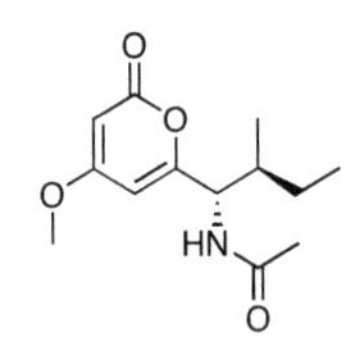

1307 (campyrone A)

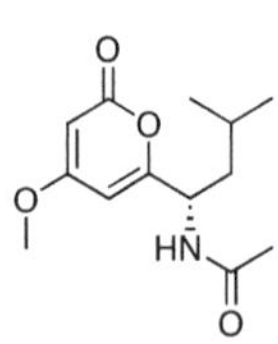

1308 (campyrone B)

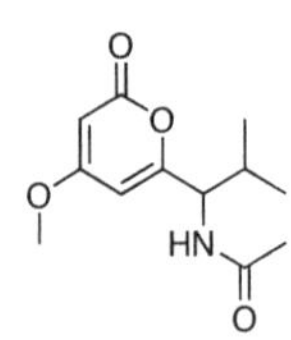

1309 (campyrone C)

Three new secondary metabolites named asperfumigatin (**1310**), isochaetominine (**1318**) and 8′-*O*-methylasterric acid (**1329**) were isolated from the endophytic fungus, *Aspergillus fumigatus*, which was obtained from the Chinese liverwort, *Heteroscyphus tener*, together with nineteen known compounds (**1311–1317**, **1319–1238**, and **1330**). The structures of the new compounds were elucidated by a combination of NMR and CD spectroscopy. The cytotoxic activity of all products was evaluated against the four cancer cell lines, PC3, PC3D, A549, and NCI-H460. Most of the isolated compounds showed weak cytotoxicity against PC3 prostate cancer cells with IC_{50} values from 19.9 to 38.9 μ*M* [472].

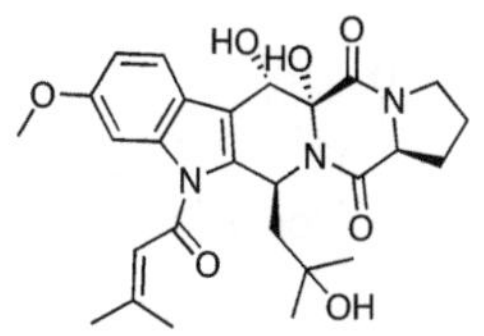

1310 (asperfumigatin)

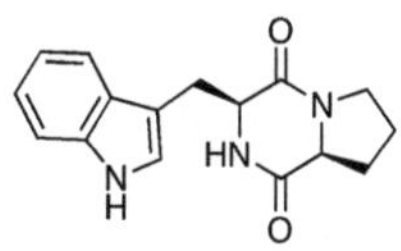

1319 (brevianamide F)

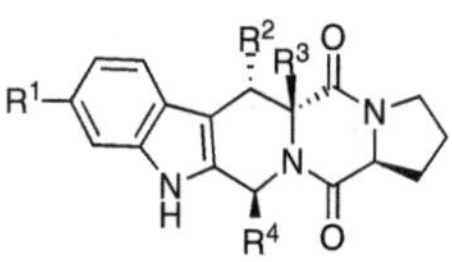

1311 (demethoxylumitremorgin C, $R^1 = R^2 = H$, $R^3 = \alpha$-H, $R^4 = A$)
1312 (fumitremorgin C, $R^1 = OMe$, $R^2 = H$, $R^3 = \alpha$-H, $R^4 = A$)
1313 (cyclotryprostatin C, $R^1 = H$, $R^2 = OH$, $R^3 = \alpha$-OH, $R^4 = A$)
1314 (12,13-dihydroxyfumitremorgin C, $R^1 = OMe$, $R^2 = OH$, $R^3 = \alpha$-OH, $R^4 = A$)
1315 (verruculogen TR-2, $R^1 = OMe$, $R^2 = OH$, $R^3 = \alpha$-OH, $R^4 = B$)
1316 (20-hydroxycyclotryprostatin B, $R^1 = R^2 = OMe$, $R^3 = \beta$-OH, $R^4 = B$)
1320 (13-dehydroxycyclotryprostatin C, $R^1 = R^2 = H$, $R^3 = \alpha$-OH, $R^4 = A$)

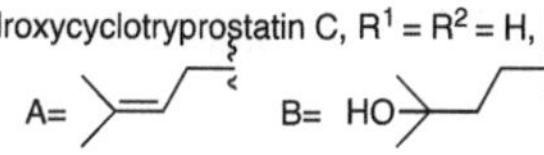

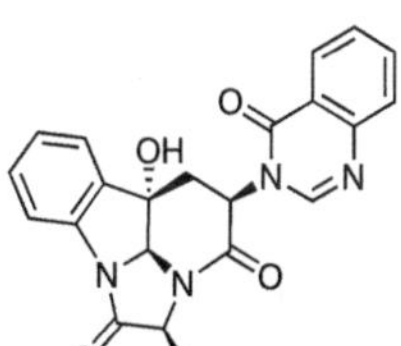

1317 (chaetominine)

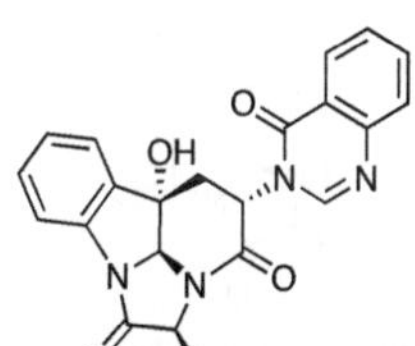

1318 (isochoetominine)

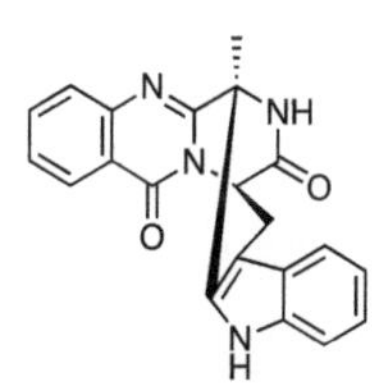

1319a (fumiquinazoline J)

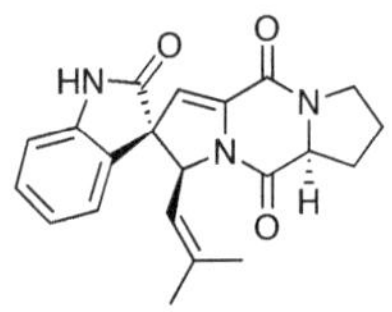

1321 (spirotryprostatin B)

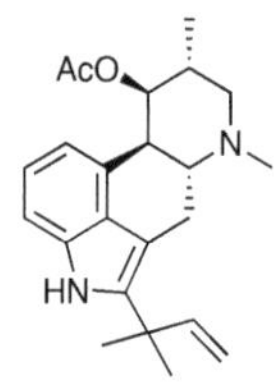

1322 (fumigaclavine C)

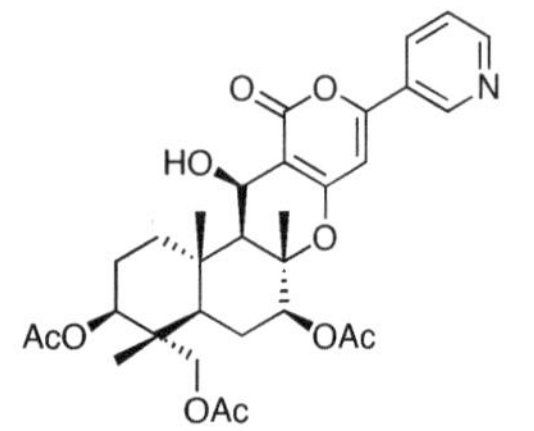

1323 (pyripyropene A)

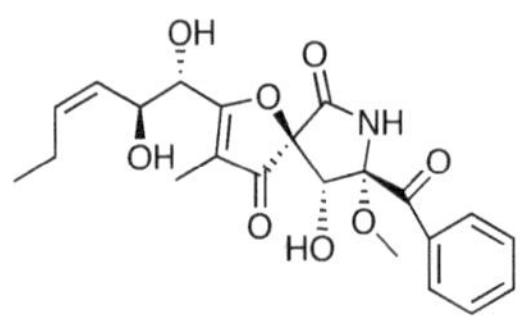

1324 (pseurotin A)

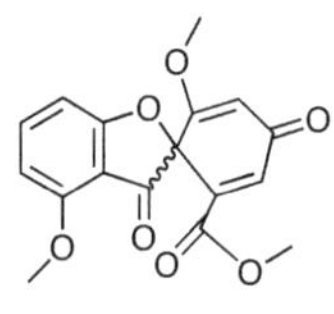

1325 (trypacidin)

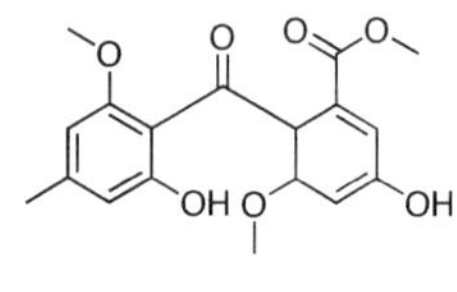

1326 (1,2-*seco*-trypacidin)

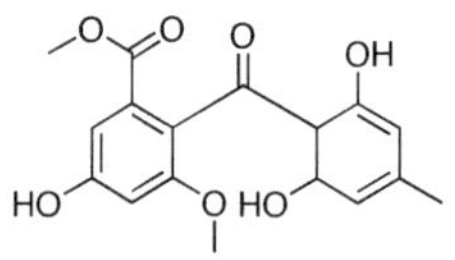

1327 (sulochirin)

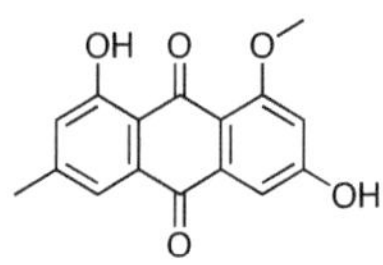

1328 (questin)

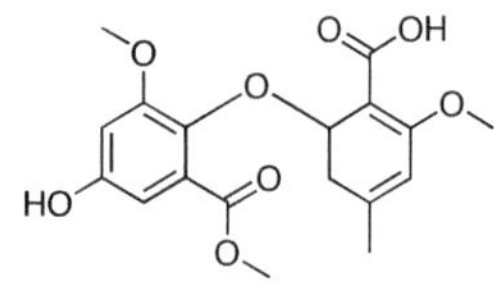

1329 (8'-*O*-methylasterric acid))

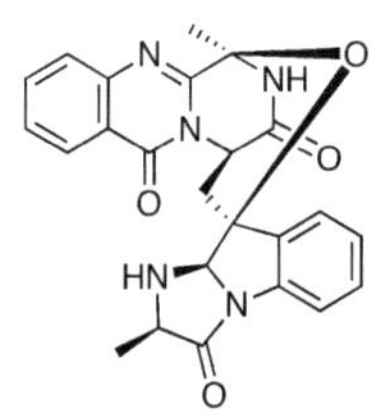

1330 (fumiquinazoline C)

An endophytic bacterium was isolated from the European liverwort *Marchantia polymorpha* and this bacterium was extracted with *n*-hexane and ethyl acetate and its secondary metabolites obtained were compared with those of the host liverwort. The ethyl acetate extract of the endophyte contained anthranilic acid (**1331**), *N*-phenylethylacetamide (**1332**), and gancidin W (**1333**), while the *n*-hexane extract showed the presence of *N*-(phenylethyl)phenyl-acetamide (**1334**), 2,2-dimethyl-*N*-phenylethylpropionamide (**1335**), propanoic acid, and cyclo(phenyl-alanylprolyl) (**1336**), which were not found in the ethyl acetate extract. The chemical profiles of the solvent extracts from the endophyte of *M. polymorpha* were quite different from that of the host liverwort since the host plant biosynthesizes mainly lipophilic sesquiterpenoids inclusive of β-chamigrene (**1337**), cuparene (**1338**), cyclopropanecuparenol (**1339**), and *epi*-cyclopropanecuparenol (**1340**), as the major components. The cytotoxicity of the ethyl acetate extract of the endophyte microorganisms was evaluated against a panel of human cancer cells comprisining FaDu pharyngeal sqaumous cell carcinoma, HeLa cervical adenocarcinoma, and SCC-25 tongue squamous cell carcinoma, in addition to VERO normal green monkey kidney fibroblasts. The selectivity index (*SI*) was calculated with reference to cytotoxicity observed for VERO cells ($SI = CC_{50}$ VERO/CC_{50} cancer cells) to assess the antitumor potential of the extract. The highest cytotoxic selectivity was observed with tHeLa cells with a CC_{50} value of 26.1 mg/cm^3 and a *SI* of 14.7. In turn, FaDu cells showed sensitivity to the ethyl acetate extract (CC_{50} 55.5 and *DI* 6.9), while SCC-25 cells were insensitive to the extract evaluated (*SI* 0.7). This was the first report of endophyte components from the liverwort, *M. polymorpha* and their cytotoxicity determination [473].

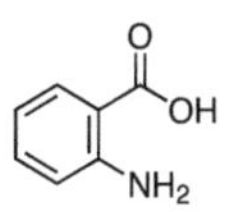

1331 (anthranilic acid)

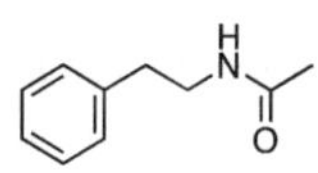

1332 (*N*-phenethyl-acetamide)

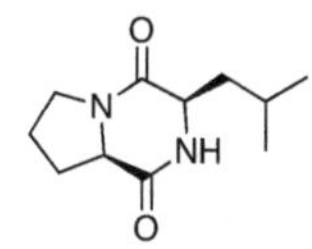

1333 (gancidin W)

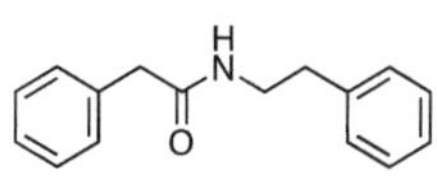

1334 (*N*-4(phenethyl)phenyl-acetamide)

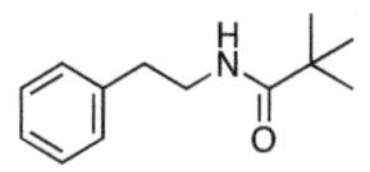

1335 (2,2-dimethyl-*N*-(phenethyl)propionamide)

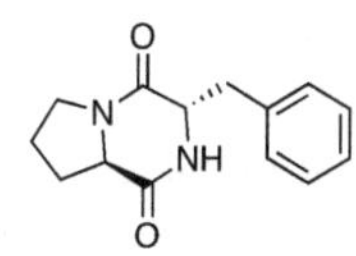

1336 (*cycb*(phenyl-alanylprolyl)

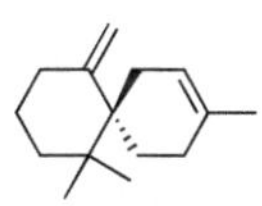

1337 (β-chamigrene)

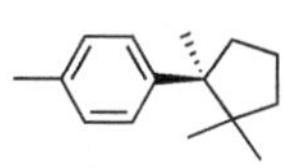

1338 (cuparene)

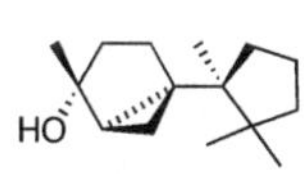

1339 (cyclopropane-cuparenol)

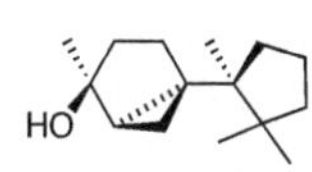

1340 (*epi*-cyclopropane-cuparenol)

The further fractionation of the ethyl acetate extract of the endophytic metabolites of *Marchantia polymorpha* resulted in the isolation of compounds **1333**, **1334**, *cyclo*(D-Leu-D-Prop) (**1341**), *cyclo*(D-Phe-D-Prop) (**1342**), *cyclo*(L-Phe-L-Prop) (**1343**) and oleic acid amide (**1344**) as the major products, together with **1331**, **1335**, 1,1-dibutoxybutane (**1345**), and pyrolidino[1,2-*a*]piperazine-3,6-dione (**1346**) as minor ones. The cytotoxic effects of the ethyl acetate extract and of its fractionated constituents were evaluated against three cancer cell lines and against VERO cells. The ethyl acetate extract exerted moderate cytotoxic effects with some cancer cell line selectivity. The fraction containing *cyclo*(L-Phe-L-Prop) (**1343**) and *cyclo*(D-Leu-D-Prop) (**1341**) showed higher CC_{50} values for the cancer cell lines, and a decrease in anticancer selectivity was also observed. The fraction containing oleamide (**1344**) and *N*-phenylethylacetamide (**1332**) instead of diketopiperazine derivatives showed the most potent cytotoxic effects, but also the lowest selectivity against the cancer cell lines used, as shown in Table 45 [474].

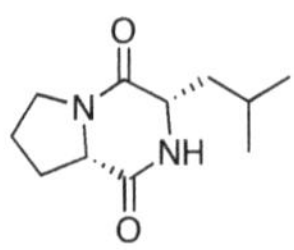

1341 (*cyclo*(D-Leu-D-Pro)

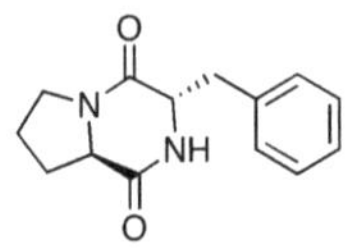

1342 (*cyclo*(D-Phe-D-Pro))

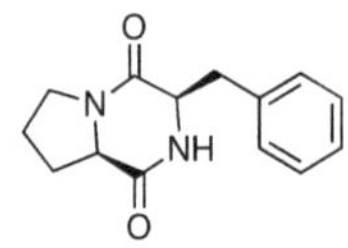

1343 (*cyclo*(L-Phe-L-Pro))

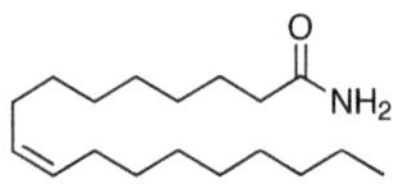

1344 (oleic acid amide)

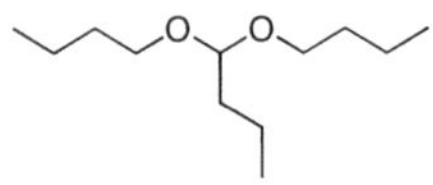

1345 (1,1-dibutoxybutane)

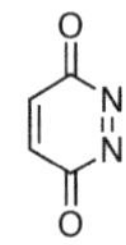

1346 (piperidin-3,6-dione)

The chloroform extract of the endophytic fungal strain *Smardaea* species AZ0432, which was obtained from the living tissue of the moss *Ceratodon purpureus*, was fractionated to afford five isopimarane diterpenoids, named smardaesidins A–F (**1347**–**1352**), and the two new 20-*nor*-isopimaranes, smardaesidins F (**1352**) and G (**1353**), along with sphaeropsidins A–F (**1354**–**1359**). Among these, compounds **1348** and **1359** were obtained as an inseparable mixture of isomers. Sphaeropsidin B (**1355**), and 7-*O*-15,16-tetrahydro-sphaeropsidin A (**1360**), and its new derivative, 7-hydroxy-6-oxo-isopimara-7-en-20-oic acid (**1361**), were obtained from sphaeropsidin A (**1354**) by chemical reduction and catalytic hydrogenation, respectively. The acetylation and methylation of **1354** gave two known compounds, 6-*O*-acetoxy-sphaeropsidin A (**1362**) and 8,14-methylene-sphaeropsidin A methyl ester (**1363**), respectively. Methylation of sphaeropsidin C (**1356**) afforded its methyl ester (**1364**). The relative configurations of **1347**–**1353** and **1361** were determined by analysis of their 1D and 2D NMR and MS spectra, while the absolute configurations of **1350** and **1352**–**1354** were derived using the modified Mosher's ester method, their CD spectra, and the comparison of specific rotation data with literature values. Among the newly isolated compounds, smardaesidin A (**1347**) possesses a rare cyclopropane ring fused to ring C of the isopimarane skeleton. Smardaesidins F (**1352**) and G (**1353**) are two polyhydroxylated 20-*nor*-isopimaranes, which are a rare group of diterpenoids. Compounds **1347**–**1364** were evaluated using human cancer cells and cells derived from normal human primary fibroblasts. Compounds **1344**, **1357**, and **1362** exhibited significant cytotoxic activity against the NCI-H460, SF268, NCF-7, MDA-MB-231, PC-3, PC-3 M, and MIAPaCa-2 cancer cells, and WI-38 normal human fibroblasts, with IC_{50} values of 1.9–2.8, 2.1–4.0, 2.0–3.0, 1.4–3.7, 2.5–9.0, 2.4–2.7, 2.0–9.0, and 3.7– > 10 μ*M*, respectively. Compound **1354** showed cancer cell type selectivity and inhibited the migration of metastatic breast adenocarcinoma (MDA-MB-231) cells at sub-cytotoxic concentrations [475].

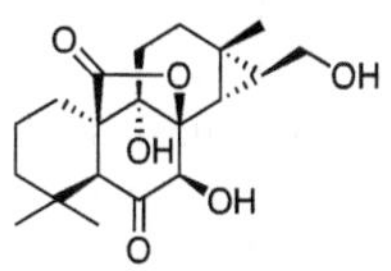

1347 (smardaesidin A)

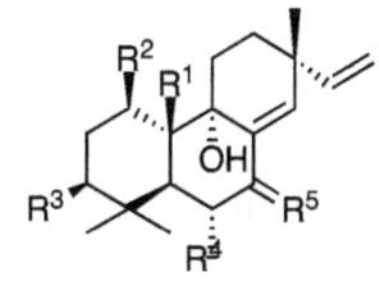

1348 (smardaesidin B, $R^1 = CH_2OH$, $R^2 = OH$, $R^3 = R^4 = H$, $R^5 = O$)
1350 (smardaesidin D, $R^1 = Me$, $R^2 = R^3 = OH$, $R^4 = H$, $R^5 = O$)
1356 (sphaeropsidin C, $R^1 = CH_2OH$, $R^2 = OH$, $R^3 = R^4 = H$, $R^5 = O$)
1359 (sphaeropsidin F, $R^1 = Me$, $R^2 = R^4 = OH$, $R^3 = H$, $R^5 = \alpha$-OH)
1364 (sphaeropsidin C methyl ester, $R^1 = COOMe$, $R^2 = R^3 = R^4 = H$, $R^5 = O$)

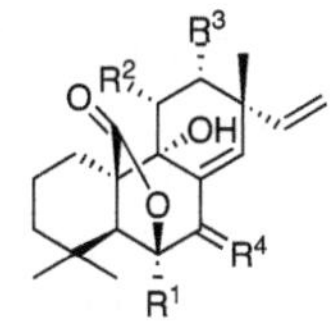

1351 (smardaesidin E, $R^1 = R^2 = H$, $R^3 = OH$, $R^4 = \alpha$-OH)
1354 (sphaeropsidin A, $R^1 = OH$, $R^2 = R^3 = H$, $R^4 = O$)
1355 (sphaeropsidin B, $R^1 = OH$, $R^2 = R^3 = H$, $R^4 = \beta$-OH)
1357 (sphaeropsidin D, $R^1 = R^2 = OH$, $R^3 = H$, $R^4 = O$)
1362 (6-*O*-acetylsphaeropsidin A, $R^1 = OAc$, $R^2 = R^3 = H$, $R^4 = O$)

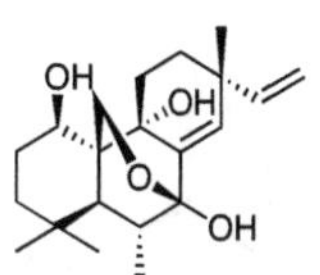

1349 (smardaesidin C)

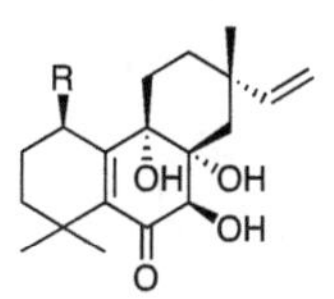

1352 (smardaesidin F, R = H)
1353 (smardaesidin G, R = OH)

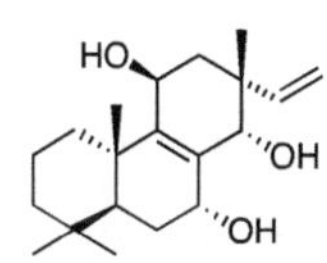

1358 (sphaeropsidin E)

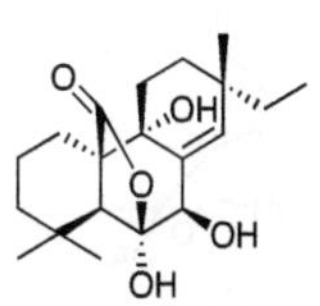

1360 (7-*O*-15,16-tetrahydro-sphaeropsidin A)

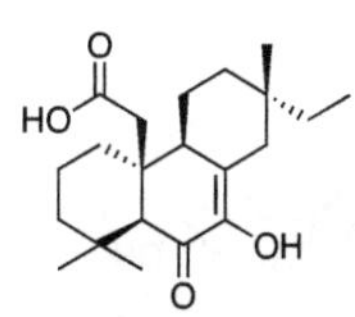

1361 (7-hydroxy-6-oxoiso-pimara-7-en-20-oic acid)

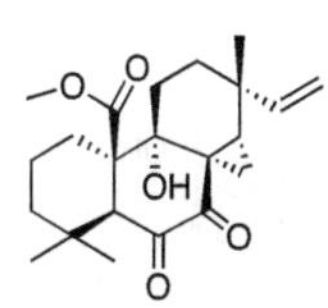

1363 (8,14-methylene-sphaeropsidin A methyl ester)

Several *Hypnum* species were studied chemically by Asakawa et al., and their crude extracts were found to possess cytotoxic activity against several cancer cell lines, as mentioned earlier [3]. Wei et al., reported in 2015 the isolation of four guaiane-type sesquiterpenoids, named xylaguaianols A–D (**1365–1368**), in addition to a cadinane-type compound, isocadinanol (**1369**), and the α-pyrone, 9-hydroxylarone (**1370**), along with the five known sesquiterpenes, epiguaidiol (**1371**), hydroheptelidic acid (**1372**), gliocladic acid (**1373**), bullatantriol (**1374**), and $1\beta,4\beta,7\alpha$-trihydroxyeudesmane (**1375**), and the four known cytochalasins, **1376–1379**, from a culture broth of *Xylaria* species NC1214, a fungal endophyte of a *Hypnum* moss. Cytochalasins C (**1376**), D (**1377**), Q (**1378**), and R (**1379**) were evaluated for their cytotoxic activity against five tumor cell lines: PC-3 M human prostate adenocarcinoma, NCI-H460 human non-small cell lungcancer, SF-268 human CNS glioma, MCF-7 human breast cancer, and MDA-MB-231 human metastatic breast adenocarcinoma. Cytochalasin D (**1377**) showed significant cytotoxic effects against all five cancer cells with IC_{50} values of 0.22–1.44 μM, while cytochalasins C (**1376**) and Q (**1378**) exhibited cytotoxic effects against all tested cancer cell lines, except for MCF-7 cells, with IC_{50} values of 0.96–1.72 and 1.31–1.53 μM [476].

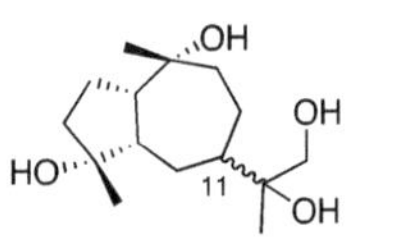

1365 (xylaguaianol A, (11*R*))
1366 (xylaguaianol B, (11*S*))

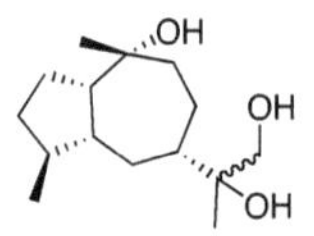

1367 (xylaguaianol C)

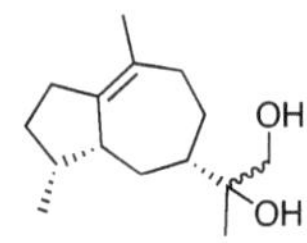

1368 (xylaguaianol D)

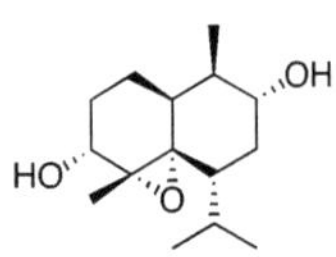

1369 (isocadinanol A)

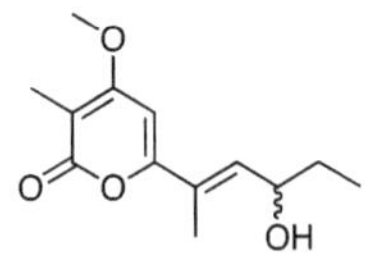

1370 (9-hydroxylarone)

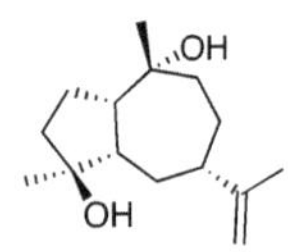

1371 (*epi*-guaidiol)

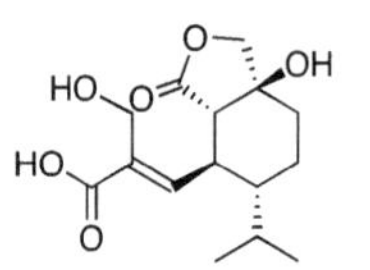

1372 (hydroheptelidic acid)

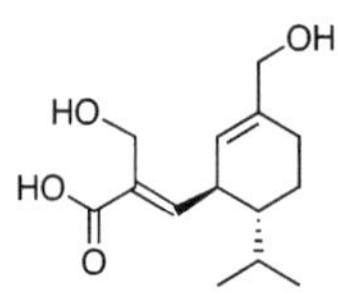

1373 (gliocladic acid)

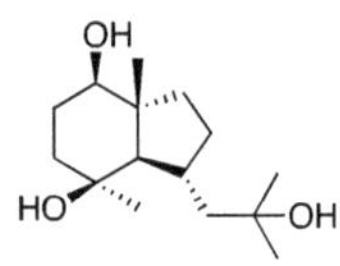

1374 (bullatantriol)

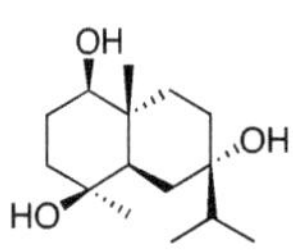

1375 (1β,4β,7α-trihydroxy-eudesmane)

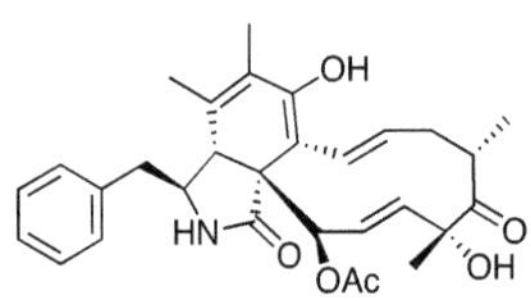

1376 (cytochalasin C)

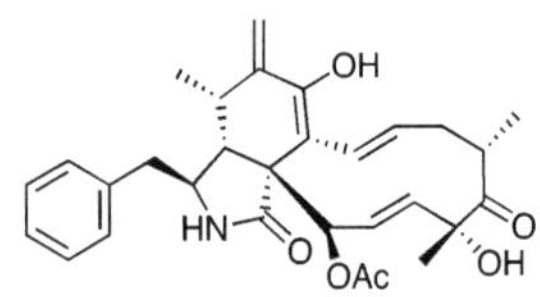

1377 (cytochalasin D)

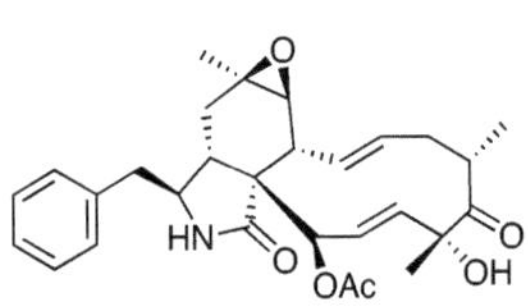

1378 (cytochalasin Q)

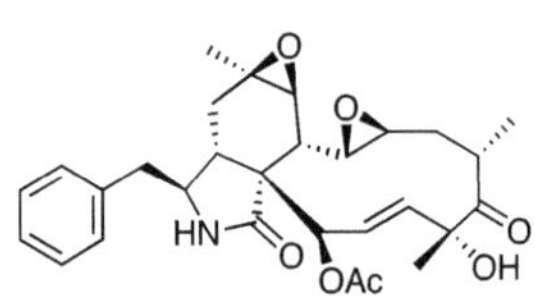

1379 (cytochalasin R)

6.2 *Antimicrobial, Antifungal, Antiviral, and Antioxidant Activities*

The prenylated indole alkaloid, *ent*-homocyclopiamine B (**1380**), possessing an acyclic nitro group, and 2-methylbutane-1,2,4-triol (**1381**), were isolated from the endophytic fungus *Penicillium concentricum*, which was obtained from the liverwort *Trichocolea tomentella* (Table 46).

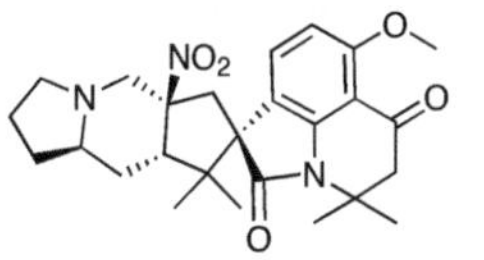

1380 (*ent*-homocyclopiamine B)

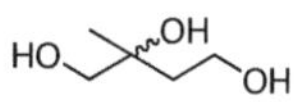

1381 (2-methylbutane-1,2,4-triol)

The antimicrobial activity of **1380** was evaluated against several Gram-positive (*Bacillus subtilis*, *Mycobacterium smegmatis*, *Rhodococcus jhostii*, and *Corynebacterium glutamicum*) and Gram-negative bacteria (*Escherichia coli*, *Salmonella* LT2, *Micrococcus luteus*, *Pseudomonas putida*, and *Serratia marcescens*). Although compound **1380** showed growth inhibition of *Bacillus subtilis* and *Mycobacterium smegmatis* on agar plates, the zones were smaller than those from an equivalent amount of the control antibacterial drug, kanamycin. *B. subtilis* and *C. glutamicum* were found to be somewhat susceptible in the microbroth dilution assays conducted. Hence, with 100 μ*M* of **1380**, this showed 30% inhibition, when compared with the control, kanamycin. No antibacterial effect of **1380** was observed against the Gram-negative bacteria used [477].

An Antarctic endophytic fungus *Mortierrella alpina* was isolated from the moss *Schistidium antarctici.* This strain elaborated high concentration levels of the polyunsaturated fatty acids, γ-linolenic and arachidonic acids, represented by 48.3% of the total fatty acid content, with the monounsaturated fatty acids, oleic, heptadecenoic, and eicosenoic acids (16.2%) and myristic, palmitic, and stearic acids (13.9%) also obtained. The fungal extract showed antioxidant activity in a DPPH radical-scavenging assay, with an ED_{50} value of 48.7 mg/cm^3 and also antimicrobial effects against *Escherichia coli* with a *MIC* value of 26.9 μg/cm^3, and against *Pseudomonas aeruginosa* and *Enterococcus faecalis*, both with a *MIC* value of 107 μg/cm^3. The GC/MS analysis of the chloroform fraction from the crude extract revealed the presence of the potential antibacterial components, 3-isobutylhexahydropyrrolo[1,2-*a*]pyrazine-1,4-dione (**1382**) and 3-benzylhexahydropyrrolo[1,2-*a*]pyrazine-1,4-dione (**1383**) as the major constituents present [478].

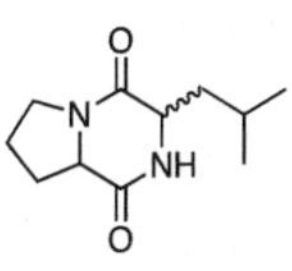

1382 (3-isobutylhexahydro-pyrrolo[1,2-*a*]pyrazine-1,4-dione)

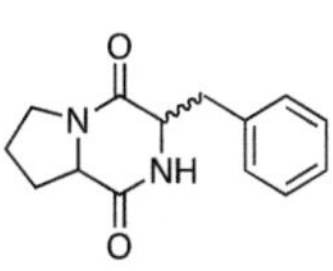

1383 (3-benzylhexahydro-pyrrolo[1,2-*a*]pyrazine-1,4-dione)

Table 46 Antimicrobial, antifungal, and antiviral active compounds and solvent extracts from endophytes of bryophytes

Species name[a] (fungi/bryophytes)	Compound or solvent extract	Microbial species	Activity value	Refs.
Chaetomium fusiforme/[L] *Scapania verrucosa*	Et_2O *C. fumigatus* (acetic acid, methyl valerate (**1289**), butane-2,3-diol (**1290**)-rich) *S. verrucosa* (aromadendrene (**816**), palmitic, acetic acid-rich)	(Fungi)	IC_{50} (μg/ cm^3)	[469]
		Aspergillus fumigatus	32/32	
		Candida albicans	32/8	
		Cryptococcus neoformans	32/64	
		Pyricularia oryzae	128/ > 128	
		Trichophyton rubrum	nd/64	
Penicillium concentricum/[L] *Trichocolea tomentella*	*ent*-Homocyclopiamine B (**1380**)	(Bacteria) *Bacillus subtilis* *Corynebacterium glutamicum*	30% inhibition/ 100 μ*M*	[477]
Mortierrella alpina/[M] *Schistidium antarctici*	CH_2Cl_2		*MIC* (μg/ cm^3)	[478]
		Enterococcus faecalis	107	
		Escherichia coli	26.9	
		Pseudomonas aeruginosa	107	
		Staphylococcus aureus	215.3	
		Klebsiella pneumoniae	315.3	
		Salmonella typhi	215.3	
	$CHCl_3$ (fraction)		Inhibition zone (mm)	
		Enterococcus faecalis	13	
		Escherichia coli	10	
		Klebsiella pneumoniae	13	
		Pseudomonas aeruginosa	10	
		Staphylococcus aureus	10	
		Salmonella typhi	15	

(continued)

Table 46 (continued)

Species name[a] (fungi/bryophytes)	Compound or solvent extract	Virus	Activity value	Refs.
Bacterial endophytes/[L] *Marchantia polymorpha*	EtOAc fraction 1 *cyclo*(L-Leu-L-Pro) = ganicidin W (**1333**), (*cyclo*(L-Phe-L-Pro) (**1343**), included)	Human herpes virus type-1	100 μg/cm^3	[474]
	EtOAc fraction 2 (*N*-phenyl-acetamide) (**1332**), oleic acid amide (**1344**) included		100	

[a] [L] liverwort, [M] moss

The further fractionation of the ethyl acetate extract of the endophytic fungal metabolites of *Marchantia polymorpha* using column chromatography resulted in the isolation of compounds **1333**, **1334**, **1341**, **1342**, **1343**, and **1344** as the major products, together with **1331**, **1335**, **1345**, and **1346** as the minor ones. Antiviral activities of the ethyl acetate extract and each fraction were tested against the human herpes virus type 1. The ethyl acetate extract and the *cyclo*(L-phenyl-alanyl-L-prolyl)-rich fraction from the endophytic extract of *M. polymorpha* diminished the formation of the HHV-1 induced cytopathic effect, and reduced the viral infectious titer by 0.61–1.16 log and the viral load by 0.93–1.03 log. They exerted a low antiviral potential against the human herpes virus type 1 replicating in VERO cells [478].

Stelmasiewicz et al., reviewed in 2023 the current knowledge on the diversity of compounds produced by endophytes isolated from bryophytes, emphasizing the biologically active molecules. In addition, the isolation procedures and biodiversity of endophytes from mosses, liverworts, and hornworts were described in detail. The authors proposed the terminology "bryendophytes" for the endophytes isolated from the Bryophyta. The secondary metabolites from the "bryendophytes" show significant structural diversity and their chemical profiles are totally different from the host bryophytes. They contain nitrogen-containing compounds, such as alkaloids and peptides, as well as phenolics, polyketides, and terpenoids. The metabolites isolated from the bryendophytes show various biological and pharmacological activities [479]. The newly isolated bryendophytes and compounds and their structures are shown in Table 46.

6.3 Immunosuppressive Activity

Two new sulfur-containing xanthone derivatives named sydoxanthones A (**1384**) and B (**1385**), and 13-*O*-acetylsydowinin B (**1386**) were isolated from an endophytic fungus, *Aspergillus sydowii*, from the Chinese liverwort *Scapania* *ciliata* (Table 47), along with the seven known related compounds, 8-hydroxy-6-methyl-9-oxo-9*H*-xanthene-1-carboxylic acid methyl ester (**1387**), pinselin (**1388**), moniliphenone (**1389**), emodin (**1390**), questin (**1391**), 1-hydroxy-6,8-dimethoxy-3-methyl-anthraquinone (**1392**), and 1,9-dihydroxy-3-(hydroxymethyl)-10-methoxy-dibenzo[*b*,*e*]-oxcepine-6,11-dione (**1393**) (Table 47). Emodin (**1390**) showed moderate immunosuppressive activities against the proliferation of mouse splenic lymphocytes that were Con-A-induced and LPS-induced, with IC_{50} values of 8.5 μg/cm^3 and 10.3 μg/cm^3, respectively. Also, questin (**1391**) showed comparable IC_{50} values of 10.1 and 14.1 μg/cm^3 [480].

Table 47 Immunosuppressive compounds from endophytes of bryophytes

Species name[a]	Compound or solvent extract	Mouse splenic lymphocytes	Refs.
Aspergillus sydowii/[L] *Scapania ciliata*		Con A-induced/ LPS-induced proliferation IC_{50}	[480]
	Sydoxanthone A (**1384**)	> 5000 / > 5000	
	Sydoxanthone B (**1385**)	22.5/15.3	
	13-*O*-Acetylsydowinin (**1386**)	172.7 /213.4	
	8-Hydroxy-6-methyl-9-oxo-9*H*-anthene-1-carboylic acid methyl ester (**1387**)	18.2/20.6	
	Pinselin (**1388**)	26.8 /10.7	
	Moniliphenone (**1389**)	> 5000 / > 5000	
	Emodin (**1390**)	8.5 /10.3	
	Questin (**1391**)	10.1 /14.1	
	1-Hydroxy-6,8-dimethoxy-3-methylanthraquinone (**1392**)	31.2 /51.2	
	1,9-Dihydroxy-3-(hydroxymethyl)-10-methoxydibenzo[*b*,*e*] oxcepine-6,11-dione (**1393**)	> 5000/ > 5000	

[a] [L] liverwort

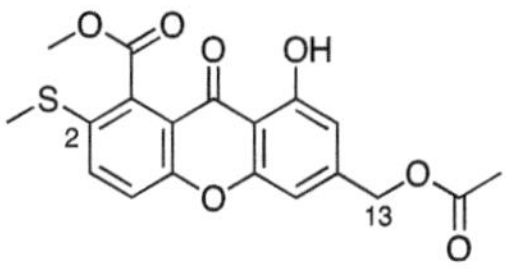

1384 (sydoxanthone A)

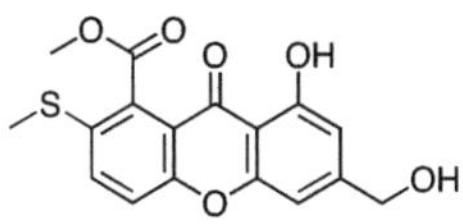

1385 (smardaesidin B)

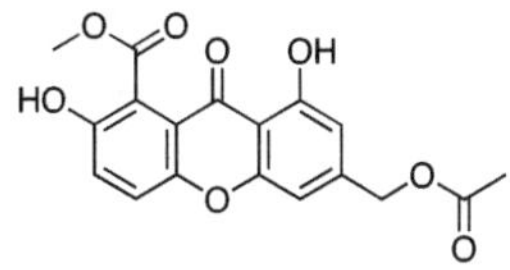

1386 (13-*O*-acetylsydowinin B)

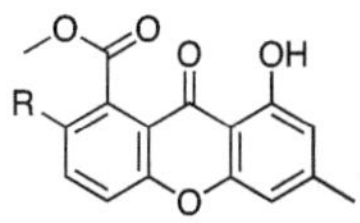

1387 (2-dehydroxy-13-deacetoxy-sydowinin B, R = H)
1388 (pinselin, R = OH)

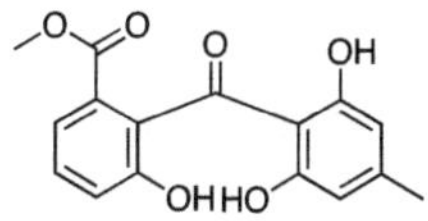

1389 (moniliphenone)

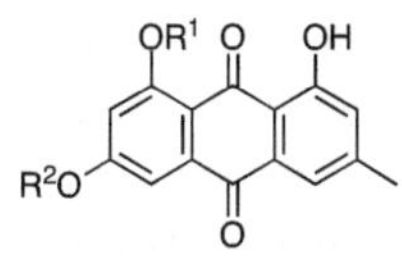

1390 (emodin, $R^1 = R^2 = H$)
1391 (questin, $R^1 = Me$, $R^2 = H$)
1392 (1-hydroxy-6,8-dimethoxy-3-methyl-anthraquinone, $R^1 = R^2 = Me$)

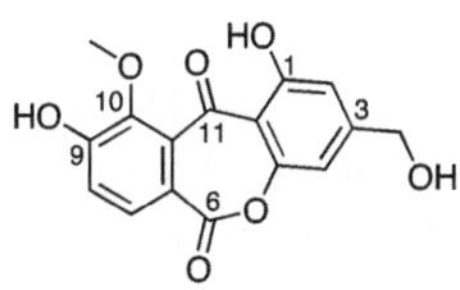

1393 (1,9-dihydroxy-3-(hydroxy-methyl)-10-methoxydibenzo[*b,e*]oxepine-6,11-dione)

7 Applications of Phytochemicals from Bryophytes

7.1 Applications of Volatile Compounds in Cosmetic Formulations

As mentioned earlier, several components isolated from liverworts, possess a commercially significant value in the cosmetic field [481]. Bicyclohumulenone (**51**) from the liverwort *Plagiochila sciophila* and tamariscol (**52**) from *Frullania tamarisci* subsp. *tamarisci* and *F. tamarisci* subsp. *obscura* are such examples, since they can be applied as fragrant components to perfumes and other cosmetics as well as in deodorant sprays. With respect to liverwort species with pleasant odors, the present author and his colleagues have obtained their essential oils and solvent extracts. The essential oils and solvent extracts from many liverworts, for example, *Marchantia polymorpha*, *Pellia endiviifolia*, *Plagiochila*, and *Porella* species possess potent antimicrobial and antifungal properties, as referred to in Tables 12 and 13. Not only the isolated compounds, but also the essential oils and solvent extracts could be used for the preservation of food and as food additives as well as in sprays and deodorants. Also, steam vapors that are emitted from humidifiers that include essential oils from the above-mentioned liverworts could play an important role in reducing infections from the COVID-19 virus and other viruses.

7.2 Applications in Foods and Beverages

The present authors and his associates have found success in the highly efficient production of the mushroom-like fragrant components of the most expensive Japanese mushroom, *Tricholoma matsutake*, from the large thalloid liverwort, *Conocephalum conicum* (Plate 35), and of perillaldehyde (**57**) from *Marchantia paleacea* subsp. *diptera*, using plant engineering systems. These have been used as sources of food additives for consommé soup, and the fresh thallus of the former liverwort has been used directly for flavoring hamburgers and tempura items in Japanese restaurants [482].

Marchantia polymorpha has been utilized for a long time as a medicinal liverwort in the Far East, because it displays multiple biological and pharmacological activities. Thus, tea-bags containing powdered *M. polymorpha* are now used as an herbal tea.

As feeds for livestock and foods for humans, bryophytes are very useful because they contain certain vitamins, minerals, as well as fiber. The mosses, *Barbella pendula, B. enervis, Floribundaria nipponica, Hypnum plumaeforme*, and *Neckeropsis nitidula,* and some of the liverworts, contain quantitatively considerable amounts of vitamin B_2. Chickens and puppies fed a diet including these powered bryophytes gained more weight than did the control animals. The supplement formulation used did not cause any sickness or distaste [483]. Since there are more than 14,000 species of mosses, further species possessing vitamin B_2 undoubtedly will be discovered.

Plate 35 *Conocephalum conicum* (liverwort) used as a food

The liverworts, *Marchantia polymorpha* and *Pellia endiviifolia,* and the mosses, *Atrichum undulatum* and *Mnium hornum*, produce vitamin E (α-tocopherol) (**15**), vitamin K (**1394**), plastoquinone (**1395**), plastohydroquinone (**1396**), and α-tocoquinone (**1397**) [38]. The last compound was also found in the moss *Racomitrium japonicum* [75]. In 2007 Nishiki et al., analyzed 700 liverworts chemically, and found that almost all of them contained α-tocopherol (**15**) and squalene [484].

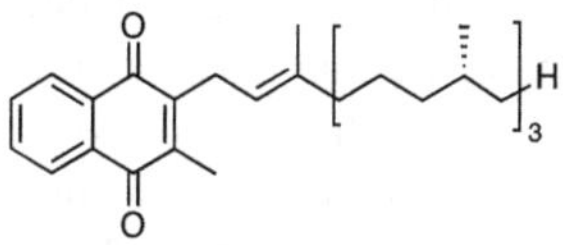

1394 (vitamin K)

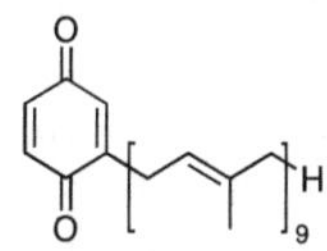

1395 (plastoquinone)

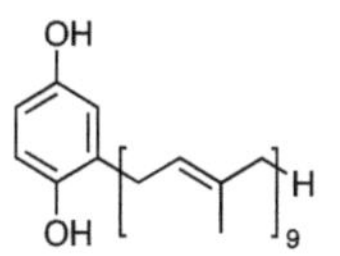

1396 (plastohydroquinone)

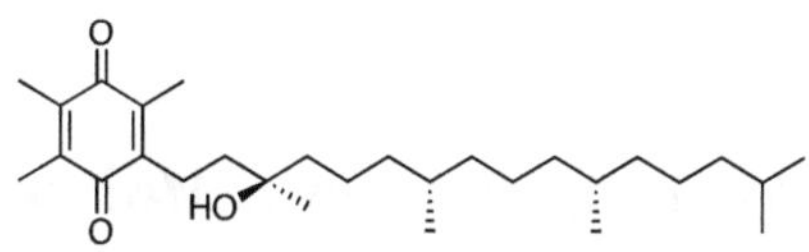

1397 (α-tocoquinone)

The prostaglandin-like fatty acids **1018**, **1019**, and **1398–1401** have been found in several mosses, such as *Dicranum scoparium*, *D. japonicum*, and *Leucobryum* species [485, 486] These, and other highly unsaturated fatty acids from bryophytes are viscous liquids, and it is thought that they are instrumental in protecting herbivorous animals living in very inhospitable places from the cold [487]. Such unsaturated fatty acids from mosses [488], like those obtained from fish oils, play an important role as antioxidants in the human body.

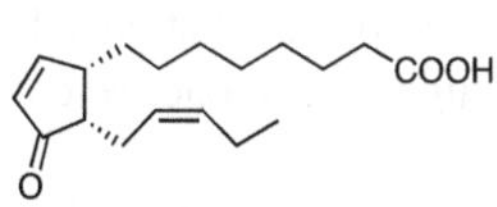

1398 (8-((1*S*,5*S*)-4-oxo-5-((*Z*)-pent-2-en-1-yl)cyclopent-2-en-1-yl)octanoic acid)

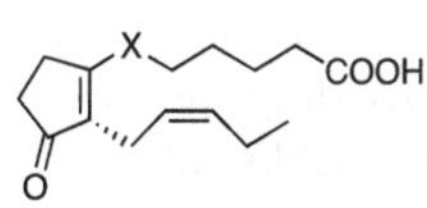

1399 ((*Z*)-6-(3-oxo-2-(pent-2-en-1-yl)cyclopent-1-en-1-yl)-6-octynoic acid, X = $-CH_2-C\equiv C-$)

1400 ((*Z*)-6-(3-oxo-2-(pent-2-en-1-yl)cyclopent-1-en-1-yl)-6-octenoic acid, X = $-CH_2CH=CH-$)

1401 ((*Z*)-6-(3-oxo-2-(pent-2-en-1-yl)cyclopent-1-en-1-yl)octanoic acid, X = $-(CH_2)_3-$)

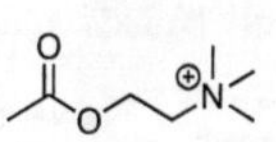

1402 (acetylcholine)

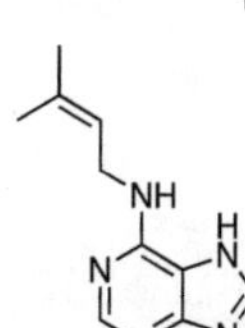

1403 (N_6-(isopentenyl)adenine)

Acetylcholine (**1402**) and the cytokinin-like compound N_6-(isopentenyl)adenine (**1403**) have been found in a callus tissue produced from a hybrid of *Funaria hygrometrica* x *Physcomitrium pyriforme* [489, 490]. Many liverworts produce hot-tasting compounds, which could be used as spices for foods and as food preservatives, especially because they also may possess potent antimicrobial and antifungal activities.

The mineral content in plants is important in foods used in the diet. Generally, the calcium ion content in liverworts is high. For example, the calcium content was found to be elevated (138.8 g/kg) in *Plagiochila asplenioides.* The second most abundant element found was potassium (91.6 g/kg), with the other metal ion contents being Al 2.0, Fe 741, Zn 130, Mg 529, Cu 17, Mn 972, Pb 349, and Cr 8.6 mg/kg. However, Co and Ni were found to be below the limit of quantification [390].

7.3 Potential Applications as Medicinal Agents

In Table 1, the medicinal bryophytes are listed and their biological and pharmacological activities indicated [36]. However, the active compounds from the mosses probably have not yet been fully elucidated, except for the presence of the terpenoids and aromatic compounds listed in Tables 9, 10, 11, 12, 13, 14, 15, 16, 17, 18, 19, 20, 21, 22, 23, 24, 25, 26, 27, and 28. Only a relatively few pre-clinically or clinically promising phytochemicals have been isolated and their structures determined.

The mosses, *Rhodobryum giganteum* (Plate 36) and *R. roseum* are fascinating species because the crude extracts process neurotrophic and cardiovascular protective effects. *Polytrichum* species are also beneficial species because of the presence of potential antitumor constituents. There are many other species that have been used as traditional crude drugs in various countries, although their bioactive compounds still remain to be established.

Plate 36 *Rhodobryum giganteum* (moss)

A number of the crude extracts and isolated compounds from bryophytes have been screened against various human cancer cell lines as shown in Table 9. Several of these possess potent cytotoxicity when evaluated in this manner, and could benefit from further laboratory evaluation.

One of the most important bryophytes is the liverwort *Marchantia polymorpha*, which contains the structurally stable marchantin A (**1**) in large amounts in the crystal state, and shows multiple biological and pharmacological activities, like cancer cell line cytotoxicity, as discussed earlier. Marchantins A (**1**) and C (**197**) display tubulin polymerization inhibitory effects and the former bis-bibenzyl possesses potent antitrypanosomal, anti-influenza and muscle relaxant activity. Generally, liverworts are a tiny group of plants, but, however, *M. polymorpha* is very easily collected in the field, or cultured in greenhouses, and can be obtained in 500–1000 kg amounts in the form of fresh material (Plates 37 and 38) [76].

The major component of *M. paleacea* subsp. *diptera* is also marchantin A (**1**), with its content being much higher than in *M. polymorpha*, and is readily collected in mountainous regions. As large amounts of the marchantins were obtained from these thalloid liverworts, it is easy to modify their original structures chemically or enzymatically to give further types of bioactive compounds. This approach was applied by the groups of Asakawa and Lou to produce the riccardin and plagiochin series of compounds, as mentioned in the section of this contribution on cancer cell line cytotoxicity. The macrocyclic bis-bibenzyls possess equivalent tubulin polymerization activity to that of paclitaxel, which is used widely clinically as an established anticancer drug. Also, **1** has shown potent antitrypanosomal activity. Although active compounds have not yet been isolated, a formulated product prepared from the *n*-hexane crude extract of the African liverwort *Marchantia debilis* was found to cure a bacterial-infected skin disease completely after two weeks of treatment [491]. In Africa and other regions of the world, many bacterial and fungal infectious skin diseases occur, so therefore, further investigation of the chemical types and their efficacy and of mechanism of action of the potentially active lipophilic compounds from *M. debilis* is desirable.

Another promising clinically used compound is *cis*-perrottetinene (**279**), which initially was isolated from Japanese and New Zealand liverworts in the genus *Radula*. The structure of **279** resembles that of (–)-*trans*-Δ^9-tetrahydrocannabinol (**281**) obtained from *Cannabis sativa*. Not surprisingly, **279** showed the same psychoactivity as **281**, and, in addition, it exhibits antiinflammatory activity. It is very easy to obtain *Radula* species by cultivation throughout the whole year, so **279** is a potentially useful clinical agent. Furthermore, riccardin C (**182**), which showed LXRα and β agonist and antagonist effects, led to elevation of high-density lipoprotein cholesterol and decreased triglycerides in vitro, and in vivo using mice. Thus, this structurally stable macrocyclic bis-bibenzyl is also a possible clinically effective drug [492].

Plate 37 *Marchantia polymorpha* (liverwort) field collection

Plate 38 *Marchantia polymorpha* (liverwort) cultured in a greenhouse

7.4 Applications as Other Useful Materials

Many *Sphagnum* species were used to treat soldiers in the First World War as a substitute for cotton, since they contain antimicrobial and antifungal compounds and absorb fifty times more water than cotton. *Sphagnum* mosses have been used in pots for culturing orchid species, which are protected by *Sphagnum.* A special pillow including *Sphagnum* is manufactured for bed-ridden elderly men in Japan and sold through the Internet. In cold countries, *Sphagnum* has been used in bedding and house building supplies. In Japanese architecture, companies use sheets of the moss *Racomitrium canescens* to spread on roof tops for urban warming prevention.

8 Conclusions

The bryophytes are terrestrial green spore-forming plants and there are 23,000 species globally. They are divided to three taxonomic classes, the Bryophyta (mosses), Marchantiophyta (liverworts), and Anthocerotophyta (hornworts), among which the Marchantiophyta species have been studied widely chemically, because they contain cellular oil bodies that are extracted readily with organic solvents to obtain large amounts of secondary metabolites. Since 1903, about one thousand literature references concerning phytochemicals of bryophytes have been published [1–3] and more than two thousand secondary metabolites have been isolated from or detected in bryophytes, with their strucures elucidated, and with many showing various biological activities, as described earlier in this volume.

The crude extracts and essential oils from liverworts are complex chemically, since they contain various lipophilic terpenoids, aromatic compounds and acetate-derived compounds, which are used as chemical markers for each liverwort genus and family. Among the secondary metabolites, bis-bibenzyls and bibenzyls are the most characteristic chemical components of liverworts. It should be noted that only two bis-bibenzyls, riccardin C (**182**) and perrottetin E (**183**), have been found elsewhere in the plant kingdom, such as in *Primula* species. Bis-bibenzyls possess not only characteristic structures but also various biological and pharmacological activities, such as muscle relaxant, antioxidant, tubulin polymerization and NO production inhibition, antiviral, antimicrobial, antifungal, and antitrypanosomal effects, among others. Since 2003, only a small percentage of the known bryophytes have been studied chemically, because bryophytes are minute in size, difficult to identify, and thought to be impractical for use in the human diet. However, owing to recent plant engineering techniques, some liverworts, such as the thalloid liverworts, *Conocephalum conicum*, *Marchantia paleacea* subsp. *diptera*, and *M. polymorpha*, and the stem-leafy species, *Bazzania japonica* and *Radula perrottetii*, may be cultivated readily for the production of large quantities of secondary metabolites including cytotoxic bis-bibenzyls and characteristic scent compounds (Plate 39). Currently, the fresh thallus of *C. conicum* has been used as a food component of hamburgers and in

Plate 39 Culture room for liverworts

consommé soup in Japan, although bryophytes in general have been neglected for a long time as food ingredients.

Some mosses produce vitamin B_2 and different types of highly unsaturated fatty acids and lipids (triglycerides) containing highly unsaturated alkane moieties. Thus, a further strategy is to focus on their secondary metabolites, which could be utilized in the diets of cattle and cultured fish. The metabolites of liverworts also may be able to lead to clinically useful drugs because they possess some potentially important metabolites, particularly bis-bibenzyls of the marchantin and isoplagiochin series, which interfere with the normal breakdown of microtubules during cell division of cancer cells, like the clinically useful anticancer drugs, paclitaxel (Taxol®) and docetaxel. A bibenzyl cannabinoid, *cis*-perrottetinene (**279**) from the peculiar liverwort, *Radula* species, possesses the potential to become a drug used clinically since it shows exactly the same psychoactivity as that of the well-known compound, (–)-Δ^9-tetrahydrocannabinol (**281**), from *Cannabis sativa*, and also shows anti-inflammatory activity.

Acknowledgments The author thanks Profs. M. Toyota, F. Nagashima, M. Tori, T. Hashimoto and Y. Fukuyama (Tokushima Bunri Univ.), the deceased G. Ourisson (Univ. Louis Pasteur), M. Zenk and T. Kutchan (Donald Danforth Plant Center), G.A. Cordell (Natural Products Inc., IL), J.M. Pezzuto (Western New England Univ.), A.D. Kinghorn (Ohio State Univ.), S. Gibbons (University of Nizwa, Nizwa, Oman), A. Bardón (Tucuman National Univ., Argentina) and Drs. L. Harrison and M. Buchanan (Univ. of Glasgow), L. Harinantenaina (Ohio State Univ.), A. Ludwiczuk (Lublin Medical Univ.), and M. Novakovic and D. Bukvicki (Univ. of Belgrade), for their collaboration, discussion, and support on the phytochemistry of bryophytes. The author is thankful to Prof. R. Gradstein (National Museum of Natural History, Paris), Drs. S. Hattori and M. Mizutani (Hattori Botanical Laboratory, Nichinan, Japan), Dr. H. Akiyama Museum of Nature and Human Activities, Hyogo, Japan, Dr. T. Furuki (Natural History Museum and Institute, Chiba, Japan), and Profs. P. Raharivelomanana, J.-P. Bianchini and Dr. A. Pham (Univ. French Polynesia, Tahiti), Dr. L. Zhang (Shenzhen Fairylake Botanical Garden, China), and Mr. M. Izawa (Saitama, Japan) for their collection, identification, and photographs of liverworts. The author is also indebted to Professor Dr. Heinz Falk (Johannes Kepler University, Linz, Austria) whose contribution to this review went far beyond his editorial duties and his revision of the manuscript as well as valuable comments and suggestions. A part of this work was supported by a Grant-in-Aid for Scientific Research (A) (No. 11309012) from the Ministry of Education, Culture, Sports, Science, and Technology, Tokyo, Japan and Matsumae International Foundation, Tokyo, Japan.

References

1. Asakawa Y (1982) Chemical constituents of the Hepaticae. In: Herz W, Grisebach H, Kirby GW (eds) Progress in the chemistry of organic natural products, vol 42. Springer, Vienna, p 1
2. Asakawa, Y (1995) Chemical constituents of bryophytes. In: Herz W, Kirby WB, Moore RE, Steglich W, Tamm Ch (eds) Progress in the chemistry of organic natural products, vol 65. Springer, Vienna, p 1
3. Asakawa Y, Ludwiczuk A, Nagashima F (2013) Chemical constituents of bryophytes. Bio- and chemical diversity, biological activity, and chemosystematics. In: Kinghorn AD, Falk H, Kobayashi J (eds) Progress in the chemistry of organic natural products, vol 95. Springer, Vienna, p 1
4. Asakawa Y, Toyota M, Uemoto M, Aratani T (1976) Sesquiterpenoids of six *Porella* species. Phytochemistry 15:1929
5. Asakawa Y, Takemoto T (1977) Sesquiterpene lactones and heterocyclic compounds of bryophytes. Heterocycles 8:563
6. Asakawa Y, Tokunaga N, Toyota M, Takemoto T, Suire C (1979) Chemosystematics of bryophytes. I. The distribution of terpenoids in bryophytes. J Hattori Bot Lab 45:395
7. Asakawa Y, Tokunaga N, Toyota M, Takemoto T, Hattori S, Mizutani M, Suire C (1979) Chemosystematics of bryophytes. II. The distribution of terpenoids in Hepaticae and Anthocerotae. J Hattori Bot Lab 46:67
8. Asakawa Y, Hattori S, Mizutani M, Tokunaga N, Takemoto T (1979) Chemosystematics of bryophytes III. Terpenoids of primitive Hepaticae, *Takakia* and *Haplomitrium*. J Hattori Bot Lab 46:77
9. Asakawa Y, Inoue H, Toyota M, Takemoto T (1980) Sesquiterpenoids of fourteen *Plagiochila* species. Phytochemistry 19:2651
10. Asakawa Y, Tokunaga N, Takemoto T, Hattori S, Mizutani M, Suire C (1980) Chemosystematics of bryophytes. IV. The distribution of terpenoids and aromatic compounds in Hepaticae and Anthocerotae. J Hattori Bot Lab 47:153
11. Asakawa Y, Suire C, Toyota M, Tokunaga N, Takemoto T (1980) Chemosystematics of bryophytes. V. The distribution of terpenoids and aromatic compounds in European and Japanese Hepaticae. J Hattori Bot Lab 48:285
12. Asakawa Y, Matsuda R, Takemoto T, Hattori S, Mizutani M, Inoue H, Suire C, Huneck S (1981) Chemosystematics of bryophytes. VII. The distribution of terpenoids and aromatic compounds in European and Japanese Hepaticae. J Hattori Bot Lab 50:107
13. Asakawa Y, Matsuda R, Toyota M, Suire C, Takemoto T, Inoue, H, Hattori S, Mizutani M (1981) Chemosystematics of bryophytes. VIII. The distribution of terpenoids and aromatic compounds in Japanese Hepaticae. J Hattori Bot Lab 50:165
14. Gradstein SR, Matsuda R, Asakawa Y (1981) Studies on Colombian cryptogams. XIII. Oil bodies and terpenoids in Lejeuneaceae and other selected Hepaticae. J Hattori Bot Lab 50:231
15. Asakawa Y, Matsuda R, Toyota M, Hattori S, Ourisson G (1981) Terpenoids and bibenzyls of 25 liverwort *Frullania* species. Phytochemistry 20:2187
16. Asakawa Y, Toyota M, Takemoto T, Mues R (1981) Aromatic esters and terpenoids of the liverworts in the genera *Trichocolea*, *Neotrichocolea* and *Trichocoleopsis*. Phytochemistry 20:2695
17. Asakawa Y, Matsuda R, Schofield WB, Gradstein SR (1982) Cuparane and isocuparane-type sesquiterpenoids in liverworts of genus *Herbertus*. Phytochemistry 21:2471
18. Asakawa Y, Campbell EO (1982) Terpenoids and bibenzyls from some New Zealand liverworts. Phytochemistry 21:2663
19. Asakawa Y, Matsuda R, Toyota M, Takemoto T, Connolly JD, Phillips WR (1983) Sesquiterpenoids from *Chiloscyphus*, *Clasmatocolea* and *Frullania* species. Phytochemistry 22:961
20. Asakawa Y, Toyota M, Matsuda R, Takikawa K, Takemoto T (1983) Distribution of novel cyclic bisbibenzyls in *Marchantia* and *Riccardia* species. Phytochemistry 22:1413

21. Takikawa K, Tori M, Asakawa Y (1989) Chemical constituents and chemosystematics of *Radula* species (Liverworts). J Hattori Bot Lab 67:365
22. Asakawa Y, Lin X, Kondo K, Fukuyama Y (1991) Terpenoids and aromatic compounds from selected East Malaysian liverworts. Phytochemistry 30:4019
23. Tori M, Nagai T, Asakawa Y, Huneck S, Ogawa K (1993) Terpenoids from six Lophoziaceae liverworts. Phytochemistry 33:1445
24. Asakawa Y, Ludwiczuk A (2008) Bryophytes—chemical diversity, bioactivity and chemosystematics. Part 1. Chemical diversity and bioactivity. Medicinal plants in Poland and in the world (Rośliny Lecznicze w Polsce i na Świecie) 2:33
25. Asakawa Y, Ludwiczuk A (2008) Bryophytes-chemical diversity, bioactivity and chemosystematics. Part 2. Chemosystematics. Medicinal plants in Poland and in the world (Rośliny Lecznicze w Polsce i Na Świecie) 3/4:43
26. Asakawa Y, Toyota M, Nagashima F, Hashimoto T (2008) Chemical constituents of selected New Zealand and Japanese liverworts. Nat Prod Commun 3:289
27. Ludwiczuk A, Asakawa Y (2008) Distribution of terpenoids and aromatic compounds in selected southern hemispheric liverworts. Fieldiana Bot 47:37
28. Ludwiczuk A, Asakawa Y (2010) Chemosystematics of the liverworts collected in Borneo. Trop Bryol 31:33
29. Asakawa Y, Ludwiczuk A (2012) Distribution of cyclic and acyclic bis-bibenzyls in the Marchantiophyta (liverworts), ferns and higher plants and their biological activity, biosynthesis, and total synthesis. Heterocycles 86:891
30. Asakawa Y, Ludwiczuk A, Nagashima F (2013) Phytochemical studies on bryophytes: bio- and chemical diversity, and biological activity. Phytochemistry 91:52
31. Coulerie R, Nour M, Toubenot L, Asakawa Y (2014) Sesquiterpene hydrocarbons from the liverwort *Treubia isignensis* var. *isignensis* with chemotaxonomic significance. Nat Prod Commun 9:1059
32. Coulerie P, Thouvenot L, Nour M, Asakawa Y (2015) Chemical originalities of New Caledonian liverworts from Lejeuneaceae family. Nat Prod Commun 10:1501
33. Santoni CJ, Asakawa Y, Nour M, Montenegro G (2017) Volatile chemical constituents of the Chilean bryophytes. Nat Prod Commun 12:1929
34. Asakawa Y, Baser KHC, Erol B, Reus SV, Konig WA, Ozenglu H, Gokler I (2018) Volatile components of some selected Turkish liverworts. Nat Prod Commun 13:899
35. Ding H (1982) Zhong guo Yao Yun Bao zi Zhi Wu, Kexue Jishu Chuban She, Shanghai, p 1
36. Glime JM (2017) Medical uses: medical conditions. In: Glime JM (ed) Bryophyte ecology, vol 5. Uses ebook. Sponsored by Michigan Technological University and the International Association of Bryologists. http://digitalcomMons.mtu.edu/bryophyte-ecology/
37. Asakawa Y (1990) Biologically active substances from bryophytes. In: Chopra RN, Bhatla SC (eds) Bryophyte development: physiology and biochemistry. CRC Press, Boca Raton, FL, USA, p 259
38. Asakawa Y (1990) Terpenoids and aromatic compounds with pharmaceutical activity from bryophytes. In: Zinsmeister DH, Mues R (eds) Bryophytes: their chemistry and chemical taxonomy. Oxford University Press, Oxford, UK, p 369
39. Asakawa Y (1993) Biologically active terpenoids and aromatic compounds from liverworts and inedible mushroom *Cryptoporus volvatus*. In: Colegate S, Molyneux RJ (eds) Bioactive natural products: detection, isolation, and structural determination. CRC Press, Boca Raton, FL, USA, p 319
40. Asakawa Y (1994) Biologically active terpenoids and aromatic compounds from liverworts and inedible mushroom *Cryptoporus volvatus*. Pure Appl Chem 66:2193
41. Asakawa Y (2001) Recent advances in phytochemistry of bryophytes: acetogenins, terpenoids and bis(bibenzyl)s from selected Japanese, Taiwanese, New Zealand, Argentinean and European liverworts. Phytochemistry 56:297
42. Asakawa Y (2004) Chemosystematics of the Hepaticae. Phytochemistry 65:623
43. Asakawa Y (2007) Biologically active compounds from Bryophytes. Chemia 9:73

44. Asakawa Y (2008) Liverworts—potential source of medicinal compounds. Curr Pharm Design 14:3067
45. Asakawa Y (2008) Recent advances of biologically active substances from the Marchantiophyta. Nat Prod Commun 3:77
46. Asakawa Y, Ludwiczuk A (2009) Bio- and chemical diversity, bioactivity and chemosystematics, Marchantiophyta. Malaysian J Sci 28:229
47. Asakawa Y (2015) Search for new liverwort constituents of biological interest. In: Chauhan AK, Pushpangadan P, Gerge V (eds) Natural products: recent advances. Write & Print, Kerala, India, p 25
48. Asakawa Y (2007) Biologically active compounds from bryophytes. Pure Appl Chem 79:557
49. Asakawa Y, Ludwiczuk A, Nagashima F, Toyota M, Hashimoto T, Tori M, Fukuyama Y, Harinantenaina L (2009) Bryophytes—bio- and chemical diversity, bioactivity and chemosystematics. Heterocycles 77:99
50. Asakawa Y, Toyota M, Tori M, Hashimoto T (2000) Chemical structures of macrocyclic bis(bibenzyls) isolated from liverworts (Hepaticae). Spectroscopy 14:149
51. Asakawa Y, Ludwiczuk A, Toyota M (2014) Chemical analysis of bryophytes. In: Hostettmann K, Stuppner H, Marston A, Chen S (eds) Handbook of chemical and biological analysis methods (II), 1st edn. Wiley, Oxford, UK, p 1
52. Zinsmeister HD, Becker H, Either T (1991) Moose, eine Quelle biologisch aktiver Naturstoffe? Angew Chem 103:134
53. Harris ES (2008) Ethnobryology: traditional uses and folk classification of bryophytes. Bryologist 111:169
54. Azuelo AG, Darianna LG, Pabuaoan MP (2011) Some medicinal bryophytes their ethnobotanical uses and morphology. Asian J Biodivers 92:49
55. Nandy S, Dey A (2012) Bibenzyls and bisbibenzyls of bryophytic origin as promising source of novel therapeutics: pharmacology, synthesis and structure-activity. Daru J Pharm Sci 28:701
56. Singh S, Srivastava K (2013) Bryophytes as green brain: unique and indispensable small creature. Int J Pharm Sci Rev Res 23:28
57. Dziwak M, Wróblewska K, Szumny A, Galek R (2022) Modern use of bryophytes as a source of secondary metabolites. Agronomy 12:1456
58. Dey A, Mukerjee A (2015) Therapeutic potential of bryophytes and derived compounds against cancer. J Acute Disease 4:236
59. Chandra S, Chandra D, Barh A, Pankaj PRK, Sarma IP (2017) Bryophytes: hoard of remedies, an ethno-medicinal review. J Trad Complement Med 7:94
60. Cianciulllo P, Maresca V, Sorbo S, Basile A (2022) Antioxidant and antibacterial properties of extracts and bioactive compounds in bryophytes. Appl Sci 12:160
61. Cianciullo P, Cimmino F, Maresca V, Sorbo S, Bontempo P (2022) Anti-tumour activities from secondary metabolites and their derivatives in bryophytes: a brief review. Appl Sci 1:73
62. Horn A, Pascal A, Loncarevic I, Marques RV, Lu Y, Miguel S, Bourgaud F, Thorsteinsdottir M, Cronberg N, Becker JD, Reski R, Simonsen HT (2021) Natural products from bryophytes: from basic biology to biotechnological applications. Crit Rev Plant Sci 40:191
63. Commisso M, Guarino F, Marchi L, Muto A, Piro A, Degola F (2021) Bryo-activities: a review on how bryophytes are contributing to the arsenal of natural active compounds against fungi. Plants 10:203
64. Sala-Carvalho WR, Peralta DF, Furlan CM (2023) A chemistry overview of the beautiful miniature forest known as mosses. Bryologist 126:341
65. Martinez-Abaigar J, Nunez-Olivera E (2021). In: Sinha RP, Häder D-P (eds) Natural bioactive compounds—technological advances. Novel biotechnological substances from bryophytes. Elsevier Inc., Philadelphia, p 233
66. Sen K, Khan MI, Paul R, Ghoshal U, Asakawa Y (2023) Recent advances in the phytochemistry of bryophytes: distribution, structure and biological activity of bibenzyls and bisbibenzyl compounds. Plants 12:4173
67. Costica M, Stratu A, Costica N (2023) Liverworts and mosses from Romania with medicinal potential. Acta Biol Marisci 6:113

68. Motti R, Palma AD, de Falco B (2023) Bryophytes used in folk medicine: an ethnobotanical overview. Horticulture 9:137
69. Singh S, Sharma R, Joshi S, Alam A (2023) Utilization of the first land plants (Bryophytes) as a source of beneficial bioactive chemicals. Res J Phytochem 17:26
70. Sabovljević M, Bijelović A, Grubisic D (2001) Bryophytes as a potential source of medicinal compound. Lakovite Sirovine (Natural Medicinal Materials) 21:17
71. Koid CW, Shaipulah LGE, Gradstein SR, Asakawa Y, Andriani Y, Mohammed A, Norhazrina N, Chia PW, Ramlee MZ (2022) Volatile organic compounds of bryophytes from peninsular Malaysia and their roles in bryophytes. Plants 11:2575
72. Das K, Kityania S, Nah R, Das S, Nath D, Talukdar D, Barukial J, Hazaka P, Valarezo E, Meneses MA, Jaramillo-Fierro X, Jain V, Ghorai M, Das T, Dey A, Lunic T, Bozic B, Nadeljkovic BB, Marques RV, Salwinski A, Enemark-Rasmussen K, Gotfredsen CH, Lu Y, Hocquigny N, Risler A, Duval RE, Miguel DS, Bourgaud F, Simmonsen HT, Barukial J, Hazarika P (2023) Bioactive compounds in bryophytes and pteridophytes. In: Murthy HN (ed) Reference series in phytochemistry. Springer Nature, Cham, Switzerland, p 1
73. Inoue H (1988) Bryophytes as an indicator of continental drift (Gondwanaland). Kagakuasahi 8:116
74. Mohamed H, Baki BB, Nasrulhaq-Boyce A, Lee PKY (2008) Bryology in the new millennium. University of Malaya, Kuala Lumpur, Malaysia, p 1
75. Asakawa Y, Ludwiczuk A (2018) Chemical constituents of bryophytes: structures and biological activity. J Nat Prod 81:641
76. Asakawa Y, Ludwiczuk A, Novakovic M, Bukvicki D, Anchang KY (2022) Bis-bibenzyls, bibenzyls and terpenoids in 33 genera of the Marchantiophyta (liverworts): structures, synthesis, and bioactivity. J Nat Prod 85:729
77. von Reuss, SH, Konig WA (2005) Olefinic isothiocyanates and iminodithiocarbonates from the liverwort *Corsinia coriandrina.* Eur J Org Chem 1184
78. Asakawa Y, Toyota M, Tanaka H, Hashimoto T, Joulain D (1995) Chemical constituents of an unidentified Malaysian liverwort *Asterella* (?) species. J Hattori Bot Lab 78:183
79. Asakawa Y (2012) Bio- and chemical diversity of bryophytes; chemical structures and biological activity of scents, and related compounds. Aroma Res 12:70
80. Mizutani M (1975) On the smell of some mosses. Misc Bryol Lichenol 6:64
81. Hayashi S, Kami T, Matsuo A, Ando H, Seki T (1977) The smell of liverworts. Bryol Soc Jap 2:38
82. Toyota M, Koyama H, Asakawa Y (1997) Volatile components of the liverworts *Archilejeunea olivacea*, *Cheilolejeunea imbricata* and *Leptolejeunea elliptica.* Phytochemistry 44:1261
83. Sakurai K, Tomiyama K, Yaguchi Y, Asakawa Y (2020) Characteristic scent from the Japanese liverwort, *Leptolejeunea elliptica.* J Oleo Sci 69:767
84. Asakawa Y, Sono M, Wakamatsu M, Kondo K, Hattori S, Mizutani M (1991) Geographical distribution of tamariscol, a mossy odorous sesquiterpene alcohol in the liverwort *Frullania tamarisci* complex and related species. Phytochemistry 30:2295
85. Sono M (1991) Odoriferant substances of liverwort *Frullania tamarisci* and structures and total synthesis of related sesquiterpene alcohols of *Conocephalum conicum.* Ph.D. thesis, Tokushima Bunri University, Tokushima, Japan, p 1
86. Tori M, Sono M, Nishigaki Y, Nakashima K, Asakawa Y (1991) Studies on liverwort sesquiterpene alcohol tamariscol. Synthesis and absolute configuration. J Chem Soc Perkin Trans I:435
87. Toyota M, Saito T, Matsunami J, Asakawa Y (1997) A comparative study on three chemotypes of the liverwort *Conocephalum conicum* using volatile constituents. Phytochemistry 44:1265
88. Ghani N, Ludwiczuk A, Ismail NH, Asakawa Y (2016) Volatile components of the stressed liverwort *Conocephalum conicum.* Nat Prod Commun 11:103
89. Sakurai K, Tomiyama K, Kawakami Y, Ochiai N, Yabe S, Nakagawa T, Asakawa Y (2016) Volatile components emitted from the liverwort *Marchantia paleacea* subsp. *diptera.* Nat Prod Commun 11:263

90. Connolly JD (1982) New terpenoids from the Hepaticae. Rev Latinoam Quím 12:121
91. Connolly JD (1990) Monoterpenoids and sesquiterpenoids from the Hepaticae. In: Zinsmeister HD, Mues R (eds) Bryophytes: their chemistry and chemical taxonomy. Oxford University Press, Oxford, UK, p 41
92. Toyota M, Asakawa Y, Frahm JP (1990) Homomono- and sesquiterpenoids from the liverwort *Lophocolea heterophylla*. Phytochemistry 29:2334
93. Spörle J (1990) Phytochemische Untersuchungen an ausgewählten panamaischen Lebermoosen. Ph.D. thesis, Universität des Saarlandes, Saarbrücken, Germany, p 1
94. Huneck S, Connolly JD, Freer AA, Rycroft DS (1988) Grimaldone. A tricyclic sesquiterpenoid from *Mannia fragrans*. Phytochemistry 27:1405
95. Asakawa Y, Toyota M, Cheminat A (1986) Terpenoids from the French liverwort *Targionia hypophylla*. Phytochemistry 25:2555
96. Geis W, Becker H (2001) Odoriferous sesquiterpenoids from the liverwort *Gackstroemia decipiens*. Flav Fragr J 16:422
97. Cullmann F, Becker H (1998) Terpenoid and phenolic constituents of sporophytes and spores from the liverwort *Pellia epiphylla*. J Hattori Bot Lab 84:28
98. Asakawa Y, Nii K, Higuchi M (2015) Identification of sesquiterpene lactones in the Bryophyta (mosses) *Takakia*: *Takakia* species are closely related chemically to the Marchantiophyta (liverworts). Nat Prod Commun 10:5
99. Asakawa Y (2007) Scent substances from bryophytes. Bryol Res 9:227
100. Ludwiczuk A, Nagashima F, Gradstein SR, Asakawa Y (2008) Volatile components from the selected Mexican, Ecuadorian, Greek, German, and Japanese liverworts. Nat Prod Commun 3:133
101. Ludwiczuk A, Komala I, Pham A, Bianchini JP, Raharivelomanana P, Asakawa Y (2009) Volatile components from selected Tahitian liverworts. Nat Prod Commun 4:1387
102. Toyota M, Asakawa Y (1994) Volatile constituents of the liverwort *Chiloscyphus pallidus* (Mitt.) Engel & Schuster. Flav Fragr J 9:237
103. Sakurai K, Tomiyama K, Kawakami Y, Yaguchi Y, Asakawa Y (2018) Characteristic scent from the Tahitian liverwort, *Cyathodium foetidissimum*. J Oleo Sci 67:1265
104. Allen NS, Santana AI, Gomez N, Chung C, Gupta MP (2017) Identification of volatile compounds from three species of *Cyathodium* (Marchantiophyta: Cyathodiaceae) and *Leiosporoceros dussii* (Anthocerotophyta: Leiosporocerotaceae) from Panama, and *C. foetidissimum* from Costa Rica. Bol Soc Argent Bot 52:357
105. Asakawa Y (2014) Scent and tasty constituents of bryophytes. Koryo 263:29
106. Inoue H (1978) *Koke no Sekai* (World of Bryophytes). Idemitsu Ltd., Tokyo, p 1
107. Asakawa Y, Aratani T (1976) Sesquiterpenes of *Porella vernicosa*. Bull Soc Chim Fr 1469
108. Asakawa Y, Ludwiczuk A, Harinantenaina L, Toyota M, Nishiki M, Bardon A, Nii K (2012) Distribution of drimane sesquiterpenoids and tocopherols in liverworts, ferns and higher plants: Polygonaceae, Canellaceae and Winteraceae species. Nat Prod Commun 7:685
109. Hashimoto T, Okumura Y, Suzuki K, Takaoka S, Kan Y, Tori M, Asakawa Y (1995) The absolute structures of new 1β-hydroxysacculatane-type diterpenoids with piscicidal activity from the liverwort *Pellia endiviifolia*. Chem Pharm Bull 43:2030
110. Asakawa Y, Harrison LJ, Toyota M (1985) Occurrence of a potent piscicidal diterpenedial in the liverwort *Riccardia lobata* var. *yakushimensis*. Phytochemistry 24:261
111. Ono K, Sakamoto T, Asakawa Y (1992) Constituents from cell suspension cultures of selected liverworts. Phytochemistry 31:124
112. Ono K, Sakamoto T, Tanaka H, Asakawa Y (1996) Sesquiterpenoids from a cell suspension culture of the liverwort *Porella vernicosa* Lindb. Flav Frag J 11:53
113. Hashimoto T, Tanaka H, Asakawa Y (1994) Stereostructure of plagiochiline A and conversion of plagiochiline A and stearoylvelutinal into hot-tasting compounds by human saliva. Chem Pharm Bull 42:1542
114. Toyota M, Ueda A, Asakawa Y (1991) Sesquiterpenoids from the liverwort *Porella acutifolia* subsp. *tosana*. Phytochemistry 30:567

115. Asakawa Y, Toyota M, Oiso Y, Braggins JE (2001) Occurrence of polygodial and 1-(2,4,6-trimethozyphenyl)-but-2-en-1-one from some ferns and liverworts: role of pungent components in bryophytes and pteridophytes evolution. Chem Pharm Bull 49:138
116. Huneck S, Asakawa Y, Taira Z, Cameron A, Connolly JD, Rycroft DS (1983) Gymnocolin, a new *cis*-clerodane diterpenoid from the liverwort *Gymnocolea inflata*. Crystal structure analysis. Tetrahedron Lett 24:115
117. Nagashima F, Toyota M, Asakawa Y (1990) Bitter kaurane-type diterpene glucosides from the liverwort *Jungermannia infusca*. Phytochemistry 29:1619
118. Rycroft DS (1990) Some recent NMR studies of diterpenoids from the Hepaticae. In: Zinsmeister DH, Mues R (eds) Bryophytes: their chemistry and chemical taxonomy. Oxford University Press, Oxford, UK, p 109
119. Huneck S, Connolly JD, Harrison LJ, Joseph R, Phillip W, Rycroft DS, Ferguson G, Parvez M (1986) New labdane ditrerpenoids from the liverwort *Scapania undulata*. J Chem Res (S):162
120. Asakawa Y, Muller JC, Ourisson G, Foussereau J, Ducombs G (1976) Nouvelles lactone sesquiterpeniques de *Frullania* (Hepaticae). Isolement, structures, proprietes allergisantes. Bull Soc Chim Fr:1465
121. Asakawa Y, Masuya T, Tori M, Campbell EO (1987) Long chain alkyl phenols from the liverwort *Schistochila appendiculata*. Phytochemistry 26:735
122. Evans FJ, Schmidt RJ (1980) Plants and plant products that induce contact dermatitis. Planta Med 38:289
123. Asakawa Y (2017) The isolation, structure elucidation, and bio- and total synthesis of bis-bibenzyls, from liverworts and their biological activity. Nat Prod Commun 12:1335
124. Oiso Y, Toyota M, Asakawa Y (1999) Occurrence of bis-bibenzyls derivative in the Japanese fern *Hymenophyllum barbatum*: first isolation and identification of perrottetin H from the Pteridophytes. Chem Pharm Bull 47:297
125. Kosenkova YS, Polovincka MP, Komarova NI, Korchagina DV, Kurochikina NY, Cheremushkina VA, Salakhutdinov NP (2007) Riccardin C, a bisbibenzyl compound from *Primula macrocalyx*. Chem Nat Compd 43:712
126. Kosenkova YS, Polovincka MP, Komarova NI, Korchagina DV, Kurochikina NY, Cheremushkina VA, Salakhutdinov NP (2007) Seasonal dynamic of riccardin accumulation in *Primula macrocalyx* Bge. Chem Sustain Dev 17:501
127. Bukvicki D, Kovtonyuk NK, Legin AA, Keppler BK, Brecker L, Asakawa Y, Velant-Vetschera K (2022) Hunting for bis-bibenzyls in *Primula veris* subsp. *macrocalyx* (Bunge) Ludi: organ-specific accumulation and cytotoxic activity. Phytochem Lett 44:90
128. Novakovic M, Ilic-Tomic T, Djordjevic I, Andjelkovic B, Tesevic V, Milosavljevic S, Asakawa Y (2023) Bisbibenzyls from Serbian *Primula veris* subsp. *columnae* (Ten.) Ludi and *P. acaulis* (L.) L. Phytochemistry 212:113719
129. Asakawa Y, Matsuda R (1982) Riccardin C, a novel cyclic bibenzyl derivative from *Reboulia hemisphaerica*. Phytochemistry 21:2143
130. Friederich S, Maier UH, Deus-Meumann B, Asakawa Y, Zenk MH (1999) Biosynthesis of cyclic bis(bibenzyls) in *Marchantia polymorpha*. Phytochemistry 50:589
131. Friederich S, Rueffer M, Asakawa Y, Zenk MH (1999) Biosynthesis of cyclic bis(bibenzyls) in *Marchantia polymorpha*. Phytochemistry 52:1192
132. Jantwal A, Rana M, Rana AJ, Upadhyay J, Durgapal S (2019) Pharmacological potential of genus *Marchantia*: a review. J Pharmacog Phytochem 8:6
133. Sabovljević MS, Vujičić M, Wang X, Garraffo HM, Bewley CA, Sabovljević A (2017) Production of the macrocyclic bis-bibenzyls in axenically farmed and wild liverwort *Marchantia polymorpha* L. subsp. *ruderalis* Bischl. et Boisselier. Plant Biosystems 151:414
134. Huang WJ, Wu CL, Lin CW, Chi LL, Chen PY, Chiu CJ, Huang CY, Chen CN (2010) Marchantin A, a cyclic bis(bibenzyl ether), isolated from the liverwort *Marchantia emarginata* subsp. *tosana* induces apoptosis in human MCF-7 breast cancer cells. Cancer Lett 291:1083
135. Jensen JSRS, Omarsdottir S, Thorsteljsdottir JB, Ogmundsdorttir HM, Olafsdottir ES (2012) Synergistic cytotoxic effect of the microtubule inhibitor marchantin A from *Marchantia polymorpha* and the aurora kinase inhibitor MLN8237 on breast cancer cells in vitro. Planta Med 78:448

136. Gawel-Beben K, Osika P, Asakawa Y, Antosiewicz B, Growniak K, Ludwiczuk A (2019) Evaluation of anti-melanoma and tyrosinase inhibitory properties of marchantin A, a natural macrocyclic bisbibenzyl isolated from *Marchantia* species. Phytochem Lett 31:192
137. Xi GM, Sun B, Jian HH, Kong F, Yan HQ, Lou HX (2010) Bisbibenzyl derivatives sensitize vincristine-resistant KB/VCR cells to chemotherapeutic agents by retarding *P*-gp activity. Bioorg Med Chem 18:6725
138. Shen J, Li G, Liu Q, He Q, Gu J, Shi Y, Lou H (2010) Marchantin C: a potential anti-invasion agent in glioma cells. Cancer Biol Ther 9:33
139. Lv Y, Song Q, Shao Q, Gao W, Mao H, Lou H, Wu Z, Li X (2012) Comparison of the effects of marchantin C and fucoidan on sFLt-1 and angiogenesis in glioma microenvironment. J Pharm Pharmacog 64:604
140. Jiang J, Sun B, Wang YY, Cui M, Zhang L, Cui CZ, Wang YF, Liu ZG, Lou HX (2012) Synthesis of macrocyclic bibenzyl derivatives and their anticancer effects as anti-tubulin agents. Bioorg Med Chem 20:2382
141. Scher JM, Burgess EJ, Lorimer SD, Perry NB (2002) A cytotoxic sesquiterpene and unprecedented sesquiterpene-bisbibenzyl compounds from the liverwort *Schistochila glaucescens*. Tetrahedron 58:7875
142. Shi YQ, Liao YX, Qu XJ, Yuan HQ, Li S, Qu JB, Lou HX (2008) Marchantin C, a macrocyclic bisbibenzyl, induces apoptosis of human glioma A172 cells. Cancer Lett 262:1732
143. Shi YQ, Zhu XJ, Yuan HQ, Li BQ, Gao J, Qu ZJ, Sun B, Cheng YN, Li S, Li X, Lou HX (2009) Marchantin C, a novel microtubule inhibitor from liverwort with anti-tumor activity both in vivo and in vitro. Cancer Lett 276:160
144. Zhang L, Ji ZT, Sun B, Qian LL, Hu XL, Lou HX (2019) Anti-cancer effect of marchantin C via inducing lung cancer cellular senescence associated with less secretory phenotype. Biochem Biophys Acta 1863:1443
145. Xu H, Qu JB, Liu SM, Syed AKA, Yuan HQ, Lou HX (2010) Cyclic bisbibenzyls induce growth arrest and apoptosis of human prostate cancer PC3 cell. Acta Pharmacol Sin 31:609
146. Zhang TW, Xing Tang JL, Lu JX, Liu CX (2015) Marchantin M induces apoptosis of prostate cancer cells through endoplasmic reticulum stress. Med Sci Monit 21:3570
147. Jian H, Sun J, Xu Q, Liu Y, Wei J, Young CYF, Yuan H, Lou HX (2013) Marchantin M: a novel inhibitor of proteasome induces autophagic cell death in prostate cancer cells. Cell Death Dis 4:e761
148. Mishra T, Sahu V, Meena S, Pal M, Asthana K, Datta D, Upreti DK (2023) A comparative study of in vitro cytotoxicity and chemical constituents of wild and cultured plants of *Marchantia polymorpha* L. S Afr J Bot 157:274
149. Li X, Wu WKK, Sun B, Cui M, Liu S, Gao J, Lou H (2011) Dihydroptychantol A, a macrocyclic bisbibenzyl derivative induces autophagy and following apoptosis associated with p53 pathway in human osteosarcoma U20S cells. Toxicol Appl Pharmacol 251:146
150. Sun B, Yuang HQ, Xi GM, Ma YD, Lou HX (2009) Synthesis and multidrug resistance reversal activity of dihydroptychantol A and its novel derivatives. Bioorg Med Chem 17:4981
151. Pang Y, Si M, Sun B, Niu L, Xu X, Lu T, Yuan H, Lou H (2014) DHA2, a synthesized derivative of bisbibenzyl, exerts antitumor activity against ovarian cancer through inhibition of XIAP and/mTOR pathway. Food Chem Toxicol 69:163
152. Novaković M, Bukvički D, Anđjelković B, Tomic TL, Veljić M, Tesevic V, Asakawa Y (2019) Cytotoxic activity of riccardin and perrottetin derivatives from the liverwort *Lunularia cruciata*. J Nat Prod 82:694
153. Qu JB, Sun LM, Lou HX (2013) Antifungal variant bis(bibenzyl)s from the liverwort *Asterella angusta*. Chin Chem Lett 24:801
154. Ivković I, Novaković M, Veljić M, Mojsin M, Stevanović M, Marin PD, Bukvički D (2021) Bis-bibenzyls from the liverwort *Pellia endiviifolia* and their biological activity. Plants 10:1063
155. Bardon A, Kamiya N, Toyota M, Takaoka S, Asakawa Y (1999) Sesquiterpenoids, hopanoids and bis(bibenzyls) from the Argentine liverwort *Plagiochasma rupestre*. Phytochemistry 52:1323

156. Ji M, Shi Y, Lou HX (2011) Overcoming of P-glycoprotein-mediated multidrug resistance in K562/A02 cells using riccardin F and pakyonol, bibenzyl derivatives from liverworts. Biosci Trends 5:192
157. Yoshida T, Hashimoto T, Takaoka S, Kan Y, Tori M, Asakawa Y (1996) Phenolic constituents of the liverwort: four novel cyclic bisbibenzyl dimers from *Blasia pusilla* L. Tetrahedron 52:14487
158. Xue X, Qu XJ, Gao ZH, Sun CC, Liu HP, Zhao CR, Cheng YN, Lou HX (2012) Riccardin D, a novel macrocyclic bisbibenzyl, induces apoptosis of human leukemia cells by targeting SHCA topoisomerase II. Invest New Drugs 30:212
159. Xue X, Sun DF, Sun CC, Liu HP, Yue B, Zhao R, Lou HX, Qu XJ (2012) Inhibitory effect of riccardin D on growth of human non-small cell lung cancer; in vitro and in vivo studies. Lung Cancer 76:300
160. Sun CC, Zhang YC, Xue X, Cheng YN, Liu HP, Zhao CR, Lou HX, Qu XJ (2011) Inhibition of angiogenesis involves in anticancer activity of riccardin D, a macrocyclic bisbibenzyl, in human lung carcinoma. Eur J Pharmacol 667:136
161. Liu HP, Gao ZH, Cui SX, Sun DF, Wang Y, Zhao CR, Lou HX, Qu XJ (2012) Inhibition of intestinal adenoma formation in $APC^{Min/+}$ mice by riccardin D, a natural product derived from liverwort plant *Dumortiera hirsuta*. PLoS One 7:e33243
162. Liu H, Li G, Zhang B, Sun D, Wu J, Chen F, Kong F, Luan Y, Jiang W, Wang R, Xue X (2018) Suppression of the NF-κB signaling pathway in colon cancer cells by the natural compound riccardin D from *Dumortiera hirsuta*. Mol Med Rep 17:5837
163. Hu Z, Zhang D, Hao J, Tia K, Wang W, Lou H, Yuan H (2014) Induction of DNA damage and P21-dependent senescence by riccardin D is novel mechanism contributing to its growth suppression in prostate cancer cells in vitro and in vivo. Cancer Chemother Pharmacol 73:397
164. Sun B, Liu J, Gao Y, Zheng HB, Li L, Hu QW, Yuan HQ, Lou HX (2017) Design, synthesis and biological evaluation of nitrogen-containing macrocyclic bisbibenzyl derivatives as potent anticancer agents by targeting the lysosome. Eur J Med Chem 136:603
165. Yue B, Zhao CXR, Xu HM, Li YY, Cheng YN, Ke HN, Yuan Y, Wang RQ, Shi YQ, Lou HX, Qu XJ (2013) Riccardin D26, a synthesized bibenzyl macrocyclic compound, inhibits human oral squamous carcinoma cell KB and KB/VCR: in vitro and in vivo studies. Biochim Biophys Acta 1830:2194
166. Speicher A, Groh M, Zapp J (2009) A synthesis-driven structure revision of "plagiochin E", a highly bioactive bisbibenzyl. Synlett 11:1852
167. Shi YQ, Qu XJ, Liao YX, Xie CF, Cheng YN, Li S, Lou HX (2008) Reversal effect of a macrocyclic bisbibenzyl plagiochin E on multidrug resistance in adriamycin-resistant K562/A02 cells. Eur J Pharmacol 584:66
168. Toyota M, Ikeda R, Kenmoku H, Asakawa Y (2013) Activity-guided isolation of cytotoxic bis-bibenzyl constituents from *Dumortiera hirsuta*. J Oleo Sci 62:105
169. Morita H, Tomizawa Y, Tsuchiya T, Hirasawa Y, Hashimoto T, Asakawa Y (2009) Antimitotic activity of two macrocyclic bis(bibenzyls), isoplagiochins A and B from the liverwort *Plagiochila fruticosa*. Bioorg Med Chem Lett 19:493
170. Ghani NA, Ismail NH, Noma Y, Asakawa Y (2017) Microbial transformation of some natural and synthetic aromatic compounds by fungi: *Aspergillus* strain and *Neurospora crassa*. Nat Prod Commun 12:1237
171. Bukvicki D, Novaković M, Ilic TT, Nikodinovic-Rumic NJ, Todorović N, Veljic M, Asakawa Y (2021) Biotransformation of perrottetin F by *Aspergillus niger*: new bioactive secondary metabolites. Rec Nat Prod 15:281
172. Novakovic M, Simić S, Koračak L, Zlatović M, Ilic-Tomic T, Asakawa Y, Nikodinovic-Runic J, Opsenica I (2020) Chemo- and biocatalytic esterification of marchantin A and cytotoxic activity of ester derivatives. Fitoterapia 142:104520
173. Asakawa Y, Nagashima F, Ludwiczuk A (2020) Distribution of bibenzyls, prenyl bibenzyls, bis-bibenzyls, and terpenoids in the liverwort genus *Radula*. J Nat Prod 83:756
174. Asakawa Y, Nagashima F (2022) Heterocyclic stilbene and bibenzyl derivatives in liverworts: distribution, structures, total synthesis and biological activity. Heterocycles 105:1174

175. Guo DX, Xiang F, Wang XN, Yuan HQ, Xi GM, Wang YY, Yu WT, Lou HX (2010) Labdane diterpenoids and highly methoxylated bibenzyls from the liverwort *Frullania inouei*. Phytochemistry 71:1573
176. Lu ZQ, Fan PH, Ji M, Lou HX (2006) Terpenoids and bisbibenzyls from Chinese liverworts *Conocephalum conicum* and *Dumortiera hirsuta*. J Asian Nat Prod Res 8:187
177. Lorimer SD, Perry NB, Tangney RS (1993) An antifungal bibenzyl from the New Zealand liverwort *Plagiochila stephensoniana*, bioactivity-directed isolation, synthesis, and analysis. J Nat Prod 56:1444
178. Zhang CY, Gao Y, Zhou JC, Qiao YN, Zhang JZ, Lou HX (2021) Diverse prenylated bibenzyl enantiomers from the Chinese liverwort *Radula apiculata* and their cytotoxic activities. J Nat Prod 84:1459
179. Zhu R, Zhou j, Li Y, Qiao Y, Zhang C, Zhang J, Gao Y, Chen W, Lou HX, (2019) Prenyl bibenzyls isolated from Chinese liverwort *Radula amoena* and their cytotoxic activities. Phytochem Lett 31:53
180. Zhang CY, Gao Y, Zhu RX, Qiao YN, Zhou JC, Zhang JZ, Li Y, Li SW, Fam SH, Lou HX (2019) Prenyl bibenzyls from the Chinese liverwort *Radula constricta* and their mitochondria-derived paraptotic cytotoxic activities. J Nat Prod 82:1741
181. Wang Z, Li L, Zhu R, Zhang J, Zhou J, Lou H (2017) Bibenzyl-based meroterpenoid enantiomers from the Chinese liverwort *Radula sumatrana*. J Nat Prod 80:3143
182. Komala I, Ito T, Nagashima F, Asakawa Y (2010) Volatile components of selected liverworts, and cytotoxic, radical scavenging and antimicrobial activities of their crude extracts. Nat Prod Commun 5:1375
183. Komala I, Ito T, Nagashima F, Yagi Y, Asakawa Y (2011) Cytotoxic bibenzyls, and germacrane- and pinguisane-type sesquiterpenoids from Indonesian, Tahitian and Japanese liverworts. Nat Prod Commun 6:303
184. Komala I, Ito T, Nagashima F, Yagi Y, Yamaguchi K, Asakawa Y (2010) Zierane sesquiterpene lactone, cembrane and fusicoccane diterpenoids from the Tahitian liverwort *Chandonanthus hirtellus*. Phytochemistry 71:1387
185. Shu YF, Wei HC, Wu CL (1994) Sesquiterpenoids from liverworts *Lepidozia vitrea* and *L. fauriana*. Phytochemistry 37:773
186. Wu C, Gunatilaka AAL, McCabe FL, Johnson RK, Spjut RW, Kingston DGI (1997) Bioactive and other sesquiterpenes from *Chiloscyphus rivularis*. J Nat Prod 60:1281
187. Zhang JZ, Qiao YN, Li L, Wang YJ, Li Y, Fei K, Zhou JZ, Wang X, Lou HX (2016) *ent*-Eudesmane-type sesquiterpenoids from the Chinese liverwort *Chiloscyphus polyanthus* var. *rivularis*. Planta Med 82:1128
188. Zhu MZ, Li Y, Zhou JZ, Lv DX, Fu XJ, Liang Z, Yuan SZ, Han JJ, Zhang JZ, Xu ZJ, Chang WQ, Lou HX (2023) *ent*-Eudesmane sesquiterpenoids from the Chinese liverwort *Chiloscyphus polyanthus*. Phytochemistry 214:113796
189. Lorimer SD, Burgess EJ, Perry NB (1997) Diplophyllolide: a cytotoxic sesquiterpene lactone from the liverworts *Clasmatocolea vermicularis* and *Chiloscyphus subporosa*. Phytomedicine 4:261
190. Baek SH, Perry NB, Lorimer SD (2003) *ent*-Costunolide from the liverwort *Hepatostolonophora paucistipula*. J Chem Res:14
191. Kim YC, da Bolzani VS, Baj N, Gunatilaka AAL, Kingston DGI (1996) A DNA-damaging sesquiterpene and other constituents from *Frullania nisquallensis*. Planta Med 62:61
192. Burgess EJ, Larsen L, Perry NB (2000) A cytotoxic sesquiterpene caffeate from the liverwort *Bazzania novae-zelandiae*. J Nat Prod 63:537
193. Liu N, Guo DX, Wang SQ, Wang YY, Zhang L, Li G, Lou HX (2012) Bioactive sesquiterpenoids and diterpenoids from the liverwort *Bazzania albifolia*. Chem Biodivers 9:2254
194. Perry NB, Foster LM (1995) Sesquiterpene/quinol from a New Zealand liverwort, *Riccardia crassa*. J Nat Prod 58:1131
195. Komala I, Ito T, Nagashima F, Yagi A, Asakawa Y (2010) Cytotoxic, radical scavenging and antimicrobial activities of sesquiterpenoids from the Tahitian liverwort *Mastigophora diclados* (Brid.) Nees (Mastigophoraceae). J Nat Med 64:41

196. Komala I, Sitorus S, Dewi FR, Nurmeillis N, Hendermin LA (2022) Cytotoxic activity of the Indonesian fern *Angiopteris angustifolia* C. Presl and liverwort *Mastigophora diclados* (Birs. ex Web) Nees against breast cancer cell lines (MCF-7). J Kim Valensi 8:79
197. Ng SY, Kamada T, Vairappan CS (2017) New pimarane-type diterpenoid and *ent*-eudesmane-type sesquiterpenoid from Bornean liverwort *Mastigophor diclados*. Rec Nat Prod 11:508
198. Nagashima F, Ohi Y, Nagai T, Tori M, Asakawa Y, Huneck S (1993) Terpenoids from some German and Russian liverworts. Phytochemistry 33:144
199. Toyota M, Tanimura K, Asakawa Y (1998) Cytotoxic 2,3-secoaromadendrane-type sesquiterpenoids from the liverwort *Plagiochila ovalifolia*. Planta Med 64:462
200. Aponte JC, Yang H, Vaisberg AJ, Castillo D, Malaga E, Verastegui M, Casson LK, Stivers N, Bates PJ, Rojas R, Fernandez I, Lewis WH, Sarasara C, Sauvain M, Gilman RH, Hammond GB (2010) Cytotoxic and anti-infective sesquiterpenes in *Plagiochila disticha* (Plagiochilaceae) and *Ambrosia peruviana* (Asteraceae). Planta Med 76:705
201. Wang S, Liu SS, Lin ZM, Li RJ, Wang XN, Zhou JC, Lou HX (2013) Terpenoids from the Chinese liverwort *Plagiochila pulcherrima* and their cytotoxic effects. J Asian Nat Prod Res 15:473
202. Pannequin A, Quetin-Leclercq J, Costa J, Tintaru A, Muselli A (2023) First phytochemical profiling and in vitro antiprotozoal activity of essential oil and extract of *Plagiochila porelloides*. Molecules 28:616
203. Stivers N, Islam A, Reyes-Reyes EM, Casson LK, Aponte JC, Vaisberg AJ, Hammond GB, Bates PJ (2018) Plagiochiline A inhibits cytokinetic abscission and induces cell death. Molecules 23:1418
204. Bailly C (2023) Discovery and anticancer activity of the plachilins from the liverwort genus *Plagiochila*. Life 13:758
205. Vergoten G, Bailly C (2023) Plagiochilins from *Plagiochila* liverworts: binding to α-tubulin and drug design perspectives. Appl Chem 3:217
206. Guo L, Wu JZ, Han T, Cao T, Rahman K, Qin LP (2008) Chemical composition, antifungal and antitumor properties of ether extracts of *Scapania verrucosa* Heeg. and endophytic fungus *Chaetomium fusiforme*. Molecules 13:2114
207. Chen X, Xiao J (2006) In vitro cytotoxic activity of extracts of *Marchantia convoluta* on human liver and lung cancer cell lines. Afr Trad CAM 3:32
208. Xiao JB, Chen XQ, Zang YW, Jiang XY, Xu M (2006) Cytotoxicity of *Marchantia convoluta* leaf extracts to human liver and lung cancer cells. Braz J Med Biol Res 39:731
209. Perry NB, Burgess EJ, Foster LM, Gerard PJ, Toyota M, Asakawa Y (2008) Insect antifeedant sesquiterpene acetals from the liverwort *Lepidolaena clavigera*. 2. Structures, artifacts and activity. J Nat Prod 71:258
210. Valarezo E, Tandazo O, Galan K, Rosales J, Benitez A (2020) Volatile metabolites in liverworts of Ecuador. Metabolites 10:92
211. Sharma R, Singh S, Marreddy NSR, Merchant N, Alam A (2023) Gas chromatography-mass spectroscopic profiling and cytotoxic activity of *Riccia billardieri* Mont. & Nees. (Bryophyta: Liverwort). Results Chem 6:101004
212. Radulović NS, Filipović SI, Nešić MS, Stojanović NM, Mitić KV, Mladenović MZ, Randelović VN (2020) Immunomodulatory constituents of *Conocephalum conicum* (snake liverwort) and the relationship of isolepidozenes to germacranes and humulanes. J Nat Prod 83:3554
213. Liu N, Wu C, Wang P, Lou H (2018) Diterpenoids from liverworts and their biological activities. Curr Org Chem 22:1847
214. Komala I (2011) Phytochemical studies on the selected Indonesian, Japanese and Tahitian liverworts. Ph.D. dissertation, Tokushima Bunri University, Tokushima, Japan, p 1
215. Wang Y, Harrison LJ, Tan BC (2009) Terpenoids from the liverwort *Chandonanthus hirtellus*. Tetrahedron 65:4035
216. Li RJ, Lin ZM, Kang YQ, Guo YX, Lv X, Zhou JC, Wang S, Lou HX (2014) Cembrane-type diterpenoids from the Chinese liverworts *Chandonanthus hirtellus* and *C. birmensis*. J Nat Prod 77:339

217. Lin ZM, Guo YX, Wang SQ, Wang XN, Chang WQ, Zhou JC, Yuan H, Lou H (2014) Diterpenoids from the Chinese liverwort *Heteroscyphus tener* and their antiproliferative effects. J Nat Prod 77:1336
218. Li RJ, Wang S, Li G, Zhou JC, Zhang JZ, Zhang YM, Shi GS, Lou HX (2016) Four new kaurane diterpenoids from the Chinese liverwort *Jungermannia comata* Nees. Chem Biodivers 13:1685
219. Nagashima F, Kasai W, Kondoh M, Fujii M, Watanabe Y, Braggins JE, Asakawa Y (2003) New *ent*-kaurene-type diterpenoids possessing cytotoxicity from the New Zealand liverwort *Jungermannia* species. Chem Pharm Bull 51:1189
220. Nagashima F, Kondoh M, Fujii M, Takaoka S, Watanabe Y, Asakawa Y (2005) Novel cytotoxic diterpenoids from the New Zealand liverwort *Jungermannia* species. Tetrahedron 51:4531
221. Nagashima F, Kondoh M, Uematsu T, Nishiyama A, Saito S, Sato M, Asakawa Y (2002) Cytotoxic and apoptosis-inducing *ent*-kaurane-type diterpenoids from the Japanese liverwort *Jungermannia truncata* Nees. Chem Pharm Bull 50:808
222. Nagashima F, Kondoh M, Kawase M, Simizu S, Osada H, Fuji M, Watanabe Y, Sato M, Asakawa Y (2003) Apoptosis-inducing properties of *ent*-kaurene-type diterpenoids from the liverwort *Jungermannia truncata*. Planta Med 69:377
223. Kondoh M, Suzuki I, Sato M, Nagashima F, Simizu S, Harada M, Fujii M, Osada H, Asakawa Y, Watanabe Y (2004) Kaurene diterpene induces apoptosis in human leukemia cells partly through a caspase-8-dependent pathway. J Pharmacol Exp Ther 311:115
224. Suzuki I, Kondoh M, Harada M, Koizumi N, Fujii M, Nagashima F, Asakawa Y, Watanabe Y (2004) An *ent*-kaurene diterpene enhances apoptosis induced by tumor necrosis factor in human leukemia cells. Planta Med 70:723
225. Suzuki I, Kondoh M, Nagashima F, Fujii M, Asakawa Y, Watanabe Y (2004) A comparison of apoptosis and necrosis induced by *ent*-kaurene-type diterpenoids in HL-60 cells. Planta Med 70:401
226. Kondoh M, Nagashima F, Suzuki I, Harada M, Fuji M, Asakawa Y, Watanabe Y (2005) Induction of apoptosis by new *ent*-kaurene-type diterpenoids isolated from the New Zealand *Jungermannia* species. Planta Med 71:1005
227. Kondoh M, Suzuki I, Harada M, Nagashima F, Fuji M, Asakawa Y, Watanabe Y (2005) Activation of p38 mitogen-activated protein kinase during *ent*-11α-hydroxy-16-kauren-15-one-induced apoptosis in human leukemia HL-60 cells. Planta Med 71:275
228. Lin Z, Guo Y, Gao Y, Wang S, Wang X, Xie Z, Niu H, Chang W, Liu L, Yuan H, Lou H (2015) *ent*-Kaurane diterpenoids from Chinese liverworts and their antitumor activities through Michael addition as detected in situ by a fluorescence probe. J Med Chem 58:3944
229. Guo YX, Lin ZM, Wang MJ, Dong YW, Niu HM, Young CYF, Lou HX (2016) Jungermannenone A and B induce ROD- and cell cycle-dependent apoptosis in prostate cancer cells in vitro. Acta Pharmacol Sin 37:813
230. Sun Y, Qiao Y, Liu Y, Zhou J, Wang X, Zheng H, Xu Z, Zhang J, Zhou Y, Qian L, Zhang C, Lou H (2021) *ent*-Kaurene diterpenoids induce apoptosis and ferroptosis through targeting redox resetting to overcome cisplatin resistance. Redox Biol 43:101977
231. Lorimer SD, Perry NB, Burgess EJ, Foster LM (1997) Hydroxy diterpenes from two New Zealand liverworts, *Paraschistochila pinnatifolia* and *Trichocolea mollissima*. J Nat Prod 60:421
232. Perry NB, Burgess E J, Tangney RS (1996) Cytotoxic 8,9-secokaurane diterpenes from a New Zealand liverwort, *Lepidolaena taylorii*. Tetrahedron Lett 37:9387
233. Perry NB, Burgess EJ, Baek SH, Weavers RT, Geis W, Mauger AB (1999) 11-Oxygenated cytotoxic 8,9-*seco*-kauranes from a New Zealand liverwort, *Lepidolaena taylorii*. Phytochemistry 50:423
234. Métoyer B, Lebouvier N, Hnawia E, Thouvenot L, Wang F, Rakotondraibe LH, Raharivelomanana P, Asakawa Y, Nour M (2021) Chemotaxonomy and cytotoxicity of the liverwort *Porella viridissima*. Nat Prod Res 35:2099
235. Chien TV, Anh NT, Loc TV, Sung TV, Thao TTO, Bakalin V, Manh NH, Sinh NV (2023) A new sacculatane and a new oplopanone sesquiterpenoid from *Porella perrottetiana* (Mont.) Trevis. collected in Sapa, Vietnam. Nat Prod Res 2023:1

236. Han J, Sun Y, Zhou J, Li Y, Jin X, Zhu M, Xu Z, Zhang Z, Lou H (2024) Sacculatane diterpenoids from the liverwort *Plagiochila nitens* collected in China. J Nat Prod 87:1124
237. Toyota M, Saito T, Asakawa Y (1998) Novel skeletal diterpenoids from the Japanese liverwort *Pallavicinia subciliata*. Chem Pharm Bull 46:178
238. Li ZJ, Lou HX, Yu WT, Fan PH, Ren DM, Ma B, Ji M (2005) Structures and absolute configurations of three 7,8-secolabdane diterpenes from the Chinese liverwort *Pallavicinia ambigua*. Helv Chim Acta 88:2637
239. Wang LN, Zhang JZ, Li X, Wang XN, Xie CF, Zhou JC, Lou HX (2012) Pallambins A and B, unprecedented hexacyclic 19-*nor*-secolabdane diterpenoids from the Chinese liverwort *Pallavicinia ambigua*. Org Lett 14:1102
240. Zhang J, Li Y, Zhu R, Li L, Wang Y, Zhou J, Qiao Y, Zhang Z, Lou H (2015) Scapairrins A-Q, labdane-type diterpenoids from the Chinese liverwort *Scapania irrigua* and their cytotoxic activity. J Nat Prod 78:2087
241. Kang YQ, Zhou JC, Fan PH, Wang SQ, Lou HX (2015) Scapaundulin C, a novel labdane diterpenoid isolated from Chinese liverwort *Scapania undulata*, inhibits acetylcholinesterase activity. Chin J Nat Med 13:933
242. Guo DX, Zhu RX, Wang XN, Wang LN, Wang SQ, Lin ZM, Lou H (2010) Scaparvin A, novel caged *cis*-clerodane with an unprecedented C-6/C-11 bond, and related diterpenoids from the liverwort *Scapania parva*. Org Lett 12:4404
243. Lou HX, Li GY, Wang FQ (2002) A cytotoxic diterpenoid and antifungal phenolic compounds from *Frullania muscicola* Steph. J Asian Nat Prod Res 4:87
244. Liu CM, Zhu RL, Liu RH, Li HL, Shan L, Xu XK, Zhang WD (2009) *cis*-Clerodane diterpenoids from the liverwort *Gottschelia schizopleura* and their cytotoxic activity. Planta Med 75:1597
245. Ng SY, Kamada T, Suleiman M, Vairappan CS (2018) Two new clerodane-type diterpenoids from Bornean liverwort *Gottschelia schizopleura* and their cytotoxic activity. Nat Prod Res 32:1832
246. Wang S, Li RJ, Zhu RX, Hu XY, Guo YX, Zhou JC, Lin MZ, Zhang JZ, Wu JY, Kang YQ, Morris-Natschke SL, Lee KH, Yuan HQ, Lou HX (2014) Notolutesins A-J, dolabrane-type diterpenoids from the Chinese liverwort *Notoscyphus lutescens*. J Nat Prod 77:2081
247. Wu JY, Wang X, Zhang JZ, Zhou JC, Li L, Lou HX (2016) Notolutesin K-P, dolabrane-type diterpenoids from the Chinese liverwort *Notoscyphus collenchymatosus*. Phytochem Lett 17:226
248. Kenmoku H, Tada H, Oogushi M, Esumi T, Takahashi H, Noji M, Sassa T, Toyota M, Asakawa Y (2014) Seed dormancy breaking diterpenoids from the liverwort *Plagiochila sciophila* and their differentiation inducing activity in human promyelocytic leukemia HL-60 cells. Nat Prod Commun 9:915
249. Ng SY, Kamada T, Suleiman M, Virappon CS (2016) A new secoclerodane-type diterpenoid from Bornean liverwort *Schistochila acuminata*. Nat Prod Commun 11:1071
250. Perry NB, Burgess EJ, Baek SH, Weavers RT (2001) The first atisane diterpenoids from a liverwort: polyols from *Lepidolaena clavigera*. Org Lett 3:4243
251. Fan S, Li Y, Zhou J, Qiao Y, Zhang C, Gao Y, Jin X, Zhang J, Chen W, Lou H (2019) Secondary metabolites from the Chinese liverwort *Diprophyllum apiculatum*. Phytochem Lett 31:92
252. Zhang CY, Chu ZJ, Zhou JC, Liu SG, Zhang JZ, Qian L, Lou HX (2021) Cytotoxic activities of 9,10-*seco*-cycloartane-type triterpenoids from the Chinese liverwort *Lepidozia reptans*. J Nat Prod 84:3020
253. Bakar MFA, Karim FA, Suleiman M, Isha A, Rahmat A (2015) Phytochemical constituents, antioxidant and antiproliferative properties of a liverwort, *Lepidozia borneensis* Stephani from Mount Kinabalu, Sabah. Malaysia. Evid-Based Compl Alt Med 4:936215
254. Neves M, Morais R, Gafner S, Stoeckli-Evans H, Hostettmann K (1999) New sesquiterpene lactones from the Portuguese liverwort *Targionia lorbeeriana*. Phytochemistry 50:967
255. Wong S-M, Oshima Y, Pezzuto JM, Fong HHS, Farnsworth NR (1986) Plant anticancer agents XXXIX. Triterpenes from *Iris missouriensis* (Iridaceae). J Pharm Sci 75:317

256. Guo DX, Du Y, Wang YY, Sun LM, Qu JB, Wang XN, Lou HX (2009) Secondary metabolites from the liverwort *Ptilidium pulcherrimum*. Nat Prod Commun 4:1319
257. Izumi S, Nishio Y, Takashima O, Hirata T (1997). Monoterpenoids, potent inducers of apoptosis in the cells of *Marchantia polymorpha*. Chem Lett:837
258. Lorimer SD, Perry NB (1994) Antifungal hydroxyacetophenones from the New Zealand liverwort, *Plagiochila fasciculata*. Planta Med 60:386
259. Perry NB, Foster LM, Lorimer SD, May BCH, Weavers RT, Toyota M, Nakaishi E, Asakawa Y (1996) Isoprenyl phenyl ethers from liverworts of the genus *Trichocolea*: cytotoxic activity, structural corrections, and synthesis. J Nat Prod 59:729
260. Baek SH, Oh HJ, Lim JA, Chun HJ, Lee HO, Ahn JW, Perry NB, Kim HM (2004) Biological activities of methyl 4[[(2*E*)-3,7-dimethyl-2,6-octadienyl]oxy]-3-hydroxybenzoate. Bull Kor Chem Soc 25:195
261. Tan TQ, Tam LT, Hoang PN, Phuong QND (2018) Biological activities of the liverwort *Marchantia polymorpha* L. collected at Da Lat, Lam Dong Province. Sci Technol Develop J 2:26
262. Phan-Duy NN, Nguyen-Vo KD, Tran-Quoe T, Quach-Ngo DP (2023) Investigation on the biological activity of the leafy liverwort *Demotarisia linguifolia* (De Nt.) Grolle. Res J Biotechnol 18:163
263. Vollár M, Gyovai A, Szűcs P, Zupkó I, Marschall M, Csupor-Löffler B, Bérdi P, Vecsermyés A, Csorba A, Lictor-Busa E, Urbán E, Csupor D (2018) Antiproliferative and antimicrobial activities of selected bryophytes. Molecules 23:1520
264. Liktor-Busa E, Urban E, Zupkó I, Szűcs P, Csupor D (2015) In vitro antibacterial and antiproliferative screening of Hungarian bryophytes. Planta Med 81:1393
265. Klavina L, Springe G, Nikolajeva V, Martsinkevich, Nakurte I, Dzabijeva D, Steinberga I (2015) Chemical composition analysis, antimicrobial activity and cytotoxicity screening of moss extract (moss phytochemistry) Molecules 20:17221
266. Nozaki H, Hayashi KI, Nishimura N, Kawaide H, Matsuo A, Takaoka D (2007) Momilactone A and B as allelochemicals from moss *Hypnum plumaeforme*: first occurrence in bryophytes. Biosci Biotechnol Biochem 71:3127
267. Kim SJ, Park HR, Park E, Lee SC (2007) Cytotoxic and antitumor activity of momilactone B from rice hulls. J Agric Food Chem 55:1702
268. Zheng G, Chang C, Stout TJ, Clardy J, Ho DK, Cassady JM (1993) Ohioensins: novel benzonaphthoxanthenones from *Polytrichum ohioense*. J Org Chem 58:366
269. Zheng GQ, Ho DK, Elder PJ, Stephens RE, Cottrell CE, Cassady JM (1994) Ohioensins and pallidisetins: novel cytotoxic agents from the moss *Polytrichum pallidiscetum*. J Nat Prod 57:32
270. Fu P, Lin S, Shan L, Lu M, Shen YH, Tang J, Liu RH, Zhang X, Zhu RL, Zhang WD (2009) Constituents of the moss *Polytrichum commune*. J Nat Prod 72:1335
271. Csupor D, Kurtan T, Vollar M, Kusz N, Kover KE, Mandi A, Szucs P, Marschall M, Senobar Tahaei SA, Zupko I, Hohmann J (2020) Pigments of the moss *Paraleucobryum longifolium*: isolation and structure elucidation of prenyl-substituted 8,8′-linked 9,10-phenanthrene-quinone dimers. J Nat Prod 83:268
272. Lunić TM, Oalđe MM, Mandić MR, Saboviljević AD, Sabovljević MS, Gasiv UM, Duletic-Lausević SN, Božić BD, Nedeljković BDR (2020) Extracts characterization and in vitro evaluation of potential immunomodulatory activities of the moss *Hypnum cupressiforme* Hedw. Molecules 25:3343
273. Li M, Wang L, Li S, Hua C, Gao H, Ning D, Li C, Zhang C, Jiang F (2022) Chemical composition, antitumor properties, and mechanism of the essential oil from *Plagiomnium acutum* T. Kop. Int J Mol Sci 23:14790
274. Toyota M, Kimura K, Asakawa Y (1998) Occurrence of *ent*-sesquiterpene in the Japanese moss *Plagiomnium acutum*: first isolation and identification of *ent*-sesqui- and dolabellane-type diterpenoids from the musci. Chem Pharm Bull 46:1488
275. Wolski GJ, Sadowska B, Fol M, Podsedek A, Kajszczak D, Kobylinska A (2021) Cytotoxicity, antimicrobial and antioxidant activities of mosses obtained from open habitats. PLoS One 16:e0257479

276. Yağlioğlu MS, Abay G, Demirtas I, Yağlioğlu AS (2017) Phytochemical screening, antiproliferative and cytotoxic activities of the mosses *Rhytidiadelphus triquetrus* (Hedw.) Warnst. and *Tortella tortuosa* (Hedw.) Limpr. Anatolian Bryol 3:31
277. Singh S, Sharma R, Singh B, Alam A (2013) Phytochemical screening and cytotoxic activity of a moss: *Barbula javanica* Dozy & Molk. Results Chem 6:101003
278. Onbasli D, Yuvali G (2020) In vitro medicinal potentials of *Bryum capillare*, a moss sample, from Turkey. Saudi J Biol Sci 28:478
279. Mandić MR, Oalđe MM, Lunić TM, Sabovljević AD, Sabovljević MS, Gašić UM, Duletić-Lauševič SN, Božić BD, Božić-Nedeljković BDB (2021) Chemical characterization and in vitro immunomodulatory effects of different extracts of moss *Hedwigia ciliata* (Hedw.) P. Beauv. from the Vrsacke Planine Mts., Serbia. PLoS One 16:e0246810
280. Klegin C, de Moura NF, de Sousa HO, Frassini R, Roesch-Ely M, Bruno AN, Bitencourt TC, Flach A, Bordin J (2021) Chemical composition and cytotoxic evaluation of the essential oil of *Phyllogonium viride* (Phyllogoniaceae, Bryophyta). Chem Biodivers 18:e2000794
281. Zych M, Urbisz K, Kimsa-Dudek M, Kamionka M, Dudek S, Raczak BK, Waclawek S, Chmura D, Kaczmarczyk-Żebrowska I, Stebel A (2023) Effects of water-ethanol extracts from four *Sphagnum* species on gene expression of selected enzymes in normal human dermal fibroblasts, and their antioxidant properties. Pharmaceuticals 16:1076
282. Li RJ, Zhao Y, Tokuda H, Yang XM, Wang YH, Shi Q, Morris-Natschke SL, Lou HX, Lee KH (2014) Total synthesis of plagiochin G and derivatives as potential cancer chemopreventive agents. Tetrahedron Lett 55:6500
283. Ivković I, Bukvicki D, Novaković M, Majstorović I, Leskovac A, Petrović S, Veljić M (2021) Assessment of the biological effects of *Pellia endiviifolia* and its constituents in vitro. Nat Prod Commun 16:1
284. Onbasli D, Yuvali CG, Altuner EM, Aslim B (2019) Investigation of pharmacological properties of bryophyte *Hypnum andoi* from Turkey. Int Pharm Nat Med 7:10
285. Kamory E, Keseru GM, Papp B (1995) Isolation and antibacterial activity of marchantin A, a cyclic bis(bibenzyl) constituent of Hungarian *Marchantia polymorpha*. Planta Med 61:387
286. Ivković IM, Bukvički DR, Novaković MM, Ivanović SG, Stanojević OJ, Nikolić IC, Veljić MM (2021) Antibacterial properties of thalloid liverworts *Marchantia polymorpha* L., *Conocephalum conicum* (L.) Dum. and *Pellia endiviifolia* (Dicks.) Dumort. J Serbian Chem Soc 86:1249
287. Lakshmi KP, Rao GMN (2023) Antimicrobial screening of the solvent extracts of *Marchantia polymorpha* against some pathogenic stains. Schol Acad J Biosci 11:385
288. Bukvicki D, Novaković M, Ilic-Tomic T, Nikodinovic-Runic J, Todorović N, Veljic M, Asakawa Y (2021) Biotransformation of perrottetin F by *Aspergillus niger*: new bioactive secondary metabolites. Rec Nat Prod 15:281
289. Sawada H, Okazaki M, Morita D, Kuroda T, Matsuno K, Hashimoto Y, Miyachi H (2012) Riccardin C derivatives as anti-MRSA agents: structure-activity relationship of a series of hydroxylated bis(bibenzyl)s. Bioorg Med Chem Lett 22:7444
290. Sawada H, Onoda K, Morita D, Ishitsubo E, Matsuno K, Tokiwa H, Kuroda T, Miyachi H (2013) Structure-anti-MRSA activity relationship of macrocyclic bis(bibenzyl) derivatives. Bioorg Med Chem Lett 23:6563
291. Onoda K, Sawada H, Morita D, Fujii K, Tokiwa H, Kuroda T, Miyachi H (2015) Anti-MRSA activity of isoplagiochin-type macrocyclic bis(bibenzyl)s is mediated through cell membrane damage. Bioorg Med Chem 23:3309
292. Harinantenaina L, Asakawa Y (2004) Chemical constituents of Malagasy liverworts, part II: mastigophoric acid methyl ester of biogenetic interest from *Mastigophora diclados* (Lepicoleaceae subf. Mastigophoraceae). Chem Pharm Bull 52:1382
293. Matsuo A, Ono K, Hamasaki K, Nozaki H (1996) Phaeophytins from a cell suspension culture of the liverwort *Plagiochila ovalifolia*. Phytochemistry 42:42
294. Scher JM, Schinkovitz A, Zapp J, Wang Y, Franzblau SG, Becker H, Lankin DC, Pauli GF (2010) Structure and anti-TB activity of trachylobanes from the liverwort *Jungermannia exsertifolia* ssp. *cordifolia*. J Nat Prod 73:656

295. Figueiredo AC, Sim-Sim, Barroso JG, Pedro LG, Santos PAG, Fontinha SS, Schripsema J, Deans SG, Scheffer JJC (2002) Comparison of the essential oil from *Marchesinia mackaii* (Hook.) S.F. Gray grown in Portugal. J Essent Oil Res 14:439
296. Xiao J, Jiang X, Chen X (2005) Antibacterial, antiinflammatory and diuretic effect of flavonoids from *Marchantia convoluta.* Afr J Trad Compl Alt Med 2:244
297. Mewari N, Kumar P (2008) Antimicrobial activity of extracts of *Marchantia polymorpha.* Pharm Biol 46:819
298. Gahtori D, Chaturvedi P (2011) Antifungal and antibacterial potential of methanol and chloroform extracts of *Marchantia polymorpha.* Arch Phytopathol Plant Protect 44:726
299. Joshi S, Singh S, Sharma R, Vats S, Nagaraju GP, Alam A (2022) Phytochemical screening and antioxidant potential of *Plagiochasma appendiculatum* Lehm. and *Sphagnum fimbriatum* Wilson. Plant Sci Today 9:986
300. Ilhan, S, Sawaroğlu F, Colak F, Iscen CF, Erdemgil FZ (2006) Antimicrobial activity of *Palustriella commutata* (Hedw.) Ochyra extracts (Bryophyta). Turk J Biol 30:149
301. Veljić M, Ciric A, Soković M, Janackovic P, Marin PD (2010) Antibacterial and antifungal activity of the liverwort (*Ptilidium pulcherrimum*) methanol extract. Arch Biol Sci Belgrade 62:381
302. Pejin B, Sabovljevic A, Soković M, Glamoclija J, Ciric A, Vujicic M, Sabovljevic M (2012) Antimicrobial activity of *Rhodobryum ontariense.* Hem Ind 66:381
303. Veljić M, Đurić A, Soković M, Ciric A, Glamoclija J, Marin PD (2009) Antibacterial activity of methanol extracts of *Fontinalis antipyretica, Hypnum cupressiforme*, and *Ctenidium molluscum.* Arch Biol Sci Belgrade 61:225
304. Kirisanth A, Nafas MNM, Dissanayake RK, Wijayabandara J (2020) Antimicrobial and *a*-amylase inhibitory activities of organic extracts of selected Sri Lankan bryophytes. Evid-Based Compl Alt Med 2020:3479851
305. Montenegro G, Portalippi M, Dalas F, Diaz M (2009) Biological properties of the Chilean native moss *Sphagnum magellanicum.* Biol Res 42:233
306. Singh M, Singh S, Nath V, Sahu V, Rawat AKS (2011) Antibacterial activity of some bryophytes used traditionally for the treatment of burn infection. Pharm Biol 49:526
307. Nikolajeva V, Liepina L, Petrina Z, Krimina G, Grube M, Muiznieks I (2012) Antibacterial activity of extracts from bryophytes. Adv Microbiol 2:345
308. Ng SY, Kamada T, Phan CS, Suleiman M, Vairappan CS (2018) New prenylated bibenzyls from Bornean liverwort *Acrobolbus saccatus.* Heterocycles 96:1958
309. Ng SY, Kamada T, Suleiman M, Vairappan CS (2016) A new cembrane-type diterpenoids from Bornean liverwort *Chandonanthus hirtellus.* J Asian Nat Prod Res 18:690
310. Basile A, Giordano S, Lopez-Saez JA, Cobianchi C (1999) Antibacterial activity of pure flavonoids isolated from mosses. Phytochemistry 52:1479
311. Manoj GS, Aswathy JM, Murugan K (2016) Bactericidal potentiality of selected bryophytes *Plagiochila beddomei, Leucobryum bowringii* and *Octoblepharum albidum.* Int J Adv Res 4:370
312. Alain A, Sharma V, Sharma SC, Kummari P (2012) Antimicrobial activity of alcoholic extracts of *Entodon nepalensis* Mizush. against some pathogenic bacteria. Rep Opin 4:44
313. Altuner EM, Canli K (2012) In vitro antimicrobial screening of *Hypnum andoi* A.J.E. Sm. Kastamonu Univ J Forestry Fac 12:97
314. Yayintas OT, Alpaslan D, Yuceer YK, Yilmaz S, Sahiner N (2017) Chemical composition, antimicrobial, antioxidant and anthocyanin activities of mosses (*Cinclidotus fontinaloides* (Hedw.) P. Beauv. and *Palustriella commutata* (Hedw.) Ochyra) gathered from Turkey. Nat Prod Res 31:2169
315. Vats S, Alam A (2013) Antibacterial activity of *Atrichum undulatum* (Hedw.) P. Beauv. against some pathogenic bacteria. J Biol Sci 13:427
316. Elibol BE, Ezer T, Kara R, Celic GY, Colak E (2011) Antifungal and antibacterial effects of some acrocarpic mosses. Afr J Biotechnol 10:886
317. Simsek O, Canli K, Benek A, Turu DM, Altuner EM (2023) Biochemical, antioxidant properties and antimicrobial activity of epiphytic leafy liverwort *Frullania dilatata* (L.) Dumort. Plants 12:1877

318. Nazir S, Murtaza G, Hameed A, Abbas G (2023) Evaluation of antibacterial activity of liverwort species of Bagh Azad Jammu and Kashmir (Western Himalaya) against pathogenic bacteria. GSC Biol Pharm Sci 22:166
319. Singh M, Govindarajan R, Nath V, Rawat AKS, Mehrotra S (2006) Antimicrobial, wound healing and antioxidant activity of *Plagiochasma appendiculatum* Lehm. et Lind. J Ethnopharmacol 107:67
320. Okan OT (2023) Antioxidant, antimicrobial and some chemical composition of *Plagiochila asplenioides* (L.) Dumort extract. Anatol Bryol 9:11
321. Karaglu SA, Yayli N, Akpinar R, Bozdeveci, Erik I, Suyabatmaz, S, Korkmaz B, Batan N, Kaya S, Nisbet C, Guler A (2023) Phytochemicals, antimicrobial, and sporicidal activities of moss, *Dicranum polysetum* Sw., against certain honey bee bacterial pathogens. Vet Res Commun 47:1445
322. Colak E, Kara R, Ezer T, Celic GY, Elibol B (2018) Investigation of antimicrobial activity of some Turkish pleurocarpic mosses. Anatol J Biotech 10:12905
323. Gilabert M, Ramos AN, Schiabone MM, Arena ME, Bardon A (2011) Bioactive sesqui- and diterpenoids from the Argentine liverwort *Porella chiliensis*. J Nat Prod 74:574
324. Guo XL, Leng P, Yeng Y, Yu LG, Lou H (2008) Plagiochin E, a botanic-derived phenolic compound, reverses fungal resistance to fluconazole relating to the efflux pump. J Appl Microbiol 104:831
325. Wu XZ, Cheng AX, Sun LM, Lou HX (2008) Effect of plagiochin E, an antifungal macrocyclic bis(bibenzyl), on cell wall chitin synthesis in *Candida albicans*. Acta Pharmacol Sin 29:147
326. Wu XZ, Cheng AX, Sun LM, Sun SJ, Lou HX (2009) Plagiochin E, an antifungal bis(bibenzyl) exerts its antifungal activity through mitochondrial disfunction-induced reactive oxygen species accumulation in *Candida albicans*. Biochim Biophys Acta 1790:770
327. Wu XZ, Chang WQ, Cheng AX, Sun LM, Lou HX (2010) Plagiochin E, an antifungal bis(bibenzyl), induced apoptosis in *Candida albicans* through a metacaspase-dependent apoptotic pathway. Biochim Biophys Acta 1800:439
328. Niu C, Qu JB, Lou HX (2006) Antifungal bis(bibenzyls) from the Chinese liverwort *Marchantia polymorpha* L. Chem Biodivers 3:34
329. Qu J, Xie C, Guo H, Yo W, Lou H (2007) Antifungal dibenzofuran bis(bibenzyl)s from the liverwort *Asterella angusta*. Phytochemistry 68:1767
330. Xie CF, Qu JB, Wu XZ, Liu N, Ji M, Lou HX (2010) Antifungal macrocyclic bis(bibenzyls) from the Chinese liverwort *Plagiochasma intermedium* L. Nat Prod Res 24:515
331. Scher JM, Speakman JB, Zapp J, Becker H (2004) Bioactivity guided isolation of antifungal compounds from the liverwort *Bazzania trilobata* (L.) S.F. Gray. Phytochemistry 65:2583
332. Gijsen HJ, Wijnberg JBPA, Stork GA, De Groot A, De Waard MA, Van Nistelrooy JGM (1992) The synthesis of mono- and dihydroxy aromadendrane sesquiterpenes starting from natural (+)-aromadendrene. Tetrahedron 48:2465
333. Burden RS, Kemp MS (1983) Sesquiterpene phytoalexin from *Ulmus glabra*. Phytochemistry 22:1039
334. Zheng S, Chang W, Zhang M, Shi H, Lou H (2018) Chiloscyphenol A derived from Chinese liverworts exerts fungicidal action by eliciting both mitochondrial dysfunction and plasma membrane destruction. Sci Rep 8:326
335. Li S, Shi H, Chang W, Li Y, Zhang M, Qiao Y, Lou H (2017) Eudesmane sesquiterpenes from Chinese liverwort are substrates of Cdrs and display antifungal activity by targeting Erg6 and Erg11 of *Candida albicans*. Bioorg Med Chem 25:5764
336. Labbe C, Faini F, Villagran C, Coll J, Rycroft DS (2007) Bioactive polychlorinated bibenzyls from the liverwort *Riccardia polyclada*. J Nat Prod 70:2019
337. Labbe C, Faini F, Villagran C, Coll J, Rycroft DS (2005) Antifungal and insect antifeedant 2-phenylethanol esters from the liverwort *Balantiopsis cancellata* from Chile. J Agric Food Chem 53:247
338. Wang XN, Yu WT, Lou HX (2005) Antifungal constituents from the Chinese moss *Homalia trichomanoides*. Chem Biodivers 2:139

339. Baek S-H, Phipps RK, Perry NB (2004) Antimicrobial chlorinated bibenzyls from the liverwort *Riccardia marginata*. J Nat Prod 67:718
340. Li Y, Xu Z, Zhu R, Zhou J, Zong Y, Zhang J, Zhu M, Jin X, Qiao Y, Zheng H, Lou H (2020) Probing the interconversion of labdane lactones from the Chinese liverwort *Pallavicinia ambigua*. Org Lett 22:510
341. Wang X, Jin XY, Zhou JC, Zhu RX, Qiao YN, Zhang JZ, Li Y, Zhang CY, Chen W, Chang WQ, Lou HX (2020) Terpenoids from the Chinese liverwort *Heteroscyphus coalitus* and their antivirulence activity against *Candida albicans*. Phytochemistry 174:112324
342. Wang F, Lou HX (2000) Chemical studies on the constituents of *Plagiochasma intermedium*. Acta Pharm Sin 35:587
343. Kamada T, Johanis ML, Ng SY, Phan CS, Suleiman M, Vairappan CS (2020) A new *ent*-neoverrucosane-type diterpenoid from the liverwort *Pleurozia subinflata* in Borneo. Nat Prod Bioprospect 10:51
344. Ng SY, Ang LP, Hau VL, Suleiman M, Vairappan CS, Kamada T (2021) Structural diversity, anti-fungal activity and chemosystematics of Bornean liverwort *Bazzania harpago* (De Not.) Schiffner. Sains Malaysia 50:101
345. Matsuo A, Kamio K, Uohama K, Yoshida K, Connolly JD, Dim G (1988) Dolabellane diterpenoids from the liverwort *Odontoschisma denudatum*. Phytochemistry 27:1153
346. Zhu MZ, Li Y, Zhou JC, Lu JH, Zhu RX, Qiao YN, Zhang JZ, Zong Y, Wang X, Jin XY, Zhang M, Chang WQ, Chen W, Lou HX (2020) Terpenoids from the Chinese liverwort *Odontoschisma grosseverrucosum* and their antifungal virulence activity. Phytochemistry 174:112341
347. Qiao YN, Jin XY, Zhou JC, Zhang JZ, Chang WQ, Li Y, Cheng W, Ren ZJ, Zhang CY, Yuan SZ, Lou HX (2020) Terpenoids from the liverwort *Plagiochila fruticosa* and their antivirulence activity against *Candida albicans*. J Nat Prod 83:1766
348. Ucuncu O, Cansu T, Zdemir T, Alpaykaraoglu S, Yayli N (2010) Chemical composition and antimicrobial activity of the essential oils of mosses (*Tortula muralis* Hedw., *Homalothecium lutescens* (Hedw.) H. Rob., *Hypnum cupressiforme* Hedw., and *Pohlia nutans* (Hedw.) Lindb.) from Turkey. Turk J Chem 34:825
349. Mewari N, Kumar P (2011) Evaluation of antifungal potential of *Marchantia polymorpha* L., *Dryopteris filix-mas* (L.) Scott and *Ephedra foliata* Boiss. against phytofungal pathogens. Archiv Phytopathol Plant Protect 44:802
350. Sabovljević A, Soković M, Glamočlija J, Ćirić A, Vujičić M, Pejin R, Sabovljević M (2011) Bio-activities of extracts from some axenically farmed and naturally grown bryophytes. J Med Plants Res 5:565
351. Asakawa Y, Ludwiczuk A, Hashimoto T (2013) Cytotoxic and antiviral components from bryophytes and inedible fungi. J Pre-Clin Clin Res 7:73
352. Asakawa Y, Miyataka H, Kenmoku H, Esumi T, Yamamoto H, Tomiyama K, Yaguchi Y (2021) 65th Symposium on chemistry of terpenes, essential oils, and aromatics, Yamaguchi, Japan, 30 Oct–1 Nov, Symposium papers, p 124
353. Xiao JM, Ren FL, Mig X (2005) Anti-hepatitis B virus activity of flavonoids from *Marchantia convoluta*. Iran J Pharmacol Therapeut 4:128
354. Iwai Y, Murakami K, Gomi Y, Hashimoto T, Asakawa Y, Okuno Y, Ishikawa I, Hatakeyama D, Echigo N, Kuzuhara T (2011) Anti-influenza activity of marchantins, macrocyclic bisbibenzyls contained in liverworts. PLoS One 6:e1982
355. Harinantenaina L, Takaoka S (2006) Cinnafragrins A-C, dimeric and trimeric drimane sesquiterpenoids from *Cinnamosma fragrans*, and structure revision of capsicodendron. J Nat Prod 69:1193
356. Perry NB, Burgess EJ, Foster LM, Gerard PJ (2003) Insect antifeedant sesquiterpene acetals from the liverwort *Lepidolaena clavigera*. Tetrahedron Lett 44:1651
357. Toyota M, Omatsu I, Braggins J, Asakawa Y (2009) Pungent aromatic compounds from New Zealand liverwort *Hymenophyton flabellatum*. Chem Pharm Bull 57:1
358. Numata A, Katsuno T, Yamamoto K, Nishida T, Takemura T, Seto K (1984) Plant constituents biologically active to insects. IV. Antifeedants for the larvae of the yellow butterfly, *Eurema hecabe mandarina*, in *Arachniodes standishii*. Chem Pharm Bull 32:325

359. Lorimer SD, Perry NB, Foster LM, Burgess EJ, Douch PGC, Hamilton MC, Donaghy MJ, McGregor RA (1996) A nematode larval motility inhibition assay for screening plant extracts and natural products. J Agric Food Chem 44:2842
360. Ainge GD, Gerard PJ, Hinkley SFR, Lorimer SD, Weavers RT (2001) Hodgsonox, a new class of sesquiterpene from the liverwort *Lepidolaena hodgsoniae*. Isolation directed by insecticidal activity. J Org Chem 66:2818
361. Shy HS, Wu CL, Paul C, Konig WA, Ean UJ (2002) Chemical constituents of two liverworts *Metacalypogeia alternifolia* and *Chandonanthus hirtellus*. J Chin Chem Soc 49:593
362. Ramirez M, Kamiya N, Popich S, Asakawa Y, Bardon A (2010) Insecticidal constituents from the Argentine liverwort *Plagiochila bursata*. Chem Biodivers 7:1855
363. Asakawa Y, Jang D, Nagashima F (2023) Azulenoids in liverworts: distribution, structures, bioactivity and chemosystematics of Calypogeiaceae. In: 67th Symposium on the chemistry of terpenes, essential oils and aromatics. Chiba, Japan, Symposium papers, p 147
364. Ramirez M, Kamiya N, Popich S, Asakawa Y, Bardon A (2017) Constituents of the Argentine liverwort *Plagiochila diversifolia* and their insecticidal activities. Chem Biodivers 14:e1700229
365. Ande AT, Wahedi JA, Fatoba PO (2010) Biocidal activities of some tropical moss extracts against maize borers. Ethnobot Leafl 14:479
366. Krishnan R, Murugan K (2015) Insecticidal potentiality of flavonoids from cell suspension culture of *Marchantia linearis* Lehm. and Lindenb. against *Spodoptera litura* F. Int J Appl Biol Pharm Technol 6:23
367. Fukada R, Kawano J, Tsuruta T, Nonaka T, Sato K, Miyajima S, Ishigami S, Ishii T, Nishikawa K, Asakawa Y, Kamada T (2023) Two new eremophilane-type sesquiterpenoids from Japanese liverwort *Bazzania japonica*. Chem Biodivers 20:e202300131
368. Romani F, Banic E, Florent SN, Kanazawa T, Goodger JQD, Mentink RA, Dierschke T, Zachgo S, Ueda T, Bowman JL, Tsiantis M, Moreno JE (2020) Oil body formation in *Marchantia polymorpha* is controlled by MpC1HDZ and serves as a defense against arthropod herbivores. Curr Biol 30:2815
369. Alves RJM, Miranda TG, Pinheiro RO, Pinheiro WBS, Andrade EHA, Tavares-Martins ACC (2022) Volatile chemical composition of *Octoblepharum albidum* Hedw. (Bryophyta) from the Brazilian Amazon. BMC Chem 16:76
370. Otoguro K, Iwatsuki M, Ishiyama A, Namatame M, Nishihara-Tukashima A, Kiyohara H, Hashimoto T, Asakawa Y, Omura S, Yamada H (2011) In vitro antitrypanosomal activity of plant terpenes against *Trypanosoma brucei*. Phytochemistry 72:2024
371. Otoguro K, Ishiyama A, Iwatsuki M, Namatame M, Nishihara-Tukashima A, Kiyohara H, Hashimoto T, Asakawa Y, Omura S, Yamada H (2012) In vitro antitrypanosomal activity of bis(bibenzyls) and bibenzyls from the liverworts against *Trypanosoma brucei*. J Nat Med 66:377
372. Jensen S, Omarsdottir S, Bwalya AG, Nielsen MA, Tasdemir D, Thorsteljsdottir JB (2012) Marchantin A, a macrocyclic bisbibenzyl ether, isolated from the liverwort *Marchantia polymorpha*, inhibits protozoal growth in vitro. Phytomedicine 19:1191
373. Toyota M, Nakamura I, Huneck S, Asakawa Y (1994) Sesquiterpene esters from the liverwort *Plagiochila porelloides*. Phytochemistry 37:1091
374. Kunz S, Becker H (1994) Bibenzyl derivatives from the liverwort *Ricciocarpos natans*. Phytochemistry 36:675
375. Roldos V, Nakayama H, Rolon M, Montero-Terres A, Trucco F, Torres S, Vega C, Marrero-Ponce Y, Heguaburu V, Yaluff G, Gómez-Barrio A, Sanabria L, Ferreira ME, Rojas de Arias A, Pandolfi E (2008) Activity of a hydroxybibenzyl bryophyte constituent against *Leishmania* spp. and *Trypanosoma cruzi*: in silico, in vitro and in vivo activity studies. Eur J Med Chem 43:1797
376. Schwartner C, Bor W, Michel C, Franck U, Muller-Jakic SB, Nenninger A, Asakawa Y, Wagner H (1995) Effect of marchantins and related compounds on 5-lipoxygenase and cyclooxygenase and their antioxidant properties: a structure activity relationship. Phytomedicine 2:113

377. Hashimoto T, Tori M, Asakawa Y (1988) Highly efficient preparation of lunularic acid and some biological activities of stilbene and dihydrostilbene derivatives. Phytochemistry 27:109
378. Seo C, Hoi YH, Sohn JH, Ahn JS, Y JH, Lee HK, Oh H (2008) Ohioensins F and G: protein tyrosine phosphatase 1B inhibitory benzonaphthoxanthones from the Antarctic moss *Polytrichum alpinum*. Bioorg Med Chem Lett 18:772
379. Tan TQ, Hoang PN, Vy LN, Anh BL, Nhut DT, Phuong QND (2020) Improving in vitro biomass and evaluating α-glucosidase inhibition activity of liverwort *Marchantia polymorpha* L. Asian J Plant Sci 19:133
380. Harinantenaina L, Asakawa Y (2007) Malagasy liverworts, source of new and biologically active compounds. Nat Prod Commun 2:701
381. Wang X, Zhang JZ, Zhou JC, Shen T, Lou HX (2016) Terpenoids from *Diplophyllum taxifolium* with quinone reductase-inducing activity. Fitoterapia 109:1
382. Han J, Li Y, Qi Z, Meng H, Zhang J, Sun Y, Qiao Y, Sun B, Lou H (2021) Dolabrane diterpenoids from the Chinese liverwort *Notoscyphus lutescens*. J Nat Prod 84:2929
383. Hashimoto T, Irita H, Yoshida M, Kikkawa A, Toyota M, Koyama H, Motoike Y, Asakawa, Y (1998) Chemical constituents of the Japanese liverworts *Odontoschisma denudatum, Porella japonica, P. acutifolia* subsp. *tosana* and *Frullania hamatiloba*. J Hattori Bot Lab 84:309
384. Qiao YN, Sun Y, Shen T, Zhang JZ, Zhou JC, Li Y, Chen W, Ren ZJ, Li YL, Wang X, Lou HX (2019) Diterpenoids from the Chinese liverwort *Frullania hamatiloba* and their Nrf2 inducing activity. Phytochemistry 158:77
385. Han J, Li Y, Zhou J, Zhang J, Qiao Y, Fang K, Zhang C, Zhu M, Lou HX (2021) Terpenoids from Chinese liverworts *Scapania* spp. J Nat Prod 84:1210
386. Filipović S, Stojanovic NM, Mitić KV, Randjelovic P, Radulović N (2022) Revisiting the effect of 3 sesquiterpenoids from *Conocephalum conicum* (snake liverwort) on rat spleen lymphocyte viability and membrane functioning. Nat Prod Commun 17:1
387. Zhang J, Wang Y, Zhu R, Li Y, Qiao Y, Zhou J, Lou H (2018) Cyperane and eudesmane-type sesquiterpenoids from Chinese liverworts and their anti-diabetic nephropathy potential. RSC Adv 8:39091
388. Wang X, Cao J, Wu Y, Wang Q, Xiao J (2016) Flavonoids, antioxidant potential, and acetylcholinesterase inhibition activity of the extracts from the gametophyte and archegoniophore of *Marchantia polymorpha*. Molecules 21:360
389. Mohandas GG, Kumaraswamy M (2018) Antioxidant activities of terpenoids from *Thuidium tamariscellum* (C. Muell.) Bosch. and Dande-Lac. A moss. Pharmacogn J 10:645
390. Basile A, Sorbo S, Conte B, Golia B, Montanari S, Castaldo-Cobianchi R, Esposito S (2011) *Leptodictyum riparium* (Bryophyta), stressed by heavy metals, heat shock, and salinity. Plant Biosystem 145:77
391. Provenzano F, Sanchez JL, Rao E, Santonocito R, Ditta LA, Borras Linares IB, Passantino R, Campisi P, Dia MG, Costa MA, Segura-Carretero A, Biagio PLS, Giacomazza D (2019) Water extract of *Cryphaea heteromalla* (Hedw.) D. Mohr bryophytes as a natural powerful source of biologically active compounds. Int J Mol Sci 20:5560
392. Ielpo MTL, Basile A, Miranda R, Moscatiello V, Nippo C, Sorbo S, Laghi E, Ricciardi MM, Ricciardi L, Vuotto ML (2000) Immunopharmacological properties of flavonoids. Fitoterapia 71:101
393. Rana M, Pant J, Jantwal A, Rana AJ, Upadhyay J, Bisht SS (2018) In-vitro anti-inflammatory and antioxidant activity of ethanol extract of *Marchantia polymorpha* in Kumauni region. World J Pharm Res 7:864
394. Gokbulut A, Satilmis B, Batcioglu K, Cetin B, Sarer E (2012) Antioxidant activity and luteolin content of *Marchantia polymorpha*. Turk J Biol 36:381
395. Hsiao G, Teng CM, Wu C, Ko FN (1996) Marchantin H, as a natural antioxidant and free radical scavenger. Archiv Biochem Biophys 334:18
396. Schwartner C, Michel C, Stettmaier K, Wagner H, Bors W (1996) Marchantins and related polyphenols from liverwort: physico-chemical studies of their radical-scavenging properties. Free Rad Biol Med 20:237

397. Sadamori M (2009) Studies on the new biologically active substances of Tahitian and Tokushima's *Plagiochila* genus. Master's thesis, Tokushima Bunri University, Tokushima, Japan, p 1
398. Bhattarai HD, Paudel B, Lee HK, Oh H, Yim JH (2009) In vitro antioxidant capacities of two benzonaphthoxanthenones: ohioensins F and G, isolated from the Antarctic *Polytristrum alpinum*. Z Naturforsch 64c:197
399. Tazaki H, Ito M, Miyoshi M, Kawabata J, Fukushi E, Fujita T, Motouri M, Furuki T, Nabeta K (2002) Subulatin, an antioxidant caffeic acid derivative isolated from the in vitro cultured liverworts, *Jungermannia subulata*, *Lophocolea heterophylla*, and *Scapania parvitexta*. Biosci Biotechnol Biochem 66:255
400. Li Y, Sun Y, Zhu M, Zhu R, Zhang J, Zhou J, Wang T, Qiao Y, Lou H (2019) Sacculatane diterpenoids from the Chinese liverwort *Pellia epiphylla* with protection against H_2O_2-induced apoptosis of PC12 cells. Phytochemistry 162:173
401. Lunić TM, Mandić MR, Pavlović MMO, Sabovljević AD, Sabovljević MS, Nedeljković BDB, Božić BD (2022) The influence of seasonality on secondary metabolite profiles and neuroprotective activities of moss *Hypnum cupressiforme* extracts: in vitro and in silico study. Plants 11:123
402. Tran-Quoc T, Nguyen-Thi MD, Tran-Quoc D, Quach-Ngo DP (2023) Study on gametophyte cultivation and bioactive evaluation on biomass of silver moss *Bryum argenteum* Hedw. Res J Biotechnol 18:22
403. Smolinska-Kondla D, Zych M, Ramos P, Waclawek S, Stebel A (2022) Antioxidant potential of various extracts from 5 common European mosses and its correlation with phenolic compounds. Herba Pol 68:54
404. Chicca A, Schafroth MA, Reynoso-Moreno I, Erni R, Petrucci V, Carreira EM, Gertsch J (2018) Uncovering the psychoactivity of a cannabinoid from liverworts associated with a legal high. Sci Adv 4:2166
405. Zhou J, Zhang J, Li R, Liu J, Fan P, Li Y, Ji M, Dong Y, Yuan H, Lou H (2016) Hapmnioides A-C, rearranged labdane-type diterpenoids from the Chinese liverwort *Haplomitrium mnioides*. Org Lett 18:4274
406. Adams DR, Brochwicz-Lewinski M, Butler AR (1999) Nitric oxide: physiological roles, biosynthesis and medical use. In: Herz W, Falk H, Kirby GW, Moore RE, Tamm Ch (eds) Progress in the chemistry of organic natural products, vol 76. Springer, Vienna, p 1
407. Marques RV, Sestito SE, Bourgaud F, Miguel S, Cailotto F, Reboul P, Jouzeau JY, Rahuel-Clermont S, Boschi-Muller S, Simonsen HT, Moulin D (2022) Antiinflammatory activity of bryophytes extracts in LPS-stimulated RA264.7 murine macrophages. Molecules 27:1940
408. Harinantenaina L, Quang DN, Nishizawa T, Hashimoto T, Kohchi C, Soma GI, Asakawa Y. (2005) Bis(bibenzyls) from liverworts inhibit lipopolysaccharide-induced inducible NOS in RAW 264.7 cells: a study of structure-activity relationships and molecular mechanism. J Nat Prod 68:1779
409. Harinantenaina L, Dang NQ, Nishizawa T, Hashimoto T, Kohchi C, Soma G, Asakawa Y (2007) Bioactive compounds from liverworts: inhibition of lipopolysaccharide-induced inducible NOS *m*RNA in RAW 264.7 cells by herbertenoids and cuparenoids. Phytomedicine 14:486
410. Quang DN, Asakawa Y (2010) Chemical constituents of the Vietnamese liverwort *Porella densifolia*. Fitoterapia 81:659
411. Harinantenaina L, Takahara Y, Nishizawa T, Kohchi C, Soma G, Asakawa Y (2006) Chemical constituents of Malagasy liverworts, part V: prenyl bibenzyls and clerodane diterpenoids with nitric oxide inhibitory activity from *Radula appressa* and *Thysananthus spathulistipus*. Chem Pharm Bull 54:1046
412. Zhu M, Li Y, Zhou J, Wang T, Li X, Zhang J, Qiao Y, Han J, Xu J, Lou H (2022) Pinguisane sesquiterpenoids from the Chinese liverwort *Trocholejeunea sandvicensis* and their anti-inflammatory activity. J Nat Prod 85:205
413. Li S, Niu H, Qiao Y, Zhu R, Sun Y, Ren Z, Yuan H, Gao Y, Li Y, Chen W, Zhou J, Lou H (2018) Terpenoids isolated from Chinese liverworts *Lepidozia reptans* and their anti-inflammatory activity. Bioorg Med Chem 26:2392

414. Tosun A, Akkol EK, Suntar I, Kiremit HO, Asakawa Y (2013) Phytochemical investigations of bioactivity evaluation of liverworts as a function of antiinflammatory and antinociceptive properties in animal models. Pharm Biol 51:1008
415. Kumar K, Singh KK, Asthana AK, Nath V (2000) Ethnotherapeutics on bryophyte *Plagiochasma appendiculatum* among the Gaddi tribes of Kangra valley, Himachal Pradesh, India. Pharm Biol 38:353
416. Zhang CY, Zhang JZ, Li YL, Xu ZJ, Qiao YN, Zhi S, Yuan Z, Tang YJ, Lou HX (2024) Heterodimers of aromadendrane sesquiterpenoid with benzoquinone from the Chinese liverwort *Mylia nuda*. J Nat Prod 87:132
417. Kim SY, Hong M, Kim TH, Lee KY, Park SJ, Hong SH, Sowndhararajan K, Kim S (2021) Antiinflammatory effect of liverwort (*Marchantia polymorpha* L.) and *Racomitrium* moss (*Racomitrium canescens* (Hedw.) Brid.) growing in Korea. Plants 10:2075
418. Zhang J, Fan P, Zhu R, Li R, Lin Z, Sun B, Zhang C, Zhou J, Lou H (2014) Marsupellins A-F, *ent*-longipinane-type sesquiterpenoids from the Chinese liverwort *Marsupella alpine* with acetylcholinesterase inhibitory activity. J Nat Prod 77:1031
419. Fukuyama Y, Asakawa Y (1991) Novel neurotrophic isocuparane-type sesquiterpene dimers, mastigophorenes A, B, C and D, isolated from the liverwort *Mastigophora diclados*. J Chem Soc Perkin Trans 1:2737
420. Fukuyama Y, Toyota M, Asakawa Y (1988) Mastigophorenes: novel dimeric isocuparane-type sesquiterpenoids from the liverwort *Mastigophora diclados*. J Chem Soc, Chem Commun: 1341
421. Fukuyama Y, Kodama M (1996) Search for novel neurotrophic factor-like substances in natural products. Food Ingred J 1:45
422. Taira Z, Takei M, Endo K, Hashimoto T, Sakiya Y, Asakawa Y (1994) Marchantin A trimethyl ether: its molecular structure and tubocurarine-like skeletal muscle relaxation activity. Chem Pharm Bull 42:52
423. Keseru GM, Nogradi M (1995) The biological activity of cyclic bis(bibenzyls): a rotational approach. Bioorg Med Chem 3:1511
424. Keseru GM, Nogradi M (1995) The chemistry of macrocyclic bis(bibenzyls). Nat Prod Rep 12:69
425. Keseru GM, Nogradi M (1996) Molecular similarity analysis on biologically active macrocyclic bis(bibenzyls) J Mol Recognit 9:133
426. Tamehiro N, Sato Y, Suzuki T, Hashimoto T, Asakawa Y, Yokoyama S, Kawanishi T, Ohno Y, Inoue K, Nagano T, Nishimaki NM (2005) Riccardin C: a natural product that functions as a liver X receptor (LXR)α agonist and an (LXR)β antagonist. FEBS Lett 579:5299
427. Hioki H, Shima N, Kawaguchi K, Harada K, Kubo M, Esumi T, Nishimaki-Mogami T, Sawada J-I, Hashimoto T, Asakawa Y, Fukuyama Y (2009) Synthesis of riccardin C and its seven analogues. Part 1: The role of their phenolic hydroxy groups as LXRα agonists. Bioorg Med Chem Lett 19:738
428. Katsunuma N (1997) Molecular mechanism of bone collagen degradation in bone resorption. J Bone Mineral Metab 15:1
429. Matsunaga Y, Saibara T, Kido H, Katunuma N (1993) Participation of cathepsin B in processing of antigen presentation to MHC class II. Fed Eur Biochem Soc Lett 324:325
430. Hashimoto T, Irita H, Yoshida M, Kikkawa A, Toyota M, Koyama H, Motoike Y, Asakawa Y (1998) Chemical constituents of the Japanese liverworts *Odontoschisma denudatum*, *Porella japonica*, *P. acutifolia* subsp. *tosana* and *Frullania hamatiloba*. J Hattori Bot Lab 84:309
431. Wei H-C, Ma SJ, Wu CL (1995) Sesquiterpenoids and cyclic bisbibenzyls from the liverwort *Reboulia hemisphaerica*. Phytochemistry 39:91
432. Lin SJ, Wu CL (1996) Isoplagiochilide from the liverwort *Plagiochila elegans*. Phytochemistry 41:1439
433. Nagashima F, Momosaki S, Watanabe Y, Toyota M, Huneck S, Asakawa Y (1996) Terpenoids and aromatic compounds from six liverworts. Phytochemistry 41:207
434. Suzuki T, Tamehiro N, Sato Y, Kobayashi T, Ishii-Watanabe A, Shinozaki Y, Nishimaki-Mogami T, Hashimoto T, Asakawa Y, Inoue K, Ohono Y, Yamaguchi T, Kawanishi T (2008)

The novel compounds that activate farnesoid X-receptor: the discovery of their effects on gene expression. J Pharmacol Sci 107:285
435. Morita H, Zaima K, Koga I, Saito A, Tamamoto H, Okazaki H, Kaneda T, Hashimoto T, Asakawa Y (2011) Vasorelaxant effects of macrocyclic bis(bibenzyls) from liverworts. Bioorg Med Chem 19:4051
436. Toyota M, Kinugawa T, Asakawa Y (1994) Bibenzyl cannabinoid and bisbibenzyl derivative from the liverwort *Radula perrottetii*. Phytochemistry 37:859
437. Toyota M, Shimamura T, Ishii H, Asakawa Y (2002) New bibenzyl cannabinoid from the New Zealand liverwort *Radula marginata*. Chem Pharm Bull 50:1390
438. Boland W (1995) The chemistry of gamete attraction: chemical strutures, biosynthesis, and (a)biotic degradation of algal pheromones. Proc Natl Acad Sci USA 92:37
439. Kajiwara T, Hatanaka A, Tanaka Y, Kawai T, Ishihara M, Tsuneya T, Fujimura T (1989) Volatile constituents from marine brown algae of Japanese *Dictyopteris*. Phytochemistry 28:636
440. Kobayashi J, Ishibashi M (1999). In: Barton D, Nakanishi K, Meth-Cohn O (eds) Comprehensive natural products chemistry, vol 8. Elsevier, Amsterdam, p 420
441. Tedone L, Komala I, Ludwiczuk A, Nagashima F, Ito T, Mondello L, Asakawa Y (2011) Volatile components of selected Japanese and Indonesian liverworts. In: 55th Symposium on the chemistry of terpenes, essential oils, and aromatics, Tsukuba, Japan, 19–21 Nov 2011, Symposium papers, p 272
442. Zhou J, Zhang J, Cheng A, Xiong Y, Liu L, Lou H (2015) Highly rigid labdane-type diterpenoids from a Chinese liverwort and light-driven structure diversification. Org Lett 17:3560
443. Pannequin A, Tintaru A, Desjiobert JM, Costa J, Muselli A (2017) New advances in the volatile metabolites of *Frullania tamarisci*. Flav Fragr J 32:409
444. Zhang W, Baudouin E, Cordier M, Frison G, Nay B (2019) One-pot synthesis of metastable 2,5-dihydrooxepines through retro-Claisen rearrangements: method and applications. Chem Eur J 25:8643
445. Thuillier S, Viola S, Lockett-Walters B, Nay B, Bailleul B, Baudouin E (2023) Mode of action of the natural herbicide radulanin A as an inhibitor of photosystem II. Pest Management Sci 2023:https://doi.org/10.1002/ps.76.hal-04124088
446. Yoshikawa H, Ichiki Y, Sakakibara K, Tamura H, Suiko M (2002) The biological and structural similarity between lunularic acid and abscisic acid. Biosci Biotechnol Biochem 66:840
447. Srikrishna A, Ramachary DB (2000) The first total synthesis of (±)-grimaldone. Tetrahedron Lett 41:2231
448. Kido F, Maruta R, Tsutsumi K, Yoshikoshi A (1979) Total synthesis of racemic frullanolide. Chem Lett:111
449. Bernasconi S, Gariboldi T, Jommi G, Montanari S, Sisti M (1981) Total synthesis of pinguisone. J Chem Soc Perkin Trans 1:2394
450. Uyehara T, Kabasawa Y, Kato T, Furuta Y (1985) Photochemical rearrangement approach to the total synthesis of (±)-pinguisone and (±)-deoxopinguisone. Tetrahedron Lett 26:2343
451. Harinantenaina L, Noma Y, Asakawa Y (2005) *Penicillium sclerotiorum* catalyzes the conversion of herbertenediol into its dimers: mastigophorenes A and B. Chem Pharm Bull 53:256
452. Fukuyama Y, Matsumoto K, Tonoi Y, Yokoyama R, Takahashi H, Minami H, Okazaki H, Mitsumoto Y (2001) Total syntheses of neuroprotective mastigophorenes A and B. Tetrahedron 57:7127
453. Blay G, Cardona L, Garcia D, Lahoz L, Pedro JR (2001) Synthesis of plagiochiline N from santonin. J Org Chem 66:7700
454. Hagiwara H, Uda H (1989) A total synthesis of (–)-sacculatal. Bull Chem Soc Jpn 62:624
455. Hagiwara H, Uda H (1990) Total synthesis of (+)-perrottetianal. J Chem Soc Perkin Trans 1:19001
456. Bhat SV, Dohadwalla AN, Bajwa BS, Dadkar NK, Dornauer H, de Souza NJ (1983) The antihypertensive and positive inotropic diterpene forskolin: effects of structural modifications on its activity. J Med Chem 26:486

457. Hashimoto T, Hori M, Toyota M, Taira Z, Takeda R, Tori T, Asakawa Y (1994) Structures of five new highly oxygenated labdane-type diterpenoids, ptychantins A-E, closely related to forskolin from liverwort *Ptychanthus striatus*. Tetrahedron Lett 35:5457
458. Hagiwara H, Takeuchi F, Hoshi T, Suzuki T, Hashimoto T, Asakawa Y (2003) First synthesis of 1,9-dideoxyforskolin from ptychantin A. Tetrahedron Lett 44:2305
459. Hagiwara H, Takeuchi F, Kudo M, Hoshi T, Suzuki T, Hashimoto T, Asakawa Y (2000) Synthetic transformation of ptychantin to forskolin and 1,9-dideoxyforskolin. J Org Chem 71:4619
460. Hagiwara H, Tsukagoshi M, Hoshi T, Suzuki T, Hashimoto T, Asakawa Y (2008) Expedient synthetic transformation of ptychantins into forskolin. Synlett 929
461. Hashimoto T, Shiki K, Tanaka M, Takaoka S, Asakawa Y (1998) Chemical conversion of labdane-type diterpenoid from the liverwort *Porella perrottetiana* into (–)-ambrox. Heterocycles 49:315
462. Song Y, Hwang S, Gong P, Kim D, Kim S (2008) Stereoselective total synthesis of (–)-perrottetinene and assignment of its absolute configuration. Org Lett 10:269
463. Park BH, Lee YR (2010) Concise synthesis of (±)-perrottetinene with bibenzyl cannabinoid. Bull Kor Chem Soc 31:2712
464. Lockell-Walters B, Thuillier S, Bauduin E, Nay B (2022) Total synthesis of phytotoxic radulanin A facilitated by the photochemical ring expansion of 2,2-dimethylchromene inflow. Org Lett 24:4029
465. Kodama M, Shiobara Y, Matsumura K, Sumitomo H (1995) Total synthesis of marchantin A, a novel cytotoxic bis(bibenzyl) isolated from liverworts. Tetrahedron Lett 26:877
466. Kodama M, Shiobara Y, Sumitomo H, Tsukamoto MK, M, Harada C, (1988) Total syntheses of marchantin A and riccardin B, cytotoxic bis(bibenzyls) isolated from liverworts. J Org Chem 53:72
467. Gottsegen A, Nogradi M, Vermes B, Kajtar-Predy M, Bihatsi-Karsai E (1990) Total syntheses of riccardins A, B, and C, cytotoxic macrocyclic bis(bibenzyls) from liverworts. J Chem Soc Perkin Trans 1:315
468. Speicher A, Holz J (2010) Synthesis of marchantin C, a novel tubulin inhibitor from liverwort. Tetrahedron Lett 51:2986
469. Guo L, Wu JZ, Han T, Cao T, Rahman K, Qin LP (2008) Chemical composition, antifungal and antitumor properties of ether extracts of *Scapania verrucosa* Heeg. and its endophytic fungus *Chaetomium fusiforme*. Molecules 13:2114
470. Li XB, Xie F, Liu SS, Li Y, Zhou JC, Liu YQ, Yuan HQ, Lou HX (2013) Naphtho-γ-pyrones from endophyte *Aspergillus niger* occurring in the liverwort *Heteroscyphus tener* (Steph.) Schiffn. Chem Biodivers 10:1193
471. Li XB, Li YL, Zhou JC, Yuan HQ, Wang N, Lou H (2015) A new diketopiperazine heterodimer from an endophytic fungus *Aspergillus niger*. J Asian Nat Prod Res 17:182
472. Xie F, Li XB, Zhou JC, Xu QQ, Wang XN, Yuan HQ, Lou HX (2015) Secondary metabolites from *Aspergillus fumigatus*, an endophytic fungus from liverwort *Heteroscyphus tener* (Steph.) Schiffn. Chem Biodivers 12:1313
473. Stelmasiewicz M, Świątek Ł, Ludwiczuk A (2022) Phytochemical profile and anticancer potential of endophytic microorganisms from liverwort species *Marchantia polymorpha* L. Molecules 27:153
474. Stelmasiewicz M, Świątek Ł, Ludwiczuk A (2022) Chemical and biological studies of endophytes isolated from *Marchantia polymorpha* L. Molecules 28:2202
475. Wang XN, Bashyal B, Wijeratne EMK, U'ren JM, Liu MX, Gunatilaka MK, Arnold AE, Gunatilaka AAL (2011) Smardaesidins A–G, isopimarane and 20-*nor*-isopimaranditerpenoids from *Smardaea* sp., a fungal endophyte of the moss *Ceratodon purpureus*. J Nat Prod 74:2052
476. Wei H, Xu YM, Espinosa-Artiles P, Liu MX, Luo JG, U'Ren JM, Arnold AE, Gunatilaka AAL (2015) Sesquiterpenes and other constituents of *Xylaria* sp. NC1214, a fungal endophyte of the moss *Hypnum* sp. Phytochemistry 18:102

477. Ali T, Pham TM, Ju KS, Rakotondraibe HL (2019) *ent*-Homocyclopiramine B, a prenylated indole alkaloid of biogenetic interest from the endophytic fungus *Penicillium concentricum*. Molecules 24:218
478. Melo IS, Santos SN, Rosa NH, Parma MM, Silva LJ, Queiroz SCN, Oellizari VH (2014) Isolation and biological activities of an endophytic *Mortierella alpina* strain from the Antarctic moss *Schistidium antarctici*. Extremophiles 18:15
479. Stelmasiewicz M, Świątek Ł, Gibbons S, Ludwiczuk A (2023) Bioactive compounds produced by endophytic microorganisms associated with bryophytes—the "bryendophytes." Molecules 28:3246
480. Song XQ, Zhang X, Han QJ, Li XB, Li G, Li RJ, Jiao Y, Zhou JC, Lou HX (2013) Xanthone derivatives from *Aspergillus sydowii*, an endophytic fungus from the liverwort *Scapania ciliata* S. Lac and their immunosuppressive activities. Phytochem Lett 6:318
481. Asakawa Y (2020) Scents and tasty substances of bryophytes and their application to cosmetics, foods and medicinal drugs (1). Aroma Res 21:68
482. Asakawa Y (2020) Scents and tasty substances of bryophytes and their application to cosmetics, foods and medicinal drugs (2). Aroma Res 21:61
483. Sugawa S (1960) Nutritive values of mosses as a food for domestic animals and fowls. Hikobia 2:119
484. Nishiki M, Toyota M, Asakawa Y (2007) Chemical constituents of the liverwort, *Radula* species and distribution of α-tocopherol in liverworts. Paper presented at 51st Symposium on chemistry of terpenes, essential oils and aromatics, Nagahama, Japan, 10–12 Nov 2007, Symposium papers, p 260
485. Ichikawa T, Namikawa M, Yamada K, Sakai K, Kondo K (1983) Novel cyclopentanoyl fatty acids from mosses, *Dicranum scoparium* and *Dicranum japonicum*. Tetrahedron Lett 24:3337
486. Ichikawa T, Yamada K, Namikawa M, Sakai K, Kondo K (1984) New cyclopentanoyl fatty acids from Japanese mosses. J Hattori Bot Lab 56:209
487. Prins HHT (1981) Why are mosses eaten in cold environments only? Okios 38:374
488. Lu Y, Eiriksson F, Thorsteinsdottir M, Simonsen HT (2019) Valuable fatty acids in bryophytes—production, biosynthesis, analysis and applications. Plants 8:525
489. Bhatla SC, Dhingra-Babbar S (1990) Growth regulating substances in mosses. In: Chopra RN, Bhatla SC (eds) Bryophyte development: physiology and biochemistry. CRC Press, Boca Raton, FL, USA, p 79
490. Wang TL, Cove DJ, Beutelmann P, Hartmann E (1980) Isopentenyladenine from mutans of the moss *Physcomitrella paten*s. Phytochemistry 19:1103
491. Yongabi KA, Novakovic M, Bukvicki D, Reeb C, Asakawa Y (2016) Management of diabetic bacterial foot infections with organic extracts of liverwort, *Marchantia debilis* from Cameroon. Nat Prod Commun 11:1333
492. Asakawa Y (2020) Scents and tasty substances of bryophytes and their application to cosmetics, foods and medicinal drugs (3). Aroma Res 21:58

Yoshinori Asakawa first studied biology at Tokushima University, and then went to graduate school at Hiroshima University in 1964 and studied organic chemistry, obtaining his Ph.D. degree there and then being appointed as a research assistant (1972–1975). He then worked as a postdoctoral fellow at the Université Louis Pasteur, Strasbourg, France for two years (1972–1974) with Prof. Guy Ourisson. In 1976 he moved to Tokushima Bunri University as an Associate Professor and was promoted to Full Professor in 1981. He served twice as Dean, and, since 1986, has been Director of the Institute of Pharmacology. He is the former Editor of "Phytomedicine" and "Spectroscopy" and serves on the editorial boards of the "Journal of Natural Products", "Phytochemistry", "Phytochemistry Letters", "Planta Medica", "Fitoterapia", "Natural Product Research", "Natural Product Communications", and the "Flavour & Fragrance Journal", among others. He is the president of Phytochemical Society of Asia and of the Chemistry of Terpenes, Essential Oils, and Aromatics. He is also a permanent committee member of International Symposium on Essential Oils. His research interests are on the bioactive secondary metabolites of medicinal plants (bryophytes and pteridophytes), fungi and insects, and on the biotransformation of secondary metabolites by fungi and mammals, and on oxidation reactions by organic peracid. To date, Prof. Asakawa has published 710 original papers, 14 reviews, and 44 books. For his outstanding research, he has been awarded the Hedwig Medal from the International Society of Bryologists, the International prize of "Phytochemistry" (Elsevier), the Jack Cannon International Gold Medal, the International Symposium of Essential Oils Award, the Gusi International Peace Prize, the Gerald Blunden Award, the International Symposium on Essential Oils Medal of Honor, the Japanese Society of Pharmacognosy Prize, and an honorary doctorate and Gold Medal from the Medical University of Lublin, Poland. His international standing has also led to him to receive the Polish Ambassador of Pharmacy Award, a Fellowship of the National Society of Ethnopharmacology of India, and an Honorary Membership of the Turkish Academy of Science. Over the years, he has been the opening, special, or invited lecturer at scientific meetings held in 52 countries. Prof. Asakawa has trained 57 postdoctoral fellows and 30 graduate students from various countries, and has organized international symposia held at his university a total of eight times as the chairperson.

MIX
Papier aus verantwortungsvollen Quellen
Paper from responsible sources
FSC® C105338

If you have any concerns about our products,
you can contact us on
ProductSafety@springernature.com

In case Publisher is established outside the EU,
the EU authorized representative is:
Springer Nature Customer Service Center GmbH
Europaplatz 3, 69115 Heidelberg, Germany

Printed by Libri Plureos GmbH
in Hamburg, Germany